# FINITE MATHEMATICS

**Eighth Edition**

**Hybrid Edition**

## HOWARD L. ROLF

*Baylor University*

BROOKS/COLE
CENGAGE Learning·

Australia • Brazil • Japan • Korea • Mexico • Singapore • Spain • United Kingdom • United States

*Finite Mathematics*, **8th Edition, Hybrid Edition**
**Howard L. Rolf**

Publisher: Richard Stratton

Senior Development Editor:
   Laura Wheel

Editorial Assistant: Danielle Hallock

Media Editor: Andrew Coppola

Brand Manager: Gordon Lee

Market Communications Manager:
   Linda Yip

Content Project Manager: Alison
   Eigel Zade

Senior Art Director: Linda May

Manufacturing Planner: Doug Bertke

Rights Acquisition Specialist:
   Shalice Shah-Caldwell

Production Service: Graphic World
   Inc.

Text Designer: Rokusek Design

Cover Designer: Jenny Willingham

Cover Image: ©Shutterstock

Compositor: Graphic World Inc.

For product information and technology assistance, contact us at
**Cengage Learning Customer & Sales Support, 1-800-354-9706**

For permission to use material from this text or product,
submit all requests online at **www.cengage.com/permissions**

Further permissions questions can be emailed to
**permissionrequest@cengage.com**

Library of Congress Control Number: 2012951494

Student Edition:

ISBN-13: 978-1-285-08464-0

ISBN-10: 1-285-08464-0

Brooks/Cole
20 Channel Center Street
Boston, MA 02210
USA

Cengage Learning is a leading provider of customized learning solutions with office locations around the globe, including Singapore, the United Kingdom, Australia, Mexico, Brazil, and Japan. Locate your local office at **international.cengage.com/region**

Cengage Learning products are represented in Canada by Nelson Education, Ltd.

For your course and learning solutions, visit **www.cengage.com.**

Purchase any of our products at your local college store or at our preferred online store **www.cengagebrain.com.**

**Instructors:** Please visit **login.cengage.com** and log in to access instructor-specific resources.

Printed in the United States of America
1  2  3  4  5  6  7  16  15  14  13  12

To Some Special People
*Julia Roseanna Rolf*
*James Howard Rolf*

# CONTENTS

# PREFACE

Why a hybrid text? Many traditional lecture-based courses are evolving into courses for which all homework and tests are delivered online. In addition, with the rapid growth of distance learning courses, there is an even greater need for course materials that blend traditional print resources with rich media-based tools. A hybrid text is designed to address the needs of these courses through the integration of both print and online components. For this hybrid edition, the end-of-section exercises have been removed from the text and are available exclusively online in Enhanced WebAssign, an easy-to-use online homework system.

## About the Course

Mathematics as we know it came into existence through an evolutionary process, and that process continues today. Occasionally a mathematical idea will fade away as it is replaced by a better idea. Old ideas become modified or new, and significant concepts are born and take their place. Some mathematical concepts have been developed in an attempt to solve problems in a particular discipline. Many disciplines have found that mathematical concepts are useful in understanding and applying the ideas of that discipline.

As technology affects more and more areas of the workplace and our culture, mathematical proficiency increases in importance. Science and engineering traditionally rely heavily on mathematics for analyzing and solving problems. Disciplines in business, the social sciences, and the life sciences have more recently applied and developed mathematical concepts in an attempt to solve problems in those disciplines. For that reason, an introductory course in finite mathematics, covering topics relevant to those disciplines, has become popular in recent decades. This text is intended to support that course, and it deals with topics, such as functions, linear systems, and matrices, that serve as useful tools in expressing and analyzing problems in areas using mathematics. Other topics, such as linear programming, probability, and mathematics of finance, have more direct applications to problems in business and industry.

## Philosophy

Because the audience for this book includes a wide range of students with varying interests, the author is sensitive to the needs of students and heeds the advice he received years ago: "Write for the student."

*Finite Mathematics* emphasizes the learning of mathematical concepts and techniques with applications of these concepts as the reason for studying them. For this reason, the examples and exercises in the text are designed to point toward these applications. Even so, the application of mathematical concepts requires an understanding of the mathematics and an expertise in the area of application. Generally, a straightforward application of the mathematics does not occur because there is a certain amount of "fuzziness" due to complications, exceptions, and variations. Thus, an application may be more difficult to accomplish than it initially appears. In spite of the difficulties in analyzing real-world problems, we can obtain an idea of the usefulness of finite mathematics using examples and exercises that are greatly simplified versions of actual applications.

While the author assumes only a background in high-school algebra, more challenging exercises are included for students who are ready to think more deeply about mathematical concepts. Although some exercises involve the use of a graphing calculator or spreadsheet, the book is designed to be used independently of them.

## Flexibility

This book provides flexibility in the choice and order of topics. Some topics must be covered in sequence. The following diagram shows prerequisite chapters.

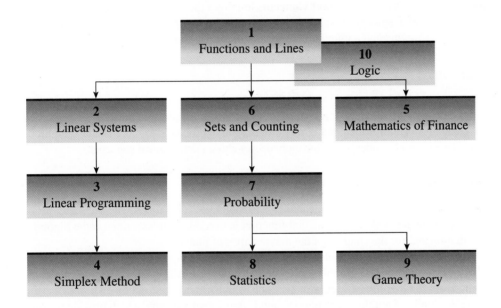

## CHANGES IN THE EIGHTH EDITION

Many examples and exercises have been updated, replaced, or added. The text has 450 examples and ample exercises in Enhanced WebAssign to provide an abundance of homework and practice problems.

The **Using Excel** feature has been updated to include instructions on the use of Excel 2007 and Excel 2010. For additional instructions, an *Excel Guide* is now available on the book's companion Website.

The end-of-chapter material has been expanded to include **Important Concepts.** Together with **Important Terms**, this new feature offers students a robust end-of-chapter summary and a valuable study tool.

# FEATURES

## Exposition

The author has concentrated on writing that is lucid, friendly, and considerate of the student. Applications are often used to motivate section concepts. Definitions, formulas, theorems, procedures, and summaries are boxed for emphasis and easy reference. Examples contain enough detail so that students can easily follow successive steps in the problem-solving process.

## Notes and Cautions

**Notes** reinforce and expand upon concepts, addressing special cases or showing alternative approaches or additional solution steps. **Cautions** point out typical problem areas and potential errors.

## End-of-Chapter Review

**Important Terms** and **Important Concepts** highlight the key terms and concepts of each chapter and offer students a valuable study tool. The comprehensive **Review Exercises** contain both routine and applied problems for additional practice and review.

## Exercises

The exercises in Enhanced WebAssign offer students a variety of problems to apply concepts and gives instructors ample problems to choose from when making assignments. The exercises are graded by level of difficulty and range from routine problems to more challenging problems that encourage students to think more deeply about mathematical concepts.

## Technology

Students are encouraged to use a calculator to facilitate calculations when working problems. A calculator with an exponential function is needed for the section on the binomial distribution and for the chapter on the mathematics of finance.

Use of a graphing calculator or Excel is optional in this book. However, for those who wish to incorporate graphing calculators or Excel in the course, the book includes **Using Your TI Graphing Calculator** sections, which provide instruction on how to use the TI-83/84 family of graphing calculators, and **Using Excel** sections, which provide instruction on how to use Excel 2007 and Excel 2010.

Several TI programs are included to (1) show students how to create programs that aid in solving problems and (2) carry out computations, sometimes tedious, to better focus on the concepts and logic of an application. In some cases, a program may give the same results as a routine built into the calculator. Where the calculator routine only requires the basic data to be entered and then the answer displayed, a program can enhance the student's understanding of the procedure and the logic of the concept by requiring student input at key decision points in the procedure.

*Appendix B* gives some additional guidance on the use of a TI graphing calculator.

# SUPPLEMENTS

## For the Instructor

**WebAssign ENHANCED WEBASSIGN®** (www.webassign.net)

Printed Access Card: 978-0-538-73810-1

Online Access Code: 978-1-285-18181-3

Exclusively from Cengage Learning, Enhanced WebAssign combines the exceptional mathematics content that you know and love with the most powerful online homework solution, WebAssign. Enhanced WebAssign engages students with immediate feedback, rich tutorial content, and interactive, fully customizable eBooks (YouBook), helping students to develop a deeper conceptual understanding of their subject matter. Online assignments can be built by selecting from thousands of text-specific problems or supplemented with problems from any Cengage Learning textbook.

### CENGAGE YOUBOOK

YouBook is an interactive and customizable eBook. Containing all the content from the eighth edition of Rolf's *Finite Mathematics*, YouBook features a text edit tool that allows instructors to modify the textbook narrative as needed. With YouBook, instructors can quickly re-order entire sections and chapters or hide any content they don't teach to create an eBook that perfectly matches their syllabus. Instructors can further customize the text by publishing web links. Additional media assets include animated figures, video clips, highlighting, notes, and more! YouBook is available in Enhanced WebAssign.

### COMPLETE SOLUTIONS MANUAL (ISBN: 978-1-285-08497-8) by Howard L. Rolf

The *Complete Solutions Manual* contains fully worked-out solutions for all exercises in the text, including *Using Your TI Graphing Calculator* exercises and *Using Excel* exercises.

### SOLUTION BUILDER (www.cengage.com/solutionbuilder)

This online instructor database offers complete, worked-out solutions to all exercises in the text. Solution Builder allows you to create customized, secure solutions printouts (in PDF format) matched exactly to the problems you assign in class.

### POWERLECTURE WITH EXAMVIEW® (ISBN: 978-1-285-08509-8)

This comprehensive CD-ROM includes Solution Builder, PowerPoint slides, and ExamView® computerized testing featuring algorithmically generated questions to create, deliver, and customize tests.

## For the Student

### STUDENT SOLUTIONS MANUAL (ISBN: 978-1-285-08469-5) by Howard L. Rolf

Giving you more in-depth explanations, this insightful resource includes fully worked-out solutions for the answers to the odd-numbered exercises.

### CENGAGEBRAIN.COM

To access additional course materials and companion resources, please visit **www.cengagebrain.com.** At the CengageBrain.com home page, search for the ISBN of your title (from the back cover of your book) using the search box at the top of the page. This will take you to the product page where free companion resources can be found.

# ACKNOWLEDGMENTS

A number of people have contributed to the writing of this book. I am most grateful to users of the book and the following reviewers, whose constructive criticism and suggestions for clarifying ideas have greatly improved the book.

Hollie Buchanan, *West Liberty University*
Jeanne Byrnes, *Northwest Community College*
Eric Ebersohl, *John A. Logan College*
Glenn Jablonski, *Triton College*
Fritz Keinert, *Iowa State University*
Peter Lee, *Santa Monica College*
Mary Peddycoart, *Lone Star College—Kenwood*
Armando Perez, *Laredo Community College*
Laurie Poe, *Santa Clara University*
Arthur Rosenthal, *Salem State University*
Mary Jane Sterling, *Bradley University*

Thanks to Revathi Narasimhan for her revision of the *Using Excel* sections, to Roger Lipsett for checking the accuracy of the book, and to Christi Verity for checking the accuracy of the solutions in the solutions manuals. Many thanks to the Brooks/Cole team: Richard Stratton, Laura Wheel, Alison Eigel Zade, Andrew Coppola, and Linda May. Thanks to Graphic World, the compositor, and Michael Ederer for a job well done.

Finally, thanks to the teachers and students who called attention to errors and made suggestions for improving the clarity of examples and exercises.

**Howard L. Rolf**

# TO THE STUDENT

Here is your first quiz in finite mathematics: What do the following people have in common?

- A banker
- A sociologist studying a culture
- A person planning for retirement
- A proud new parent
- A young couple buying their first house
- A politician assessing the chances of winning an election
- A casino manager
- A feedlot operator caught in a highly competitive cattle market
- A marketing manager of a corporation who wants to know if the company should invest in marketing a new product

The answer: All of these persons directly or indirectly use or can use some area of finite mathematics to help determine the best course of action. *Finite Mathematics* helps to analyze problems in business and the social sciences and provides methods that help determine the implications and consequences of various choices available.

This book gives an introduction to mathematics that is useful in a variety of disciplines. Mathematics can help you understand the underlying concepts of a discipline. It can help you organize information into a more useful form. Predictions and trends can be obtained from mathematical models. A mathematical analysis can provide a basis for making a good decision.

How can this course benefit you? First, you must understand your discipline, and second, you must understand this course. It is up to you to learn your discipline. This book is written to help you understand the mathematics. Perhaps these suggestions will be helpful.

1. You must study the material on a regular basis. Do your homework.
2. Study to understand the concepts. Read the explanations and study the examples.
3. Work the exercises and relate them to the concepts presented.
4. After you work an exercise, take a minute to review what you have done and make sure you understand how you worked the problem.

Your general problem-solving skills can improve because of this course. As you analyze problems, consider which method to use, and work through the solution, you are experiencing a simple form of the kind of process you will use throughout your life in your job or in daily living when you respond to the question "Hey, we have a problem, what should we do?" Most of life's problems are word problems, and so are those in this course.

## STUDY FEATURES

Several features in the text will assist you in your study of the concepts.

**Boldface** words indicate new terms.

**Boxes** emphasize definitions, theorems, procedures, and summaries.

**Notes** reinforce and expand upon concepts, addressing special cases or showing alternative approaches or additional solution steps. **Cautions** point out typical problem areas and potential errors.

**Important Terms** and **Important Concepts** are summarized at the end of each chapter to provide you with a convenient study tool.

**Review Exercises** at the end of each chapter help you practice skills and review key problem types.

**Answers** to the odd-numbered exercises are provided at the back of the text. In addition, the ***Student Solutions Manual*** contains worked-out solutions to the odd-numbered exercises. The *Student Solutions Manual* is available for purchase at CengageBrain.com.

# FUNCTIONS AND LINES

**1**

Mathematics is a powerful tool used in the design of automobiles, electronic equipment, and buildings. It helps in solving problems of business, industry, environment, science, and social sciences. Mathematics helps to predict sales, population growth, the outcome of elections, and the location of black holes. Some techniques of mathematics are straightforward; others are complicated and difficult. Nearly all practical problems involve two or more quantities that are related in some manner. For example, the amount withheld from a paycheck for FICA is related to an employee's salary; the area of a rectangle is related to the length of its sides (area = length × width); the amount charged for sales tax depends on the price of an item; and UPS shipping costs depend on the weight of the package and the distance shipped.

One of the simplest relationships between variables can be represented by a straight line, a basic geometric concept encountered by people in different experiences and situations. We refer to a line of people waiting to buy a ticket to a movie, the line formed by a string between two stakes to plant a straight row of vegetables, the stripes on a parking lot, the boundaries of a basketball court, and the representation of streets and highways on a map.

We realize that the "lines" mentioned are not really lines. At best they form an approximation of a line segment. They may not be exactly straight, a basketball court boundary is really a stripe, and the most carefully drawn line has "ragged edges" when enlarged sufficiently.

We give the ancient Greeks credit for the generalization and abstraction of geometric concepts such as the ideal line. We mention this because the ideal line — the line that expresses the essence of the stretched string, the mark in the sand, or the artist's finely drawn line — can be extended to certain relationships between two variables and can be expressed mathematically and used to give insight into the behavior of phenomena and activities in science, business, and some daily activities. In this chapter we discuss linear equations and their applications.

| **1.1** | # FUNCTIONS |

Mathematicians formalize certain kinds of relationships between quantities and call them **functions.** Nearly always in this book, the quantities involved are measured by real numbers.

A function consists of three parts: two sets and a rule. The **rule of a function** describes the relationship between a number in the first set (called the **domain**) and a number in the second set (called the **range**). The rule is often stated in the form of an equation like $A = \text{length}^2$ for the relationship between the area of a square and the length of a side.

In this case, the domain consists of the set of numbers representing lengths and the range consists of the set of numbers representing areas.

| **DEFINITION** | A **function** is a rule that assigns to each number from the first set (domain) exactly one |
| **Function** | number from the second set (range). |

We generally use the letter $x$ to represent a number from the first set (domain) and the letter $y$ to represent a number from the second set. Thus, for each value of $x$ the rule assigns exactly one value of $y$ to $x$.

Generally, a number may be arbitrarily selected from the domain, so $x$ is called an **independent variable.** Once a value of $x$ is selected, the rule determines the corresponding value of $y$. Because $y$ depends on the value of $x$, we call $y$ a **dependent variable.**

**Example 1**     Mr. Riggs consults for a trailer manufacturing company. His fee is $300 for miscellaneous expenses plus $50 per hour. What is the domain of this function? The number of hours worked determines the fee, so the domain consists of the number of hours worked and the range consists of the fees charged. It makes sense to say that hours worked must be a positive number, and the minimum fee is $300. Thus, positive numbers make the domain, and numbers 300 or larger make the range. The rule that determines the fee that corresponds to the number of consulting hours is given by the formula

$$y = 50x + 300$$

where $x$ represents the number of hours consulted and $y$ represents the total fee in dollars. Note that the formula gives exactly one fee for each number of consulting hours. So, this determines a function where the domain consists of the set of numbers representing hours worked, and the range consists of the set of numbers representing the dollar amount of fees.

In some cases, the quantities in the domain and range of a function may be limited to a few values, as illustrated in the next example.

**Example 2**   When a family goes to a concert, the amount paid depends on the number attending; that is, the total admission is a function of the number attending. The ticket office may have a chart giving total admission, so for them the rule is a chart something like this:

| Number of Tickets | Total Admission |
|---|---|
| 1 | $ 6.50 |
| 2 | $13.00 |
| 3 | $19.50 |
| 4 | $26.00 |
| 5 | $32.50 |
| 6 | $39.00 |

From the chart, the domain consists of the set of numbers representing the number of tickets sold. Because we never sell a fractional number of tickets, we restrict the domain to positive integers. In this case, the domain is the set $\{1, 2, 3, 4, 5, 6\}$.

Mathematicians have a standard notation for functions. For example, the equation $y = 50x + 300$ is often written as

$$f(x) = 50x + 300$$

$f(x)$ is read "$f$ of $x$," indicating "$f$ is a function of $x$." The notation $f(x)$ is a way of naming a function $f$ and indicating that the variable used is $x$. The notation $g(t)$ indicates another function named $g$ using the variable $t$. The $f(x)$ notation is especially useful to indicate the substitution of a number for $x$. $f(3)$ looks as though 3 has been put in place of $x$ in $f(x)$. This is the correct interpretation. $f(3)$ represents the **value** of the function when 3 is substituted for $x$ in

$$f(x) = 50x + 300$$

That is,

$$f(3) = 50(3) + 300$$
$$= 150 + 300$$
$$= 450$$

The next three examples illustrate some uses of the $f(x)$ notation.

**Example 3**   **(a)** If $f(x) = -7x + 22$, then

$$f(2) = -7(2) + 22 = 8$$
$$f(-1) = -7(-1) + 22$$
$$= 7 + 22$$
$$= 29$$

**(b)** If $f(x) = 4x - 11$, then

$$f(5) = 4(5) - 11$$
$$= 20 - 11$$
$$= 9$$

$$f(0) = 4(0) - 11$$
$$= 0 - 11$$
$$= -11$$

**(c)** If $f(x) = x(4 - 2x)$, then

$$f(6) = 6\big(4 - 2(6)\big)$$
$$= 6(4 - 12)$$
$$= 6(-8)$$
$$= -48$$
$$f(a) = a(4 - 2a)$$
$$f(a + 3) = (a + 3)\big(4 - 2(a + 3)\big)$$
$$= (a + 3)(4 - 2a - 6)$$
$$= (a + 3)(-2a - 2)$$
$$= -2a^2 - 8a - 6$$

Note that sometimes we are given an equation that relates numbers in the domain to those in the range, but does not specify the domain and range. In such cases, we define the domain to be all real numbers that can be substituted for $x$ and that yield a real number for $y$. The range is the set of values of $y$ so obtained. For example, $x = 9$ is in the domain of $y = \sqrt{x}$ because it yields the real number $y = 3$. However, $x = -4$ is not in the domain because $\sqrt{-4}$ is not a real number.

For $f(x) = \dfrac{3x + 2}{x - 5}$, 2 is in the domain because $f(2) = \frac{8}{-3}$ but 5 is not in the domain because $f(5) = \frac{17}{0}$, which is undefined. In fact, the domain consists of all real numbers except $x = 5$.

When applying a function, the nature of the application may restrict the domain or range. For example, the function that determines the amount of postage depends on the weight of the letter. It makes no sense for the domain to contain negative values of weight.

In summary: When a domain consists of real numbers, some equations automatically exclude certain real numbers from the domain.

**(a)** When the equation contains the independent variable, $x$, in the denominator of a fraction, that excludes the values of $x$ that make the denominator zero.
**(b)** If the equation contains a square root, this excludes the value of $x$ that makes the expression within the square root negative.
**(c)** An application problem usually restricts the values used for $x$ depending on the nature of the variables represented by $x$.
We conclude this section with some applications.

**Example 4**   From 1980 to 2000, the population of the United States can be estimated with the function

$$p(t) = 2.745t - 5210.35$$

where

$$t = \text{the year}$$
$$p(t) = \text{the population in millions}$$

**(a)** Based on this function, find $p(1970)$. Find $p(2010)$.
**(b)** Estimate when the population will reach 325 million.

**Solution**

**(a)** $p(1970) = 2.745(1970) - 5210.35 = 5407.65 - 5210.35 = 197.3$ million.
$p(2010) = 2.745(2010) - 5210.35 = 307.1$ million

*Note*: The 2010 census reported a U.S. population of 308.7 million.

**(b)** If $p(t) = 325$, then

$$325 = 2.745t - 5210.35$$
$$5535.35 = 2.745t$$
$$t = \frac{5535.35}{2.745} = 2016.5$$

The function estimates that the population will reach 325 million in the year 2016.

**Example 5**   Andy works at Papa Rolla's Pizza Parlor. He makes $8 per hour and time-and-a-half for all hours over 40 in a week. Thus, his weekly salary is $S(h) = 12h + 320$, where $h$ is the number of hours overtime and $S(h)$ is his weekly salary.

**(a)** What is $S(5.25)$?

**(b)** Find his weekly salary when he works 44.5 hours.

**(c)** One week Andy's salary was $362. How many hours overtime did he work?

**Solution**

**(a)** $S(5.25) = 12(5.25) + 320 = 383$

**(b)** In this case, $h = 44.5 - 40 = 4.5$, so $S(h) = 12(4.5) + 320 = 374$.
His salary was $374.

**(c)** $S(h) = 362$, so

$$362 = 12h + 320$$
$$12h = 362 - 320 = 42$$
$$h = \tfrac{42}{12} = 3.5$$

Andy worked 3.5 hours overtime.

Observe that we use letters other than $f$ and $x$ to represent functions and variables. We might refer to the cost of producing $x$ items as $C(x) = 5x + 540$; the price of $x$ pounds of steak as $p(x) = 5.19x$; the area of a circle of radius $r$ as $A(r) = \pi r^2$; and the distance in feet an object falls in $t$ seconds as $d(t) = 16t^2$.

A function requires that each number in the domain be associated with *exactly* one number in the range. Sometimes a rule assigns more than one number in the range to a number in the domain. In such a case, the relationship is *not* a function. Here is an example.

**Example 6**   A grocery store sells apples for $1.29 a pound. When Sarah buys six apples ($x = 6$), the checker determines the weight in order to know the cost. The six apples of the customer behind Sarah likely will correspond to a different weight. Thus, the cost "function" of $x$ apples is not a function of the number of apples because a number in the domain (number of apples) may be related to more than one number in the range (weight of $x$ apples).

If we let $x$ = the weight of the apples, then the cost relation to $x$ pounds is a function because there is a unique cost for a given weight.

**Example 7**    The domain of a linear function often includes all real numbers, but a function can have a limited domain or range. The function may not even be described by an equation. For example, a weather station monitors temperatures that are plotted to yield a graph like this.

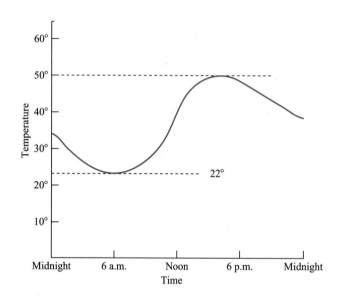

The domain covers the time from midnight through the day to 11:59 p.m. The range consists of the set of temperatures from 22 degrees to 50 degrees, written as [22°, 50°] in interval notation.

## 1.1    EXERCISES

Access end-of-section exercises online at **www.webassign.net**

**Using Your TI Graphing Calculator**

The TI graphing calculator can be used to calculate values of a function $y = f(x)$ for a single or several values of $x$.

*Note: The notation using a box such as* $\boxed{\text{ENTER}}$ *indicates the key to be pressed.*

## Example

For $y = 7x - 5$, calculate $y$ for $x = 3, 7, 2$, and $-12$. Here's how:

**(a)** You may calculate the value of the function by substituting the values of $x$ one by one, such as $7 \times 3 - 5$

**(b)** You may use a table that will calculate $y$ automatically when you enter a value of $x$:

    **1.** Select $\boxed{Y =}$ and enter $7x - 5$ as the $Y_1$ function.

        Next, we set up a table that will calculate the values of $y$ for values of $x$ listed in the table.

    **2.** Press $\boxed{\text{Tbl Set}}$ (i.e., $\boxed{\text{2nd}}$ $\boxed{\text{WINDOW}}$) to display the TABLE SETUP screen.

        Enter 0 for **TblStart** and press $\boxed{\text{ENTER}}$.

        Enter 1 for **ΔTbl** and press $\boxed{\text{ENTER}}$.

        Select the **Ask** option for **Indpnt** and press $\boxed{\text{ENTER}}$.

        Select **Auto** for **Depend** and press $\boxed{\text{ENTER}}$.

        You will then see the first screen shown below.

    **3.** Enter the values of $x$ in the table:

        Press $\boxed{\text{TABLE}}$ (i.e., $\boxed{\text{2nd}}$ $\boxed{\text{GRAPH}}$) and enter the values of $x$ in the list headed X. As you enter the values of $x$, the values of $7x - 5$ will appear in the list under $Y_1$ as shown in the second screen below.

 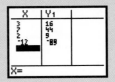

## Exercises

  **1.** Calculate $y = 6x - 3$ for $x = 1, 5$, and 9.

  **2.** Calculate $y = 17x + 4$ for $x = 23$.

  **3.** Calculate $y = x^2 + 5$ for $x = -2, 2, 3$, and 5.

  **4.** Calculate $y = (x - 2)^2$ for 2, 3, 5, and 6.

  **5.** Calculate $y = \dfrac{2x + 1}{x - 4}$ for $x = -1/2, 0, 1.5$, and 5.3.

  **6.** Calculate $y = 1.98x - 3.11$ for $x = 6.16, 8.25$, and 9.80.

## Using Excel

A cell in a spreadsheet may be used to record either a number or alphabetic information. For example, to enter 17.5 in cell B3, select the cell and type 17.5.

To enter the current date in cell C2, select the cell and type the current date.

|   | A | B | C |
|---|---|---|---|
| 1 |   |   |   |
| 2 |   |   | Sept. 18 |
| 3 |   | 17.5 |   |
| 4 |   |   |   |
| 5 |   |   |   |

## Formulas

A cell may contain a number, or a formula that uses numbers from other cells.

Here's how a spreadsheet adds the numbers in cells B3 and C3 with the answer stored in cell D3.

1. Enter 3 in cell B3 and 5 in cell C3.

2. Select the cell D3.

3. Type an =.

4. Type B3 + C3. Press the return key or click on the check mark in the top bar. The value 8 (3 + 5 in this case) appears in D3. If you change the numbers in B3 or C3, the new result will appear in D3.

|   | A | B | C | D |
|---|---|---|---|---|
| 1 |   |   |   |   |
| 2 |   |   |   |   |
| 3 |   | 3 | 5 | 8 |
| 4 |   |   |   |   |

## Example

Calculate $y = 3x + 4$ for $x$ found in A2. Store the result in B2.

1. Type =3*A2+4 in B2

| SUM | ▾ | × ✓ $f_x$ | =3*A2+4 | | |
|---|---|---|---|---|---|
|   | A | B | C | D | |
| 1 |   |   |   |   | |
| 2 |   | =3*A2+4 |   |   | |
| 3 |   |   |   |   | |

2. Press return or click on the check mark in the top bar.

3. For $x = 5$ in A2, the result $3 \times 5 + 4$ appears in B2.

|   | A | B |   |
|---|---|---|---|
| 1 |   |   |   |
| 2 | 5 | 19 |   |
| 3 |   |   |   |
| 4 |   |   |   |

You may calculate values for $y = 3x + 4$ using several values of $x$ as follows:

1. Enter the values of $x$ (5 values in this case) in A2 through A6.
2. Select cell B2 where =3*A2+4 is stored. Notice that the dark rectangle outlining B2 has a small square hanging on the lower right corner.
3. Place the cursor on the small square, click, and hold down while dragging the cell down to B6.

| | A | B | C | D | E | F |
|---|---|---|---|---|---|---|
| 1 | | | | | | |
| 2 | 5 | 19 | =3*A2+4 | | | |
| 3 | 7 | 25 | =3*A3+4 | | | |
| 4 | -3 | -5 | =3*A4+4 | | Dragging B2 puts these formulas in B2 through B6. | |
| 5 | 13 | 43 | =3*A5+4 | | | |
| 6 | 44 | 136 | =3*A6+4 | | | |
| 7 | | | | | | |

The B column now shows the values of $y$ corresponding to the $x$ values in column A.

The next screen shows cell B3 selected and the bar at the top shows the formula in B3 is the same as the one entered in B2, except it uses cell A3 instead of A2. The formulas in B4 through B6 use cells A4 through A6.

| B3 | ▾ | $f_x$ =3*A3+4 | | |
|---|---|---|---|---|
| | A | B | C | D |
| 1 | | | | |
| 2 | 5 | 19 | | |
| 3 | 7 | 25 | | |
| 4 | -3 | -5 | | |
| 5 | 13 | 43 | | |
| 6 | 44 | 136 | | |

A formula may be written in standard mathematical notation using $+$, $-$, $*$, and $/$ for addition, subtraction, multiplication, and division.

**Example:** =(A2−B2)/C2.

Exponentiation is indicated with ^.

**Example:** =A1^3 indicates $x^3$ where $x$ is in A1.

## Exercises

Write the Excel formulas for the calculations described in the exercises.

1. Add the numbers in A4 and B4 with the result in C4.
2. Add the numbers in A1, B1, and C1 with the result in D1.
3. Add the numbers in C4 and C5 with the result in C6.
4. Subtract the number in B4 from the number in A4 with the result in C4.
5. Multiply the numbers in B2 and B3 with the result in B4.
6. Divide the number in C2 by the number in D2 with the result in E2.
7. Divide the sum of the numbers in B1 and B2 by 2 with the result in B3.
8. Calculate $2x + 6$ where $x$ is in B3 and the result is in C3.
9. Calculate $2.1x - 1.8$ where $x$ is in A5 and the result is in B5.
10. For each of these values of $x$: 2, 5, $-1$, and 8 ($x$'s in A1 through A4), calculate $2x - 3$ with the results in B1 through B4.
11. For each of the values of $x$: 5, 8, $-4$, 0, 1.6 and 2.9 ($x$'s in A1 through A6), calculate $1.5x + 3.25$ with results in B1 through B6.

## **1.2**  GRAPHS AND LINES

- Definition of a Graph
- Linear Functions and Straight Lines
- Slope and Intercept
- Horizontal and Vertical Lines
- Slope-Intercept Equation

- Point-Slope Equation
- Two-Point Equation
- The *x*-Intercept
- Parallel Lines
- Perpendicular Lines

### Definition of a Graph

"A picture is worth a thousand words" may be an overworked phrase, but it does convey an important idea. You may even occasionally use the expression, "Oh, I see!" when you really grasp a difficult concept. A **graph** of a function shows a picture of a function and can help you to understand the behavior of the function.

A graph often makes it easier to notice trends and to draw conclusions. Let's look at an elementary example. Jason kept a record of the number of text messages he received. He told his roommate that he averaged 12 messages per day over a two-week period. "Tell me more," was his roommate's response. "Were there any days when you received none? What's the most you got? How many times did you get 12 messages? On what days, if any, did you receive 20 messages?"

To satisfy his roommate's curiosity and obtain the information of interest, Jason made a graph showing the number of messages for each day. Figure 1–1 shows the graph.

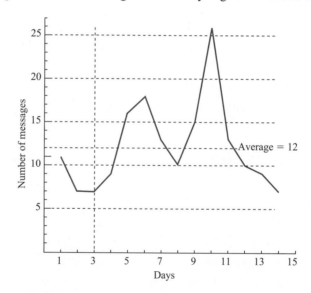

**FIGURE 1–1**

The graph shows that Jason received at least 7 messages each day. The largest number of messages was 26 on the tenth day. The graph shows no consistent pattern of the number of messages but indicates considerable variation from day to day. He never received exactly 12 messages, although the graph does coincide with 12 messages at about days 4.3, 7.5, 8.5, and 11.5. This doesn't make much sense, but it does illustrate an important point.

The daily number of messages really should be unconnected dots. However, it is much harder to get information from a graph drawn with just dots. Remember those drawings you made as a child by connecting the dots? The picture made a lot more sense after you drew in the connecting lines. In Figure 1–1, we sacrificed some technical accuracy by "connecting the dots" but got a better picture of what happened by doing so. Professional users of mathematics do the same thing; an accountant may let $C(x)$ represent the cost of manufacturing $x$ items, or a history professor may let $P(x)$ represent the class attendance in American history for day $x$ of the course. In reality, the domains of these functions involve only positive integers. But many methods of mathematics require the domain of the function to be an interval or intervals, rather than isolated points. These methods have proven so powerful in solving problems that people set up their functions using such domains. They then use some common sense in interpreting their answer. If the manager finds that the most efficient number of people to assign to a project is 54.87, she will probably end up using either 54 or 55 people.

Because the picture of a function might help convey the information that the function represents, we might ask how the picture of a function, its graph, relates to the rule, or equation, of the function. We obtain a point on a graph from the value of a number in the domain, $x$, and the corresponding value of the function, $f(x)$. We obtain the complete graph by using all numbers in the domain. Here is the definition.

| **DEFINITION** **Graph of a Function** | The **graph of a function** $f$ is the set of points $(x, y)$ in the plane that satisfies the equation $y = f(x)$. |
| --- | --- |

Generally, it is impossible to plot *all* points of the graph of a function. Sometimes we can find several points on the graph, and that suffices to give us the general shape of the graph.

**Example 1**  We write the function

$$f(x) = 2x^2 - 3 \quad \text{as}$$
$$y = 2x^2 - 3$$

so that we can relate the $x$-coordinates and $y$-coordinates of points on its graph to the equation of the function.

When $x = 2$, we find from $y = 2(2^2) - 3$ that $y = 5$, so the ordered pair $(2, 5)$, is a **solution** to $y = 2x^2 - 3$.

Thus, the point $(2, 5)$ lies on the graph of $y = 2x^2 - 3$. Other solutions include the ordered pairs (points on the graph) $(-2, 5), (0.5, -2.5), (3, 15), (0, -3), (-1, -1)$, and $(5, 47)$. When we plot these points and all other points that are solutions, we have the graph of the function $y = 2x^2 - 3$ (Figure 1–2). Since the graph of $y = 2x^2 - 3$ extends upward indefinitely, we cannot show all points on the graph. We can, however, show enough to convey the shape and location of the graph.

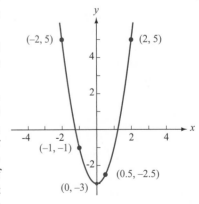

**FIGURE 1–2**   Graph of $y = 2x^2 - 3$.

As we develop different mathematical techniques throughout this text, we will use some concrete applications. This in turn will require some familiarity with the functions involved and some idea of the shape of their graphs. We start with the simplest functions and graphs.

## Linear Functions and Straight Lines

**DEFINITION**
**Linear Function**

A function is called a **linear function** if its rule—its defining equation—can be written $f(x) = mx + b$. Such a function is called linear because its graph forms a straight line.

From geometry we learned that two points determine a line. One point and the direction of a line also determine a line. We will learn how to find the equation of a line in each of these situations.

**Example 2**     Let's see how we can draw the graph of $f(x) = 2x + 5$.

**Solution**
The graph will be a straight line, and it takes just two points to determine a straight line. If we let $x = 1$, then we have $f(1) = 7$; if we let $x = 4$, then $f(4) = 13$. This means that the points $(1, 7)$ and $(4, 13)$ lie on the graph of $f(x) = 2x + 5$. Because we also use $y$ for $f(x)$, we could also say that these points lie on the line $y = 2x + 5$. By plotting the points $(1, 7)$ and $(4, 13)$ and drawing the line through them, we obtain Figure 1–3. (We can use any pair of $x$-values to get two points on the line.) It is usually a good idea to plot a third point to help catch any error. Because $f(0) = 5$, the point $(0, 5)$ also lies on the graph of $f$.

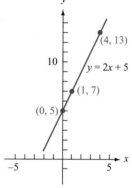

**FIGURE 1–3**   Graph of $f(x) = 2x + 5$.

**CAUTION**

The coefficient of $x$ in a general linear equation does not automatically give you the slope of the line. When the equation of the line is in the form $y = mx + b$, the coefficient of $x$ is the slope of the line, and the constant term is the $y$-intercept. If the equation is in another form, it is a good idea to change to this form to determine the slope and $y$-intercept.

## Slope and Intercept

The equations such as $y = 3x + 8$, $y = -2.5x + 17$, and $y = 12.1x - 62$ are equations of lines, and all are of the form $y = mx + b$. In that form, the constants $m$ and $b$ give key information about the line. The constant $b$ is the value of $y$ that corresponds to $x = 0$, so $(0, b)$ is a point on the line. We call $(0, b)$ the **$y$-intercept** of the line, because it tells where the line intercepts the $y$-axis. A common practice shortens the $y$-intercept notation of $(0, b)$ to just the letter $b$. Thus, in the linear equation $y = mx + b$, $b$ is called the $y$-intercept of the line. The other constant, $m$, determines the direction, or slant, of a line. We call $m$ the **slope** of the line. It measures the relative steepness of a line and will be discussed in detail later.

Note the equation of the line in Example 2 can be written $y = 2x + 5$, so it has slope $m = 2$, a $y$-intercept of 5, and passes through $(0, 5)$.

$y = mx + b$ For a linear function written in the form $y = mx + b$, we call $b$ the **y-intercept** of the line and $(0, b)$ is on the line.

$m$ is called the **slope** of the line and determines the direction and steepness of the line.

**Example 3**  Find the slope and $y$-intercept of each of the following lines.

(a) $y = 3x - 5$          (b) $y = -6x + 15$

**Solution**
(a) For the line $y = 3x - 5$, the slope $m = 3$ and the $y$-intercept $b = -5$.
(b) For the line $y = -6x + 15$, the slope is $-6$ and the $y$-intercept is 15.

Some equations may represent a line even though they are not in the form $y = mx + b$. The next example illustrates how we can still find the slope and $y$-intercept of those lines.

**Example 4**  Find the slope and $y$-intercept of the line $3x + 2y - 4 = 0$.

**Solution**
We can rewrite the equation $3x + 2y - 4 = 0$ in the slope-intercept form, $y = mx + b$, by solving the given equation for $y$:

$$3x + 2y - 4 = 0$$
$$2y = -3x + 4$$
$$y = -\frac{3}{2}x + 2$$

Thus, we have the slope-intercept form $y = -\frac{3}{2}x + 2$. Now we can say that the slope is $-\frac{3}{2}$ and the $y$-intercept is 2.

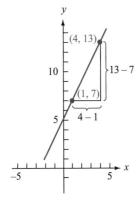

**FIGURE 1–4**  Graph of $y = 2x + 5$.

Now let's use the linear function $y = 2x + 5$ to illustrate the way the slope relates to the direction of a line. Select two points on the line $y = 2x + 5$, such as $(1, 7)$ and $(4, 13)$. (See Figure 1–4.) Compute the difference in the $y$-coordinates of the two points: $13 - 7 = 6$. Now compute the difference in $x$-coordinates: $4 - 1 = 3$. The quotient $\frac{6}{3} = 2$ is $m$, and the slope of the line $y = 2x + 5$. Following this procedure with any other two points on the line $y = 2x + 5$ will also yield the answer 2.

Examples 3 and 4 illustrate the following general formula that shows how to compute the slope of a line.

**Slope Formula**  Choose two points $P$ and $Q$ on the line. Let $(x_1, y_1)$ be the coordinates of $P$ and $(x_2, y_2)$ be the coordinates of $Q$. The **slope** of the line, $m$, is given by the equation

$$m = \frac{y_2 - y_1}{x_2 - x_1} = \frac{\text{change in } y}{\text{change in } x} \quad \text{where} \quad x_2 \neq x_1$$

Obtain the slope using the difference in the *y*-coordinates divided by the difference in the *x*-coordinates.

Figure 1–5 shows the geometric meaning of this quotient. The slope tells how fast *y* changes for each unit change in *x*.

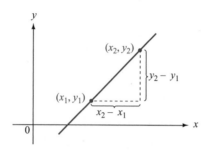

**FIGURE 1–5**   Geometric meaning of slope quotient.

We now use the slope formula to obtain the slope of a line through two points.

**Example 5**   Find the slope of a line through the points (2, 5) and (6, 1).

**Solution**

Let the point $(x_1, y_1)$ be (2, 5) and $(x_2, y_2)$ be (6, 1). Substituting these values into the definition of *m*,

$$m = \frac{5 - 1}{2 - 6} = \frac{4}{-4} = -1$$

Figure 1–6 shows the geometric relationship.

**NOTE**

It doesn't matter which point we call $(x_1, y_1)$ and which one we call $(x_2, y_2)$; it doesn't affect the computation of *m*. If we label the points differently in Example 5, the computation becomes

$$m = \frac{1-5}{6-2} = \frac{-4}{4} = -1$$

The answer is the same. Just be sure to subtract the *x*- and *y*-coordinates in the same order.

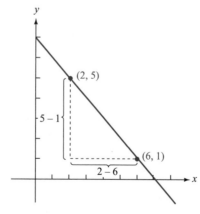

**FIGURE 1–6**   Geometric relationship of the slope between the points (2, 5) and (6, 1).

**Example 6** Determine the slope of a line through the points (2, 2) and (7, 10).

**Solution**
For these points

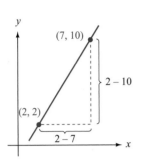

$$m = \frac{2 - 10}{2 - 7} = \frac{-8}{-5} = \frac{8}{5}$$

The geometric relationship is shown in Figure 1–7.

**FIGURE 1–7** Geometric relationship of the slope between the points (2, 2) and (7, 10).

The two preceding examples suggest the direction a line takes when the slope is positive and when it is negative. For a line with positive slope, the line slopes upward to the right. For a negative slope, the line slopes downward to the right.

Now we consider situations when the slope is neither positive nor negative.

## Horizontal and Vertical Lines

**Example 7** Find the equation of the line through the points (3, 4) and (7, 4).

**Solution**
The slope of the line is

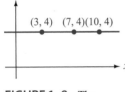

$$m = \frac{4 - 4}{7 - 3} = \frac{0}{4} = 0$$

Whenever $m = 0$, we can write the equation $f(x) = 0x + b$ more simply as $f(x) = b$; $f$ is called the **constant function**. The graph of a constant function is a line parallel to the $x$-axis; such a line has an equation of the form $y = b$ and is called a **horizontal line**. (See Figure 1–8.) Because all points on a horizontal line have the same $y$-coordinates, the value of $b$ can be determined from the $y$-coordinate of any point on the line. The equation of the line in this example is $y = 4$.

**FIGURE 1–8** The horizontal line $y = 4$. Note that the $y$-coordinate of a point on the line is always 4.

When a line has slope zero, the line is a horizontal line. Conversely, a horizontal line has slope zero.

**Horizontal Line**   A **horizontal line** has slope zero.

The next example illustrates a line that has no slope.

**Example 8** Determine the equation of the line through the points (4, 1) and (4, 3).

**Solution**
We can try to use the rule for computing the slope, but we obtain the quotient

$$\frac{3 - 1}{4 - 4} = \frac{2}{0}$$

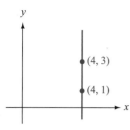

which doesn't make sense because division by zero is not defined. The slope is not defined. When we plot the two points, however, we have no difficulty in drawing the line through them. (See Figure 1–9.)

The line, parallel to the $y$-axis, is called a **vertical line**. A point lies on this line when the first coordinate of the point is 4, so the equation of the line is $x = 4$.

**FIGURE 1–9** The vertical line $x = 4$. Note that the $x$-coordinate of a point on the line is always 4.

When the slope of a line is undefined, the line is a vertical line. Conversely, a vertical line has an undefined slope.

| **Vertical Line** | The slope of a **vertical line** is undefined. |
|---|---|

**CAUTION**

A vertical line does *not* have a slope, but it does have an equation.

Whenever $x_2 = x_1$, you get a 0 in the denominator when computing the slope, so we say that the *slope does not exist* for such a line.

The slope of a line can be positive, negative, zero, or even not exist. These situations are depicted in Figure 1–10. This figure shows the relationship between the slope and the slant of the line. If $m > 0$, the graph slants up as $x$ moves to the right. If $m < 0$, the graph slants down as $x$ moves to the right. If $m = 0$, the graph remains at the same height. If $m$ does not exist, the line is vertical, and the line is *not* the graph of a linear function. The equation of a vertical line cannot be written in the form $f(x) = mx + b$ because there is no $m$.

We conclude this section by showing how to find equations of lines. Linear functions arise in many applied settings. When an application provides the appropriate two pieces of information, you can find an equation of the corresponding line. This information may be given either as:

**1.** the slope and a point on the line or

**2.** two points on the line.

We now show each form in a particular application and then give the general method of solving the problem.

### Slope-Intercept Equation

When we know the slope of a line and the point given is the $y$-intercept, the equation of the line is given directly by the $y = mx + b$ form.

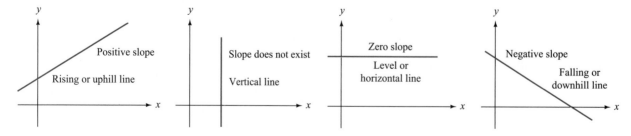

**FIGURE 1–10** Relationship of the direction of a line to its slope (reading from left to right).

**Example 9**   Determine an equation of the line with slope 3 and $y$-intercept 2.

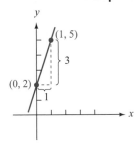

**Solution**
This information gives $m = 3$ and $b = 2$ in the equation $y = mx + b$, so the equation is

$$y = 3x + 2$$

The graph of this equation is shown in Figure 1–11.

**FIGURE 1–11**   Graph of a line with slope 3 and $y$-intercept 2.

When the slope of a line and a point on the line are given, we can find the values of $m$ and $b$ in the slope-intercept form $y = mx + b$.

**Example 10**   Determine an equation of the line that has slope $-2$ and passes through the point $(-3, 5)$.

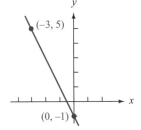

**Solution**
The value of $m = -2$, so the line has an equation of form

$$y = -2x + b$$

To complete the solution, we find the value of $b$. Since the point $(-3, 5)$ lies on the line, $x = -3$ and $y = 5$ must be a solution to $y = -2x + b$. Just substitute those values into $y = -2x + b$ to obtain

$$5 = (-2)(-3) + b$$
$$5 = 6 + b$$
$$5 - 6 = b$$

**FIGURE 1–12**   Graph of a line with slope $-2$ passing through $(-3, 5)$.

Thus, $b = -1$ and the equation of the line is $y = -2x - 1$. The graph of this equation is shown in Figure 1–12.

The following table displays some values of $x$ and the corresponding values of $y$ for the line $y = -2x - 1$ from Example 10. The values of $x$ listed increase in steps of 2 in order to point out that $y$ also changes in equal steps $(-4)$. Although the step size for $y$ differs from the step size of $x$, they are related. Observe that the change in $y$ equals the slope times the change in $x\left(-4 = -2(2)\right)$. This relationship holds in general for linear functions. That is, the step size of $y$ is $m$ times the step size of $x$.

| $x$ | $y$ |
|---|---|
| $-2$ | 3 |
| 0 | $-1$ |
| 2 | $-5$ |
| 4 | $-9$ |
| 6 | $-13$ |

## Point-Slope Equation

We can use another method for writing the equation of a line when we know the slope and a point. We call this method the **point-slope equation.**

We will work with the information in Example 10, $m = -2$ and the point $(-3, 5)$. Use the slope formula with $(-3, 5)$ as $(x_1, y_1)$ and use an arbitrary point, $(x, y)$, as $(x_2, y_2)$. Because $m = -2$, we can write

$$-2 = \frac{y - 5}{x - (-3)} = \frac{y - 5}{x + 3}$$

Multiply both sides by $x + 3$ to obtain

$$-2(x + 3) = y - 5$$

The formula is usually written as

$$y - 5 = -2(x + 3)$$

Check that this has the same slope-intercept form as we obtained in Example 10.

Note what we did in this example and observe that the procedure generally holds. When we are given the slope of a line, $m$, and a specific point on the line, $(x_1, y_1)$, then for any other point on the line, $(x, y)$, the slope of the line can be obtained as $\frac{y - y_1}{x - x_1}$, so we write

$$\frac{y - y_1}{x - x_1} = m$$

Now multiply through by $x - x_1$ to obtain

$$y - y_1 = m(x - x_1)$$

an equation of a line with a given slope and point.

---

**Point-Slope Equation**   If a line has slope $m$ and passes through $(x_1, y_1)$, an equation of the line is

$$y - y_1 = m(x - x_1)$$

---

Now we use the point-slope equation to find the equation of a line.

---

**Example 11**   Find an equation of the line with slope 4 that passes through $(-1, 5)$.

**Solution**
$x_1 = -1$ and $y_1 = 5$; $m = 4$, so the point-slope formula gives us

$$y - 5 = 4(x - (-1)) = 4(x + 1)$$

which can be simplified to $y = 4x + 9$.

---

Now let's look at a simple application of a linear function that illustrates the role of the slope and the $y$-intercept.

**Example 12**    Carlos started a paper route and decided to add $7 each week to his savings account. By the eighth week, he had $242 in savings. Write his total savings as a linear function of the number of weeks since he started his paper route.

**Solution**

This example gives the slope and one point on the line.

> Let $x$ be the number of weeks and $y$ be the total in savings. Each time $x$ increases by 1 (one week), $y$ increases by 7 ($7 deposit). Thus, the slope equals

**NOTE**

The slope of the line is the rate of change of $y$ per unit change in $x$.

$$m = \frac{\text{change in } y}{\text{change in } x} = \frac{7}{1} = 7$$

In this case the slope gives the change of dollars per week, so it represents $7 per week.

> At 8 weeks, the savings totaled $242, so the point $(8, 242)$ lies on the line. Using the point-slope formula, we have

$$y - 242 = 7(x - 8)$$

Solving for $y$, we have $y = 7x + 186$.

> Note that the $y$-intercept, 186, indicates that Carlos had $186 when he decided to start his periodic savings.

## Two-Point Equation

Just as we have a point-slope formula, we have a formula for finding an equation of a line using the coordinates of two points on the line.

**Two-Point Equation**    If a line passes through the points $(x_1, y_1)$ and $(x_2, y_2)$, with $x_1 \neq x_2$, an equation of the line is

$$y - y_1 = m(x - x_1)$$

where

$$m = \frac{y_2 - y_1}{x_2 - x_1}$$

> Note that the **two-point equation** is a variation of the point-slope equation. The two given points enable us to find the slope. Then use the slope and one of the given points in the point-slope equation. (It doesn't matter which one of the two points we use.)

**Example 13**

Determine an equation of the straight line through the points (1, 3) and (4, 7).

**NOTE**

There is no one way that the equation of a line must be written. This equation, $y - 3 = \frac{4}{3}(x - 1)$, can be simplified to $y = \frac{4}{3}x + \frac{5}{3}$. Another form is $4x - 3y = -5$. Each form is correct, but one form may be preferred depending on the situation.

**Solution**

Let $(x_1, y_1)$ be the point (1, 3) and let $(x_2, y_2)$ be the point (4, 7); then

$$m = \frac{7 - 3}{4 - 1} = \frac{4}{3}$$

The point-slope formula gives $y - 3 = \frac{4}{3}(x - 1)$ as an equation of this line, whose graph is shown in Figure 1–13.

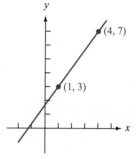

**FIGURE 1–13**    Graph of a line through (1, 3) and (4, 7).

**Example 14**

An electric utility computes the monthly electric bill for residential customers with a linear function of the number of kilowatt hours (kWh) used. One month, a customer used 1560 kWh, and the bill was $118.82. The next month, the bill was $102.26 for 1330 kWh used. Identify the variables $x$ and $y$, and find the equation relating kWh used and the monthly bill.

**Solution**

Let $x$ = the number of kWh used and $y$ = the monthly bill. The information provided gives two points on the line, (1560, 118.82) and (1330, 102.26). The slope of the line is

$$m = \frac{118.82 - 102.26}{1560 - 1330} = \frac{16.56}{230} = 0.072$$

In this case $m$ represents the cost per kWh.

The equation can now be written as

$$y - 102.26 = 0.072(x - 1330)$$
$$y - 102.26 = 0.072x - 95.76$$
$$y = 0.072x + 6.50$$

**Alternate Method**

Given two points on a line, you can find the equation of the line in the form $y = mx + b$ as follows:

**(a)** Compute $m$ using the two given points and subtitute for $m$ in the equation.

**(b)** Subtitute a given point into the equation $y = mx + b$.

**(c)** Calculate $b$.

**(d)** Use the value of $m$ and the value of $b$ found above to obtain the equation $y = mx + b$.

## The *x*-Intercept

The **x-intercept** of a line is similar to the *y*-intercept. It is the *x*-coordinate of the point where the line crosses the *x*-axis. To find the *x*-intercept, set $y = 0$ in the linear equation and solve for *x*.

**Example 15**   Find the *x*-intercept of the line $4x - 9y = 30$.

**Solution**
Let $y = 0$. We have $4x - 9(0) = 30$ and $x = \dfrac{30}{4} = 7.5$. The *x*-intercept is 7.5, and the line crosses the *x*-axis at $(7.5, 0)$. (See Figure 1–14.)

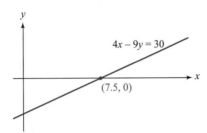

**FIGURE 1-14**   The *x*-intercept of $4x - 9y = 30$.

We conclude this section with a discussion of parallel and perpendicular lines.

## Parallel Lines

**DEFINITION**
**Parallel Lines**
Two lines are parallel if they have the same slope or if they are both vertical lines.

**Example 16**   Determine if the line through $(2, 3)$ and $(6, 8)$ is parallel to the line through $(3, 1)$ and $(11, 11)$.

**Solution**
Let $L_1$ be the line through $(2, 3)$ and $(6, 8)$ and let $m_1$ be its slope. Then,

$$m_1 = \frac{8 - 3}{6 - 2} = \frac{5}{4}$$

Let $L_2$ be the line through $(3, 1)$ and $(11, 11)$ and let $m_2$ be its slope. Then,

$$m_2 = \frac{11 - 1}{11 - 3} = \frac{10}{8} = \frac{5}{4}$$

The slopes are identical, so the lines are parallel. (See Figure 1–15.)

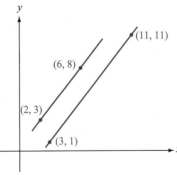

**FIGURE 1-15**   Lines that have the same slope are parallel.

**Example 17**   Is the line through $(5, 4)$ and $(-1, 2)$ parallel to the line through $(3, -2)$ and $(6, 4)$?

**Solution**

The slope of the first line is

$$m_1 = \frac{4 - 2}{5 - (-1)} = \frac{2}{6} = \frac{1}{3}$$

The slope of the second line is

$$m_2 = \frac{-2 - 4}{3 - 6} = \frac{-6}{-3} = 2$$

Since the slopes are not equal, the lines are not parallel.

**Example 18**   When we have the equation of a line, we can find the equation of a parallel line, provided that we know a point on the second line. We illustrate this by finding the line through the point $(8, 5)$ that is parallel to the line $4x - 3y = 12$. We know a point on the line—namely, $(8, 5)$—and we can find the slope from the given line $4x - 3y = 12$ by putting it in the slope-intercept form:

$$4x - 3y = 12$$

$$y = \frac{4}{3}x - 4$$

The desired slope is $\frac{4}{3}$, so the equation of the parallel line is found from

$$y - 5 = \frac{4}{3}(x - 8)$$

$$y = \frac{4}{3}x - \frac{17}{3}$$

We can put this in another standard form by multiplying by 3:

$$3y = 4x - 17 \quad \text{or} \quad 4x - 3y = 17$$

Note that the coefficients of $x$ and $y$ are the same as those in the given line. This generally holds, giving us another way to find the equation of the parallel line using the knowledge that its equation is of the form

$$4x - 3y = \text{some constant}$$

Since the point $(8, 5)$ lies on the line, it must give the desired constant when the coordinates are used for $x$ and $y$; that is,

$$4(8) - 3(5) = \text{the constant} = 17 \text{ in this case}$$

and the equation is

$$4x - 3y = 17$$

### Perpendicular Lines

**DEFINITION**
**Perpendicular Lines**

Two lines are perpendicular when they intersect in right angles (90°).

We can tell when two lines are perpendicular by the way their slopes are related.

**Perpendicular Lines**

Two lines with slopes $m_1$ and $m_2$ are perpendicular if and only if

$$m_1 m_2 = -1$$

or in another form

$$m_2 = -\frac{1}{m_1}$$

**Example 19**  The lines

$$y = -\frac{2}{3}x + 17 \quad \text{and} \quad y = \frac{3}{2}x - 9$$

are perpendicular because the product of their slopes is

$$-\frac{2}{3} \times \frac{3}{2} = -1$$

The lines

$$3x + 5y = 7 \quad \text{and} \quad 11x - 3y = 15$$

are not perpendicular because the product of their slopes

$$-\frac{3}{5} \times \frac{11}{3} = -\frac{11}{5}$$

is not $-1$.

---

**Summary of Equations of a Linear Function**

**Slope:** $m = \dfrac{y_2 - y_1}{x_2 - x_1}$, where $(x_1, y_1)$ and $(x_2, y_2)$ are points on the line with $x_2 \neq x_1$.

**Standard Equation:** $Ax + By = C$ where $(x, y)$ is any point on the line and at least one of $A$, $B$ is not zero.

**Slope-Intercept Equation:** $y = mx + b$, where $m$ is the slope and $b$ is the $y$-intercept.

**Point-Slope Equation:** $y - y_1 = m(x - x_1)$, where $m$ is the slope and $(x_1, y_1)$ is a given point on the line.

**Horizontal Line:** $y = k$, where $(h, k)$ is a point on the line and all points on the line have the same $y$-coordinate, $k$. The slope is zero.

**Vertical Line:** $x = h$, where $(h, k)$ is a point on the line and all points on the line have the same $x$-coordinate, $h$. The slope is undefined.

## 1.2   EXERCISES

Access end-of-section exercises online at **www.webassign.net**

### Using Your TI Graphing Calculator

### Graphs of Lines and Evaluating Points on a Line

The graphing of a line, or lines, on a graphics calculator requires these steps:

**A.** Set the $x$ and $y$ ranges for the window.

**B.** Enter the linear equation(s).

**C.** Select the GRAPH command.

Here are examples of the TI graphing calculator screens resulting from these steps using the linear equations

$$y = 2x + 5$$
$$3x + 4y = 12$$

*Note:* The second equation must be solved for $y$ to enter it. It may be written as

$$y = (12 - 3x)/4 \qquad \text{or} \qquad y = -.75x + 3$$

**A.** Settings for the window to show $x$ and $y$ ranges from $-10$ to $10$, with tick marks 2 units apart. The $\boxed{\text{WINDOW}}$ key will select the screen.

```
WINDOW
 Xmin=-10
 Xmax=10
 Xscl=2
 Ymin=-10
 Ymax=10
 Yscl=2
 Xres=1
```

**B.** The $\boxed{\text{Y} =}$ key selects the screen for entering the equations $y = 2x + 5$ and $y = (12 - 3x)/4$.

**C.** The $\boxed{\text{GRAPH}}$ key activates the graphing of the equations.

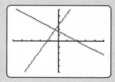

## Exercises

Graph the lines in Exercises 1 through 8. Use –10 to 10 for the range of both $x$ and $y$.

**1.** $y = 5x + 4$         **2.** $y = 0.5x - 6$         **3.** $y = -1.4x + 8.2$

**4.** $y = -0.65x + 7.3$         **5.** $5x - 2y = 12$         **6.** $3x + y = 8$

**7.** $2.4x + 5.3y = 15.6$         **8.** $3.3x - 7.2y = 22.8$

You can evaluate a function $y = f(x)$ at values of $x$ using the **value** option from the **CALC** menu. We illustrate with $y = 2x - 3$ and $x = 4.7$.

Press $\boxed{Y=}$ and enter the equation.

Press $\boxed{\text{CALC}}$ (using $\boxed{\text{2nd}}$ $\boxed{\text{TRACE}}$). You will obtain the screen

Select <1:value> and press $\boxed{\text{ENTER}}$ to obtain the screen

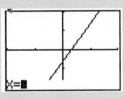

Enter 4.7 for the value of $x$ and press ENTER to get the screen

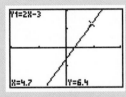

where you see $y = 6.4$ when $x = 4.7$.

To obtain the value of $y$ for another value of $x$, enter the value of $x$ and press $\boxed{\text{ENTER}}$.

**9.** For the line $y = 3.86x + 1.22$, use **value** to find the value of $y$ when $x = 4.5$.

**10.** For the line $y = 1.98x - 3.11$, use **value** to find the value of $y$ when $x = 6.16$.

**11.** For the line $y = 4.6x + 7.2$, find the values of $y$ for $x = 6.16, -3.2,$ and $4.1$.

**12.** For the line $3.8x + 5.4y = 29.7$, find the values of $y$ for $x = -1.1, 3.9,$ and $7.8$.

# Using Excel

We now illustrate how to find the slope of a line through two points and the slope-intercept equation of the line.

## Example 1

Find the slope of the line through the points (3, 8) and (4, 5).

### Solution

We enter the coordinates of the first point on line 2, and the coordinates of the second point on line 3, and the slope formula in cell C3. Line 1 is used to identify the entries in the columns. The screen that we get follows.

|   | A | B | C | D | E |
|---|---|---|---|---|---|
| 1 | X | Y |  Slope |  |  |
| 2 |  3 |  8 |  |  | The formula |
| 3 |  4 |  5 |  -3 |  | =(B3-B2)/(A3-A2) is entered in C3 |
| 4 |  |  |  |  |  |
| 5 |  |  |  |  |  |

## Example 2

Find the slope-intercept equation of the line through the points (2, 5) and (4, 9).

### Solution

The $y$-intercept of a line, $b$, can be found from the slope of the line, $m$, and a point on the line, $(x_1, y_1)$ using

$$b = y_1 - mx_1$$

Here is the spreadsheet using Line 1 to identify the entries in each column, entering the coordinates of the first point in Line 2, the coordinates of the second point in Line 3, the calculated slope in C3, and the $y$-intercept in D3.

|   | A | B | C | D | E | F |
|---|---|---|---|---|---|---|
| 1 | X | Y | Slope | Y-intercept |  |  |
| 2 |  2 |  5 |  |  | The formula |  |
| 3 |  4 |  9 |  2 |  1 ← | =B2-C3*A2 is entered in D3 |  |
| 4 |  |  |  |  |  |  |
| 5 |  |  |  |  |  |  |
| 6 |  |  | The formula =(B3-B2)/(A3-A2) |  |  |  |
| 7 |  |  | is entered in C3 |  |  |  |
| 8 |  |  |  |  |  |  |

The slope-intercept equation is

$$y = 2x + 1$$

## Exercises

1.  Find the slope of the line through the points (6, 5) and (−2, 3).
2.  Find the slope of the line through the points (3, 8) and (6, −2).
3.  Find the slope of the line through the points (5.3, 1.9) and (9.7, 13.9).
4.  Find the slope of the line through the points (156.2, 298.5) and (644.4, 903.8).
5.  Find the equation of the line through the points (2, 3) and (8, −1).
6.  Find the equation of the line through the points (5, 4) and (10, 16).
7.  Find the equation of the line through the points (1.5, 3.4) and (4.7, 1.6).

## Finding the Graph and Equation of a Line through Two Points

A series of Excel menu selections will give the graph of a line through two points and find the equation of the line.

### Example 3

Graph the line through the points $(1, 4)$ and $(3, 8)$.

**Solution**
Here are the steps to find the graph and equation.

- Enter the first point in cells A2:B2 ($x$ in A2, $y$ in B2) and the second point in A3:B3.
- Select the cells, A2:B3, containing the points.
- Click on the **Insert** tab. Move to the **Charts** group. Select **Scatter** with the scatter subtype.

- A graph will be inserted in your worksheet. The worksheet will be in the **Chart Tools** environment.
- Click on **Select Data** under the **Data group** in **Chart Tools**. If necessary, click on **Switch Row/Column** so that the correct $x$ values, 1 and 3, appear in the horizontal axis label.
- Click on **OK**.
- Right-click on one of the data points in the graph.
- Click on **Add Trendline**.
- In the dialog box, click on **Linear** and check the box to **Display equation on chart**.

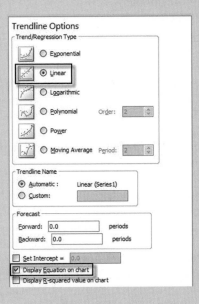

- Click on **Close**.

*(continued)*

You will get the screen

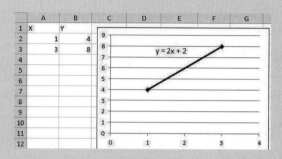

Once you have this sequence set up on your spreadsheet, you can change the points, and the graph and equation will automatically give the new graph and equation.

## Exercises

1. Find the graph and equation of the line through the points (3, 5) and (8, 12).
2. Find the graph and equation of the line through the points (1, 3) and (5, −11).
3. Find the graph and equation of the line through the points (−2.2, −1.4) and (4.2, 6.6).

## 1.3    MATHEMATICAL MODELS AND APPLICATIONS OF LINEAR FUNCTIONS

- Applications
- Cost–Volume Function
- Revenue Function
- Break-Even Analysis
- Straight-Line Depreciation
- Applications of Linear Inequalities

Nat Hambrick is a contractor in a competitive construction market. He needs to answer such questions as, "We want to add another floor to the top of the Amico Building for an elite restaurant. Will the present building support it?" "Will this roof design support a 2-foot snowfall?" "Will the proposed office tower withstand 100-mile-per-hour winds?"

The building trade knows how to handle many problems from past experience of successes and failures. A builder may avoid a failure by deliberately overdesigning, but that may drive up costs until the builder cannot compete. Because a trial-and-error approach may result in costly failures or expensive successes, an approach that uses proper mathematical analyses can provide valuable answers. Engineers have developed mathematical equations and formulas, called **mathematical models**, that can be used to answer "what if?" questions of building design.

Mathematical models abound in numerous other disciplines. They describe scientific phenomena, population growth, sociological trends, economic growth, and product costs. Politicians depend on mathematically based opinion polls to determine campaign strategy.

One of the more famous mathematical models is the equation $E = mc^2$, which was chalked on sidewalks of campuses across the nation on April 19, 1955, the day after Albert Einstein died. Written in tribute to his profound work on the theory of relativity, this equation relates energy, $E$, the mass of a body, $m$, and the speed of light, $c$. This equation does not tell all about energy, but it describes the energy released when matter is fully transformed into energy. It helps predict the energy released from uranium and the energy stored in the Sun.

Although mathematical models generally only approximate the real situations, they can be useful in making decisions and estimating the consequences of these decisions: A mathematical analysis can help a hospital administrator determine the most economical order for the quantity of surgical supplies, can give a proud new grandparent an estimate of the worth of an investment when the grandchild begins college, or can help the phone company plan for the growing demand for cell phones.

Sometimes a mathematical model may help us get a glimpse into the future, but don't expect a mathematical model to tell you something like the content of the next test. A model may provide a glimpse into the future when a recent trend is known. For example, parents of young children wishing to save for their college expenses find it helpful to have an estimate of college expenses when their children become college age. A company that considers expanding could make better decisions if they had an estimate of sales for the next five years. An estimate of future trends of the growth, or decline, of the economy, and the unemployment rate might give you a better idea of your job prospects upon graduation.

One simple mathematical model for estimating trends assumes that the trend is linear and can be explained with a linear equation. It uses past and present data to determine the trend. In this section we will determine linear equations to estimate future trends. *Warning*: When a trend is *not* linear, you may still find a linear equation, but don't expect it to give you reliable estimates.

## Applications

In practical problems, the relationship between the **variables** can be quite complicated. For example, the variables and their relationships that affect the stock market still defy the best analysts. However, many times a linear relationship can be used to provide a reasonable and useful model for solving practical problems. We now look at several applications of the linear function.

## Cost–Volume Function

The manufacturer of a home theater system conducted a study of production costs and found that fixed costs averaged $5600 per week and component costs averaged $759 per system. This information can be stated as

$$C = 759x + 5600$$

where $x$ represents the number of systems produced per week, also called the **volume**, and $C$ is the total weekly cost of producing $x$ systems. A linear function like this is used when:

1. there are **fixed costs**—such as rent, utilities, and salaries—that are the same each week, independent of the number of items produced;
2. there are **variable costs** that depend on the number of items produced, such as the cost of materials for the items, packaging, and shipping costs.

The home theater system example illustrates a linear **cost–volume function** (often simply called the *cost function*). A linear function is appropriate when the general form of the cost function $C$ is given by

$$C(x) = ax + b$$

where

> $x$ is the number of items produced (**volume**),
>
> $b$ is the **fixed cost** in dollars,
>
> $a$ is the **unit cost** (the cost per item) in dollars,
>
> $C(x)$ is the **total cost** in dollars of producing $x$ items.

Note the form of the cost function. It is essentially the slope-intercept form of a line, where the unit cost gives the slope and the fixed cost gives the $y$-intercept.

**Example 1**    If the cost of manufacturing $x$ home theater systems per week is given by

$$C(x) = 415x + 5200$$

then:

**(a)** Determine the unit cost and the fixed cost.

**(b)** Determine the cost of producing 700 systems per week.

**(c)** Determine how many systems were produced if the production cost for one week was $230,130.

**Solution**

**(a)** The unit cost is $415, and the fixed cost is $5200 per week.

**(b)** Substitute $x = 700$ into the cost equation to obtain

$$C(700) = 415(700) + 5200$$
$$= 290,500 + 5200$$
$$= 295,700$$

So, the total cost is $295,700. (See Figure 1–16.)

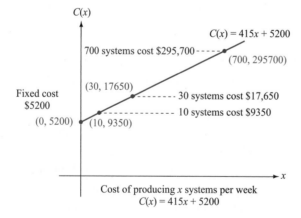

**FIGURE 1–16**

**(c)** This information gives $C(x) = 230,130$, so we have

$$230,130 = 415x + 5200$$

We need to solve this for $x$:

$$230{,}130 - 5200 = 415x$$
$$224{,}930 = 415x$$
$$x = \frac{224{,}930}{415} = 542$$

So, 542 systems were produced that week.

The unit cost and fixed costs may not be known directly, but the cost function can be obtained if the information gives two points on the line, as illustrated in the next example.

**Example 2**  A company made a cost study and found that it cost $10,170 to produce 800 pairs of running shoes and $13,810 to produce 1150 pairs.

**(a)** Determine the cost–volume function.

**(b)** Find the fixed cost and the unit cost.

**Solution**

**(a)** Let $x =$ the number of pairs. The information gives two points on the cost–volume line: (800, 10170) and (1150, 13810). The slope of the line is

$$m = \frac{13{,}810 - 10{,}170}{1150 - 800} = \frac{3640}{350} = 10.40$$

Using the point (800, 10170) in the point-slope equation, we have

$$y - 10{,}170 = 10.40(x - 800)$$
$$y = 10.40x + 1850$$

Therefore, $C(x) = 10.40x + 1850$.

**(b)** From the equation $C(x) = 10.40x + 1850$, the fixed cost is $1850 per week, and the unit cost is $10.40.

## Revenue Function

T's and More specializes in T-shirts for fraternities and sororities. They sell their T-shirts for $13.50 each. The total income (**revenue**) in dollars is 13.50 times the number of shirts sold. This illustrates the general concept of a **revenue function**; it gives the total revenue from the sale of $x$ items. This example gives the revenue function as

$$R(x) = 13.50x$$

where $x$ represents the number of T-shirts sold, 13.50 is the selling price in dollars for each item, and $R(x)$ is the total revenue in dollars from $x$ items.

**Example 3**    T's and More has a sale on T-shirts at $12.25 each.

(a) Give the revenue function.

(b) The store sold 213 T-shirts. What was the total revenue?

(c) One sorority bought $575.75 worth of T-shirts. How many did they buy?

**Solution**

(a) The revenue function is given by

$$R(x) = 12.25x$$

(b) The revenue for 213 shirts is obtained from the revenue function when $x = 213$:

$$R(213) = 12.25(213) = 2609.25$$

The total revenue was $2609.25.

(c) This gives $R(x) = 575.75$, so

$$575.75 = 12.25x$$
$$x = \frac{575.75}{12.25} = 47$$

The sorority bought 47 T-shirts.

## Break-Even Analysis

**Break-even analysis** answers a common management question: At what sales volume will we break even? When do revenues equal costs? Greater sales will induce a profit, whereas lesser sales will show a loss.

The **break-even point** occurs when the cost equals the revenue, so the cost and revenue functions can be used to determine the break-even point. Using function notation, we write this as $C(x) = R(x)$.

**Break-Even Point**    The point at which cost equals revenue:

$$C(x) = R(x)$$

**Example 4**    Cox's Department Store pays $99 each for DVD players. The store's monthly fixed costs are $1250. The store sells the DVD players for $189.95 each.

(a) What is the cost–volume function?

(b) What is the revenue function?

(c) What is the break-even point?

**Solution**

Let $x$ represent the number of DVD players sold.

(a) The cost function is given by

$$C(x) = 99x + 1250$$

**(b)** The revenue function is defined by

$$R(x) = 189.95x$$

**(c)** The break-even point occurs when cost equals revenue,

$$C(x) = R(x)$$

Writing out the function gives

$$99x + 1250 = 189.95x$$

The solution of this equation gives the break-even point:

$$
\begin{aligned}
99x + 1250 &= 189.95x \\
1250 &= 189.95x - 99x \\
1250 &= 90.95x \\
x &= \frac{1250}{90.95} \\
&= 13.74
\end{aligned}
$$

Because $x$ represents the number of DVD players, we use the next integer, 14, as the number sold per month to break even. If 14 or more are sold, there will be a profit. If fewer than 14 are sold, there will be a loss.

The break-even point occurs at $x = 14$ DVD players, with a revenue of $2659.30.

Geometrically (see Figure 1–17), the break-even point occurs where the cost function and revenue function intersect. A profit occurs for the values of $x$ when the graph of the revenue function is above the cost function. At a given value of $x$, the profit is the vertical distance between the graphs. Similarly, a loss occurs when the graph of the revenue function is below the graph of the cost function and the vertical distance between the graphs represents the amount of loss.

**NOTE**

As shown in Figure 1–17, using $x = 13.74$ at the break-even point gives a revenue of $2610. This ignores the restriction of $x$ to integer values.

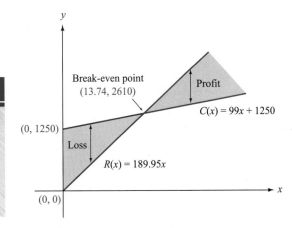

**FIGURE 1–17** A graph showing
Cost function:        $C(x) = 99x + 1250$
Revenue function:   $R(x) = 189.95x$
Break-even point:    $(13.74, 2610)$

**Example 5**   A temporary secretarial service has a fixed weekly cost of $896. The wages and benefits of the secretaries amount to $7.65 per hour. A firm that employs a secretary pays Temporary Service $10.40 per hour. How many hours per week of secretarial service must Temporary Service place to break even?

**Solution**

First, write the cost and revenue functions. The fixed cost is $896, and the unit cost is $7.65, so the cost function is given by

$$C(x) = 7.65x + 896$$

where $x$ is the number of hours placed each week. The revenue function is given by

$$R(x) = 10.40x$$

Equating cost and revenue, we have

$$7.65x + 896 = 10.40x$$

This equation reduces to

$$
\begin{aligned}
10.40x - 7.65x &= 896 \\
2.75x &= 896 \\
x &= \frac{896}{2.75} \\
&= 325.8
\end{aligned}
$$

Temporary Service must place secretaries for a total of 326 hours per week (rounded up) to break even.

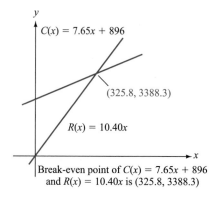

Break-even point of $C(x) = 7.65x + 896$ and $R(x) = 10.40x$ is (325.8, 3388.3)

**FIGURE 1–18**

## Profit Function

The break-even point gives the boundary between profit and loss. To determine the profit (or loss) for a given level of production, we can use the profit function.

**DEFINITION**   The **profit function** is the difference between revenue and cost:
$$P(x) = R(x) - C(x)$$
where

$$
\begin{aligned}
P(x) &= \text{profit function} \\
R(x) &= \text{revenue function} \\
C(x) &= \text{cost function}
\end{aligned}
$$

**Example 6**  Find the profit function when

$$R(x) = 21.6x$$
$$C(x) = 8.5x + 95$$

**Solution**
The profit function, $P(x)$, is

$$P(x) = R(x) - C(x)$$
$$= 21.6x - 8.5x - 95$$
$$= 13.1x - 95$$

**Example 7**  For the revenue and cost functions, let

$$R(x) = 15x$$
$$C(x) = 6.40x + 142$$

(a) Find the profit function.

(b) Graph the revenue, cost, and profit functions.

(c) Find the profit for $x = 10, 30, 95,$ and $220$.

**Solution**
(a) $P(x) = R(x) - C(x)$
$$= 15x - 6.40x - 142$$
$$= 8.60x - 142$$

(b)

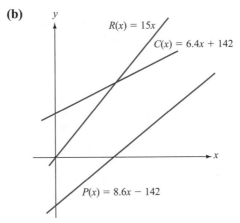

**FIGURE 1–19**

(c) At $x = 10$, $P(10) = -56$ (a loss of 56)

At $x = 30$, $P(30) = 116$

At $x = 95$, $P(95) = 675$

At $x = 220$, $P(220) = 1750$

## Straight-Line Depreciation

To prepare a tax return and report the financial condition of a company to its stockholders, a company needs to estimate the value of buildings, equipment, and so on. Part of this procedure sometimes uses a linear function to estimate the value of equipment. For instance, when a corporation buys a fleet of cars, it expects them to decline in value because of wear and tear. For example, if the corporation purchases new cars for $17,500 each, they may be worth only $5500 each three years later, due to normal wear and tear. This decline in value is called **depreciation**. The value of an item after deducting depreciation is called its **book value**. In three years, each car depreciated $12,000, and its book value at the end of three years was $5500. For tax and accounting purposes, each year during the life of an item, a company will report depreciation as an expense and book value as an asset. The Internal Revenue Service allows several methods of depreciation. The simplest is **straight-line depreciation**. This method assumes that the book value is a linear function of time—that is,

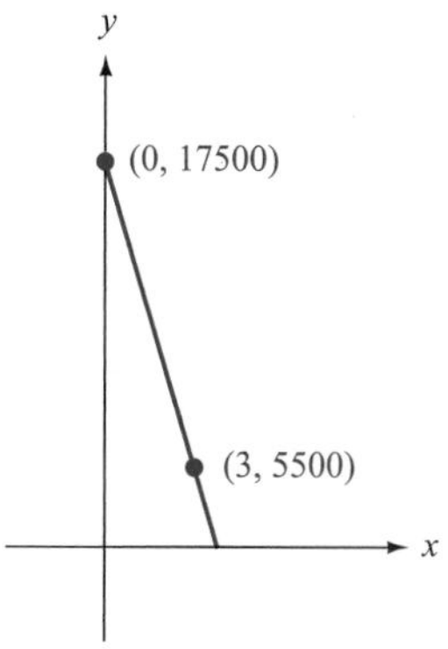

$$B = mx + b$$

where $B$ is the book value and $x$ is the number of years. For example, each car had a book value of $17,500 when $x = 0$ ("brand new" occurs at zero years). When $x = 3$, the car's book value declined to $5500. This information is equivalent to giving two points (0, 17500) and (3, 5500) on the straight line representing book value. (See Figure 1–20.)

We obtain the linear equation of the book value by finding the equation of a line through these two points. The slope of the line is

$$m = \frac{y_2 - y_1}{x_2 - x_1} = \frac{5500 - 17,500}{3 - 0} = \frac{-12,000}{3} = -4000$$

and the $y$-intercept is 17,500, so the equation is

$$B = -4000x + 17,500$$

The book value at the end of two years is

$$B = -4000(2) + 17,500 = -8000 + 17,500 = 9500$$

The *negative* value of the slope indicates a *decrease* in the book value of $4000 each year. This annual decrease is the **annual depreciation**.

Generally, a company will estimate the number of years of useful life of an item. The estimated value of the item at the end of its useful life is called its **scrap value**. We restrict the values of $x$ to $0 \le x \le n$, where $n$ represents the number of years of useful life.

Here is a simple example.

**FIGURE 1–20**
Straight-line depreciation.

---

**Example 8**    Acme Manufacturing Co. purchases a piece of equipment for $28,300 and estimates its useful life as eight years. At the end of its useful life, its scrap value is estimated at $900.

**(a)** Find the linear equation expressing the relationship between book value and time.

**(b)** Find the annual depreciation.

**(c)** Find the book value at the end of the first, fifth, and seventh years.

### Solution

**(a)** The line passes through the two points (0, 28300) and (8, 900). Therefore, the slope is

$$m = \frac{900 - 28{,}300}{8 - 0} = -3425$$

and the $y$-intercept is 28,300, giving the equation

$$B = -3425x + 28{,}300$$

(See Figure 1–21.)

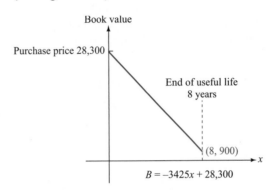

Book value

Purchase price 28,300

End of useful life
8 years

(8, 900)

$x$

$B = -3425x + 28{,}300$

**FIGURE 1–21**   Book value.

**(b)** The annual depreciation is obtained from the slope and is $3425.

**(c)** For year 1, $B = -3425(1) + 28{,}300 = 24{,}875$. For year 5, $B = -3425(5) + 28{,}300 = 11{,}175$. For year 7, $B = -3425(7) + 28{,}300 = 4325$. Thus, after seven years the book value of the equipment is $4325.

## Linear Trends

Many people like to ask "what if" questions such as:

> If worldwide temperatures increase two degrees, how much coastal area will be flooded?
>
> If I don't do my homework for tomorrow, will we have a quiz?
>
> If our quarterback can't play, can we win?

A variation of this question finds people observing trends and making projections such as:

> Will the price of textbooks continue to increase, and by how much?
>
> What will the job market be like when I graduate?
>
> The per capita federal debt increased from $21,600 in 2002 to $45,000 in 2011.
>
> If this trend continues, what will be my share of the federal debt in 2020?

Mathematical models can help analyze some trends and make projections based on the trends. Linear functions sometimes describe a trend quite well and sometimes describe a trend poorly.

**Example 9**    Studies indicate that the number of overweight people in the United States has increased, posing many health concerns. In 1994, 55% of the adults in the United States were overweight; this increased to 67% by 2008.

(a) Find the linear equation that describes this trend.

(b) Use the equation to estimate the percentage of people overweight in 2000.

(c) Use the equation to estimate when the percentage reaches 100%.

**Solution**

(a) We let $x$ = number of years since 1994, so $x = 0$ for 1994 and $x = 14$ for 2008. The variable $y$ represents the percentage of the population that is overweight. We then have the points $(0, 55)$ and $(14, 67)$. We find the slope

$$m = \frac{67 - 55}{14 - 0} = \frac{12}{14} = 0.86$$

Since the $y$-intercept is 55, the equation becomes

$$y = 0.86x + 55$$

(b) For 2000, $x = 6$ so

$$y = 0.86(6) + 55 = 60$$

This estimates that 60% of U.S. adults were overweight in 2000. The National Center for Health Statistics found the percentage to be 65%. Thus, the equation gave a poor estimate.

(c) The proportion of overweight adults reaches 100% when $y = 100$.

$$100 = 0.86x + 55$$
$$0.86x = 45$$
$$x = \frac{45}{0.86} = 52$$

So, in the year $1994 + 52 = 2046$, it is predicted that all adults will be overweight. However, we don't expect there will ever be a time when everyone is overweight. This illustrates that estimates of trends become less reliable as you move further away from the years upon which it is based.

**Example 10**    In the past three years donations to the Brazos Valley College Scholarship Fund have totaled $8.55 million, $9.22 million, and $7.95 million, respectively. In planning the next year's budget (fourth year), the Scholarship Committee wants an estimate of the fourth-year donations. To do so, let $x$ = the year ($x = 1, 2, 3$) and $y$ = the donations in millions of dollars. Determine the linear equation using the points in (a) through (c):

(a) $(1, 8.55)$ and $(2, 9.22)$

(b) $(1, 8.55)$ and $(3, 7.95)$

(c) $(2, 9.22)$ and $(3, 7.95)$

(d) Use each linear function in (a) through (c) to estimate the fourth year's donations.

**Solution**

(a) $m = \dfrac{9.22 - 8.55}{2 - 1} = \dfrac{0.67}{1}$

$y - 8.55 = 0.67\,(x - 1)$

$y = 0.67x + 7.88$

(b) $m = \dfrac{7.95 - 8.55}{3 - 1} = \dfrac{-0.60}{2} = -0.30$

$y - 8.55 = -0.30\,(x - 1)$

$y = -0.30x + 8.85$

(c) $m = \dfrac{7.95 - 9.22}{3 - 2} = \dfrac{-1.27}{1}$

$y - 9.22 = -1.27\,(x - 2)$

$y = -1.27x + 11.76$

(d) For (a), $x = 4$ so $y = 0.67(4) + 7.88 = 10.56$

This estimated fourth-year donations total $10.56 million.

For (b), $x = 4$ so $y = -0.30(4) + 8.85 = 7.65$

This estimated fourth-year donations total $7.65 million.

For (c), $x = 4$ so $y = -1.27(4) + 11.76 = \$6.68$

This estimated fourth-year donations total $6.68 million.

The variation in the estimated fourth-year donations of $7.65 million, $10.56 million, and $6.68 million suggest that a linear equation does not adequately estimate the trend.

## Applications of Linear Inequalities

For some applications, an interval of values—instead of one specific value—gives an appropriate answer to the problem. These situations can often be represented by inequalities. We give two examples using linear inequalities. (See Appendix A.4 for a review of the properties of linear inequalities.)

**Example 11**   A doughnut shop sells doughnuts for $3.29 per dozen. The shop has fixed weekly costs of $650 and unit costs of $1.55 per dozen. How many dozens of doughnuts must be sold weekly for the shop to make a profit?

**Solution**
Let $x =$ number of dozens of doughnuts sold per week. The revenue and cost functions are

$$R(x) = 3.29x$$
$$C(x) = 1.55x + 650$$

The shop makes a profit when the revenue exceeds the costs—that is, when $R(x) > C(x)$. Therefore, we want to solve

$$3.29x > 1.55x + 650$$
$$3.29x - 1.55x > 650$$
$$1.74x > 650$$
$$x > \dfrac{650}{1.74} = 373.56$$

Thus, at least 374 dozen doughnuts must be sold in order to make a profit. The interval notation for this solution is $[374, \infty)$. (See Figure 1–22.)

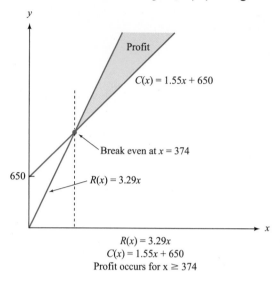

$$R(x) = 3.29x$$
$$C(x) = 1.55x + 650$$
Profit occurs for x ≥ 374

**FIGURE 1–22**

The following example illustrates how an analysis of an inequality can help decide on the best course of action.

**Example 12**     A quick-copy store can select from two plans to lease a copy machine. Plan A costs $75 per month plus five cents per copy. Plan B costs $200 per month plus two cents per copy. When will it be to the copy shop's advantage to lease under plan A?

**Solution**
Let $x$ = number of copies per month. Then the monthly costs are as follows:

$$\text{Plan A:} \quad CA(x) = \;\; 75 + 0.05x$$
$$\text{Plan B:} \quad CB(x) = 200 + 0.02x$$

Plan A is better when the cost of Plan A is less than the cost of Plan B, $CA(x) < CB(x)$.

We find the solution by solving

$$75 + 0.05x < 200 + 0.02x$$
$$0.05x - 0.02x < 200 - 75$$
$$0.03x < 125$$
$$x < \frac{125}{0.03} = 4166.7$$

Plan A is better when the number of copies per month is fewer than 4167 copies. The interval notation for this answer is $(0, 4167)$.

## **1.3**    EXERCISES

Access end-of-section exercises online at **www.webassign.net**    WebAssign

 ### **Using Your TI Graphing Calculator**

### Intersection of Lines

Finding the break-even point requires finding the intersection of two lines. We illustrate by finding the intersection of the lines

$$y = 2x + 5$$
$$3x + 4y = 12$$

Graph the two equations using Y= and GRAPH :

The intersection of the two lines can then be found using the **intersect** command. With the graph of the two lines on the screen, press CALC (2nd TRACE) and select <5 : intersect> from the menu. Press ENTER three times, and the coordinates of the intersection will appear as in the following window:.

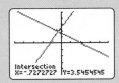

*Note:* Notation such as <5 : intersect> indicates a command to be selected from a menu.

### Exercises

Use the **intersect** command to find the intersection of the following pairs of lines.

**1.** $y = 2x - 1$
   $y = 1.75x$

**2.** $y = 3x - 11.5$
   $y = 2x - 5$

**3.** $y = -4x + 18$
   $y = 1.5x - 5.5$

**4.** $y = 6.2x + 1.4$
   $y = -2.1x + 9.3$

**5.** $3x + 4y = 30$
   $5x - 2y = -2$

**6.** $3.1x + 1.4y = 19.46$
   $5.7x - 2.3y = 6.46$

# Using Excel

## Example 1

Let's use Excel to find the profit for the cost and revenue functions

$$C(x) = 38x + 648$$
$$R(x) = 95x$$

where $x$ = number of items. Find the profit for selling 10, 58, 72, 109, and 223 items.

Here is the Excel template to calculate $C(x)$, $R(x)$, and the profit, $P(x)$.

| | A | B | C | D |
|---|---|---|---|---|
| 1 | X | C(x) | | R(x) | P(x) |
| 2 | 10 | 1028 | 950 | -78 |
| 3 | 58 | 2852 | 5510 | 2658 |
| 4 | 72 | 3384 | 6840 | 3456 |
| 5 | 109 | 4790 | 10355 | 5565 |
| 6 | 223 | 9122 | 21185 | 12063 |
| 7 | | | | |
| 8 | | Enter the formula | Enter the formula | Enter the formula |
| 9 | | =38*A2+648 in B2 | =95*A2 in C2 and | =C2-B2 in D2 and |
| 10 | | and drag down to B6 | drag down to C6 | drag down to D6 |
| 11 | | | | |

Note that for $x = 10$ there is a loss of 78.

Excel has a tool called **Goal Seek** that can be used to find the intersection of two linear functions such as when you want to find the break-even point.

## Example 2

Find the break-even point for the cost and revenue functions

$$C(x) = 3.6x + 9895$$
$$R(x) = 12.6x$$

We set up the spreadsheet as shown below with entries in Line 1 to identify the entries in Line 2 that are used in the analysis. The entry of 10 for $x$ is an arbitrary value.

Assign the cell A2 for the location of $x$ and put the equation of $C(x)$ in B2, the equation for $R(x)$ in C2, and the equation for $R(x) - C(x)$ in D2.

| | A | B | C | D |
|---|---|---|---|---|
| 1 | X | C(x) | R(x) | R(x)-C(x) |
| 2 | 10 | =3.6*A2+9895 | =12.6*A2 | =C2-B2 |
| 3 | | | | |
| 4 | | | | |

The calculations for this are

| | A | B | C | D |
|---|---|---|---|---|
| 1 | X | C(x) | R(x) | R(x)-C(x) |
| 2 | 10 | 9931 | 126 | -9805 |
| 3 | | | | |
| 4 | | | | |

We will use the formula in D2, $R(x) - C(x)$, to find the intersection, because this difference is zero at the intersection. Now we are ready to use this spreadsheet to find the intersection of the cost and revenue lines.

1. Select **Goal Seek** under the What-If Analysis option in the Data tab. This will bring up the following screen:

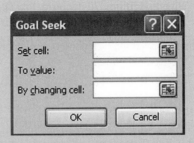

2. Place the cursor in the blank to the right of **Set cell:** and then select the cell D2 on the spreadsheet.
3. Move the cursor to the blank to the right of **To value:** and enter the number zero.
4. Move the cursor to the blank to the right of **By changing cell:** and then select cell A2 in the spreadsheet. This gives you the following screen:

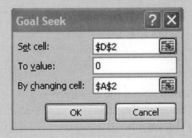

5. Click on OK and the spreadsheet shows

| | A | B | C | D |
|---|---|---|---|---|
| 1 | X | C(x) | R(x) | R(x)-C(x) |
| 2 | 1099.44444 | | 13853 | 13853 | 0 |
| 3 | | | | |
| 4 | | | | |

Note that the value of $x$, 1099.4444, gives the same cost and revenue value, 13,853, so we have found the break-even point (1099.4444, 13853). When $x$ represents a number of items, we would round it to 1099.

The **Goal Seek** procedure we have just used seeks the value of $x$ that makes the value of D2, $R(x) - C(x)$, equal to the value placed in **To value,** namely zero. If you use the same procedure and enter 1000 in **To value,** **Goal Seek** will seek the value of $x$ that makes $R(x) - C(x)$ equal to 1000, that is, find the $x$ that yields a profit of 1000. Try it. You should get $x = 1210.55556$.

You can use the same spreadsheet to find the loss when $x = 900$. Enter 900 in A2, do not use **Goal Seek,** and D2 shows $-1795$, a loss of $1795.

## Exercises

1. $C(x) = 10x + 535$ and $R(x) = 35x$
   Calculate profit for $x = 20, 30, 45, 62,$ and 81.
2. $C(x) = 54x + 1250$ and $R(x) = 97.50x$
   Calculate profit for $x = 225, 315, 450,$ and 620.

*(continued)*

3. $C(x) = 537.60x + 7431$ and $R(x) = 779x$

Calculate profit for $x = 25, 100, 250, 350$, and $560$.

4. Find the break-even point for the cost and revenue functions.

$$C(x) = 5x + 450 \quad \text{and} \quad R(x) = 12.5x$$

5. Find the break-even point for the cost and revenue functions.

$$C(x) = 22.2x + 1165 \quad \text{and} \quad R(x) = 48.6x$$

6. **(a)** Find the break-even point for the cost and revenue functions.

$$C(x) = 3.6x + 224.64 \quad \text{and} \quad R(x) = 18x$$

**(b)** Find the value of $x$ for which profit $= 1500$.

**(c)** Find the profit, or loss, when $x = 10$, when $x = 33$.

# IMPORTANT TERMS

**1.1**

Function
Rule of a Function
Domain
Range
Independent Variable
Dependent Variable
Function Value

**1.2**

Graph
Graph of a Function
Solution
Linear Function
$y$-Intercept

Slope
Constant Function
Horizontal Line
Vertical Line
Slope-Intercept Equation
Point-Slope Equation
Two-Point Equation
$x$-Intercept
Parallel Lines
Perpendicular Lines

**1.3**

Mathematical Model
Variables
Volume

Fixed Costs
Variable Costs
Cost–Volume Function
Unit Costs
Total Cost
Revenue
Revenue Function
Break-Even Analysis
Break-Even Point
Profit Function
Depreciation
Book Value
Straight-Line Depreciation
Annual Depreciation
Scrap Value

# IMPORTANT CONCEPTS

**Function:**    A rule, often an equation in $x$ and $y$, that assigns exactly one value of $y$ to each $x$ in the domain.

**Graph of a Function**    The set of points $(x, y)$ in the plane that satisfy the equation $y = f(x)$.

**Linear Function**    A function that can be written $f(x) = mx + b$ and whose graph is a straight line.

**Slope Formula**    The slope, $m$, of a line through two points $(x_1, y_1)$ and $(x_2, y_2)$ is given by the equation

$$m = \frac{y_2 - y_1}{x_2 - x_1}, \; x_2 \neq x_1$$

| | |
|---|---|
| **Equations of a Linear Function** | *Standard Equation:* $Ax + By = C$ where $(x, y)$ is any point on the line and at least one of $A$, $B$ is not zero. |
| | *Slope-intercept Equation:* $y = mx + b$ where $m$ is the slope and $b$ the $y$-intercept. |
| | *Point-slope Equation:* $y - y_1 = m(x - x_1)$ where $m$ is the slope and $(x_1, y_1)$ is a given point on the line. |
| | *Horizontal Line:* $y = k$ where $(h, k)$ is on the line. The slope is zero. |
| | *Vertical Line:* $x = h$ where $(h, k)$ is on the line. The slope is undefined. |
| | *Parallel Lines:* Lines with the same slope or are vertical lines. |
| | *Perpendicular Lines:* Two lines with slopes $m_1$ and $m_2$ where $m_1 m_2 = -1$. |
| **Cost Function** | $C(x) = ax + b$ where $x$ is the number of items produced, $b$ is the fixed cost, and $a$ is the unit cost. |
| **Revenue Function** | $R(x) = ax$ where $x$ is the number of items sold and $a$ is the unit price. |
| **Break-Even Point** | The point at which cost equals revenues: $R(x) = C(x)$ |
| **Profit Function** | The difference between revenue and cost: $P(x) = R(x) - C(x)$ |

## REVIEW EXERCISES

**1.** If $f(x) = \dfrac{7x - 3}{2}$, find

(a) $f(5)$  (b) $f(1)$

(c) $f(4)$  (d) $f(b)$

**2.** If $f(x) = 8x - 4$, find

(a) $f(2)$  (b) $f(-3)$

(c) $f\left(\dfrac{1}{2}\right)$  (d) $f(c)$

**3.** If $f(x) = \dfrac{x + 2}{x - 1}$ and $g(x) = 5x + 3$, find $f(2) + g(3)$.

**4.** If $f(x) = (x + 5)(2x - 1)$, find

(a) $f(1)$  (b) $f(0)$

(c) $f(-5)$  (d) $f(a - 5)$

**5.** Apples cost $1.20 per pound, so the price of a bag of apples is $f(x) = 1.20x$, where $x$ is the weight in pounds and $f(x)$ is the purchase price in dollars.

(a) What is $f(3.5)$?

(b) A bag of apples costs $3.30. How much did it weigh?

**6.** Tuition and fees charges at a university are given by

$$f(x) = 135x + 450$$

where $x$ is the number of semester hours enrolled and $f(x)$ is the total cost of tuition and fees.

(a) Find $f(15)$.

(b) A student's bill for tuition and fees was $2205; for how many semester hours was she enrolled?

**7.** Write an equation of the function described by the following statements.

(a) All the shoes on this table are $29.95 per pair.

(b) A catering service charges $40 plus $2.25 per person to cater a reception.

**8.** Sketch the graph of

(a) $f(x) = 2x - 5$  (b) $6x + 10y = 30$

**9.** Graph the following lines:

(a) $y = 3x - 5$  (b) $y = -7$

(c) $x = 5.5$  (d) $y = x$

**10.** Graph the following lines:

(a) $y = 6.5$  (b) $x = -4.75$

(c) $y = -1.3$  (d) $x = 7$

**11.** Find the slope and $y$-intercept for the following lines:

(a) $y = -2x + 3$  (b) $y = \dfrac{2}{3}x - 4$

(c) $4y = 5x + 6$  (d) $6x + 7y + 5 = 0$

**12.** Find the slope of the line through the following pairs of points:

  **(a)** $(2, 7)$ and $(-3, 4)$     **(b)** $(6, 8)$ and $(-11, 8)$

  **(c)** $(4, 2)$ and $(4, 6)$

**13.** For the line $6x + 5y = 15$, find

  **(a)** the slope          **(b)** the $y$-intercept

  **(c)** the $x$-intercept

**14.** For the line $-2x + 9y = 6$, find

  **(a)** the slope          **(b)** the $y$-intercept

  **(c)** the $x$-intercept

**15.** Find an equation of the following lines:

  **(a)** With slope $-\dfrac{3}{4}$ and $y$-intercept 5

  **(b)** With slope 8 and $y$-intercept $-3$

  **(c)** With slope $-2$ and passing through $(5, -1)$

  **(d)** With slope 0 and passing through $(11, 6)$

  **(e)** Passing through $(5, 3)$ and $(-1, 4)$

  **(f)** Passing through $(-2, 5)$ and $(-2, -2)$

  **(g)** Passing through $(2, 7)$ and parallel to $4x - 3y = 22$

**16.** Find an equation of the line with the given slope and passing through the given point:

  **(a)** $m = 5$ and point $(2, -1)$

  **(b)** With slope $-\dfrac{2}{3}$ and point $(5, 4)$

  **(c)** With $m = 0$ and point $(7, 6)$

  **(d)** With slope 1 and point $(-2, -2)$

**17.** Find an equation of the line through the given points:

  **(a)** $(6, 2)$ and $(-3, 2)$

  **(b)** $(-4, 5)$ and $(-4, -2)$

  **(c)** $(5, 0)$ and $(5, 10)$

  **(d)** $(-7, 6)$ and $(7, 6)$

**18.** Determine whether the following pairs of lines are parallel:

  **(a)** $7x - 4y = 12$  and  $-21x + 12y = 17$

  **(b)** $3x + 2y = 13$  and  $2x - 3y = 28$

**19.** Is the line through $(5, 19)$ and $(-2, 7)$ parallel to the line through $(11, 3)$ and $(-1, -5)$?

**20.** Is the line through $(4, 0)$ and $(7, -2)$ parallel to the line through $(7, 4)$ and $(10, 2)$?

**21.** Determine whether the line through $(8, 6)$ and $(-3, 14)$ is parallel to the line $8x + 4y = 34$.

**22.** Determine whether the line through $(-2.5, 0)$ and $(-1, 4.5)$ is parallel to the line $3x - y = 19$.

**23.** Determine whether the line through $(9, 10)$ and $(5, 6)$ is parallel to the line $3x - 2y = 14$.

**24.** Determine whether the following pairs of lines are parallel:

  **(a)** $y = 5x + 13$       **(b)** $6x + 2y = 15$
      $y = 5x - 24$            $15x + 5y = -27$

  **(c)** $-8x + 9y = 41$     **(d)** $12x - 5y = 60$
      $9x - 8y = 13$            $6x + y = 15$

**25.** A manufacturer has fixed costs of $12,800 per month and a unit cost of $36 per item produced. What is the cost function?

**26.** The weekly cost function of a manufacturer is

$$C(x) = 83x + 960$$

  **(a)** What are the weekly fixed costs?

  **(b)** What is the unit cost?

**27.** The cost function of producing $x$ bags of Hi-Gro fertilizer per week is

$$C(x) = 3.60x + 2850$$

  **(a)** What is the cost of producing 580 bags per week?

  **(b)** If the production costs for one week amounted to $5208, how many bags were produced?

**28.** The Shoe Center has a special sale in which all jogging shoes are $28.50 per pair. Write the revenue function for jogging shoes.

**29.** A T-shirt shop pays $6.50 each for T-shirts. The shop's weekly fixed expenses are $675. It sells the T-shirts for $11.00 each.

  **(a)** What is the revenue function?

  **(b)** What is the cost function?

  **(c)** What is the break-even point?

**30.** Midstate Manufacturing sells calculators for $17.45 each. The unit cost is $9.30, and the fixed cost is $17,604. Find the quantity that must be sold to break even.

**31.** The comptroller of Southern Watch Company wants to find the company's break-even point. She has the following information: The company sells the watches for $19.50 each. One week it produced

1840 watches at a production cost of $25,260.00. Another week it produced 2315 watches at a production cost of $31,102.50. Find the weekly volume of watches the company must produce to break even.

**32.** Find an equation of the line described in the following:

    **(a)** Through the point (5, 7) and parallel to $6x - y = 15$

    **(b)** Through the point $(-2, -5)$ and parallel to the line through $(4, -2)$ and $(9, 5)$

    **(c)** With $y$-intercept 6 and passing through $(2, -5)$

**33.** Norton's Inc. purchases a piece of equipment for $17,500. The useful life is eight years, and the scrap value at the end of eight years is $900.

    **(a)** Find the equation relating book value and its age using straight-line depreciation.

    **(b)** What is the annual depreciation?

    **(c)** What is the book value for the fifth year?

**34.** The function for the book value of a truck is

$$f(x) = -2300x + 16,500$$

where $x$ is its age and $f(x)$ is its book value.

    **(a)** What did the truck cost?

    **(b)** If its useful life is seven years, what is its scrap value?

**35.** An item cost $1540, has a useful life of five years, and has a scrap value of $60. Find the equation relating book value and number of years using straight-line depreciation.

**36.** Find the $x$- and $y$-intercepts of the line $8x + 6y = 24$ and sketch its graph.

**37.** A line passes through the point (2, 9) and is parallel to $4x - 5y = 10$. Find the value of $k$ so that $(-3, k)$ is on the line.

**38.** Lawn Care Manufacturing estimates that the material for each lawnmower costs $85. The manufacturer's fixed operating costs are $4250 per week. Find the company's weekly cost function.

**39.** Nguyen's Auto Parts bought a new delivery van for $22,000. Nguyen expects the van to be worth $3000 in five years. Find the value of the van as a linear function of its age.

**40.** A manufacturer found that it cost $48,840 to produce 940 items in one week. The next week it cost $42,535 to produce 810 items. Find the linear cost–volume function.

**41.** A hamburger place estimates that the materials for each hamburger cost $0.67. One day it made 1150 hamburgers, and the total operating costs were $1250.50. Find the cost function.

**42.** A fast-food franchise owner must pay the parent company $1200 per month plus 4.1% of receipts. Find the monthly franchise cost function.

**43.** A company offers an inventor two royalty options for her product. The first is a one-time-only payment of $17,000. The second is a payment of $2000 plus 75 cents for each item sold. Determine when the second option is better than the first.

**44.** A university estimates that 92% of the applicants who pay the admissions deposit will enroll. How many applicants who pay their admissions deposit are required in order to have a class of 2300 students?

**45.** Find the value of $k$ so that the points (9, 4) and $(-2, k)$ lie on a line with slope $-2$.

**46.** A caterer's fee to cater a wedding reception is a linear function based on a fixed fee and an amount per person. Stephanie and Roy's wedding was planned for 350 guests and cost $3475. Jennifer and Brett's wedding was planned for 290 guests and cost $2695. Find the cost function.

**47.** A manufacturer of bobble-head dolls uses a linear cost function. One week, the company produced 1730 dolls for a total cost of $12,813.60. The unit cost is known to be $6.82. Find the fixed cost.

**48.** Jones established a small business with a reserve of $12,000 to cover operating expenses in the early stages when a loss is expected. The reserve is reduced $620 each week to cover operating costs.

    **(a)** Write the function giving the amount remaining in the reserve fund.

    **(b)** How much is in the fund after eight weeks?

    **(c)** How long will it take to deplete the fund?

**49.** Juan is a salesman for L.L. Bowers Corp. He has a choice of three compensation plans. Plan 1 pays $2500 per month. Plan 2 pays $2000 per month plus 15% commission. Plan 3 pays $1700 per month plus 30% commission. Graph the three plans and determine which is best.

**50.** The Grounds, a coffee shop, has average daily fixed costs of $215 per day and unit costs of $0.85 per cup of coffee. The coffee sells for $2.25 per cup.

   **(a)** Find the daily average number of cups of coffee that must be sold to break even.

   **(b)** After several months fixed costs rise to $265 per day. At that time, the manager introduces a small cup of coffee that contains half as much coffee and is priced at $1.25 per cup. The regular size cup remains at $2.25. At the break-even point, find the relationship between the number of small and regular sizes that must be sold.

**51.** The Homewood branch library opened with 8400 books. The budget provides for 120 additional books each month.

   **(a)** Write the equation that gives the number of books $x$ months after the library opened.

   **(b)** When will the library have 10,000 books?

**52.** The Uganda population in 1996 was 20.2 million and was 33.4 million in 2010.

   **(a)** Use this information to find a linear equation that estimates population as a function of years since 1996.

   **(b)** Use the equation to estimate Uganda population in 2025 and 2050.

   **(c)** The U.S. Census Bureau projects a Uganda population of 56.7 million in 2025 and 128.0 million in 2050. How well does the equation in part (a) estimate the Census Bureau projections?

**53.** Pete sells used cars at a profit of $500 per car, not taking into account monthly overhead expenses of $1700. Taking overhead expense into account, write the equation relating monthly profit ($y$) to monthly sales of cars ($x$).

**54.** In 1994, 47.5% of individuals in the 20–34 age group were overweight. By 2002, the rate rose to 52.8%.

   **(a)** Find the linear equation that describes this trend.

   **(b)** Use the equation to estimate the percentage of overweight people in 2015.

**55.** In 1990/1991, the percentage of students with disabilities was 11.43%. In 1997/1998 the portion increased to 12.80%.

   **(a)** Find a linear function that describes the percentage of disabled individuals as a function of years since 1990/1991.

   **(b)** Use the linear function to estimate the percentage with disabilities in 1999/2000. How does your result compare to the actual figure of 13.33%?

   **(c)** Use the linear function to estimate when the percentage of students with disabilities will reach 15%.

**56.** The Medicare Hospital Insurance Fund pays for some hospital services for those in the Medicare program. Based on income and disbursements during 1990–2003, the linear trends for the fund's income and disbursements are:

$$y = 8.2x + 75.7 \quad \text{for income}$$
$$y = 5.9x + 81.0 \quad \text{for disbursements}$$

where $x$ = number of years since 1990 and $y$ = billions of dollars.

   **(a)** Find the intersection of these trends.

   **(b)** Based on the linear trends, when will the fund's disbursements exceed income?

**57.** *USA Today* reported that the average winter heating cost of a U.S. home was $564 in 1999 and $989 in 2005.

   **(a)** Use this information to find the linear trend of average heating costs as a function of years since 1999.

   **(b)** Assume the linear trend continues and use the linear equation to estimate the average winter heating cost in 2015.

   **(c)** Estimate when average winter heating costs will be quadruple the 1999 cost—that is, $2256.

**58.** In 2000 the percentage of individuals 18 and over in the United States who had never married was 23.9%. In 2009 it was 26.1%.

   **(a)** Assume that this is a linear trend. Find the linear equation expressing the percentage never married as a function of the number of years since 2000.

   **(b)** Use the linear function found in (a) to estimate the percentage who have never married in 2015.

   **(c)** Based on the function in (a), when will the percentage reach 50%?

# LINEAR SYSTEMS

In Chapter 1, you learned that a linear equation may represent or model a situation reasonably well. You also learned that an important business concept, the break-even point, is determined from two equations, a cost function and a revenue function. The intersection of the two lines determines the break-even point.

Other, more complex situations may require two or more linear equations to describe. As with the break-even problem, we want to find a point common to all equations, their intersection. We illustrate with a simple example.

The Bluebonnet Campfire Kids sold cookies and candy to raise money for summer camp. They sold a total of 325 boxes of candy and cookies. The candy sold for $4 per box, and the cookies sold for $3 per box. They made a profit of $2.10 for each box of candy and $1.80 for each box of cookies. Their sales totaled $1165, with a profit of $642. This information can be represented mathematically in the following way. Let $x = $ the number of boxes of candy sold and let $y = $ the number of boxes of cookies sold. Then, $x + y = 325$ represents the total number of boxes sold. Because each box of candy sells for $4 and each box of cookies sells for $3, $4x + 3y = 1165$ represents total sales. The profit from sales is

$$2.10x + 1.80y = 642$$

To answer the question of the number of boxes of candy and the number of boxes of cookies sold, we want the solution to the system

$$
\begin{aligned}
x + \phantom{4}y &= \phantom{1}325 \\
4x + \phantom{4}3y &= 1165 \\
2.10x + 1.80y &= \phantom{1}642
\end{aligned}
$$

To find a solution to the system, we want to find a value of $x$ and a value of $y$ that makes *all three* equations true. We can easily find a solution to the first equation, such as $x = 100$ and $y = 225$, but these values make neither of the last two equations true. The values $x = 265$ and $y = 35$ make the second equation true but not the other two. Thus, the point (265, 35) lies on the second line but not on the other two. We have not found the solution of the system until we find a point common to all three lines, their point of intersection. We do not solve this system here. The purpose of this chapter is to demonstrate some methods of finding a solution to a system of equations; then you will be able to solve this system. Some of the methods used can be applied to systems with a large number of variables and equations.

## 2.1    SYSTEMS OF TWO EQUATIONS

- Substitution Method
- Elimination Method
- Applications

- Inconsistent Systems
- Systems That Have Many Solutions
- Application: Supply and Demand Analysis

We call the equations used in the introduction a **system of equations**. A pair of numbers, one a value of $x$ and the other a value of $y$, that makes *all* equations true is called a **solution of the system of equations**.

We now look at some methods used to find the solution to a system of equations. The first method is the substitution method.

### Substitution Method

**The Substitution Method**    To solve a system of two linear equations in two variables by the substitution method:

1. Solve for a variable in one of the equations (say $x$ in terms of $y$).
2. Substitute for $x$ in the other equation.
3. You now have an equation in one variable (say $y$). Solve for that variable.
4. Substitute the value of the variable ($y$) just obtained into the first equation and solve for the other variable ($x$).

**Example 1**    Solve the system

$$2x - y = 3$$
$$x + 2y = 4$$

by substitution.

**Solution**
In this case, it is easy to solve for $x$ in the second equation because its coefficient is 1. We obtain

$$x = 4 - 2y$$

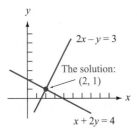

**FIGURE 2–1**
$2x - y = 3$ and
$x + 2y = 4$ intersect
at the point $(2, 1)$.

Substitute this expression for $x$ in the first equation, $2x - y = 3$:

$$2(4 - 2y) - y = 3$$
$$8 - 4y - y = 3$$
$$8 - 5y = 3$$
$$8 = 3 + 5y$$
$$5 = 5y$$
$$y = 1$$

Now substitute 1 for $y$ in $x = 4 - 2y$ to obtain $x = 4 - 2 = 2$. You could also substitute in $2x - y = 3$. Thus, the solution to the system is $(2, 1)$. You may also solve for $y$ in the first equation and substitute it in the second equation. This will yield the same solution. (Try it.) To be sure you have made no errors, you should check your solution in *both* equations. Figure 2–1 shows the graph of the solution.

---

**Example 2**   Solve the system

$$5x - 3y = 18$$
$$4x + 2y = 10$$

**Solution**
In this case, it is a little easier to solve for $y$ in the second equation.

$$2y = 10 - 4x$$
$$y = 5 - 2x$$

Substitute the expression for $y$ in the first equation:

$$5x - 3(5 - 2x) = 18$$
$$5x - 15 + 6x = 18$$
$$11x = 33$$
$$x = 3$$

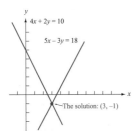

**FIGURE 2–2**
$5x - 3y = 18$ and
$4x + 2y = 10$ intersect
at the point $(3, -1)$.

Now substitute $x = 3$ into one of the equations. We use $y = 5 - 2x$.

$$y = 5 - 2(3) = -1$$

The pair $x = 3$, $y = -1$ gives the solution to the system. Check it in both equations. Figure 2–2 shows the graph of the solution.

---

The method of substitution to solve systems of equations has limited usefulness. It may work quite well for systems with two variables, but they do not extend well to systems with more than two variables. We now introduce a method that can be extended to systems with a larger number of variables.

## Elimination Method

The **elimination method** finds the solution by systematically modifying the system to simpler systems. It does so in a manner that algebraically modifies the system without disturbing the solution. The goal is to modify the system until one of the equations contains just one unknown. The simpler system so obtained is called an **equivalent system** because it has exactly the same solution as the original system. The elimination method is especially useful because it can be used with systems with several variables and equations. The elimination method produces a series of systems of equations by eliminating a variable from an equation or equations, to obtain a simpler system. As you study the examples, observe that the operations used to transform the system (eliminate a variable) into a simpler yet equivalent system uses the following operations:

**Equivalent Linear Systems**

To transform one system of linear equations into an **equivalent linear system**, use one or more of the following:

1. Interchange two equations.
2. Multiply or divide one or more equations by a nonzero constant.
3. Multiply one equation by a constant and add the result to or subtract it from another equation.

The procedure of converting a linear system to an equivalent linear system will be used again later in this chapter and in Chapter 4, so be sure you understand the process. We illustrate this method with the system

$$3x - y = 3$$
$$x + 2y = 8$$

The arithmetic is a little easier if we eliminate a variable that has 1 as a coefficient. It is usually more convenient to use the first equation to eliminate $x$, so we interchange the two equations to get

$$x + 2y = 8$$
$$3x - y = 3$$

Now eliminate $x$ from the second equation as follows:

$$-3x - 6y = -24 \quad \text{(multiply the first equation by } -3 \text{ because it gives } -3x, \text{ the negative of the } x\text{-term in the second equation)}$$

$$\underline{3x - y = 3}$$
$$-7y = -21 \quad \text{(now add it to the second equation)}$$
$$\text{(the new second equation)}$$

Since the new equation came from equations in the system, it holds true whenever the system is true. It replaces the equation $3x - y = 3$ to give the system

$$x + 2y = 8$$
$$-7y = -21$$

Notice that the second equation has been modified so that the variable $x$ has been *eliminated* from it. Simplify the second equation further by dividing by $-7$:

$$x + 2y = 8$$
$$y = 3$$

This system has the same solution as the original system, but it has the advantage of giving the value of $y$ at the common solution, namely, 3. Now substitute 3 for $y$ in the first equation to obtain

$$x + 2(3) = 8$$

(Actually, you can substitute $y$ into either of the original equations.) This simplifies to $x = 2$, so the solution to the system is $(2, 3)$.

When a system has no coefficient equal to 1, you can still solve such a system by elimination, as illustrated in the next example.

**Example 3**  Solve this system by elimination:

$$2x - 3y = -19$$
$$5x + 7y = \phantom{-}25$$

**Solution**

We want to eliminate $x$ from the second equation and find the value of $y$ for the common solution. We modify the system of equations to obtain coefficients of $x$, so one coefficient is the negative of the other. Do this:

$$-10x + 15y = \phantom{0}95 \quad \text{(multiply the first equation by } -5\text{)}$$
$$10x + 14y = \phantom{0}50 \quad \text{(multiply the second equation by 2)}$$
$$29y = 145 \quad \text{(now add)}$$
$$y = \phantom{00}5 \quad \text{(divide by 29 to find } y\text{)}$$

Replace the second equation to obtain the modified, equivalent system:

$$2x - 3y = -19$$
$$y = \phantom{-1}5$$

Now substitute this value of $y$ into one of the equations. We use the first to obtain

$$2x - 3(5) = -19$$
$$2x = \phantom{-}{-4}$$
$$x = \phantom{-}{-2}$$

The solution to the original system is $(-2, 5)$. Figure 2–3 shows the graph of the solution.

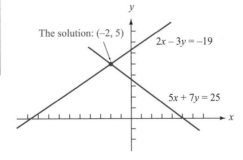

**FIGURE 2–3**  $2x - 3y = -19$ and $5x + 7y = 25$ intersect at the point $(-2, 5)$.

Observe that the elimination method follows this general procedure.

**Elimination Method** To solve a system of two linear equations in two unknowns by the **elimination method:**

1. Multiply one or both equations by appropriate constants so that a variable in one equation has a coefficient $c$ and the same variable in the other equation has a coefficient $-c$.

2. Add the equations to eliminate this variable.

3. The resulting equation replaces one of the two equations.

## Applications

Before mathematics can be used to solve an application, the information given must be converted to a mathematical form. Depending on the application, one of many forms may be appropriate. For our purposes in this chapter, we look at some applications that can be represented with a system of equations and solved by elimination.

**Example 4** A woman must control her diet. She selects milk and bagels for breakfast. How much of each should she serve in order to consume 700 calories and 28 grams of protein? Each cup of milk contains 170 calories and 8 grams of protein. Each bagel contains 138 calories and 4 grams of protein.

**Solution**
Let $m$ be the number of cups of milk and $b$ the number of bagels. Then the total number of calories is

$$170m + 138b$$

and the total protein is

$$8m + 4b$$

So, we need to solve the system

$$8m + 4b = 28$$
$$170m + 138b = 700$$

Divide the top equation by 4 and the bottom equation by 2 to simplify somewhat:

$$2m + b = 7$$
$$85m + 69b = 350$$

Next, multiply the top equation by $-69$ and add it to the bottom equation in order to eliminate $b$ from the second equation:

$$
\begin{array}{rl}
-138m - 69b = & -483 \\
85m + 69b = & 350 \\
\hline
-53m \phantom{+ 69b} = & -133
\end{array}
$$

$$m = \frac{133}{53} = 2.509 \quad \text{(rounded)}$$

Now substitute and solve for $b$:

$$2(2.509) + b = 7$$
$$5.018 + b = 7$$
$$b = 1.982$$

It is reasonable to round these answers to 2.5 cups of milk and 2 bagels.

**Example 5**   A health food company has two nutritional drinks prepared. One contains 6.25 grams of carbohydrates per ounce, and the second contains 5.125 grams of carbohydrate per ounce. A customer wants 400 ounces of a nutritional drink containing 5.625 grams of carbohydrate per ounce. The company dietitian will mix the two drinks on hand to provide the requested drink. How much of each should be used?

**Solution**

Let $x$ = number of ounces of the first drink and let $y$ = number of ounces of the second drink. Then,

$$x + y = 400 \quad \text{(number of ounces of the mixture)}$$
$$6.25x = \text{grams of carbohydrate from the first drink}$$
$$5.125y = \text{grams of carbohydrate from the second drink}$$
$$5.625(400) = 2250 = \text{grams of carbohydrate in the mixture}$$

We summarize this information with the system

$$\begin{aligned} x + \quad\quad y &= \quad 400 \\ 6.25x + 5.125y &= 2250 \end{aligned}$$

We can use the substitution method to solve the system. Substitute $x = 400 - y$ into the second equation:

$$\begin{aligned} 6.25(400 - y) + 5.125y &= 2250 \\ 2500 - 6.25y + 5.125y &= 2250 \\ -1.125y &= -250 \\ 1.125y &= 250 \\ y &= 222.22 \end{aligned}$$

Rounded to the nearest ounce, the dietitian uses 222 ounces of the second drink and $400 - 222 = 178$ ounces of the first drink.

## Inconsistent Systems

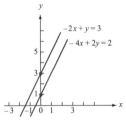

**FIGURE 2–4** Each line has a slope 2, and they do not intersect, so no solution exists for this system.

Each system in the preceding examples has exactly one solution. Do not expect this always to be the case. If the equations represent two parallel lines, they have no points in common, and a solution to the system does not exist. We say that this is an **inconsistent system**. Figure 2–4 shows the graph of the two lines

$$\begin{aligned} -2x + \quad y &= 3 \\ -4x + 2y &= 2 \end{aligned}$$

Each line has slope 2, and they do not intersect, so no solution exists for the system.

Note that the equations in this system have the slope-intercept forms

$$y = 2x + 3 \quad\quad \text{and} \quad\quad y = 2x + 1$$

In an inconsistent system of two equations, the lines have the same slope but different $y$-intercepts.

**Example 6**    The equations of two parallel lines give a system of equations. Let's see what happens when we try to solve such a system:

$$3x - 2y = \;\;\;5$$
$$6x - 4y = -6$$

We can eliminate $x$ from an equation by multiplying the first equation by $-2$ and adding the two equations:

$$
\begin{array}{rcl}
-6x + 4y &=& -10 \\
6x - 4y &=& \;\;-6 \\
\hline
0x + 0y &=& -16 \qquad \text{or} \\
0 &=& -16
\end{array}
$$

The process of solving the system leads to an inconsistency, $0 = -16$, so the system has *no* solution. You may expect such an inconsistency when attempting to solve a system that represents two different parallel lines.

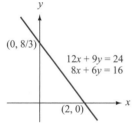

**FIGURE 2–5**  The graphs of $12x + 9y = 24$ and $8x + 6y = 16$ coincide.

## Systems That Have Many Solutions

When you graph the lines

$$12x + 9y = 24$$
$$8x + 6y = 16$$

you will find that the graphs coincide, so they represent the same line. When you put both equations in the slope-intercept form, you find that both have slopes $-\frac{4}{3}$ and $y$-intercept $\frac{8}{3}$, so the lines and their graphs are identical. Because the lines coincide, *every* point on this line is a solution to the given system (see Figure 2–5). Let's look at an example that illustrates what happens when we try to solve such a system.

**Example 7**    Solve the system

$$12x + 9y = 24$$
$$8x + 6y = 16$$

**Solution**
First, eliminate $x$ from an equation:

$$
\begin{array}{rcll}
24x + 18y &=& \;\;\;48 & \text{(multiply the first equation by 2)} \\
-24x - 18y &=& -48 & \text{(multiply the second equation by } -3) \\
\hline
0 &=& \;\;\;0 & \text{(add the equations)}
\end{array}
$$

This transforms the system into

$$
\begin{array}{rcl}
12x + 9y &=& 24 \\
0 &=& 0
\end{array}
$$

Any solution to the equation $12x + 9y = 24$ gives a solution to the system. Since an infinite number of points lie on this line, the system has an infinite number of solutions that may be represented as

$$y = -\frac{4}{3}x + \frac{8}{3}$$

where $x$ may be any number. For example, when $x = 1$, 5, and $-2$, we have the solutions

$$x = 1, \quad y = \frac{4}{3}$$
$$x = 5, \quad y = -4$$
$$x = -2, \quad y = \frac{16}{3}$$

**NOTE**

Because $k$ can be any real number, an infinite number of solutions exists.

In this example, the variable $y$ can be expressed in terms of $x$. When this occurs, we call $x$ a **parameter**. In practice, we often use another letter—say, $k$—to represent an arbitrary value of $x$. We obtain the corresponding value of $y$ when $x = k$ ($k =$ any real number) is substituted into the equation. When doing so, we can express the infinity of solutions as

$$x = k \quad \text{and} \quad y = -\frac{4}{3}k + \frac{8}{3}$$

We call this the **parametric form** of the solution, which we can also write as $(k, -\frac{4}{3}k + \frac{8}{3})$.

## Application: Supply and Demand Analysis

Department stores are well aware that they can sell large quantities of goods if they advertise a reduction in price. The lower the price, the more they sell. Retailers understand this relationship between the price of a commodity and the consumer **demand** (the amount consumers buy). They also know that there may be more than one relationship between price and demand, depending on circumstances. In times of shortages, a different psychology takes effect, and prices tend to *increase* when demand *increases*.

In the competitive situation, a decrease in price can cause an increase in demand (when a store has a sale). This suggests that demand is a function of price. On other occasions, a store may lower prices because an item is in great demand and it expects to increase profits by a greater volume. This suggests that price is a function of demand. Because the cause and effect relationship between price and demand can go either way—a change in price causes a change in demand or a change in demand can cause a change in price—we need to decide how to write the demand equation. The analysis is easier if we write the demand equation (and the supply equation in the next example) so that the price is a function of demand (or supply).

**Example 8**  The Bike Shop held an annual sale. The consumer price and demand relationship for the Ten-Speed Special was

$$p = -2x + 179$$

where $x$ is the number of bikes in demand at the price $p$. The negative slope, $-2$, indicates that when an *increase* occurs in one of the variables, price or demand, a decrease occurs in the other. This relationship between price and demand is a linear function. Its graph illustrates the decrease in price with an increase in demand (see Figure 2–6 on the next page). When demand increases from 10 to 40, prices drop from \$159 to \$99.

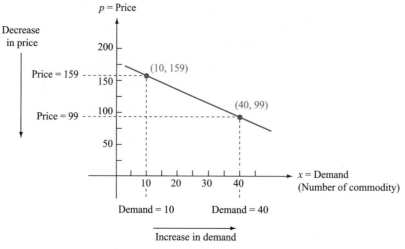

The price–demand function $p = -2x + 179$

**FIGURE 2–6** A decrease in price leads to a higher demand.

The Bike Shop cannot lower prices indefinitely because the supplier wants to make a profit also. In a competitive situation, a price increase gives the supplier incentive to produce more. When prices fall, the supplier tends to produce less. The quantity produced by the supplier is called **supply**. Suppose Bike Manufacturing produces the Ten-Speed Special, and the relationship between supply and price is given by the linear function

$$p = 1.5x + 53$$

The graph of this equation illustrates that an increase in price leads to a higher supply (see Figure 2–7). When the price increases from \$83 to \$128, the supply increases from 20 to 50 units.

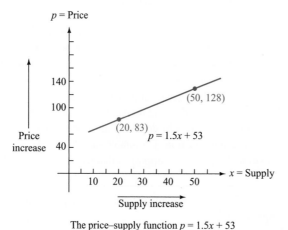

The price–supply function $p = 1.5x + 53$

**FIGURE 2–7** An increase in price leads to a greater supply.

Supply and demand are two sides of a perfect competitive market. They interact to determine the price of a commodity. The price of a commodity settles down in the market to one at which the amount willingly supplied and the amount willingly demanded are equal. This price is called the **equilibrium price**. The equilibrium price may be determined by solving a system of equations. In our example, we have the system

$$p = -2x + 179$$
$$p = 1.5x + 53$$

Solve the system to obtain the equilibrium solution by either the substitution method or the elimination method. The solution is $x = 36$ and $p = 107$. (Be sure you can find this solution.) The equilibrium price is $107 when the supply and demand are 36 bikes (see Figure 2–8).

Figure 2–8 also helps us see when there is a surplus or shortage of bikes. For example, draw a horizontal line at $p = 150$. Note that it intersects the demand equation at about $x = 15$ and it intersects the supply equation at about $x = 65$. Thus, when the price is $150, there is a demand for 15 bikes, but there are 65 supplied, so a surplus exists. Similarly, when $p = 75$, the demand is about 50, and the supply is about 10, so a shortage exists.

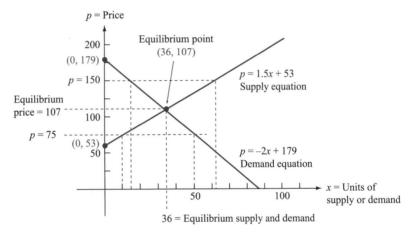

**FIGURE 2–8** Equilibrium solution of supply and demand equations.

Observe that the $p$-intercept for the demand equation is $179. This indicates that the price must be less than $179 if the shop expects to sell any bikes. Similarly, the $p$-intercept for the supply equation is 53, so the price must be more than $53 for the supplier to manufacture bikes.

**Example 9**  The Catalog Store found that the supply and demand quantities for two prices of their student backpack were:

| Demand (1000's) | Supply (1000's) | Price |
|---|---|---|
| 14.0 | 19.2 | $32 |
| 22.0 | 16.8 | $26 |

From this information, find

(a) points in the demand and supply linear equations.

(b) the linear demand equation.

(c) the linear supply equation.

(d) the equilibrium quantity and price.

### Solution

Let $x =$ quantity (in thousands) of backpacks and let $p =$ price of the backpack.

(a) Points on the demand equation are (14.0, 32) and (22.0, 26). Points on the supply equation are (19.2, 32) and (16.8, 26).

(b) We use the points (14.0, 32) and (22.0, 26) to find the linear demand equation.

$$m = \frac{32 - 26}{14.0 - 22.0} = \frac{6}{-8} = -0.75$$
$$p - 32 = -0.75(x - 14)$$
$$p = -0.75x + 10.5 + 32 = -0.75x + 42.5$$

The demand equation is $p = -0.75x + 42.5$.

(c) We use the points (19.2, 32) and (16.8, 26) to find the linear supply equation.

$$m = \frac{32 - 26}{19.2 - 16.8} = \frac{6}{2.4} = 2.5$$
$$p - 32 = 2.5(x - 19.2)$$
$$p = 2.5x - 48 + 32 = 2.5x - 16$$

The supply equation is $p = 2.5x - 16$.

(d) Equilibrium occurs when the demand price = the supply price, that is,

$$-0.75x + 42.5 = 2.5x - 16$$
$$-3.25x = -58.5$$
$$x = \frac{-58.5}{-3.25} = 18$$
$$p = 2.5(18) - 16 = 29$$

Equilibrium occurs at a price of $29 and a quantity of 18,000.

## 2.1     EXERCISES

Access end-of-section exercises online at **www.webassign.net**

  **Using Excel**

See **Using Excel** in Section 1.3 for the procedure to graph two lines and determine their intersection.

## 2.2 SYSTEMS WITH THREE VARIABLES: AN INTRODUCTION TO A MATRIX REPRESENTATION OF A LINEAR SYSTEM OF EQUATIONS

- Elimination Method
- Matrices
- Matrices and Systems of Equations
- Gauss-Jordan Method
- Application

The tuition–fee cost of courses at Mountainview Community College consists of a general fee of $250 and tuition of $38 per semester hour that can be represented by a linear equation in two variables:

$$y = 38x + 250$$

where $x$ is the number of semester hours and $y$ is the total tuition–fee cost.

At Valley Junior College, the tuition–fee cost consists of a general fee of $225, a building use fee of $25 per course, and tuition of $41 per semester hour. The tuition–fee cost can be represented by a linear equation

$$y = 41h + 25x + 225$$

where $h$ is the number of semester hours, $x$ is the number of courses, and $y$ is the total tuition–fee cost.

In the first case, the tuition–fee cost can be represented with a linear equation with two variables. For Valley Junior College, the tuition–fee function is also linear, but it requires the use of three variables.

This illustrates that, as in Section 2.1, many applications can be modeled using a linear equation with two variables, but some linear models require more than two variables.

### Elimination Method

Some applications require the solution of two or more linear equations in three variables (or more), so we will use some examples with "nice" numbers to illustrate the procedure for solving such a system.

**Example 1** Cutter's mother opened Amy's Online Store so that he could purchase some books from the Bargain Baskets for his cousin Andrew's birthday. He was allowed to select from the $1 basket, the $2 basket, and the $3 basket. Based on the following information, determine how many books Cutter selected from each basket:

**(a)** He selected five books at a total cost of $10.

**(b)** Shipping costs were $2.00 for each $1 book and $1.00 for each $2 and $3 book.

**(c)** The total shipping cost was $6.00.

**Solution**
Let's state this information in mathematical form and determine the number of books chosen from each table. Let

$$x = \text{number of books from the \$1 basket}$$
$$y = \text{number of books from the \$2 basket}$$
$$z = \text{number of books from the \$3 basket}$$

The given information may be written as follows:

$$\begin{aligned} x + y + z &= 5 \quad \text{(total number of books)} \\ x + 2y + 3z &= 10 \quad \text{(total cost of books)} \\ 2x + y + z &= 6 \quad \text{(total shipping costs)} \end{aligned}$$

The solution to this system of three equations in three variables gives the number of books chosen from each basket. Before we solve this system, let's look at some basic ideas.

A set of values for $x$, $y$, $z$ that satisfies all three equations is called a **solution** to the system. *Be sure* you understand that a solution consists of three numbers, one each for $x$, $y$, and $z$.

You know that a linear equation in two variables represents a line in two-dimensional space. However, a linear equation in three variables does not represent a line, it represents a plane in three-dimensional space. A solution to a system of three equations in three variables corresponds to a point that lies in all three planes. Figure 2–9 (on the next page) illustrates some possible ways in which the planes might intersect. If the three planes have just one point in common, the solution will be unique. If the planes have no points in common, there will be no solution to the system. If the planes have many points in common, the system will have many solutions.

Note that two planes can have points in common, such as plane $A$ and $B$ in Figure 2–9(c) and (d), but if those points do not also lie in plane $C$, then they do not represent a solution.

We need not stop with a system of three variables. Applications exist that require larger systems with more variables. Larger systems become more difficult to interpret geometrically and are more tedious to solve, but they also can have a unique solution, no solution, or many solutions.

The method of elimination used to solve systems of two equations can be adapted to larger systems. We now solve the system in Example 1. First, we make a change in notation. Replace the variables $x$, $y$, and $z$ with $x_1$, $x_2$, and $x_3$. We do this to emphasize that we can run out of letters when more variables are needed. (Some applications use dozens of variables.) The use of $x_1, x_2, x_3, x_4, \ldots$ for variables is another way to identify different variables, and it has the advantage of providing a notation for any number of variables. So, here is the system of Example 1 using $x_1$, $x_2$, and $x_3$:

$$\begin{aligned} x_1 + x_2 + x_3 &= 5 \\ x_1 + 2x_2 + 3x_3 &= 10 \\ 2x_1 + x_2 + x_3 &= 6 \end{aligned}$$

Now let's proceed with the solution. First, eliminate $x_1$ from the second equation by subtracting the first equation from the second equation to obtain a new second equation:

$$\begin{array}{rl} x_1 + 2x_2 + 3x_3 = 10 & \text{(Equation 2)} \\ \underline{x_1 + x_2 + x_3 = 5} & \text{(Equation 1)} \\ x_2 + 2x_3 = 5 & \text{(this will replace the second equation in the system)} \end{array}$$

As we work through the example, note that we usually multiply one equation by a constant before adding it to another. The choice of the multiple is determined by the coefficients of the variable we want to eliminate.

Unique solution:

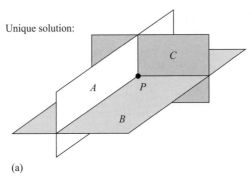

(a)

Three planes $A$, $B$, $C$ intersect at a single point $P$;
$P$ corresponds to a unique solution.

No solutions:

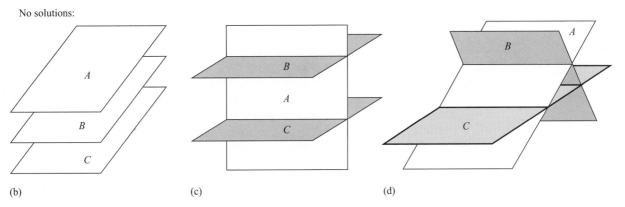

(b)　　　　　　　　　　　(c)　　　　　　　　　　　(d)

Planes $A$, $B$, $C$ have no point of common intersection, no solution.

Many solutions:

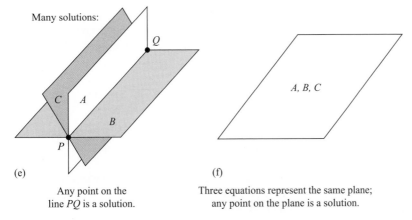

(e)　　　　　　　　　　　(f)

Any point on the
line $PQ$ is a solution.

Three equations represent the same plane;
any point on the plane is a solution.

**FIGURE 2–9**  Six possible ways three planes can intersect. (a) Unique solution: Planes $A$, $B$, and $C$ intersect at a single point $P$; $P$ corresponds to a unique solution. (b, c, and d) No solutions: Planes $A$, $B$, and $C$ have no point of common intersection; no solution. (e and f) Many solutions: (e) Any point on the line $PQ$ is a solution. (f) Three equations represent the same plane, and any point on the plane is a solution.

**NOTE**

Observe that the reason we multiplied the first equation by $-2$ was to give a coefficient of $x_1$ that was the negative of the coefficient of $x_1$ in the third equation. Adding the two equations then eliminated $x_1$ from the third equation.

Now eliminate $x_1$ from the third equation by multiplying the first equation by $-2$ and adding it to the third:

$$
\begin{array}{rrrrrl}
2x_1 & + & x_2 & + & x_3 & = & 6 \qquad \text{(Equation 3)}\\
-2x_1 & - & 2x_2 & - & 2x_3 & = & -10 \qquad (-2 \times \text{Equation 1})\\
\hline
& & -x_2 & - & x_3 & = & -4 \qquad \text{(this will replace the third equation in the system)}
\end{array}
$$

This gives the new and simpler equivalent system

$$
\begin{array}{rrrrrl}
x_1 & + & x_2 & + & x_3 & = & 5\\
& & x_2 & + & 2x_3 & = & 5\\
& & -x_2 & - & x_3 & = & -4
\end{array}
$$

Observe that the variable $x_1$ has been eliminated from both the second and third equations. This system of three equations becomes the current system from which we eliminate another variable.

To describe the operations performed in a concise manner, we will use the notation Eq. 2 $-$ Eq. 1 to mean the first equation is subtracted from the second. $-2(\text{Eq. 1}) + \text{Eq. 3}$ means to multiply the first equation by $-2$ and add it to the third. We continue the process by eliminating more variables from the new system.

Now we can eliminate the variable $x_2$ from the first and third equations of the current system by using the second equation:

| **Current System** | | **Next System** |
|---|---|---|
| $x_1 + x_2 + x_3 = 5$ | Eq. 1 $-$ Eq. 2 yields | $x_1 \qquad - \; x_3 = 0$ |
| $x_2 + 2x_3 = 5$ | remains | $x_2 + 2x_3 = 5$ |
| $-x_2 - x_3 = -4$ | Eq. 3 $+$ Eq. 2 yields | $x_3 = 1$ |

The system on the right is equivalent to the original system and is now the current system. We can complete the solution by eliminating $x_3$ from the first and second equations of this latest system:

| **Current System** | | **Next System** |
|---|---|---|
| $x_1 \qquad - \; x_3 = 0$ | Eq. 1 $+$ Eq. 3 yields | $x_1 = 1$ |
| $x_2 + 2x_3 = 5$ | Eq. 2 $-$ 2(Eq. 3) yields | $x_2 = 3$ |
| $x_3 = 1$ | remains | $x_3 = 1$ |

The solution to this system and, thus, the solution to the original system is $x_1 = 1$, $x_2 = 3, x_3 = 1$, which is also written $(1, 3, 1)$. This solution tells us that Cutter chose 1 book from the \$1 basket, 3 books from the \$2 basket, and 1 book from the \$3 basket. (Check the solution in the original system.)

Note that when we have reached the system

$$
\begin{array}{rrrrl}
x_1 & - & x_3 & = & 0\\
x_2 & + & 2x_3 & = & 5\\
& & x_3 & = & 1
\end{array}
$$

we can finish the solution by using what is called **back substitution**. We substitute $x_3 = 1$ into the first two equations to obtain

$$x_1 - 1 = 0 \quad \text{(which gives } x_1 = 1)$$

and

$$x_2 + 2 = 5 \quad \text{(which gives } x_2 = 3)$$

## Matrices

The method of elimination may be used to solve larger systems, but it is tedious, with errors easily made. To make the elimination method more efficient, we convert it to an equivalent procedure that can still be tedious, but less so. The good news is that it can be used to obtain computer solutions of a system. We will learn the **Gauss-Jordan Method**, which uses matrices, a notation that keeps track of the variables in the system but does not write them.

When you worked some solution by elimination exercises, you likely noticed that the coefficients of the variables determined the multipliers used in the operations. Thus, some labor can be saved if we can avoid writing the variables and concentrate on the coefficients. The matrix notation allows that.

Before we show you the Gauss-Jordan Method, we introduce you to matrices. The formal definition of a matrix is the following:

**DEFINITION**
**Matrix**

A **matrix** is a rectangular array of numbers. The numbers in the array are called the **elements of the matrix**. The array is enclosed with brackets.

An array composed of a single row of numbers is called a **row matrix**.
An array composed of a single column of numbers is called a **column matrix**.

Some examples of matrices are

$$\begin{bmatrix} 1 & 2 & 3 \\ 0 & -1 & 1 \end{bmatrix} \quad \begin{bmatrix} 2 & 3 \\ 1 & 1 \\ 4 & 1 \end{bmatrix} \quad \begin{bmatrix} 1 & 2 & 3 \\ 4 & 5 & 6 \\ 0 & 1 & 2 \end{bmatrix} \quad [2.5 \; 8.3] \quad \begin{bmatrix} 3 \\ -5 \\ 8 \end{bmatrix}$$

The location of each element in a matrix is described by the row and column in which it lies. Count the rows from the top of the matrix and the columns from the left.

$$\begin{array}{cccc} & \text{Col. 1} & \text{Col. 2} & \text{Col. 3} & \text{Col. 4} \\ \text{Row 1} & 4 & 6 & -3 & 2 \\ \text{Row 2} & 1 & 5 & 9 & -2 \\ \text{Row 3} & 7 & 8 & 3 & 4 \end{array}$$

The element 9 is in row 2 and column 3 of the matrix. We call this location the $(2, 3)$ location; the row is indicated first and the column second. A standard notation for the number in that location is $a_{23}$ (read as "$a$ sub two-three"), so $a_{23} = 9$. The element 7, designated $a_{31} = 7$, is in the $(3, 1)$ location; and $-3$, designated $a_{13} = -3$, is in the $(1, 3)$ location.

**Example 2**    For the matrix

$$\begin{bmatrix} 2 & -6 & -5 & -1 & 0 \\ 1 & 7 & 6 & -4 & 4 \\ 9 & 5 & -8 & 3 & -2 \end{bmatrix}$$

find the following:

**(a)** The $(1, 1)$ element $(a_{11})$

**(b)** The $(2, 5)$ element $(a_{25})$

**(c)** The $(3, 3)$ element $(a_{33})$

**(d)** The location of $-4$

**(e)** The location of $0$

**Solution**

**(a)** 2 is the $(1, 1)$ element $(a_{11} = 2)$.

**(b)** 4 is the $(2, 5)$ element $(a_{25} = 4)$.

**(c)** The $(3, 3)$ element is $-8$ $(a_{33} = -8)$.

**(d)** $-4$ is in the $(2, 4)$ location $(a_{24} = -4)$.

**(e)** 0 is in the $(1, 5)$ location $(a_{15} = 0)$.

Let's pause to point out that you often encounter information that can be summarized in matrix form. An article reporting on the outcome of a basketball game will include a summary that resembles a matrix where a row represents a player and columns that represent minutes played, field goals, free throws, fouls, assists, and total points. A calendar resembles a matrix with each row representing a week and each column representing a day of the week.

The rows and columns of a matrix often represent categories. A grade book shows a rectangular array of numbers. Each column represents a test, and each row represents a student.

## Matrices and Systems of Equations

Let's use the following system to see how matrices relate to systems of equations.

$$\begin{array}{rcrcrcr} 2x_1 & + & x_2 & - & x_3 & = & 5 \\ 3x_1 & + & 5x_2 & + & 2x_3 & = & 11 \\ x_1 & - & 2x_2 & + & x_3 & = & -1 \end{array}$$

We form one matrix by using only the coefficients of the system. This gives the **coefficient matrix**:

$$\begin{bmatrix} 2 & 1 & -1 \\ 3 & 5 & 2 \\ 1 & -2 & 1 \end{bmatrix}$$

Column 1 lists the coefficients of $x_1$, column 2 lists the coefficients of $x_2$, and column 3 lists the coefficients of $x_3$. Notice that the entries in each column are listed in the order first, second, and third equations, so the first row represents the left-hand side of the first equation, and so on.

A matrix that also includes the numbers on the right-hand side of the equation is called the **augmented matrix** of the system:

$$\begin{bmatrix} 2 & 1 & -1 & 5 \\ 3 & 5 & 2 & 11 \\ 1 & -2 & 1 & -1 \end{bmatrix}$$

The augmented matrix gives complete information, in a compact form, about a system of equations, provided that we agree that each row represents an equation and each column, except the last, consists of the coefficients of a variable.

Generally, we place a vertical line between the coefficients and the column of constant terms. This gives a visual reminder of the location of the equal sign in the equations.

**Example 3**    Write the coefficient matrix and the augmented matrix of the system

$$\begin{aligned} 5x_1 - 7x_2 + 2x_3 &= 17 \\ -x_1 + 3x_2 + 8x_3 &= 12 \\ 6x_1 + 9x_2 - 4x_3 &= -23 \end{aligned}$$

**Solution**

The coefficient matrix is

$$\begin{bmatrix} 5 & -7 & 2 \\ -1 & 3 & 8 \\ 6 & 9 & -4 \end{bmatrix}$$

and the augmented matrix is

$$\begin{bmatrix} 5 & -7 & 2 & 17 \\ -1 & 3 & 8 & 12 \\ 6 & 9 & -4 & -23 \end{bmatrix}$$

**Example 4**    Write the system of linear equations represented by the augmented matrix

$$\begin{bmatrix} 3 & 7 & 2 & -3 & 8 \\ 4 & 0 & -5 & 7 & -2 \end{bmatrix}$$

**Solution**

The system is

$$\begin{aligned} 3x_1 + 7x_2 + 2x_3 - 3x_4 &= 8 \\ 4x_1 \qquad\quad - 5x_3 + 7x_4 &= -2 \end{aligned}$$

We can solve a system of linear equations by using its augmented matrix. Since each row in the matrix represents an equation, we perform the same kinds of operations on rows of the matrix as we do on the equations in the system. Here are the **row operations**:

---

**Row Operations**    1. Interchange two rows.

2. Multiply or divide a row by a nonzero constant.

3. Multiply a row by a constant and add it to or subtract it from another row.

Two **augmented matrices are equivalent** if one is obtained from the other by using row operations.

---

These row operations are the same kind of operations that we used on the equations in Section 2.1 to convert a system of equations to an equivalent system.

## Gauss-Jordan Method

To illustrate the matrix technique, we solve a system of linear equations using the augmented matrix and row operations. This technique is basically a simplification of the elimination method. You will soon find that even this "simplified" method can be tedious and subject to arithmetic errors. However, a method such as this is widely used on computers to solve systems of equations. Whereas the examples and exercises use equations that yield relatively simple answers, actual applications don't always have nice, neat, unique solutions.

We cannot give you a manual method for solving systems of equations that is quick, simple, and foolproof. We want you to understand and apply systems of equations. When a "messy" system needs to be solved, you need to be aware that solutions are available using computers or a TI graphing calculator.

First, we solve a system of two equations in two variables and then a system of three equations in three variables. In each case, we show the solutions of the systems by both the elimination method and the method using matrices. We show them in parallel so that you can see the relationship between the two methods.

**Example 5**    Solve the system of equations

$$x + 3y = 11$$
$$2x - 5y = -22$$

| **Sequence of Equivalent Systems of Equations** | **Corresponding Equivalent Augmented Matrices** |
|---|---|
| *Original system:* | *Original augmented matrix:* |

$$x + 3y = 11$$
$$2x - 5y = -22$$

$$\begin{bmatrix} 1 & 3 & | & 11 \\ 2 & -5 & | & -22 \end{bmatrix}$$

Eliminate $x$ from the second equation by multiplying the first equation by $-2$ and adding to the second:

Get a 0 in the second row, first column by multiplying the first row by $-2$ and adding it to the second row:

$$x + 3y = 11$$
$$-11y = -44$$

$$\begin{bmatrix} 1 & 3 & | & 11 \\ 0 & -11 & | & -44 \end{bmatrix}$$

Simplify the second equation by dividing by $-11$:

$$
\begin{aligned}
x + 3y &= 11 \\
y &= 4
\end{aligned}
$$

Eliminate $y$ from the first equation by multiplying the second equation by $-3$ and adding it to the first:

$$
\begin{aligned}
x &= -1 \\
y &= 4
\end{aligned}
$$

Simplify the second row by dividing by $-11$:

$$
\left[\begin{array}{cc|c} 1 & 3 & 11 \\ 0 & 1 & 4 \end{array}\right]
$$

Get a 0 in the first row, second column by multiplying the second row by $-3$ and adding it to the first:

$$
\left[\begin{array}{cc|c} 1 & 0 & -1 \\ 0 & 1 & 4 \end{array}\right]
$$

Read the solution from this augmented matrix. The first row gives $x = -1$, and the second row gives $y = 4$.

Some of the details that arise in larger systems do not show up in a system of two equations with two variables, so we now solve a system of three equations with three variables.

**Example 6**   Solve the system of equations.

$$
\begin{aligned}
x_1 + x_2 + x_3 &= 5 \\
x_1 + 2x_2 + 3x_3 &= 10 \\
2x_1 + x_2 + x_3 &= 6
\end{aligned}
$$

**Solution**
We solved this system earlier in Example 1. We use it again so that you can concentrate on the procedure. We show the solution of this system by both the elimination method and the method using matrices. We show them in parallel so that you can see the relationship between the two methods.

| **Sequence of Equivalent Systems of Equations** | **Corresponding Equivalent Augmented Matrices** |
|---|---|

*Original system:*

$$
\begin{aligned}
x_1 + x_2 + x_3 &= 5 \\
x_1 + 2x_2 + 3x_3 &= 10 \\
2x_1 + x_2 + x_3 &= 6
\end{aligned}
$$

*Original augmented matrix:*

$$
\left[\begin{array}{ccc|c} 1 & 1 & 1 & 5 \\ 1 & 2 & 3 & 10 \\ 2 & 1 & 1 & 6 \end{array}\right]
$$

Eliminate $x_1$ from the second equation by multiplying the first equation by $-1$ and adding to the second.

Get 0 in the second row, first column by multiplying the first row by $-1$ and adding to the second.

Eliminate $x_1$ from the third equation by multiplying the first equation by $-2$ and adding to the third:

$$
\begin{array}{rrrcr}
x_1 & + x_2 & + x_3 & = & 5 \\
    & x_2 & + 2x_3 & = & 5 \\
    & -x_2 & - x_3 & = & -4
\end{array}
$$

Eliminate $x_2$ from the first equation by multiplying the second equation by $-1$ and adding to the first.

Eliminate $x_2$ from the third equation by adding the second equation to the third:

$$
\begin{array}{rrrcr}
x_1 & & - x_3 & = & 0 \\
    & x_2 & + 2x_3 & = & 5 \\
    & & x_3 & = & 1
\end{array}
$$

Eliminate $x_3$ from the first equation by adding the third equation to the first.

Eliminate $x_3$ from the second equation by multiplying the third equation by $-2$ and adding to the second:

$$
\begin{array}{rrrcr}
x_1 & & & = & 1 \\
    & x_2 & & = & 3 \\
    & & x_3 & = & 1
\end{array}
$$

Get 0 in the third row, first column by multiplying the first row by $-2$ and adding to the third:

$$
\left[\begin{array}{ccc|c}
1 & 1 & 1 & 5 \\
0 & 1 & 2 & 5 \\
0 & -1 & -1 & -4
\end{array}\right]
$$

Get 0 in the first row, second column by multiplying the second row by $-1$ and adding to the first.

Get 0 in the third row, second column by adding the second row to the third:

$$
\left[\begin{array}{ccc|c}
1 & 0 & -1 & 0 \\
0 & 1 & 2 & 5 \\
0 & 0 & 1 & 1
\end{array}\right]
$$

Get 0 in the first row, third column by adding the third row to the first.

Get 0 in the second row, third column by multiplying the third row by $-2$ and adding to the second:

$$
\left[\begin{array}{ccc|c}
1 & 0 & 0 & 1 \\
0 & 1 & 0 & 3 \\
0 & 0 & 1 & 1
\end{array}\right]
$$

Read the solution from this augmented matrix. The first row gives $x_1 = 1$, the second row $x_2 = 3$, and the third row $x_3 = 1$.

This technique of using row operations to reduce an augmented matrix to a simple matrix is called the **Gauss-Jordan Method**. The form of the final matrix is such that the solution to the original system can easily be read from the matrix. Note that the final matrix in the example preceding was

$$
\left[\begin{array}{ccc|c}
1 & 0 & 0 & 1 \\
0 & 1 & 0 & 3 \\
0 & 0 & 1 & 1
\end{array}\right]
$$

For the moment, ignore the last column of the matrix. The remaining columns have zeros everywhere except in the $(1, 1)$, $(2, 2)$, and $(3, 3)$ locations. These are called the **diagonal locations**. The Gauss-Jordan Method attempts to reduce the augmented matrix until there are 1's in the diagonal locations and 0's elsewhere (except in the last column). This procedure of obtaining a 1 in one position of a column and making all other entries in that column equal to 0 is called **pivoting**.

For a matrix reduced to this diagonal form, each row easily shows the value of a variable in the solution. In the matrix on the previous page, the rows represent

$$x_1 = 1$$
$$x_2 = 3$$
$$x_3 = 1$$

which gives the solution.

We now look at another example and focus attention on the procedure for arriving at this desired diagonal form. In this section, we focus our attention on augmented matrices that can be reduced to this diagonal form. The cases in which the diagonal form is not possible are studied in the next section.

We introduce a new notation in the next example to reduce the writing involved. When we are reducing a matrix and you see

$$\tfrac{1}{4}R1 \qquad \text{gives} \qquad [1 \quad 2 \quad -3 \quad 11] \to R1$$

this means that row 1 of the current matrix is divided by 4 and gives the new row $[1 \quad 2 \quad -3 \quad 11]$, which is placed in row 1 of the next matrix. The notation $-2R2 + R3 \to R3$ means that row 2 of the current matrix is multiplied by $-2$ and added to row 3. The result becomes row 3 of the next matrix.

**Example 7** Solve the following system by reducing the augmented matrix to the diagonal form.

$$2x_1 - 4x_2 + 6x_3 = 20$$
$$3x_1 - 6x_2 + x_3 = 22$$
$$-2x_1 + 5x_2 - 2x_3 = -18$$

The augmented matrix of this system is

$$\left[\begin{array}{ccc|c} 2 & -4 & 6 & 20 \\ 3 & -6 & 1 & 22 \\ -2 & 5 & -2 & -18 \end{array}\right]$$

We now use row operations to find the solution to the system.

| | **Matrix** | **This Operation on Present Matrix** | **Put in New Row** |
|---|---|---|---|
| Need 1 here | $\left[\begin{array}{ccc|c} 2 & -4 & 6 & 20 \\ 3 & -6 & 1 & 22 \\ -2 & 5 & -2 & -18 \end{array}\right]$ | $\tfrac{1}{2}R1$ gives $[1 \; -2 \; 3 \; 10] \to R1$ | |
| Need 0 here | $\left[\begin{array}{ccc|c} 1 & -2 & 3 & 10 \\ 3 & -6 & 1 & 22 \\ -2 & 5 & -2 & -18 \end{array}\right]$ | $-3R1 + R2$ gives $[0 \; 0 \; -8 \; -8] \to R2$  $2R1 + R3$ gives $[0 \; 1 \; 4 \; 2] \to R3$ | |
| | $\left[\begin{array}{ccc|c} 1 & -2 & 3 & 10 \\ 0 & 0 & -8 & -8 \\ 0 & 1 & 4 & 2 \end{array}\right]$ | Interchange R2 and R3, R2 $\leftrightarrow$ R3 | |

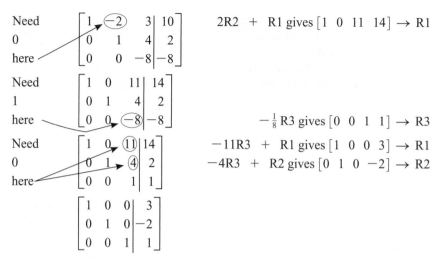

Need
0
here

$$\begin{bmatrix} 1 & -2 & 3 & | & 10 \\ 0 & 1 & 4 & | & 2 \\ 0 & 0 & -8 & | & -8 \end{bmatrix}$$

$2R2 + R1$ gives $\begin{bmatrix} 1 & 0 & 11 & 14 \end{bmatrix} \rightarrow R1$

Need
1
here

$$\begin{bmatrix} 1 & 0 & 11 & | & 14 \\ 0 & 1 & 4 & | & 2 \\ 0 & 0 & -8 & | & -8 \end{bmatrix}$$

$-\frac{1}{8} R3$ gives $\begin{bmatrix} 0 & 0 & 1 & 1 \end{bmatrix} \rightarrow R3$

Need
0
here

$$\begin{bmatrix} 1 & 0 & 11 & | & 14 \\ 0 & 1 & 4 & | & 2 \\ 0 & 0 & 1 & | & 1 \end{bmatrix}$$

$-11R3 + R1$ gives $\begin{bmatrix} 1 & 0 & 0 & 3 \end{bmatrix} \rightarrow R1$
$-4R3 + R2$ gives $\begin{bmatrix} 0 & 1 & 0 & -2 \end{bmatrix} \rightarrow R2$

$$\begin{bmatrix} 1 & 0 & 0 & | & 3 \\ 0 & 1 & 0 & | & -2 \\ 0 & 0 & 1 & | & 1 \end{bmatrix}$$

The last matrix is in diagonal form and represents the system

$$x_1 = 3 \qquad x_2 = -2 \qquad x_3 = 1$$

so the solution is $(3, -2, 1)$.

We point out that a system of linear equations can be solved without reducing the augmented matrix to diagonal form. Reducing the matrix to a form with zeros below the main diagonal suffices. For example, we use the same system of Example 7 and reduce as follows:

$$\begin{bmatrix} 2 & -4 & 6 & | & 20 \\ 3 & -6 & 1 & | & 22 \\ -2 & 5 & -2 & | & -18 \end{bmatrix}$$

$$\begin{bmatrix} 1 & -2 & 3 & | & 10 \\ 0 & 0 & -8 & | & -8 \\ 0 & 1 & 4 & | & 2 \end{bmatrix}$$

$$\begin{bmatrix} 1 & -2 & 3 & | & 10 \\ 0 & 1 & 4 & | & 2 \\ 0 & 0 & -8 & | & -8 \end{bmatrix}$$

From the last row of this matrix, we have

$$-8x_3 = -8 \qquad \text{or} \qquad x_3 = 1$$

From the second row, we have

$$x_2 + 4x_3 = 2$$

Since $x_3 = 1$, we have

$$x_2 + 4 = 2$$
$$x_2 = -2$$

From the first row, we have

$$x_1 - 2x_2 + 3x_3 = 10$$

Substituting $x_2 = -2$ and $x_3 = 1$, we have

$$x_1 + 4 + 3 = 10$$
$$x_1 = 3$$

Thus, the solution is $(3, -2, 1)$.

Generally, you can reduce the matrix to a form with zeros below the diagonal and then work from the bottom row up substituting the values of the variables.

**Example 8**   Solve the system

$$\begin{array}{rrrrcr} x_1 & -\ x_2 & +\ x_3 & +\ 2x_4 & = & 1 \\ 2x_1 & -\ x_2 & & +\ 3x_4 & = & 0 \\ -x_1 & +\ x_2 & +\ x_3 & +\ x_4 & = & -1 \\ & x_2 & & +\ x_4 & = & 1 \end{array}$$

**Solution**

The augmented matrix of this system is

$$\left[\begin{array}{rrrr|r} 1 & -1 & 1 & 2 & 1 \\ 2 & -1 & 0 & 3 & 0 \\ -1 & 1 & 1 & 1 & -1 \\ 0 & 1 & 0 & 1 & 1 \end{array}\right]$$

Performing the indicated row operations produces the following sequence of equivalent augmented matrices:

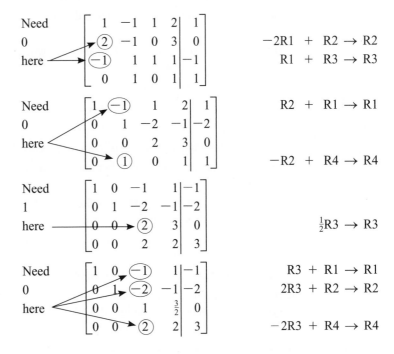

**NOTE**

In some systems, it might be impossible to get a 1 in some of the diagonal positions. The next section deals with that situation.

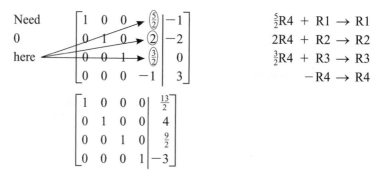

$$\frac{5}{2}R4 + R1 \rightarrow R1$$
$$2R4 + R2 \rightarrow R2$$
$$\frac{3}{2}R4 + R3 \rightarrow R3$$
$$-R4 \rightarrow R4$$

$$\begin{bmatrix} 1 & 0 & 0 & 0 & \frac{13}{2} \\ 0 & 1 & 0 & 0 & 4 \\ 0 & 0 & 1 & 0 & \frac{9}{2} \\ 0 & 0 & 0 & 1 & -3 \end{bmatrix}$$

This matrix is in the reduced diagonal form, so we can read the solution to the system:

$$x_1 = \frac{13}{2} \quad x_2 = 4 \quad x_3 = \frac{9}{2} \quad x_4 = -3$$

The examples in this section all have unique solutions. In such cases, the last column of the diagonal form gives the unique solution.

**Summary**    **Solve a System with the Gauss-Jordan Method Using Augmented Matrices**

Solve a system of $n$ equations in $n$ variables by using row operations on the augmented matrix.

**1.** Write the augmented matrix of the system.

**2.** Get a 1 in the $(1, 1)$ position of the matrix by

   **(a)** rearranging rows and/or

   **(b)** dividing row 1 by the $(1, 1)$ entry, $a_{11}$.

**3.** Get a zero in the other positions of column 1.

   **(a)** Take the negative of the number in the position to be made zero and multiply row 1 by it.

   **(b)** Add this multiple of row 1 to the row where the zero is needed. The result becomes a new row there.

**4.** Get a 1 in the $(2, 2)$ location by

   **(a)** rearranging the rows below row 1 and/or

   **(b)** dividing row 2 by the $(2, 2)$ entry, $a_{22}$.

**5.** Get a zero in all other positions of column 2 (leave 1 in the $(2, 2)$ position).

   **(a)** Take the negative of the number in the position to be made zero and multiply row 2 by it.

   **(b)** Add this multiple of row 2 to the row where the zero is needed. The result becomes the new row there.

**6.** Get a 1 in the $(3, 3)$, $(4, 4)$, ... positions, and in each case get zeros in the other positions of that column.

**7.** Each row now gives the value of a variable $x_1 = c_1, x_2 = c_2$, and so on, in the solution.

**8.** If some step in the process produces a row with all zeros except the last column, the system has produced an equation of the form $0 = c$ $(c \neq 0)$, which is a contradiction. Thus, the system has no solution.

## Application

**Example 9**    The Midway High School band sold citrus fruit for a fund-raiser. They sold grapefruit for $16 per box, oranges for $13 per box, and a grapefruit–orange combination for $15 per box. Rhonda, a French horn player, was pleased with her sales. "I sold 57 boxes, and here is the money, $845." The band director insisted he needed to know the number of boxes of each kind of fruit to process the order. "I don't remember the number of each, but I did notice that if we increased the price of a box of grapefruit by $2 and the price of a box of oranges by $1, I would have an even $900."

"That's odd information to notice, but perhaps we can use it to determine the number of each box."

Help Rhonda and the band director determine the number of boxes of each kind of fruit sold.

**Solution**

Let $x_1$ = number of boxes of grapefruit
$x_2$ = number of boxes of oranges
$x_3$ = number of boxes of grapefruit–orange

We can state the given information as follows:

$$
\begin{array}{rcl}
x_1 + x_2 + x_3 &=& 57 \quad \text{(number of boxes sold)} \\
16x_1 + 13x_2 + 15x_3 &=& 845 \quad \text{(amount Rhonda received)} \\
18x_1 + 14x_2 + 15x_3 &=& 900 \quad \text{(amount if prices were increased)}
\end{array}
$$

The solution to this system gives the desired number of boxes for each kind of fruit. Find the solution by using augmented matrices. The following sequence of matrix reductions leads to the solution:

$$
\left[\begin{array}{ccc|c}
1 & 1 & 1 & 57 \\
16 & 13 & 15 & 845 \\
18 & 14 & 15 & 900
\end{array}\right]
\quad
\begin{array}{l}
-16R1 + R2 \rightarrow R2 \\
-18R1 + R3 \rightarrow R3
\end{array}
$$

$$
\left[\begin{array}{ccc|c}
1 & 1 & 1 & 57 \\
0 & -3 & -1 & -67 \\
0 & -4 & -3 & -126
\end{array}\right]
\quad
-\tfrac{1}{3}R2 \rightarrow R2
$$

$$
\left[\begin{array}{ccc|c}
1 & 1 & 1 & 57 \\
0 & 1 & \frac{1}{3} & \frac{67}{3} \\
0 & 4 & 3 & 126
\end{array}\right]
\quad
\begin{array}{l}
-R2 + R1 \rightarrow R1 \\
\\
-4R2 + R3 \rightarrow R3
\end{array}
$$

$$
\left[\begin{array}{ccc|c}
1 & 0 & \frac{2}{3} & \frac{104}{3} \\
0 & 1 & \frac{1}{3} & \frac{67}{3} \\
0 & 0 & \frac{5}{3} & \frac{110}{3}
\end{array}\right]
\quad
\tfrac{3}{5}R3 \rightarrow R3
$$

$$
\left[\begin{array}{ccc|c}
1 & 0 & \frac{2}{3} & \frac{104}{3} \\
0 & 1 & \frac{1}{3} & \frac{67}{3} \\
0 & 0 & 1 & 22
\end{array}\right]
\quad
\begin{array}{l}
-\tfrac{2}{3}R3 + R1 \rightarrow R1 \\
-\tfrac{1}{3}R3 + R2 \rightarrow R2
\end{array}
$$

$$
\left[\begin{array}{ccc|c}
1 & 0 & 0 & 20 \\
0 & 1 & 0 & 15 \\
0 & 0 & 1 & 22
\end{array}\right]
$$

The last matrix gives $x_1 = 20$, $x_2 = 15$, and $x_3 = 22$.

Rhonda sold 20 boxes of grapefruit, 15 boxes of oranges, and 22 boxes of grapefruit–orange.

**Example 10**    The procedure of solving a system of equations by reducing the augmented matrix by a series of pivots can lead to some messy fractions. A variation of the pivot process can avoid fractions until the last step. We illustrate with a simple example.

Solve the system

$$x + 2y + 3z = 142$$
$$5x + 4y + 2z = 292$$
$$4x + 4y + 5z = 316$$

Set up the augmented matrix

$$\begin{bmatrix} 1 & 2 & 3 & | & 142 \\ 5 & 4 & 2 & | & 292 \\ 4 & 4 & 5 & | & 316 \end{bmatrix} \quad \begin{matrix} \\ -5R1 + R2 \to R2 \\ -4R1 + R3 \to R3 \end{matrix}$$

We can pivot on the $(1, 1)$ entry in the usual manner by using the row operations just indicated, which yields the matrix

$$\begin{bmatrix} 1 & 2 & 3 & | & 142 \\ 0 & -6 & -13 & | & -418 \\ 0 & -4 & -7 & | & -252 \end{bmatrix}$$

Next, we could pivot on the $-6$ in the $(2, 2)$ position by dividing the row by $-6$, giving the row $[0 \quad 1 \quad 13/6 \quad 418/6]$.

Using this row as the pivot row will involve arithmetic using fractions. Let's get the required 0's in the second column without reducing the pivot element to 1. Here's how. Use the following row operations, much like the way you solved by elimination.

$$\begin{bmatrix} 1 & 2 & 3 & | & 142 \\ 0 & -6 & -13 & | & -418 \\ 0 & -4 & -7 & | & -252 \end{bmatrix} \quad \begin{matrix} 2R2 + 6R1 \to R1 \\ \\ -4R2 + 6R3 \to R3 \end{matrix}$$

This gives

$$\begin{bmatrix} 6 & 0 & -8 & | & 16 \\ 0 & -6 & -13 & | & -418 \\ 0 & 0 & 10 & | & 160 \end{bmatrix} \quad \begin{matrix} \frac{1}{2}R1 \to R1 \\ -R2 \to R2 \\ \frac{1}{10}R3 \to R3 \end{matrix}$$

You can simplify the arithmetic with the row operations indicated above, giving

$$\begin{bmatrix} 3 & 0 & -4 & | & 8 \\ 0 & 6 & 13 & | & 418 \\ 0 & 0 & 1 & | & 16 \end{bmatrix} \quad \begin{matrix} 4R3 + R1 \to R1 \\ -13R3 + R2 \to R2 \end{matrix}$$

The following result gives the solution when we divide by the pivot elements:

$$\begin{bmatrix} 3 & 0 & 0 & | & 72 \\ 0 & 6 & 0 & | & 210 \\ 0 & 0 & 1 & | & 16 \end{bmatrix} \quad \begin{matrix} \frac{1}{3}R1 \rightarrow R1 \\ \frac{1}{6}R2 \rightarrow R2 \end{matrix} \quad \begin{bmatrix} 1 & 0 & 0 & | & 24 \\ 0 & 1 & 0 & | & 35 \\ 0 & 0 & 1 & | & 16 \end{bmatrix}$$

$$x = 24 \quad y = 35 \quad z = 16$$

## 2.2   EXERCISES

Access end-of-section exercises online at **www.webassign.net**    **WebAssign**

### Using Your TI Graphing Calculator

**Solving a System of Equations Using Intersect**

You can solve a system of two equations with two variables using the **intersect** command. See the TI Graphing Calculator section at the end of Section 1.3.

**Solving a System of Equations Using Row Operations**

Let's see how we can solve the following system of equations using row operations on the TI graphing calculator.

$$\begin{aligned} 2x_1 - 3x_2 + x_3 &= 25 \\ 6x_1 + x_2 + 5x_3 &= 33 \\ x_1 + 4x_2 - 3x_3 &= -29 \end{aligned}$$

The TI graphing calculator allows ten matrices named [A] through [J]. We will enter the augmented matrix of the system of equations in [A] and perform row operations on that matrix.

To enter matrix [A], select $\boxed{\text{MATRIX}}$ <EDIT>, select [A] from the list shown, and then press $\boxed{\text{ENTER}}$.

Enter the size of the matrix and the matrix entries, row by row, and you will see the following screens displayed:

```
MATRIX[A] 3 ×4
[ 2   -3   1    .
[ 6    1   5    .
[ ▆    4   -3   .

3,1=1
```

```
MATRIX[A] 3 ×4
 -3   1    25  ]
 1    5    33  ]
 4    -3   -29 ]

3,4=-29
```

Note that the screen does not show the entire matrix. The left screen shows the left-hand portion of the matrix. The rest of the matrix is viewed by scrolling the screen to the right. Use the $\boxed{>}$ key.

At any time, you can view matrix [A] by the sequence $\boxed{\text{MATRIX}}$ <1:[A]3x4> $\boxed{\text{ENTER}}$ $\boxed{\text{ENTER}}$.

```
[A]
 [[2  -3   1   25 ]
  [6   1   5   33 ]
  [1   4  -3  -29]]
```

*(continued)*

An augmented matrix can be reduced to a form that gives the solution to its system of equations. The operations and their TI Graphing Calculator instructions are shown below. We also include the notation for a row operation used in this chapter.

| Operation on Matrix [A] | Notation | TI Graphing Calculator Instruction |
|---|---|---|
| • Interchange rows 1 and 2 | R1 ↔ R2 | rowSwap([A],1,2) |
| • Multiply row 4 by 3.5 | 3.5R4 → R4 | *row(4,[A],3.5) |
| • Add row 1 and row 3 with the result in row 3 | R1 + R3 → R3 | row+([A],1,3) |
| • Multiply row 3 by −4 and add the result to row 5 | −4R3 + R5 → R5 | *row+(−4,[A],3,5) |

*Note:* Other TI matrices may be used and other rows as appropriate. Here is an example of how you build the instruction R2 + R4 → R4 to operate on matrix [B]:

Select row+ (with the sequence MATRIX <MATH>, and then scroll down to <D: row+ ( > and press ENTER. When row+ ( appears, you will fill in the matrix name and row numbers, in this case [B], 2, and 4, to obtain row+ ([B], 2, 4).

You enter the matrix name, [B], with the sequence: MATRIX , <2: [B]> under NAME, ENTER. Then, to complete the instruction, type ,2,4) in the usual way. Note that the second row number, 4 in this case, is the row location of the result of adding the two rows. Likewise, in the *row+ ( instruction, the second row number indicates the location of the result.

We now solve the system using the following sequence of row operations. We use the matrix

$$[A] = \begin{bmatrix} 2 & -3 & 1 & 25 \\ 6 & 1 & 5 & 33 \\ 1 & 4 & -3 & -29 \end{bmatrix}$$

1. Multiply row 1 by 0.5, (0.5R1 → R1): MATRIX <MATH> <E:*row(.5,[A],1)> ENTER , which displays the following screens:

```
*row(.5,[A],1)
[[1  -1.5  .5  12.…
 [6  1      5   33.…
 [1  4     -3  -29.…
```

```
*row(.5,[A],1)
…  -1.5  .5  12.5]
…  1      5   33 ]
…  4     -3  -29 ]]
```

The matrix shown is in working memory, not in [A]. To place it in [A], use ANS STO [A] ENTER .

2. Next, obtain a zero in row 2, column 1, by multiplying row 1 by −6 and adding to row 2 (−6R1 + R2 → R2). Use MATRX <MATH> <F:*row+(−6,[A],1,2)> ENTER . You should get the following:

```
*row+(-6,[A],1,2
[[1  -1.5  .5  12.…
 [0  10    2   -42…
 [1  4    -3  -29…
```

```
*row+(-6,[A],1,2
…  -1.5  .5  12.5]
…  10    2   -42 ]
…  4    -3  -29 ]]
```

ANS STO [A] ENTER stores the result in [A].

3. To get a zero in the (3, 1) position, multiply the first row by $-1$ and add to the third row ($-1R1 + R3 \rightarrow R3$). You should get the following:

```
*row+(-1,[A],1,3
[[1  -1.5  .5   1…
 [0  10    2
 [0  5.5  -3.5  -…
■
```

```
*row+(-1,[A],1,3
….5  .5   12.5 ]
…2        -42  ]
…5  -3.5  -41.5]]
```

ANS STO [A] ENTER stores the result in [A].

4. Now proceed to obtain a 1 in row 2, column 2, and zero in the rest of column 2.

5. Then obtain a 1 in row 3, column 3, and zero in the rest of column 3. The final matrix is

```
*row+(-.2,[A],3,
2
   [[1 0 0  3 ]
    [0 1 0 -5]
    [0 0 1  4 ]]
```

This matrix shows the solution of the system to be $(3, -5, 4)$.

## Exercises

Use row operations to solve the following systems:

1. $2x_1 + x_2 + 4x_3 = 12$
   $-x_1 + 3x_2 + 5x_3 = 8$
   $3x_1 \quad\quad + 2x_3 = 9$

2. $x_1 + 3x_2 - 2x_3 = 19$
   $2x_1 + x_2 + x_3 = 13$
   $5x_1 - 2x_2 + 4x_3 = 7$

3. $4x_1 + 3x_2 + x_3 = 26$
   $-5x_1 + 2x_2 + 2x_3 = -22$
   $3x_1 + x_2 + 5x_3 = 42$

4. $5x_1 - 3x_2 + x_3 = 31$
   $2x_1 + 4x_2 - 3x_3 = 33$
   $3x_1 + x_2 + x_3 = 23$

 ## Using Excel

## Solving Systems of Equations Using Solver

In Section 1.3, we used Goal Seek to solve an equation with one variable. When we are given a system of equations, we have to use Excel Solver instead. Click on the **Data** tab and see if there is an **Analysis** group with **Solver**. If so, then Solver is already installed, and ready to use. Otherwise, it has to be installed. The instructions for installation are given at the end of this section.

*(continued)*

Assuming that the Solver add-in is installed, let us see how we can solve the following system of linear equations using Excel Solver.

$$
\begin{aligned}
2x_1 - 3x_2 + x_3 &= 25 \\
6x_1 + x_2 + 5x_3 &= 33 \\
x_1 + 4x_2 - 3x_3 &= -29
\end{aligned}
$$

1. Make sure the equations are in a form with the variable expressions on the left-hand side and the constants on the right-hand side.

2. We have three variables, $x_1, x_2,$ and $x_3,$ and we store their values in cells B2, B3, and B4, respectively, with initial values of 0.

|   | A | B |
|---|---|---|
| 1 | Variables | |
| 2 | x1 | 0 |
| 3 | x2 | 0 |
| 4 | x3 | 0 |

3. The left-hand side of each equation is then entered as a formula in each of the cells B7, B8, and B9. For example, $2x_1 - 3x_2 + x_3$ is entered as = **2\*B2 − 3\*B3 + B4** in cell B7. Similarly, the left-hand sides of the other two equations are entered in cells B8 and B9, respectively.

4. Next, enter the corresponding right-hand side of each equation in cells C7, C8, and C9.

|   | A | B | C |
|---|---|---|---|
| 1 | Variables | | |
| 2 | x1 | 0 | |
| 3 | x2 | 0 | |
| 4 | x3 | 0 | |
| 5 | | | |
| 6 | | Left hand side of equation | Right hand side of equation |
| 7 | $2x_1 - 3x_2 + x_3 = 25$ | =2*B2-3*B3+B4 | 25 |
| 8 | $6x_1 + x_2 + 5x_3 = 33$ | =6*B2+B3+5*B4 | 33 |
| 9 | $x_1 + 4x_2 - 3x_3 = -29$ | =B2+4*B3-3*B4 | -29 |
| 10 | | | |

5. The equality portion of the equation is imposed by using Solver. Click on the **Data** tab. Move to the **Analysis** group and click on **Solver**.

6. In Excel 2007, the first equation will be our Target. Type **B7** in the **Set Target Cell** box, choose the **Value of** option, and then type 25.

7. In Excel 2010, the first equation will be our Objective. Type **B7** in the **Set Objective** box, choose the **Value of** option, then type 25.

8. Move cursor to the **By Changing Cells** entry box. Enter the cell references for the variables by selecting the cells **B2 : B4**. The dialog box should now look like one of the following, depending on the version of Excel in use.

**Excel 2007**

Set Target Cell:    $B$7

Equal To:    ○ Max    ○ Min    ⦿ Value of:    25
By Changing Cells:

$B$2:$B$4

**Excel 2010**

Set Objective:    $B$7

To:    ○ Max    ○ Min    ⦿ Value Of:    25

By Changing Variable Cells:

$B$2:$B$4

9. Click cursor into **Subject to the Constraints** entry box. This is where we enter the equations:
   a. Press the Add button to add the second equation. You will get a new dialog box for the constraint.
   b. Click cursor to the left entry box and click into cell **B8** containing the formula for the left-hand side of the first equation.
   c. The middle entry box should be set to =.
   d. Click cursor to the right entry box and click into the cell **C8** containing the right-hand side. The constraint dialog box should resemble the following.

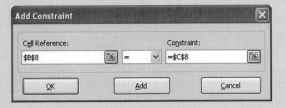

   e. Click the **Add** button to add the third equation and repeat Steps (b)–(d), with cells B9 and C9.
   f. Click **OK**.
10. The completed Solver box should resemble one of the following:

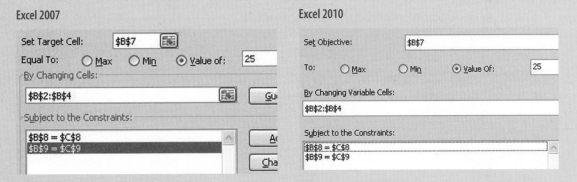

11. Now set the options for Solver, depending on the version in use.
    a. In Excel 2007, click on the options box. **Assume Non-Negative** option must be **unchecked** and the **Assume Linear Model** option should be **checked**.
    b. In Excel 2010, below the constraints box, non-negative variables option must be **unchecked** and the Solving Method should be **Simplex LP**.

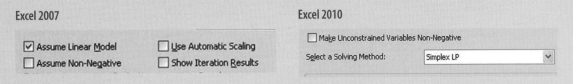

12. Click **Solve** in the Solver dialog box. A new dialog box will appear stating that Solver found a solution.

*(continued)*

13. Check the **Keep Solver Solution** button. Click **OK**. Go back and examine the cells B2, B3, and B4. They should now contain the solution, $x_1 = 3$, $x_2 = -5$, $x_3 = 4$. Note that the left- and right-hand side of the equations now have equal values, confirming that we have indeed found the solution.

| | A | B | C |
|---|---|---|---|
| 1 | Variables | | |
| 2 | x1 | 3 | |
| 3 | x2 | -5 | |
| 4 | x3 | 4 | |
| 5 | | | |
| 6 | | Left hand side of equation | Right hand side of equation |
| 7 | $2x_1 - 3x_2 + x_3 = 25$ | 25 | 25 |
| 8 | $6x_1 + x_2 + 5x_3 = 33$ | 33 | 33 |
| 9 | $x_1 + 4x_2 - 3x_3 = -29$ | -29 | -29 |

**Installation of Solver Add-In**

In Excel 2010, click on the **File** menu and then choose **Options**.

In Excel 2007, click on the **Office Button**, and then click on **Excel Options**.

Afterward, for either version of Excel, click on **Add-ins**, and then select **Excel Add-ins** in the **Manage** box. Click **Go** and then select the **Solver Add-in** from the **Add-Ins** available box. If you are prompted to install the add-in, click **Yes**.

## Exercises

Use Solver to solve the following systems.

**1.** $\begin{aligned} 2x_1 + x_2 + 4x_3 &= 12 \\ -x_1 + 3x_2 + 5x_3 &= 8 \\ 3x_1 \qquad\quad + 2x_3 &= 9 \end{aligned}$

**2.** $\begin{aligned} x_1 + 3x_2 - 2x_3 &= 19 \\ 2x_1 + x_2 + x_3 &= 13 \\ 5x_1 - 2x_2 + 4x_3 &= 7 \end{aligned}$

**3.** $\begin{aligned} 4x_1 + 3x_2 + x_3 &= 26 \\ -5x_1 + 2x_2 + 2x_3 &= -22 \\ 3x_1 + x_2 + 5x_3 &= 42 \end{aligned}$

**4.** $\begin{aligned} 5x_1 - 3x_2 + x_3 &= 31 \\ 2x_1 + 4x_2 - 3x_3 &= 33 \\ 3x_1 + x_2 + x_3 &= 23 \end{aligned}$

**5.** $\begin{aligned} x_1 + 2x_2 - x_3 + 3x_4 &= 12 \\ 2x_1 - x_2 + 2x_3 - x_4 &= 10 \\ x_1 + 3x_2 - 2x_3 + 2x_4 &= -3 \\ 3x_1 - 2x_2 + 3x_3 - 2x_4 &= 13 \end{aligned}$

| **2.3** | # GAUSS-JORDAN METHOD FOR GENERAL SYSTEMS OF EQUATIONS |

- Reduced Echelon Form
- Systems with No Solution
- Application

In this section, we expand on the Gauss-Jordan Method presented in Section 2.2. In that section, the systems had the same number of equations as variables, dealt mainly with three variables, and had a unique solution. In general, a system may have many variables, it may have more or fewer equations than variables, and it may have many solutions or no solution at all. In any case, the Gauss-Jordan Method can be used to solve the system by starting with an augmented matrix as before. Using a sequence of row operations eventually gives a simpler form of the matrix, which yields the solutions. In Section 2.2, the simplified matrices reduced to a diagonal form that gave a unique solution. Some augmented matrices will not reduce to a diagonal form, but they can always be reduced to another standard form called the **reduced echelon form**.

## Reduced Echelon Form

We give the general definition of a reduced echelon form. The diagonal forms of the preceding section also conform to this definition.

| **DEFINITION** Reduced Echelon Form | A matrix is in **reduced echelon form** if all the following are true: |

1. All rows consisting entirely of zeros are grouped at the bottom of the matrix.
2. The leftmost nonzero number in each row is 1. This element is called the *leading 1 of the row.*
3. The leading 1 of a row is to the right of the leading 1 of the rows above.
4. All entries above and below a leading 1 are zeros.

The following matrices are all in reduced echelon form. Check the conditions in the definition to make sure you understand why.

$$\begin{bmatrix} 1 & 0 & 0 & 5 \\ 0 & 1 & 0 & -3 \\ 0 & 0 & 1 & 7 \end{bmatrix} \quad \begin{bmatrix} 1 & 0 & 3 & 0 & 8 \\ 0 & 1 & -1 & 0 & 2 \\ 0 & 0 & 0 & 1 & 7 \end{bmatrix} \quad \begin{bmatrix} 1 & 5 & 0 & 2 \\ 0 & 0 & 1 & 3 \\ 0 & 0 & 0 & 0 \end{bmatrix} \quad \begin{bmatrix} 1 & 0 & -2 & 0 & 0 & 3 & 7 \\ 0 & 1 & 4 & 0 & 0 & -5 & 6 \\ 0 & 0 & 0 & 1 & 0 & 4 & 11 \\ 0 & 0 & 0 & 0 & 1 & 2 & 2 \end{bmatrix}$$

The following matrices are *not* in reduced echelon form.

$$\begin{bmatrix} 1 & 0 & 0 & 2 & 1 \\ 0 & 1 & 0 & 2 & 3 \\ 0 & 0 & 0 & 0 & 0 \\ 0 & 0 & 1 & 0 & 2 \end{bmatrix} \qquad\qquad \begin{bmatrix} 1 & 0 & 0 & 6 \\ 0 & 1 & 0 & 5 \\ 0 & 0 & 4 & 7 \end{bmatrix}$$

The row of zeros is not at the bottom of the matrix.

The leftmost nonzero entry in row 3 (4) is not 1.

$$\begin{bmatrix} 1 & 0 & 0 & 0 & 5 \\ 0 & 1 & 0 & 0 & 3 \\ 0 & 0 & 0 & 1 & 4 \\ 0 & 0 & 1 & 0 & -2 \end{bmatrix}$$

$$\begin{bmatrix} 1 & 0 & 0 & 0 & 7 \\ 0 & 1 & 3 & 0 & -2 \\ 0 & 0 & 1 & 0 & 5 \\ 0 & 0 & 0 & 1 & -1 \end{bmatrix}$$

The leading 1 in row 4 is not to the right of the leading 1 in row 3.

The entry in row 2 above the leading 1 in row 3 is not zero.

We now work through the details of modifying a matrix until we obtain the reduced echelon form. As we work through it, notice how we use row operations to obtain the leading 1 in row 1, row 2, and so on, and then get zeros in the rest of a column with a leading 1.

We use the same row operations that are used in reducing an augmented matrix to obtain a solution to a linear system. Generally, we want to obtain a reduced matrix with 1's on the diagonal.

**Example 1**   Find the reduced echelon form of the matrix

$$\begin{bmatrix} 0 & 1 & -3 & 2 \\ 2 & 4 & 6 & -4 \\ 3 & 5 & 2 & 2 \end{bmatrix}$$

**Solution**

| | Matrices | Row Operations | Comments |
|---|---|---|---|
| Need 1 here | $\begin{bmatrix} ⓪ & 1 & -3 & 2 \\ 2 & 4 & 6 & -4 \\ 3 & 5 & 2 & 2 \end{bmatrix}$ | $R2 \leftrightarrow R1$ | Interchange row 1 and row 2 to get nonzero number at top of column 1. |
| Need 1 here | $\begin{bmatrix} ② & 4 & 6 & -4 \\ 0 & 1 & -3 & 2 \\ 3 & 5 & 2 & 2 \end{bmatrix}$ | $\frac{1}{2}R1 \to R1$ | Divide row 1 by 2 to get a 1. |
| Need 0 here | $\begin{bmatrix} 1 & 2 & 3 & -2 \\ 0 & 1 & -3 & 2 \\ ③ & 5 & 2 & 2 \end{bmatrix}$ | $-3R1 + R3 \to R3$ | Get zeros in rest of column 1. |
| Leading 1 here | $\begin{bmatrix} 1 & 2 & 3 & -2 \\ 0 & ① & -3 & 2 \\ 0 & -1 & -7 & 8 \end{bmatrix}$ | | Now get leading 1 in row 2. No changes necessary this time. |
| Need 0 here | $\begin{bmatrix} 1 & ② & 3 & -2 \\ 0 & 1 & -3 & 2 \\ 0 & -① & -7 & 8 \end{bmatrix}$ | $-2R2 + R1 \to R1$  $R2 + R3 \to R3$ | Zero entries above and below leading 1 of row 2. |
| Need leading 1 here | $\begin{bmatrix} 1 & 0 & 9 & -6 \\ 0 & 1 & -3 & 2 \\ 0 & 0 & ⟨-10⟩ & 10 \end{bmatrix}$ | $-\frac{1}{10}R3 \to R3$ | Get a leading 1 in the next row. |

Need $\qquad$ $\begin{bmatrix} 1 & 0 & 9 & -6 \\ 0 & 1 & -3 & 2 \\ 0 & 0 & 1 & -1 \end{bmatrix}$ $\qquad$ $-9R3 + R1 \rightarrow R1$ $\qquad$ Zero entries above,
0 here $\qquad$ $3R3 + R2 \rightarrow R2$ $\qquad$ leading 1 in row 3.

$$\begin{bmatrix} 1 & 0 & 0 & 3 \\ 0 & 1 & 0 & -1 \\ 0 & 0 & 1 & -1 \end{bmatrix}$$ $\qquad$ This is the reduced echelon form.

**Example 2** Find the reduced echelon form of this matrix:

$$\begin{bmatrix} 0 & 0 & 2 & -2 & 2 \\ 3 & 3 & -3 & 9 & 12 \\ 4 & 4 & -2 & 11 & 12 \end{bmatrix}$$

**Solution**
Again we show much of the detailed row operations.

| | Matrices | Row Operations | Comments |
|---|---|---|---|
| Need 1 here | $\begin{bmatrix} 0 & 0 & 2 & -2 & 2 \\ 3 & 3 & -3 & 9 & 12 \\ 4 & 4 & -2 & 11 & 12 \end{bmatrix}$ | $R1 \leftrightarrow R2$ | |
| Need 1 here | $\begin{bmatrix} 3 & 3 & -3 & 9 & 12 \\ 0 & 0 & 2 & -2 & 2 \\ 4 & 4 & -2 & 11 & 12 \end{bmatrix}$ | $\frac{1}{3}R1 \rightarrow R1$ | |
| Need 0 here | $\begin{bmatrix} 1 & 1 & -1 & 3 & 4 \\ 0 & 0 & 2 & -2 & 2 \\ 4 & 4 & -2 & 11 & 12 \end{bmatrix}$ | $\frac{1}{2}R2 \rightarrow R2$ $-4R1 + R3 \rightarrow R3$ | |
| Leading 1 row 2 | $\begin{bmatrix} 1 & 1 & -1 & 3 & 4 \\ 0 & 0 & 1 & -1 & 1 \\ 0 & 0 & 2 & -1 & -4 \end{bmatrix}$ | | The leading 1 of row 2 must come from row 2 or below. Since all entries in column 2 are zero in rows 2 and 3, go to column 3 for leading 1. |
| Need 0 here | $\begin{bmatrix} 1 & 1 & -1 & 3 & 4 \\ 0 & 0 & 1 & -1 & 1 \\ 0 & 0 & 2 & -1 & -4 \end{bmatrix}$ | $R2 + R1 \rightarrow R1$ $-2R2 + R3 \rightarrow R3$ | |
| Need 0 here | $\begin{bmatrix} 1 & 1 & 0 & 2 & 5 \\ 0 & 0 & 1 & -1 & 1 \\ 0 & 0 & 0 & 1 & -6 \end{bmatrix}$ | $-2R3 + R1 \rightarrow R1$ $R3 + R2 \rightarrow R2$ | |

$$\begin{bmatrix} 1 & 1 & 0 & 0 & | & 17 \\ 0 & 0 & 1 & 0 & | & -5 \\ 0 & 0 & 0 & 1 & | & -6 \end{bmatrix}$$

This is the reduced
echelon form.

We now solve various systems of equations to illustrate the Gauss-Jordan Method of elimination. As you work through the examples, note that solving a system of equations using the Gauss-Jordan Method is basically a process of manipulating the augmented matrix to its reduced echelon form.

**Example 3**    Solve, if possible, the system

$$2x_1 - 4x_2 + 12x_3 = 20$$
$$-x_1 + 3x_2 + 5x_3 = 15$$
$$3x_1 - 7x_2 + 7x_3 = 5$$

**Solution**
We start with the augmented matrix and convert it to reduced echelon form:

$$\begin{bmatrix} 2 & -4 & 12 & | & 20 \\ -1 & 3 & 5 & | & 15 \\ 3 & -7 & 7 & | & 5 \end{bmatrix} \qquad \tfrac{1}{2}\text{R1} \rightarrow \text{R1}$$

$$\begin{bmatrix} 1 & -2 & 6 & | & 10 \\ -1 & 3 & 5 & | & 15 \\ 3 & -7 & 7 & | & 5 \end{bmatrix} \qquad \begin{array}{l} \text{R1} + \text{R2} \rightarrow \text{R2} \\ -3\text{R1} + \text{R3} \rightarrow \text{R3} \end{array}$$

$$\begin{bmatrix} 1 & -2 & 6 & | & 10 \\ 0 & 1 & 11 & | & 25 \\ 0 & -1 & -11 & | & -25 \end{bmatrix} \qquad \begin{array}{l} 2\text{R2} + \text{R1} \rightarrow \text{R1} \\ \\ \text{R2} + \text{R3} \rightarrow \text{R3} \end{array}$$

$$\begin{bmatrix} 1 & 0 & 28 & | & 60 \\ 0 & 1 & 11 & | & 25 \\ 0 & 0 & 0 & | & 0 \end{bmatrix}$$

This matrix is the reduced echelon form of the augmented matrix. It represents the system of equations

$$x_1 \qquad + 28x_3 = 60$$
$$x_2 + 11x_3 = 25$$

When the reduced echelon form gives equations containing more than one variable, such as $x_1 + 28x_3 = 60$, the system has many solutions. Many sets of $x_1$, $x_2$, and $x_3$ satisfy these equations. Usually, we solve the first equation for $x_1$ and the second for $x_2$ to get

$$x_1 = 60 - 28x_3$$
$$x_2 = 25 - 11x_3$$

This represents the general solution with $x_1$ and $x_2$ expressed in terms of $x_3$. When some variables are expressed in terms of another variable, $x_3$ in this case, we call $x_3$ a **parameter**. We find specific solutions to the system by substituting values for $x_3$. For example, one specific solution is found by assigning $x_3 = 1$. Then $x_1 = 60 - 28 = 32$, and $x_2 = 25 - 11 = 14$. In general, we assign the arbitrary value $k$ to $x_3$ and solve for $x_1$ and $x_2$. The arbitrary solution can then be expressed as $x_1 = 60 - 28k$, $x_2 = 25 - 11k$, and $x_3 = k$. As $k$ ranges over the real numbers, we get all solutions. In such a case, $k$ is called a *parameter*. For example, when $k = 2$, we get $x_1 = 4$, $x_2 = 3$, and $x_3 = 2$. When $k = -1$, we get the solution $x_1 = 88$, $x_2 = 36$, and $x_3 = -1$. In summary, the solutions to this example may be written in two ways:

$$x_1 = 60 - 28x_3$$
$$x_2 = 25 - 11x_3$$

or

$$x_1 = 60 - 28k$$
$$x_2 = 25 - 11k$$
$$x_3 = k$$

The latter is sometimes written as $(60 - 28k, 25 - 11k, k)$.

---

The reduction of an augmented matrix can be tedious. However, this method reduces the solution of a system of equations to a routine. This routine can be carried out by a computer or a graphing calculator. When dozens of variables are involved, a computer is the only practical way to solve such a system. We want you to be able to perform this routine, so we have two more examples to help you.

---

**Example 4**   Solve the system by reducing the augmented matrix.

$$
\begin{aligned}
x_1 + 3x_2 - 5x_3 + 2x_4 &= -10 \\
-2x_1 + x_2 + 3x_3 - 4x_4 + 7x_5 &= -22 \\
3x_1 - 7x_2 - 3x_3 - 2x_4 + 4x_5 &= -18
\end{aligned}
$$

We write the augmented matrix and start the process of reducing to echelon form:

$$
\left[\begin{array}{ccccc|c}
1 & 3 & -5 & 2 & 0 & -10 \\
-2 & 1 & 3 & -4 & 7 & -22 \\
3 & -7 & -3 & -2 & 4 & -18
\end{array}\right]
\quad
\begin{array}{l}
\\
2\text{R}1 + \text{R}2 \to \text{R}2 \\
-3\text{R}1 + \text{R}3 \to \text{R}3
\end{array}
$$

$$
\left[\begin{array}{ccccc|c}
1 & 3 & -5 & 2 & 0 & -10 \\
0 & 7 & -7 & 0 & 7 & -42 \\
0 & -16 & 12 & -8 & 4 & 12
\end{array}\right]
\quad
\begin{array}{l}
\\
\frac{1}{7}\text{R}2 \to \text{R}2 \\
-\frac{1}{4}\text{R}3 \to \text{R}3
\end{array}
$$

Simplify rows 2 and 3:

$$
\left[\begin{array}{ccccc|c}
1 & 3 & -5 & 2 & 0 & -10 \\
0 & 1 & -1 & 0 & 1 & -6 \\
0 & 4 & -3 & 2 & -1 & -3
\end{array}\right]
\quad
\begin{array}{l}
-3\text{R}2 + \text{R}1 \to \text{R}1 \\
\\
-4\text{R}2 + \text{R}3 \to \text{R}3
\end{array}
$$

$$
\left[\begin{array}{ccccc|c}
1 & 0 & -2 & 2 & -3 & 8 \\
0 & 1 & -1 & 0 & 1 & -6 \\
0 & 0 & 1 & 2 & -5 & 21
\end{array}\right]
\quad
\begin{array}{l}
2\text{R}3 + \text{R}1 \to \text{R}1 \\
\text{R}3 + \text{R}2 \to \text{R}2
\end{array}
$$

$$
\left[\begin{array}{ccccc|c}
1 & 0 & 0 & 6 & -13 & 50 \\
0 & 1 & 0 & 2 & -4 & 15 \\
0 & 0 & 1 & 2 & -5 & 21
\end{array}\right]
$$

We now have the reduced echelon form of the augmented matrix. This matrix represents the system:

$$
\begin{aligned}
x_1 \quad\quad\; + 6x_4 - 13x_5 &= 50 \\
x_2 + 2x_4 - \;\;4x_5 &= 15 \\
x_3 + 2x_4 - \;\;5x_5 &= 21
\end{aligned}
$$

Solving for $x_1$, $x_2$, and $x_3$ in terms of $x_4$ and $x_5$, we get

$$
\begin{aligned}
x_1 &= 50 - 6x_4 + 13x_5 \\
x_2 &= 15 - 2x_4 + \;\;4x_5 \\
x_3 &= 21 - 2x_4 + \;\;5x_5
\end{aligned}
$$

Here, $x_1$, $x_2$, and $x_3$ are solved in terms of $x_4$ and $x_5$, so this solution has two parameters, $x_4$ and $x_5$. We can select arbitrary values for $x_4$ and $x_5$ and use them to obtain values for $x_1$, $x_2$, and $x_3$. We denote this by assigning the arbitrary value $k$ to $x_4$ and $m$ to $x_5$. We use them to obtain solutions

$$
\begin{aligned}
x_1 &= 50 - 6k + 13m \\
x_2 &= 15 - 2k + \;\;4m \\
x_3 &= 21 - 2k + \;\;5m \\
x_4 &= k \\
x_5 &= m
\end{aligned}
$$

which we can also write as $(50 - 6k + 13m, 15 - 2k + 4m, 21 - 2k + 5m, k, m)$. Specific solutions are obtained by selecting specific values for $k$ and $m$, such as $k = 5$, $m = 1$, which yields the solution $(33, 9, 16, 5, 1)$. Because we can choose any real numbers for $k$ and $m$, this system has an infinite number of solutions.

It is possible for a system to have **no solution**. We illustrate this in the following example.

## Systems with No Solution

**Example 5**    The following system has no solution. Let's see what happens when we try to solve it.

$$
\begin{aligned}
x_1 + 3x_2 - 2x_3 &= 5 \\
4x_1 - \;\;x_2 + 3x_3 &= 7 \\
2x_1 - 7x_2 + 7x_3 &= 4
\end{aligned}
$$

**Solution**

$$
\left[\begin{array}{ccc|c}
1 & 3 & -2 & 5 \\
4 & -1 & 3 & 7 \\
2 & -7 & 7 & 4
\end{array}\right]
\qquad
\begin{array}{l}
-4\text{R}1 + \text{R}2 \rightarrow \text{R}2 \\
-2\text{R}1 + \text{R}3 \rightarrow \text{R}3
\end{array}
$$

$$
\left[\begin{array}{ccc|c}
1 & 3 & -2 & 5 \\
0 & -13 & 11 & -13 \\
0 & -13 & 11 & -6
\end{array}\right]
\qquad
-\text{R}2 + \text{R}3 \rightarrow \text{R}3
$$

$$
\left[\begin{array}{ccc|c}
1 & 3 & -2 & 5 \\
0 & -13 & 11 & -13 \\
0 & 0 & 0 & 7
\end{array}\right]
$$

The last matrix is not yet in reduced echelon form. However, we need proceed no further because the last row represents the equation $0 = 7$. When we reach an inconsistency like this, we know that the system has no solution.

Usually, you cannot look at a system of equations and tell whether there is no solution, just one solution, or many solutions. When a system has fewer equations than variables, we generally expect many solutions. Here is such an example.

**Example 6**  Solve the system

$$x_1 + 2x_2 - x_3 = -3$$
$$4x_1 + 3x_2 + x_3 = 13$$

**Solution**

Set up the augmented matrix and solve:

$$\begin{bmatrix} 1 & 2 & -1 & -3 \\ 4 & 3 & 1 & 13 \end{bmatrix} \qquad -4R1 + R2 \rightarrow R2$$

$$\begin{bmatrix} 1 & 2 & -1 & -3 \\ 0 & -5 & 5 & 25 \end{bmatrix} \qquad -\tfrac{1}{5}R2 \rightarrow R2$$

$$\begin{bmatrix} 1 & 2 & -1 & -3 \\ 0 & 1 & -1 & -5 \end{bmatrix} \qquad -2R2 + R1 \rightarrow R1$$

$$\begin{bmatrix} 1 & 0 & 1 & 7 \\ 0 & 1 & -1 & -5 \end{bmatrix}$$

This matrix in reduced echelon form represents the equations

$$x_1 = 7 - x_3$$
$$x_2 = -5 + x_3$$

Since $x_3$ can be chosen arbitrarily, this system has many solutions. Letting $x_3 = k$, the parametric form of this solution is

$$x_1 = 7 - k$$
$$x_2 = -5 + k$$
$$x_3 = k$$

which may be written $(7 - k, -5 + k, k)$.

You should not conclude from the preceding example that a system with fewer equations than variables will always yield **many solutions**. In some cases, the system contains an inconsistency and therefore has no solution. Here is such an example.

**Example 7**  Attempt to solve the following system:

$$x_1 + x_2 - x_3 + 2x_4 = 4$$
$$-2x_1 + x_2 + 3x_3 + x_4 = 5$$
$$-x_1 + 2x_2 + 2x_3 + 3x_4 = 6$$

**Solution**

$$\begin{bmatrix} 1 & 1 & -1 & 2 & 4 \\ -2 & 1 & 3 & 1 & 5 \\ -1 & 2 & 2 & 3 & 6 \end{bmatrix} \qquad \begin{array}{l} 2R1 + R2 \rightarrow R2 \\ R1 + R3 \rightarrow R3 \end{array}$$

$$\begin{bmatrix} 1 & 1 & -1 & 2 & 4 \\ 0 & 3 & 1 & 5 & 13 \\ 0 & 3 & 1 & 5 & 10 \end{bmatrix} \qquad -R2 + R3 \rightarrow R3$$

$$\begin{bmatrix} 1 & 1 & -1 & 2 & 4 \\ 0 & 3 & 1 & 5 & 13 \\ 0 & 0 & 0 & 0 & -3 \end{bmatrix}$$

The last row of this matrix represents $0 = -3$, an inconsistency, so the system has **no solution**.

A system with more equations than variables may have a **unique solution**, no solution, or many solutions. The following examples illustrate these cases.

**Example 8**   If possible, solve the system

$$\begin{array}{rcl} x + 3y &=& 11 \\ 3x - 4y &=& -6 \\ 2x - 7y &=& -17 \end{array}$$

**Solution**

$$\begin{bmatrix} 1 & 3 & 11 \\ 3 & -4 & -6 \\ 2 & -7 & -17 \end{bmatrix} \qquad \begin{array}{l} -3R1 + R2 \rightarrow R2 \\ -2R1 + R3 \rightarrow R3 \end{array}$$

$$\begin{bmatrix} 1 & 3 & 11 \\ 0 & -13 & -39 \\ 0 & -13 & -39 \end{bmatrix} \qquad -R2 + R3 \rightarrow R3$$

$$\begin{bmatrix} 1 & 3 & 11 \\ 0 & -13 & -39 \\ 0 & 0 & 0 \end{bmatrix} \qquad -\tfrac{1}{13}R2 \rightarrow R2$$

$$\begin{bmatrix} 1 & 3 & 11 \\ 0 & 1 & 3 \\ 0 & 0 & 0 \end{bmatrix} \qquad -3R2 + R1 \rightarrow R1$$

$$\begin{bmatrix} 1 & 0 & 2 \\ 0 & 1 & 3 \\ 0 & 0 & 0 \end{bmatrix}$$

This reduced echelon matrix gives the solution $x = 2$, $y = 3$, a unique solution.

**Example 9**    If possible, solve the system

$$x_1 - x_2 + 2x_3 = 2$$
$$2x_1 + 3x_2 - x_3 = 14$$
$$3x_1 + 2x_2 + x_3 = 16$$
$$x_1 + 4x_2 - 3x_3 = 12$$

**Solution**

$$\begin{bmatrix} 1 & -1 & 2 & | & 2 \\ 2 & 3 & -1 & | & 14 \\ 3 & 2 & 1 & | & 16 \\ 1 & 4 & -3 & | & 12 \end{bmatrix} \qquad \begin{array}{l} -2R1 + R2 \rightarrow R2 \\ -3R1 + R3 \rightarrow R3 \\ -R1 + R4 \rightarrow R4 \end{array}$$

$$\begin{bmatrix} 1 & -1 & 2 & | & 2 \\ 0 & 5 & -5 & | & 10 \\ 0 & 5 & -5 & | & 10 \\ 0 & 5 & -5 & | & 10 \end{bmatrix} \qquad \begin{array}{l} -R2 + R3 \rightarrow R3 \\ -R2 + R4 \rightarrow R4 \end{array}$$

$$\begin{bmatrix} 1 & -1 & 2 & | & 2 \\ 0 & 5 & -5 & | & 10 \\ 0 & 0 & 0 & | & 0 \\ 0 & 0 & 0 & | & 0 \end{bmatrix} \qquad \frac{1}{5}R2 \rightarrow R2$$

$$\begin{bmatrix} 1 & -1 & 2 & | & 2 \\ 0 & 1 & -1 & | & 2 \\ 0 & 0 & 0 & | & 0 \\ 0 & 0 & 0 & | & 0 \end{bmatrix} \qquad R2 + R1 \rightarrow R1$$

$$\begin{bmatrix} 1 & 0 & 1 & | & 4 \\ 0 & 1 & -1 & | & 2 \\ 0 & 0 & 0 & | & 0 \\ 0 & 0 & 0 & | & 0 \end{bmatrix}$$

This matrix represents the system

$$x_1 + x_3 = 4$$
$$x_2 - x_3 = 2$$

which gives an infinite number of solutions of the form

$$x_1 = 4 - x_3$$
$$x_2 = 2 + x_3$$

or $(4 - k, 2 + k, k)$ in parametric form.

**Example 10**    We now attempt to solve a system having no solutions, to observe the effect on the reduced matrix.

We use the system

$$\begin{aligned} x_1 + 2x_2 - 2x_3 &= 5 \\ 3x_1 + x_2 + 4x_3 &= 10 \\ x_1 - 2x_2 - 2x_3 &= -7 \\ 2x_1 \qquad\quad - 4x_3 &= 9 \end{aligned}$$

**Solution**

The augmented matrix of the system is

$$\left[\begin{array}{ccc|c} 1 & 2 & -2 & 5 \\ 3 & 1 & 4 & 10 \\ 1 & -2 & -2 & -7 \\ 2 & 0 & -4 & 9 \end{array}\right]$$

We now use row operations to reduce the matrix.

$$\left[\begin{array}{ccc|c} 1 & 2 & -2 & 5 \\ 3 & 1 & 4 & 10 \\ 1 & -2 & -2 & -7 \\ 2 & 0 & -4 & 9 \end{array}\right] \qquad \begin{array}{l} -3R1 + R2 \to R2 \\ -R1 + R3 \to R3 \\ -2R1 + R4 \to R4 \end{array}$$

$$\left[\begin{array}{ccc|c} 1 & 2 & -2 & 5 \\ 0 & -5 & 10 & -5 \\ 0 & -4 & 0 & -12 \\ 0 & -4 & 0 & -1 \end{array}\right] \qquad \begin{array}{l} -\frac{1}{5}R2 \to R2 \\ -\frac{1}{4}R3 \to R3 \end{array}$$

$$\left[\begin{array}{ccc|c} 1 & 2 & -2 & 5 \\ 0 & 1 & -2 & 1 \\ 0 & 1 & 0 & 3 \\ 0 & -4 & 0 & -1 \end{array}\right] \qquad \begin{array}{l} -2R2 + R1 \to R1 \\ \\ -R2 + R3 \to R3 \\ 4R2 + R4 \to R4 \end{array}$$

$$\left[\begin{array}{ccc|c} 1 & 0 & 2 & 3 \\ 0 & 1 & -2 & 1 \\ 0 & 0 & 2 & 2 \\ 0 & 0 & -8 & 3 \end{array}\right] \qquad \begin{array}{l} -R3 + R1 \to R1 \\ R3 + R2 \to R2 \\ \frac{1}{2}R3 \to R3 \\ 4R3 + R4 \to R4 \end{array}$$

$$\left[\begin{array}{ccc|c} 1 & 0 & 0 & 1 \\ 0 & 1 & 0 & 3 \\ 0 & 0 & 1 & 1 \\ 0 & 0 & 0 & 11 \end{array}\right]$$

The last matrix is not quite in the reduced echelon form because the last column has not been reduced to a 1 in the last row and zeros elsewhere. However, we need not proceed any further because the matrix in the present form represents the system

$$\begin{aligned} x_1 \qquad\qquad &= 1 \\ x_2 \qquad &= 3 \\ x_3 &= 1 \\ 0 &= 11 \end{aligned}$$

Because the system includes a false equation, $0 = 11$, the system cannot be satisfied by any values of $x_1$, $x_2$, and $x_3$. Thus, we must conclude that the original system has no solution.

———————●

Each nonzero row in the reduced echelon matrix gives the value of one variable—either as a number or expressed in terms of another variable or variables. When the system reduces to fewer equations than variables, not enough rows in the matrix exist to give a row for each variable. This means that you can solve only for some of the variables, and they will be expressed in terms of the remaining variables (parameters). Examples 3, 4, and 6 illustrate the relationship between the number of variables solved in terms of parameters. Note the following:

> Example 3 reduces to two equations and three variables. Two of the variables, $x_1$ and $x_2$, were solved in terms of one variable, $x_3$. We call $x_3$ a parameter.
>
> In Example 4, which reduces to three equations and five variables, three of the variables, $x_1$, $x_2$, and $x_3$, were solved in terms of two variables, $x_4$ and $x_5$. We have two parameters in this case, $x_4$ and $x_5$.
>
> In Example 6, which has two equations and three variables, two variables, $x_1$ and $x_2$, were solved in terms of one variable, $x_3$. This solution has one parameter, $x_3$.

The relationship between the number of equations and variables is as follows: If there are $k$ equations with $n$ variables in the reduced echelon matrix, and $n > k$, then $k$ of the variables can be solved in terms of $n - k$ parameters. The system has many solutions.

Whenever a row in a reduced matrix becomes all zeros, then an equation is eliminated from the system, and the number of equations is reduced by one. In the reduced echelon matrix, count the nonzero rows to count the number of equations in the solution.

## Application

**Example 11**  The Mt. Pleasant School District proposes to renovate and enlarge the school library and the football stadium. The Student Council polls the sophomore, junior, and senior classes to judge student support of the proposals. The student secretary reported the results: 790 voted, 504 supported the library renovation, and 538 supported the stadium renovation. The sophomore class representative asked, "How did the classes vote?"

"The sophomore class voted 60% for the library and 70% for the stadium."
"The junior class voted 80% for the library and 60% for the stadium."
"The senior class voted 40% for the library and 80% for the stadium."
"But how many voted in each class?"
"Oops, I don't know. I didn't write it down."
The vice president suggested that the information be given to the Math Club for them to determine the class totals.
"No big deal, we have been studying systems of equations, we will do it."
Now, assume you are in the Math Club and determine class totals.

### Solution
Let $x$ = sophomore class total, $y$ = junior class total, and $z$ = senior class total. This information can be stated as a system of equations.

| | |
|---|---|
| Total vote | $x + y + z = 790$ |
| Votes for the library | $0.60x + 0.80y + 0.40z = 504$ |
| Votes for the stadium | $0.70x + 0.60y + 0.80z = 538$ |

The augmented matrix is

$$\begin{bmatrix} 1 & 1 & 1 & 790 \\ 0.6 & 0.8 & 0.4 & 504 \\ 0.7 & 0.6 & 0.8 & 538 \end{bmatrix}$$

The matrix reduces to the following (show that it does):

$$\begin{bmatrix} 1 & 0 & 2 & 640 \\ 0 & 1 & -1 & 150 \\ 0 & 0 & 0 & 0 \end{bmatrix}$$

This represents the system

$$\begin{aligned} x + 2z &= 640 \\ y - z &= 150 \end{aligned}$$

which has multiple solutions of the form

$$\begin{aligned} x &= 640 - 2z \\ y &= 150 + z \end{aligned}$$

Since the variables represent the number of students, $x$, $y$, and $z$ must be integers and cannot be negative. Thus, $2z \le 640$ or else $x$ would be negative. Thus, $z$ can be any integer 320 or less. The best that the Math Club can report on class size follows:

| | |
|---|---|
| Seniors: | From 0 through 320 |
| Juniors: | 150 plus the number of seniors |
| Sophomores: | 640 minus twice the number of seniors |

**Summary**    The nonzero rows of the reduced echelon matrix give the needed information about the solutions to a system of equations. Three situations are possible.

1. **No solution.**    At least one row has all zeros in the coefficient portion of the matrix (the portion to the left of the vertical line) and a nonzero entry to the right of the vertical line.

$$\begin{bmatrix} 1 & 0 & 0 & 3 \\ 0 & 1 & 0 & 2 \\ 0 & 0 & 0 & 5 \end{bmatrix} \quad \text{(No solution)}$$

Two more possibilities arise when a solution exists.

2. **The solution is unique.**    The number of nonzero rows equals the number of variables in the system.

$$\begin{bmatrix} 1 & 0 & 5 \\ 0 & 1 & -2 \end{bmatrix} \qquad \begin{bmatrix} 1 & 0 & 0 & -4 \\ 0 & 1 & 0 & 3 \\ 0 & 0 & 1 & 2 \\ 0 & 0 & 0 & 0 \end{bmatrix} \quad \text{(Unique solution)}$$

**NOTE**

In the Gauss-Jordan Method, we start by obtaining a 1 in the (1, 1) position of the augmented matrix. Actually, this is not essential. We do not need to conform exactly to this sequence, and we could begin with another column or row.

**3. Infinite number of solutions.** The number of nonzero rows is less than the number of variables in the system.

$$\left[\begin{array}{ccc|c} 1 & 0 & 0 & 2 & 3 \\ 0 & 1 & 0 & 5 & 2 \\ 0 & 0 & 1 & 3 & 4 \end{array}\right] \quad \left[\begin{array}{ccc|c} 1 & 0 & 1 & 2 \\ 0 & 1 & 2 & 4 \\ 0 & 0 & 0 & 0 \end{array}\right] \quad \text{(Infinite number of solutions)}$$

Note that we solve for the variable in each row where a leading 1 occurs, and we write that variable in terms of the other variables in that row.

## **2.3**  EXERCISES

Access end-of-section exercises online at **www.webassign.net**   Web**Assign**

## Using Your TI Graphing Calculator

### Obtaining the Reduced Echelon Form of a Matrix

The process of solving a system of equations by using row operations on the augmented matrix seeks to reduce the columns to a single entry of 1's and 0's in the rest of the column. The same process is used to obtain the reduced echelon form of a matrix.

This process of modifying a matrix so a column contains a single entry with 1's and 0's in the rest of the column is called *pivoting*. A system of three equations can require up to nine row operations to solve the system. These row operations can be accomplished by the TI Graphing Calculator program called PIVOT, which can be used to pivot on specified entries of the matrix stored in [A]. The pivot row and column determine where the pivot, the 1 entry, is located.

For each pivot, the user indicates the pivot row and pivot column.

```
PIVOT
: [A] → [B]                          : Goto 4
: 2 → dim(L1)                        : *Row + (–[B](K,J),[B],I,K) → [B]
: dim([A]) → L1                      : Lbl 4
: Lbl 1                              : 1 + K → K
: Disp "PIVOT ROW"                   : If K ≤ L1(1)
: Input I                            : Goto 5
: Disp "PIVOT COL"                   : round([B],2) → [C]
: Input J                            : Pause [C]
: *row(1/[B](I,J),[B],I) → [B]       : Goto 1
: 1 → K                              : End
: Lbl 5
: If K = I
```

*(continued)*

We now show some screens that occur when using the PIVOT program. The matrix used is

$$[A] = \begin{bmatrix} 2 & 4 & 6 & -4 \\ 0 & 1 & -3 & 2 \\ 3 & 5 & 2 & 2 \end{bmatrix}$$

Begin the program with $\boxed{\text{PRGM}}$ <PIVOT> $\boxed{\text{ENTER}}$ $\boxed{\text{ENTER}}$ and enter the row and column numbers where the pivot occurs (row 1, column 1, here). The screen shows the result of the pivot:

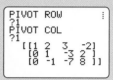

Press $\boxed{\text{ENTER}}$ to enter next pivot (row 2, column 2).

```
PIVOT ROW        :
?2
PIVOT COL
?2
  [[1 0 9   -6]
   [0 1 -3  2 ]
   [0 0 -10 10]]
```

Finally, pivot using row 3, column 3, to obtain the reduced echelon matrix:

```
PIVOT ROW        :
?3
PIVOT COL
?3
  [[1 0 0 3 ]
   [0 1 0 -1]
   [0 0 1 -1]]
```

If the original matrix represents the system

$$\begin{aligned} 2x_1 + 4x_2 + 6x_3 &= -4 \\ x_2 - 3x_3 &= 2 \\ 3x_1 + 5x_2 + 2x_3 &= 2 \end{aligned}$$

then the reduced matrix above gives the solution $x_1 = 3$, $x_2 = -1$, and $x_3 = -1$.

## Exercises

Use the PIVOT program to solve the following systems:

**1.** $\begin{aligned} x_1 + 3x_2 + 2x_3 &= 1 \\ 6x_1 - x_2 + 4x_3 &= 31 \\ 2x_1 + x_2 + 2x_3 &= 9 \end{aligned}$

**2.** $\begin{aligned} x_1 + 4x_2 + 2x_3 &= 15 \\ 3x_1 + x_2 - x_3 &= 4.7 \\ 2x_1 - 2x_2 + 3x_3 &= 11.6 \end{aligned}$

**3.** $\begin{aligned} x_1 - 2x_2 + x_3 &= 1 \\ 2x_1 + x_2 - 2x_3 &= -2 \\ -x_1 + x_2 + 2x_3 &= 13 \end{aligned}$

**4.** $\begin{aligned} 1.2x_1 + 3.1x_2 - 4.5x_3 &= -6.71 \\ 2.3x_1 - 1.8x_2 + 2.5x_3 &= 6.82 \\ 4.1x_1 + 2.6x_2 - 3.4x_3 &= -0.99 \end{aligned}$

**5.** $\begin{aligned} 5x_1 + 11x_2 + 7x_3 + 2x_4 &= 207.2 \\ 6x_1 - x_2 + 4x_3 + 5x_4 &= 104.6 \\ 9x_1 + 12x_2 - 3x_3 - 4x_4 &= 168.6 \end{aligned}$

**6.** $\begin{aligned} 3x_1 + 7x_2 - 9x_3 &= -112.90 \\ 4.1x_1 + 22.3x_2 + 17.5x_3 &= -145.31 \\ 1.1x_1 + 15.3x_2 + 26.5x_3 &= 42.6 \end{aligned}$

## The rref Command

We now show you another way to obtain the reduced echelon matrix for solving a system. Use a command that finds the reduced echelon form of a matrix, the **rref** command.

Let's illustrate **rref** using the matrix [A] shown on the screen:

```
[A]
 [[1 2 1 0 9 ]
  [2 3 0 1 7 ]
  [3 5 2 1 20]]
```

Access the **rref** command with $\boxed{\text{MATRX}}$ <MATH> <B:rref (> and **rref** will show on the screen). Enter the matrix name, [A], and press $\boxed{\text{ENTER}}$. The reduced echelon matrix is

```
rref([A])
 [[1 0 0 2  -1]
  [0 1 0 -1 3 ]
  [0 0 1 0  4 ]]
```

## Exercises

1. Use **rref** to find the reduced echelon form of the matrix

$$\begin{bmatrix} 1 & 3 & 2 & 1 & 5 \\ 2 & 2 & 0 & 4 & 6 \\ 4 & -4 & 2 & 0 & 9 \end{bmatrix}$$

3. $\quad x_1 + 3x_2 + 2x_3 = \quad 7$
$-4x_1 + 7x_2 + 5x_3 = \quad 28$
$\quad 2x_1 + \quad x_2 + 2x_3 = -5$

5. $5.1x_1 + 4.2x_2 - 7.3x_3 = \quad 58.86$
$2.2x_1 - 5.4x_2 + 9.5x_3 = -21.95$
$3.0x_1 + 6.9x_2 + 4.4x_3 = \quad 108.15$

Use **rref** to solve the following systems:

2. $5x_1 + 4x_2 - \quad x_3 = \quad 2$
$2x_1 + \quad x_2 + 3x_3 = 19$
$4x_1 + 7x_2 + 2x_3 = \quad 8$

4. $1.2x_1 + 3.6x_2 + 2.4x_3 = 24.0$
$2.3x_1 + 1.1x_2 + 4.3x_3 = 32.9$
$3.5x_1 + 2.2x_2 + 5.1x_3 = 43.9$

6. $5x_1 + \quad 3x_2 - \quad 2x_3 = 13.5$
$4x_1 - \quad 7x_2 + \quad 8x_3 = 33.0$
$\quad x_1 + 10x_2 - 10x_3 = 17.2$

## Using Excel

A system of equations may be solved using Solver (See Section 2.2) with one of the equations used in the Target Cell and the other equations used as constraints. In cases where the system has multiple solutions or no solution, Solver will indicate that no feasible solution was found.

## Exercises

Solve the following systems using Solver.

1. $4x_1 - 2x_2 + \quad x_3 = -18$
$\quad x_1 + \quad x_2 + 3x_3 = \quad 15$
$3x_1 + 7x_2 + 2x_3 = \quad 11$

3. $2.1x_1 + 1.5x_2 - 3.4x_3 = -8.2$
$1.7x_1 + 4.1x_2 + 0.4x_3 = \quad 21.2$
$2.6x_1 + 5.3x_2 + 0.8x_3 = \quad 10.6$

2. $\quad x_1 + 2x_2 - \quad x_3 = \quad 13$
$4x_1 - \quad x_2 + 3x_3 = \quad 10$
$2x_1 - 5x_2 + 6x_3 = -13$

4. $\quad x_1 + 2x_2 - \quad x_3 + 3x_4 = \quad 7$
$2x_1 + 3x_2 + \quad x_3 - 5x_4 = \quad 8$
$5x_1 + 2x_2 - 4x_3 + \quad x_4 = 15$
$\quad x_1 + \quad x_2 + \quad x_3 + \quad x_4 = \quad 8$

## 2.4 MATRIX OPERATIONS

- Additional Uses of Matrices
- Equal Matrices
- Addition of Matrices
- Scalar Multiplication

### Additional Uses of Matrices

You have used augmented matrices to represent a system of linear equations (Section 2.3) and then used row operations to solve the system of equations. This method reduces the procedure to a more straightforward sequence of steps. The procedure can also be performed on a computer that handles the computational drudgery and reduces errors.

A casual observer might view an augmented matrix as simply a table of numbers. In one sense that is correct. The interpretation of the table and an understanding of the row operations gives significant meaning to "just a table of numbers." Matrices are mathematically interesting because they can be used to represent more than a system of equations. Many applications in business, social, and biological sciences represent information in tables or rectangular arrays of numbers. The analyses of such data can often be done using matrices and matrix operations. We add, multiply, and perform other operations with matrices. We will study these operations and suggest ways in which they are useful.

We begin with a simple application using a matrix to summarize information in tabular forms. For example, a mathematics professor records grade information for two sections of a mathematics course in the following matrix:

$$
\begin{array}{c}
 \\
 \\
\begin{array}{r}
\text{Homework} \\
\text{Quizzes} \\
\text{Exams}
\end{array}
\end{array}
\begin{array}{c}
\text{Section} \\
\begin{array}{cc} 1 & 2 \end{array} \\
\begin{bmatrix} 76 & 79 \\ 71 & 70 \\ 73 & 74 \end{bmatrix}
\end{array}
$$

Each row represents a different type of grade, and each column represents a section.

Because an endless variety of ways exist in which someone might want to break information into categories and summarize it, matrices come in various shapes and sizes. Independent of the source of the information summarized in a matrix, we can classify matrices by the number of rows and columns they have. For example,

$$
\begin{bmatrix} 3 & -1 & 4 \\ 2 & 1 & 5 \end{bmatrix}
$$

is a $2 \times 3$ matrix because it has two rows and three columns.

$$
\begin{bmatrix} 5 & 0 & 1 \\ 2 & 1 & 4 \\ 3 & 2 & 2 \\ -1 & 6 & -2 \end{bmatrix}
$$

is a $4 \times 3$ matrix, having four rows and three columns. The convention used in describing the size of a matrix states the number of rows first, followed by the number of columns.

Two matrices are the **same size** if they have the same dimensions; that is, the number of rows is the same for both matrices, and the number of columns is also the same. For example,

$$\begin{bmatrix} -2 & 8 & -3 \\ 1 & 0 & 1 \end{bmatrix} \text{ and } \begin{bmatrix} 2 & 5 & 9 \\ 3 & 6 & 7 \end{bmatrix}$$

are matrices of the same size; they are both $2 \times 3$ matrices.

Here is another example of using a matrix to summarize information.

**Example 1**  The Campus Bookstore carries spirit shirts in white, green, and gold. In September it sold 238 white, 317 green, and 176 gold shirts. In October it sold 149 white, 342 green, and 369 gold shirts. In November it sold 184 white, 164 green, and 201 gold shirts. Summarize this information in a matrix.

**Solution**
Let each of three columns represent a month and each of three rows represent a color of shirts. Label the columns and rows.

$$\begin{array}{c} \\ \text{White} \\ \text{Green} \\ \text{Gold} \end{array} \begin{array}{ccc} \text{Sept.} & \text{Oct.} & \text{Nov.} \\ \begin{bmatrix} 238 & 149 & 184 \\ 317 & 342 & 164 \\ 176 & 369 & 201 \end{bmatrix} \end{array}$$

A matrix in which the number of rows equals the number of columns is called a **square matrix**.

## Equal Matrices

Two matrices of the same size are **equal matrices** if and only if their corresponding components are equal. If the matrices are not the same size, they are not equal.

**Example 2**
$$\begin{bmatrix} 3 & 7 \\ 5-1 & 4 \times 4 \end{bmatrix} = \begin{bmatrix} \frac{6}{2} & 7 \\ 4 & 16 \end{bmatrix}$$

because corresponding components are equal.

$$\begin{bmatrix} 1 & 2 & 5 \\ 3 & 6 & 4 \end{bmatrix} \neq \begin{bmatrix} 1 & 2 & 5 \\ 3 & -1 & 4 \end{bmatrix}$$

because the entries in row 2, column 2 are different; that is, the $(2, 2)$ entries are not equal.

**Example 3**  Find the value of $x$ such that

$$\begin{bmatrix} 3 & 4x \\ 2.1 & 7 \end{bmatrix} = \begin{bmatrix} 3 & 9 \\ 2.1 & 7 \end{bmatrix}$$

**Solution**

For the matrices to be equal, the corresponding components must be equal, so $4x = 9$ and $x = \frac{9}{4}$.

## Addition of Matrices

A businesswoman has two stores. She wants a record of the daily sales of the regular size and the giant economy size of laundry soap. She can use matrices to record this information. These two matrices show sales for two days.

|         | Store | | | Store | |
|---------|:-----:|:-----:|---|:-----:|:-----:|
|         | **1** | **2** | | **1** | **2** |
| Regular | 8 | 12 | | 6 | 5 |
| Giant   | 9 | 7  | | 11 | 4 |
|         | Day 1 | | | Day 2 | |

The position in the matrix identifies the store and package size. For example, store 2 sold 7 giant sizes and 12 regular on day 1, and so on.

The total sales, by store and package size, can be obtained by adding the sales in each individual category to get the total in that category—that is, by adding corresponding entries of the two matrices. We indicate this procedure with the notation

$$\begin{bmatrix} 8 & 12 \\ 9 & 7 \end{bmatrix} + \begin{bmatrix} 6 & 5 \\ 11 & 4 \end{bmatrix} = \begin{bmatrix} 14 & 17 \\ 20 & 11 \end{bmatrix}$$

This procedure applies generally to the **addition of matrices**.

> **CAUTION**
>
> To add two matrices, they must be the same size.

> **DEFINITION**
> **Matrix Addition**
>
> Obtain the **sum** of two matrices of the same size by adding corresponding elements. If two matrices are not of the same size, they cannot be added; we say that their sum does not exist. **Subtraction** is performed on matrices of the same size by subtracting corresponding elements.

Now we apply the definition of matrix addition in the next two examples.

**Example 4**     For the following matrices:

$$A = \begin{bmatrix} 2 & 1 & -1 \\ 0 & 5 & 2 \end{bmatrix} \quad B = \begin{bmatrix} 1 & 3 & 1 \\ 2 & 1 & 4 \end{bmatrix} \quad C = \begin{bmatrix} 4 & 1 \\ -1 & 2 \end{bmatrix}$$

determine the sums $A + B$ and $B + C$ if possible.

**Solution**

$$A + B = \begin{bmatrix} 2 & 1 & -1 \\ 0 & 5 & 2 \end{bmatrix} + \begin{bmatrix} 1 & 3 & 1 \\ 2 & 1 & 4 \end{bmatrix}$$

$$= \begin{bmatrix} 2 + 1 & 1 + 3 & -1 + 1 \\ 0 + 2 & 5 + 1 & 2 + 4 \end{bmatrix} = \begin{bmatrix} 3 & 4 & 0 \\ 2 & 6 & 6 \end{bmatrix}$$

Neither the sum $A + C$ nor $B + C$ exists, because matrices $A$ and $C$ and matrices $B$ and $C$ are not of the same size. (Try adding these matrices using the rule.)

Let's extend our definition to enable us to add more than just two matrices. For example, define the sum of three matrices as

$$\begin{bmatrix} 1 & 2 \\ 0 & -1 \end{bmatrix} + \begin{bmatrix} 3 & 4 \\ 2 & 1 \end{bmatrix} + \begin{bmatrix} 5 & 2 \\ -1 & 0 \end{bmatrix} = \begin{bmatrix} 1 + 3 + 5 & 2 + 4 + 2 \\ 0 + 2 - 1 & -1 + 1 + 0 \end{bmatrix}$$

$$= \begin{bmatrix} 9 & 8 \\ 1 & 0 \end{bmatrix}$$

We add a string of matrices that are the same size by adding corresponding elements. The following example illustrates a use of this rule.

**Example 5** The Green Earth Recycling Center has three locations. The Center recycles aluminum, plastic, and newspapers. Each location keeps a daily record in matrix form. Here is an illustration of one week's records. The entries represent the number of pounds collected.

Location 1

|  | Alum. | Plastic | Paper |
|------|------|------|------|
| Mon. | 920 | 140 | 1840 |
| Tue. | 640 | 96 | 1260 |
| Wed. | 535 | 80 | 955 |
| Thu. | 768 | 32 | 1030 |
| Fri. | 420 | 55 | 1320 |
| Sat. | 1590 | 205 | 2340 |

Location 2

|  | Alum. | Plastic | Paper |
|------|------|------|------|
| Mon. | 435 | 60 | 2840 |
| Tue. | 620 | 45 | 2665 |
| Wed. | 240 | 22 | 3450 |
| Thu. | 195 | 38 | 1892 |
| Fri. | 530 | 52 | 1965 |
| Sat. | 895 | 74 | 3460 |

Location 3

|  | Alum. | Plastic | Paper |
|------|------|------|------|
| Mon. | 634 | 110 | 1565 |
| Tue. | 423 | 86 | 948 |
| Wed. | 555 | 142 | 1142 |
| Thu. | 740 | 93 | 1328 |
| Fri. | 883 | 135 | 1476 |
| Sat. | 976 | 234 | 1928 |

A summary of the total collected at the three locations can be obtained by matrix addition.

$$
\text{Location 1 + Location 2 + Location 3} =
\begin{array}{c}
 \\ \\
\text{Mon.} \\ \text{Tue.} \\ \text{Wed.} \\ \text{Thu.} \\ \text{Fri.} \\ \text{Sat.}
\end{array}
\overset{\displaystyle\text{All Locations}}{
\overset{\text{Alum.} \quad \text{Plastic} \quad \text{Paper}}{
\begin{bmatrix}
1989 & 310 & 6245 \\
1683 & 227 & 4873 \\
1330 & 244 & 5547 \\
1703 & 163 & 4250 \\
1833 & 242 & 4761 \\
3461 & 513 & 7728
\end{bmatrix}}}
$$

Thus, the total aluminum collected on Monday was 1989 pounds, the amount of newspaper collected on Saturday was 7728 pounds, and so on.

Although this analysis and others like it can be carried out without the use of matrices, the handling of large quantities of data is often most efficiently done on computers using matrix techniques.

## Scalar Multiplication

Another matrix operation multiplies a matrix by a number like

$$
4\begin{bmatrix} 3 & 2 \\ 1 & 7 \end{bmatrix}
$$

This product is defined to be

$$
4\begin{bmatrix} 3 & 2 \\ 1 & 7 \end{bmatrix} = \begin{bmatrix} 12 & 8 \\ 4 & 28 \end{bmatrix}
$$

Note that this operation multiplies each entry in the matrix by 4. This illustrates the procedure for **scalar multiplication**, so called because mathematicians often use the term *scalar* to refer to a *number.*

| **DEFINITION** | **Scalar multiplication** is the operation of multiplying a matrix by a number (scalar). |
|---|---|
| Scalar Multiplication | Each entry in the matrix is multiplied by the scalar. |

**Example 6**

$$
-3\begin{bmatrix} 5 & 2 & 1 \\ 0 & 1 & 4 \\ -1 & 3 & 6 \end{bmatrix} = \begin{bmatrix} -15 & -6 & -3 \\ 0 & -3 & -12 \\ 3 & -9 & -18 \end{bmatrix}
$$

**Example 7**  A class of ten students had five tests during the semester. A perfect score on each of the tests is 50. The scores are listed in this table.

|          | Test 1 | Test 2 | Test 3 | Test 4 | Test 5 |
|----------|--------|--------|--------|--------|--------|
| Anderson | 40     | 45     | 30     | 48     | 42     |
| Boggs    | 20     | 15     | 30     | 25     | 10     |
| Chittar  | 40     | 35     | 25     | 45     | 46     |
| Diessner | 25     | 40     | 45     | 40     | 38     |
| Farnam   | 35     | 35     | 38     | 37     | 39     |
| Gill     | 50     | 46     | 45     | 48     | 47     |
| Homes    | 22     | 24     | 30     | 32     | 29     |
| Johnson  | 35     | 27     | 20     | 41     | 30     |
| Schomer  | 28     | 31     | 25     | 27     | 31     |
| Wong     | 40     | 35     | 36     | 32     | 38     |

We can express these scores as column matrices:

$$\begin{bmatrix} 40 \\ 20 \\ 40 \\ 25 \\ 35 \\ 50 \\ 22 \\ 35 \\ 28 \\ 40 \end{bmatrix} \begin{bmatrix} 45 \\ 15 \\ 35 \\ 40 \\ 35 \\ 46 \\ 24 \\ 27 \\ 31 \\ 35 \end{bmatrix} \begin{bmatrix} 30 \\ 30 \\ 25 \\ 45 \\ 38 \\ 45 \\ 30 \\ 20 \\ 25 \\ 36 \end{bmatrix} \begin{bmatrix} 48 \\ 25 \\ 45 \\ 40 \\ 37 \\ 48 \\ 32 \\ 41 \\ 27 \\ 32 \end{bmatrix} \begin{bmatrix} 42 \\ 10 \\ 46 \\ 38 \\ 39 \\ 47 \\ 29 \\ 30 \\ 31 \\ 38 \end{bmatrix}$$

To obtain each person's average, we use matrix addition to add the matrices and then scalar multiplication to multiply by $\frac{1}{5}$ (dividing by the number of tests). We get

$$\frac{1}{5} \begin{bmatrix} 205 \\ 100 \\ 191 \\ 188 \\ 184 \\ 236 \\ 137 \\ 153 \\ 142 \\ 181 \end{bmatrix} = \begin{bmatrix} 41.0 \\ 20.0 \\ 38.2 \\ 37.6 \\ 36.8 \\ 47.2 \\ 27.4 \\ 30.6 \\ 28.4 \\ 36.2 \end{bmatrix} \qquad \text{(Column matrix given each person's average score)}$$

Row matrices are also useful; a person's complete set of scores corresponds to a row matrix. For example, the row matrix

$$[25 \quad 40 \quad 45 \quad 40 \quad 38]$$

gives Diessner's scores.

This approach to analyzing test scores has the advantage of lending itself to implementation on the computer. A computer program can be written that will perform the desired matrix additions and scalar multiplications.

## 2.4    EXERCISES

Access end-of-section exercises online at **www.webassign.net**

# Using Your TI Graphing Calculator

## Matrix Operations

The notation used to add matrices and to do scalar multiplication resembles that used for the addition and multiplication of numbers. The following matrices are stored in [A] and [B].

$$[A] = \begin{bmatrix} 1 & 2 & 3 \\ 4 & 5 & 6 \\ 7 & 8 & 9 \end{bmatrix} \quad \text{and} \quad [B] = \begin{bmatrix} 10 & 11 & 12 \\ 13 & 14 & 15 \\ 16 & 17 & 18 \end{bmatrix}$$

Their sum is obtained by
[A] $\boxed{+}$ [B] $\boxed{\text{ENTER}}$,

and 3[A] is obtained by
$\boxed{3}$ $\boxed{\times}$ [A] $\boxed{\text{ENTER}}$

```
[A]+[B]
   [[11 13 15]
    [17 19 21]
    [23 25 27]]
```

```
3*[A]
   [[3  6  9 ]
    [12 15 18]
    [21 24 27]]
```

## Exercises

**1.** $[A] = \begin{bmatrix} 1 & 3 & 2 \\ 4 & 5 & 7 \end{bmatrix}$ and $[B] = \begin{bmatrix} 3 & -2 & 1 \\ 6 & 8 & -5 \end{bmatrix}$. Find the matrices:

  **(a)** $[A] + [B]$          **(b)** $2[A]$          **(c)** $[A] - [B]$          **(d)** $3[A] + 2[B]$

**2.** $[A] = \begin{bmatrix} 4 & 2 \\ 3 & 9 \end{bmatrix}$ and $[B] = \begin{bmatrix} 2 & 6 \\ 5 & 12 \end{bmatrix}$. Find the matrices:

  **(a)** $[A] + [B]$          **(b)** $-4[A]$          **(c)** $[A] - [B]$          **(d)** $2[A] - 5[B]$

## Using Excel

### Matrix Addition and Scalar Multiplication

We now show how to perform matrix addition and scalar multiplication using Excel.

### Example

$$A = \begin{bmatrix} 1 & 2 & 3 \\ 4 & 5 & 6 \end{bmatrix} \quad B = \begin{bmatrix} -1 & 3 & 1 \\ 2 & -3 & 4 \end{bmatrix}$$

Find $A + B$, $5A$, $3A - 2B$.

### Solution

We enter A in cells A2:C3 and B in cells E2:G3. We will put $A + B$ in cells A6:C7, $5A$ in cells E6:G7, and $3A - 2B$ in A10:C11.

The formulas to perform the desired operations follow much like the same operations on numbers.

For $A + B$: Enter =A2+E2 in A6. Then copy the formulas in the rest of the matrix by dragging A6. Press ENTER to activate the operation.

For $5A$: Enter =5*A2 in E6 and drag to other cells in E6:G7.

For $3A - 2B$: Enter =3*A2-2*E2 in A10 and drag to the other cells in A10:C11.

Here is the spreadsheet with the results of the operation.

| | A | B | C | D | E | F | G |
|---|---|---|---|---|---|---|---|
| 1 | Matrix A in A2:C3 | | | | Matrix B in E2:G3 | | |
| 2 | 1 | 2 | 3 | | -1 | 3 | 1 |
| 3 | 4 | 5 | 6 | | 2 | -3 | 4 |
| 4 | | | | | | | |
| 5 | A+B in A6:C7 | | | | 5A in E6:G7 | | |
| 6 | 0 | 5 | 4 | | 5 | 10 | 15 |
| 7 | 6 | 2 | 10 | | 20 | 25 | 30 |
| 8 | | | | | | | |
| 9 | 3A-2B in A10:C11 | | | | | | |
| 10 | 5 | 0 | 7 | | | | |
| 11 | 8 | 21 | 10 | | | | |

### Exercises

For Exercises 1 through 5 use

$$A = \begin{bmatrix} 1 & 3 & -2 \\ 5 & 9 & 7 \\ -4 & 0 & 6 \end{bmatrix} \quad B = \begin{bmatrix} 8 & 2 & 0 \\ -3 & 5 & 4 \\ 2 & 9 & 1 \end{bmatrix}$$

**1.** Find $A + B$.       **2.** Find $A - B$.       **3.** Find $4A$.

**4.** Find $-0.35B$.       **5.** Find $4A + 6B$.

**6.** Repeat Exercises 1 through 5 using

$$A = \begin{bmatrix} 2 & 3 & 2 \\ 5 & 1 & 5 \\ 6 & 9 & 3 \end{bmatrix} \quad B = \begin{bmatrix} 4 & 1 & 7 \\ 2 & -3 & 6 \\ 3 & 3 & 2 \end{bmatrix}$$

## 2.5 MULTIPLICATION OF MATRICES

- Dot Product
- Identity Matrix
- Matrix Multiplication
- Row Operations Using Matrix Multiplication

You have learned to add matrices and multiply a matrix by a number. You may naturally ask whether one can multiply two matrices together and whether this helps to solve problems. Mathematicians have devised a way of multiplying two matrices. It might seem rather complicated, but it has many useful applications. For example, you will learn how to use matrix multiplication to solve a problem like the following.

A manufacturer makes tables and chairs. The time, in hours, required to assemble and finish the items is given by the matrix

$$
\begin{array}{cc}
& \text{Chair} \quad \text{Table} \\
\begin{array}{c} \text{Assemble} \\ \text{Finish} \end{array} &
\begin{bmatrix} 2 & 3 \\ 2.5 & 4.75 \end{bmatrix}
\end{array}
$$

The total assembly and finishing time required to produce 950 chairs and 635 tables can be obtained by an appropriate matrix multiplication. The procedure of multiplying matrices may appear strange. Following a description of how to multiply two matrices are some examples of the uses of matrix multiplication. But first, here's an overview of the process:

1. For two matrices $A$ and $B$, we will find their product denoted by $AB$. Their product is a matrix called $C$, that is, $AB = C$.

2. $C$ is a matrix. The problem is to find the entries of $C$.

3. Each entry in $C$ will depend on a *row* from matrix $A$ and a *column* from matrix $B$.

We call the entry in row $i$ and column $j$ the $(i, j)$ entry of a matrix. The $(i, j)$ entry in $C$ is a number obtained using all entries of row $i$ in $A$ and using all entries of column $j$ in $B$. For example, the $(2, 3)$ entry in $C$ depends on row 2 of $A$ and column 3 of $B$. We show you how to find an entry in $C$ by using the *dot product* of a row in $A$ and a column in $B$.

### Dot Product

We use the two matrices

$$
A = \begin{bmatrix} 1 & 3 \\ 2 & -1 \end{bmatrix} \quad \text{and} \quad B = \begin{bmatrix} 4 & -5 \\ 1 & 6 \end{bmatrix}
$$

to illustrate matrix multiplication $AB$. We use the first row of $A$, $R1 = [1 \quad 3]$, and the first column of $B$,

$$
C1 = \begin{bmatrix} 4 \\ 1 \end{bmatrix}
$$

to find the $(1, 1)$ entry of the product. To do so, we need to find what we call the **dot product**, $R1 \cdot C1$, of the row and column. It is

$$
R1 \cdot C1 = [1 \quad 3] \cdot \begin{bmatrix} 4 \\ 1 \end{bmatrix} = 1(4) + 3(1) = 7
$$

Notice the following:

1. The dot product of a row and a column gives a single number.
2. Obtain the dot product by multiplying the first numbers from both the row and column, then the second numbers from both, and so on, and then adding the results.

There are three other dot products possible using a row from $A$ and a column from $B$. They are:

$$R1 \cdot C2 = [1 \quad 3] \cdot \begin{bmatrix} -5 \\ 6 \end{bmatrix} = 1(-5) + 3(6) \quad = 13$$

$$R2 \cdot C1 = [2 \; -1] \cdot \begin{bmatrix} 4 \\ 1 \end{bmatrix} \quad = 2(4) \quad + (-1)(1) = 7$$

$$R2 \cdot C2 = [2 \; -1] \cdot \begin{bmatrix} -5 \\ 6 \end{bmatrix} = 2(-5) + (-1)(6) = -16$$

The general form of the dot product of a row and column is

$$[a_1 \quad a_2 \cdots a_n] \cdot \begin{bmatrix} b_1 \\ b_2 \\ \vdots \\ b_n \end{bmatrix} = a_1 b_1 + a_2 b_2 + \cdots + a_n b_n$$

**NOTE**

The dot product is defined only when the row and column matrices have the same number of entries.

The total cost of a purchase at the grocery store can be determined by using the dot product of a price matrix and a quantity matrix as illustrated in the next example.

**Example 1**    Let the row matrix [2.48    2.99    2.78] represent the prices of a loaf of bread, a six-pack of soft drinks, and a package of granola bars, in that order. Let

$$\begin{bmatrix} 5 \\ 3 \\ 4 \end{bmatrix}$$

represent the quantity of bread (5), soft drinks (3), and granola bars (4) purchased in that order.

Then the dot product

$$[2.48 \quad 2.99 \quad 2.78] \cdot \begin{bmatrix} 5 \\ 3 \\ 4 \end{bmatrix} = 2.48(5) + 2.99(3) + 2.78(4)$$

$$= 12.40 + 8.97 + 11.12 = 32.49$$

gives the total cost of the purchase.

## Matrix Multiplication

Recall that we said the entries in $C = AB$ depend on a row of $A$ and a column of $B$. The entries are actually the dot product of a row and column. In the product

$$C = AB = \begin{bmatrix} 1 & 3 \\ 2 & -1 \end{bmatrix} \begin{bmatrix} 4 & -5 \\ 1 & 6 \end{bmatrix}$$

the $(1, 2)$ entry in $C$ is the dot product R1 · C2, for example. In the product $AB$,

$$C = \begin{bmatrix} R1 \cdot C1 & R1 \cdot C2 \\ R2 \cdot C1 & R2 \cdot C2 \end{bmatrix}$$

$$= \begin{bmatrix} [1 \ 3] \cdot \begin{bmatrix} 4 \\ 1 \end{bmatrix} & [1 \ 3] \cdot \begin{bmatrix} -5 \\ 6 \end{bmatrix} \\ [2 \ -1] \cdot \begin{bmatrix} 4 \\ 1 \end{bmatrix} & [2 \ -1] \cdot \begin{bmatrix} -5 \\ 6 \end{bmatrix} \end{bmatrix}$$

$$= \begin{bmatrix} 1(4) + 3(1) & 1(-5) + 3(6) \\ 2(4) + (-1)(1) & 2(-5) + (-1)(6) \end{bmatrix}$$

$$= \begin{bmatrix} 7 & 13 \\ 7 & -16 \end{bmatrix}$$

**Example 2**    Find the product $AB$ of

$$A = \begin{bmatrix} 1 & 3 & 2 \\ -1 & 0 & 4 \end{bmatrix} \quad \text{and} \quad B = \begin{bmatrix} 7 & 5 \\ -2 & 6 \\ -3 & -4 \end{bmatrix}$$

**Solution**

$$AB = \begin{bmatrix} R1 \cdot C1 & R1 \cdot C2 \\ R2 \cdot C1 & R2 \cdot C2 \end{bmatrix}$$

$$= \begin{bmatrix} 1(7) + 3(-2) + 2(-3) & 1(5) + 3(6) + 2(-4) \\ (-1)(7) + 0(-2) + 4(-3) & -1(5) + 0(6) + 4(-4) \end{bmatrix}$$

$$= \begin{bmatrix} -5 & 15 \\ -19 & -21 \end{bmatrix}$$

**Example 3**    Try to multiply the matrices

$$A = \begin{bmatrix} 1 & 3 \\ 5 & 4 \end{bmatrix} \quad \text{and} \quad B = \begin{bmatrix} -1 & 6 \\ 2 & 7 \\ 0 & 8 \end{bmatrix}$$

**Solution**
The product $AB$ is not possible because a row–column dot product can occur only when the rows of $A$ and the columns of $B$ have the same number of entries.

This example illustrates that two matrices may or may not have a product. There must be the same number of columns in the first matrix as there are rows in the second in order for multiplication to be possible.

As you work with the product of two matrices, *AB,* note how the size of *AB* relates to the size of *A* and the size of *B.* You will find that

- The number of rows of *A* equals the number of rows of *AB.*
- The number of columns of *B* equals the number of columns of *AB.*
- The number of columns of *A* must equal the number of rows of *B.*

---

**Multiplication of Matrices**  Given matrices *A* and *B,* to find $AB = C$ (**matrix multiplication**):

1. Check the number of columns of *A* and the number of rows of *B.* If they are equal, the product is possible. If they are not equal, no product is possible.

2. Form all possible dot products using a row from *A* and a column from *B.* The dot product of row *i* with column *j* gives the entry for the $(i, j)$ position in *C.*

3. The number of rows in *C* is the same as the number of rows in *A.* The number of columns in *C* is the same as the number of columns in *B.*

---

We now return to the problem at the beginning of the section and show a simple use of matrix multiplication.

---

**Example 4**  The time, in hours, required to assemble and finish a table and a chair is given by the matrix

$$
\begin{array}{cc}
& \text{Chair} \quad \text{Table} \\
\begin{array}{c} \text{Assemble} \\ \text{Finish} \end{array} &
\begin{bmatrix} 2 & 3 \\ 2.5 & 4.75 \end{bmatrix}
\end{array}
$$

How long will it take to assemble and finish 950 chairs and 635 tables?

**Solution**
Matrix multiplication gives the answer when we let

$$
\begin{bmatrix} 950 \\ 635 \end{bmatrix}
$$

be the column matrix that specifies the number of chairs and tables produced. Multiply the matrices:

$$
\begin{bmatrix} 2 & 3 \\ 2.5 & 4.75 \end{bmatrix}
\begin{bmatrix} 950 \\ 635 \end{bmatrix}
=
\begin{bmatrix} 2(950) & + & 3(635) \\ 2.5(950) & + & 4.75(635) \end{bmatrix}
=
\begin{bmatrix} 3805 \\ 5391.25 \end{bmatrix}
$$

The rows of the result correspond to the rows in the *first* matrix; the first row in each represents assembly time, and the second represents finishing time. In the final matrix, 3805 is the *total* number of hours of assembly, and 5391.25 is the *total* number of hours for finishing required for 950 chairs and 635 tables.

The next example illustrates that $AB$ and $BA$ may both exist but are not equal.

**Example 5**    Find $AB$ and $BA$:

$$A = \begin{bmatrix} 1 & 3 \\ 5 & -2 \end{bmatrix} \quad \text{and} \quad B = \begin{bmatrix} 2 & 1 \\ 3 & -4 \end{bmatrix}$$

**Solution**

$$AB = \begin{bmatrix} 1 & 3 \\ 5 & -2 \end{bmatrix}\begin{bmatrix} 2 & 1 \\ 3 & -4 \end{bmatrix} = \begin{bmatrix} 11 & -11 \\ 4 & 13 \end{bmatrix}$$

$$BA = \begin{bmatrix} 2 & 1 \\ 3 & -4 \end{bmatrix}\begin{bmatrix} 1 & 3 \\ 5 & -2 \end{bmatrix} = \begin{bmatrix} 7 & 4 \\ -17 & 17 \end{bmatrix}$$

This example shows that $AB$ and $BA$ are not always equal. In fact, sometimes one of them may exist and the other not. The following example illustrates this.

**Example 6**    Find $AB$ and $BA$, if possible.

$$A = \begin{bmatrix} 1 & 2 & 3 \\ -4 & 0 & -2 \\ 1 & 1 & 1 \end{bmatrix} \quad \text{and} \quad B = \begin{bmatrix} 5 & -2 \\ 1 & 4 \\ 2 & 3 \end{bmatrix}$$

**Solution**

$$AB = \begin{bmatrix} 1 & 2 & 3 \\ -4 & 0 & -2 \\ 1 & 1 & 1 \end{bmatrix}\begin{bmatrix} 5 & -2 \\ 1 & 4 \\ 2 & 3 \end{bmatrix} = \begin{bmatrix} 13 & 15 \\ -24 & 2 \\ 8 & 5 \end{bmatrix}$$

$$BA = \begin{bmatrix} 5 & -2 \\ 1 & 4 \\ 2 & 3 \end{bmatrix}\begin{bmatrix} 1 & 2 & 3 \\ -4 & 0 & -2 \\ 1 & 1 & 1 \end{bmatrix} = 5(1) + (-2)(-4) + ?(1)$$

When we attempt to use row 1 from $B$ and column 1 from $A$ to find the (1, 1) entry of $BA$, we find no entry in row 1 of matrix $B$ to multiply by the bottom entry, 1, of column 1 of $A$. Therefore, we, cannot complete the computation. $BA$ does not exist.

Matrix multiplication can be used in a variety of applications, as illustrated in the next two examples.

**Example 7**    The Kaplans have 150 shares of Acme Corp., 100 shares of High Tech, and 240 shares of ABC in an investment portfolio. The closing prices of these stocks one week follow:

| | |
|---|---|
| Monday: | Acme, $56; High Tech, $132; ABC, $19 |
| Tuesday: | Acme, $55; High Tech, $133; ABC, $19 |
| Wednesday: | Acme, $55; High Tech, $131; ABC, $20 |
| Thursday: | Acme, $54; High Tech, $130; ABC, $22 |
| Friday: | Acme, $53; High Tech, $128; ABC, $21 |

Summarize the closing prices in a matrix. Write the number of shares in a matrix and find the value of the Kaplans' portfolio each day by matrix multiplication.

**Solution**

Set up the matrix of closing prices by letting each column represent a stock and each row a day:

$$
\begin{array}{c}
\\
\\
\text{Mon.} \\
\text{Tue.} \\
\text{Wed.} \\
\text{Thur.} \\
\text{Fri.}
\end{array}
\begin{array}{ccc}
 & \text{High} & \\
\text{Acme} & \text{Tech} & \text{ABC} \\
\left[\begin{array}{ccc}
56 & 132 & 19 \\
55 & 133 & 19 \\
55 & 131 & 20 \\
54 & 130 & 22 \\
53 & 128 & 21
\end{array}\right]
\end{array}
$$

We point out that using rows to represent stocks and columns to represent days is also acceptable. The matrix showing the number of shares of each company could be either a row matrix or a column matrix. Which of the matrices giving the number of shares,

$$
[150 \quad 100 \quad 240] \qquad \text{or} \qquad \begin{bmatrix} 150 \\ 100 \\ 240 \end{bmatrix}
$$

should be used to find the daily value of the portfolio? First of all, note that the products

$$
[150 \quad 100 \quad 240]\begin{bmatrix} 56 & 132 & 19 \\ 55 & 133 & 19 \\ 55 & 131 & 20 \\ 54 & 130 & 22 \\ 53 & 128 & 21 \end{bmatrix}
$$

$$
\begin{bmatrix} 150 \\ 100 \\ 240 \end{bmatrix}\begin{bmatrix} 56 & 132 & 19 \\ 55 & 133 & 19 \\ 55 & 131 & 20 \\ 54 & 130 & 22 \\ 53 & 128 & 21 \end{bmatrix}
$$

$$
\begin{bmatrix} 56 & 132 & 19 \\ 55 & 133 & 19 \\ 55 & 131 & 20 \\ 54 & 130 & 22 \\ 53 & 128 & 21 \end{bmatrix}[150 \quad 100 \quad 240]
$$

are not possible.

The product

$$
\begin{bmatrix} 56 & 132 & 19 \\ 55 & 133 & 19 \\ 55 & 131 & 20 \\ 54 & 130 & 22 \\ 53 & 128 & 21 \end{bmatrix}\begin{bmatrix} 150 \\ 100 \\ 240 \end{bmatrix} = \begin{bmatrix} 26{,}160 \\ 26{,}110 \\ 26{,}150 \\ 26{,}380 \\ 25{,}790 \end{bmatrix}
$$

is possible. Does it give the desired result? Note that the first entry in the answer, 26,160, is obtained by

$$56(150) + 132(100) + 19(240)$$

which is

(price of Acme) × (no. shares of Acme)
+ (price of High Tech) × (no. shares of High Tech)
+ (price of ABC) × (no. shares of ABC)

This gives exactly what is needed to find the total value of the portfolio on Monday. The other entries are correct for the other days.

When you use a matrix product in an application, check to see which order of multiplication makes sense.

If you ask a person to set up the stock data in matrix form, they might well let the columns represent days of the week and the rows the stocks. The matrix then takes the form

|           | Mon | Tue | Wed | Thu | Fri |
|-----------|-----|-----|-----|-----|-----|
| Acme      | 56  | 55  | 55  | 54  | 53  |
| High Tech | 132 | 133 | 131 | 130 | 128 |
| ABC       | 19  | 19  | 20  | 22  | 21  |

In this case, the matrix representing the number of shares must assume a form that makes sense when the matrices are multiplied to obtain the portfolio value. Use the form [150  100  240], and the product

$$\begin{bmatrix} 150 & 100 & 240 \end{bmatrix} \begin{bmatrix} 56 & 55 & 55 & 54 & 53 \\ 132 & 133 & 131 & 130 & 128 \\ 19 & 19 & 20 & 22 & 21 \end{bmatrix}$$
$$= \begin{bmatrix} 26{,}160 & 26{,}110 & 26{,}150 & 26{,}380 & 25{,}790 \end{bmatrix}$$

gives daily portfolio values.

Thus, a matrix may be written in different ways as long as it is understood what the rows and columns represent, and the rows and columns of each matrix are arranged so the multiplication makes sense.

## Identity Matrix

You are familiar with the number fact

$$1 \times a = a \times 1 = a$$

where $a$ is any real number. We call 1 the **identity** for multiplication.

In general, we have no similar property for multiplication of matrices; there is no one matrix $I$ such that $AI = IA = A$ for all matrices $A$. However, there is such a matrix for square matrices of a given size. For example, if

$$A = \begin{bmatrix} 4 & 3 \\ 7 & 2 \end{bmatrix} \quad \text{and} \quad I = \begin{bmatrix} 1 & 0 \\ 0 & 1 \end{bmatrix}$$

then

$$AI = \begin{bmatrix} 4 & 3 \\ 7 & 2 \end{bmatrix}\begin{bmatrix} 1 & 0 \\ 0 & 1 \end{bmatrix} = \begin{bmatrix} 4 & 3 \\ 7 & 2 \end{bmatrix} = A$$

and

$$IA = \begin{bmatrix} 1 & 0 \\ 0 & 1 \end{bmatrix}\begin{bmatrix} 4 & 3 \\ 7 & 2 \end{bmatrix} = \begin{bmatrix} 4 & 3 \\ 7 & 2 \end{bmatrix} = A$$

Furthermore, for any $2 \times 2$ matrix $A$, the matrix $I$ has the property that $AI = A$ and $IA = A$. This can be justified by using a $2 \times 2$ matrix with arbitrary entries

$$A = \begin{bmatrix} a & b \\ c & d \end{bmatrix}$$

Now,

$$AI = \begin{bmatrix} a & b \\ c & d \end{bmatrix}\begin{bmatrix} 1 & 0 \\ 0 & 1 \end{bmatrix}\begin{bmatrix} a \times 1 + b \times 0 & a \times 0 + b \times 1 \\ c \times 1 + d \times 0 & c \times 0 + d \times 1 \end{bmatrix}$$

$$= \begin{bmatrix} a & b \\ c & d \end{bmatrix} = A$$

You should now multiply $IA$ to verify that it is indeed $A$. Thus,

$$\begin{bmatrix} 1 & 0 \\ 0 & 1 \end{bmatrix}$$

is the **identity matrix** for all $2 \times 2$ matrices. If we try to multiply the $3 \times 3$ matrix

$$A = \begin{bmatrix} 1 & 2 & 3 \\ 5 & 7 & 12 \\ 8 & 4 & -2 \end{bmatrix} \quad \text{by} \quad \begin{bmatrix} 1 & 0 \\ 0 & 1 \end{bmatrix}$$

we find we are unable to multiply at all because $A$ has 3 columns and $I$ has only two rows. So,

$$\begin{bmatrix} 1 & 0 \\ 0 & 1 \end{bmatrix}$$

is *not* the identity matrix for $3 \times 3$ matrices. However, the matrix

$$I = \begin{bmatrix} 1 & 0 & 0 \\ 0 & 1 & 0 \\ 0 & 0 & 1 \end{bmatrix}$$

is an identity matrix for the set of all $3 \times 3$ matrices:

$$\begin{bmatrix} a & b & c \\ d & e & f \\ g & h & i \end{bmatrix}\begin{bmatrix} 1 & 0 & 0 \\ 0 & 1 & 0 \\ 0 & 0 & 1 \end{bmatrix} = \begin{bmatrix} a & b & c \\ d & e & f \\ g & h & i \end{bmatrix}$$

$$\begin{bmatrix} 1 & 0 & 0 \\ 0 & 1 & 0 \\ 0 & 0 & 1 \end{bmatrix}\begin{bmatrix} a & b & c \\ d & e & f \\ g & h & i \end{bmatrix} = \begin{bmatrix} a & b & c \\ d & e & f \\ g & h & i \end{bmatrix}$$

In general, if we let $I$ be the $n \times n$ matrix with ones on the *main diagonal* and zeros elsewhere, we have the identity matrix for the class of all $n \times n$ matrices. (The **main diagonal** runs from the upper left to the lower right corner.)

### Row Operations Using Matrix Multiplication

We can perform row operations on a matrix by multiplying by a modified identity matrix. Let's illustrate.

$$\text{Let } A = \begin{bmatrix} 1 & 2 & 3 & 4 \\ 5 & 6 & 7 & 8 \\ 9 & 10 & 11 & 12 \end{bmatrix} \quad \text{and} \quad I = \begin{bmatrix} 1 & 0 & 0 \\ 0 & 1 & 0 \\ 0 & 0 & 1 \end{bmatrix}$$

We know that $IA = A$. Now interchange rows 1 and 2 of $I$ and multiply times $A$.

$$\begin{bmatrix} 0 & 1 & 0 \\ 1 & 0 & 0 \\ 0 & 0 & 1 \end{bmatrix} \begin{bmatrix} 1 & 2 & 3 & 4 \\ 5 & 6 & 7 & 8 \\ 9 & 10 & 11 & 12 \end{bmatrix} = \begin{bmatrix} 5 & 6 & 7 & 8 \\ 1 & 2 & 3 & 4 \\ 9 & 10 & 11 & 12 \end{bmatrix}$$

Note that this multiplication interchanges rows 1 and 2 of $A$. This illustrates a general property.

---

**Interchange Rows by Matrix Multiplication**

If a matrix $A$ has $n$ rows and $I$ is the $n \times n$ identity matrix, then modify $I$ by interchanging two rows, giving matrix $I_M$. The product $I_M A$ interchanges the corresponding rows of $A$.

---

Next, modify $I$ by adding row 3 to row 1 giving $I_M = \begin{bmatrix} 1 & 0 & 1 \\ 0 & 1 & 0 \\ 0 & 0 & 1 \end{bmatrix}$.

$$\text{The product } I_M A \text{ is } \begin{bmatrix} 1 & 0 & 1 \\ 0 & 1 & 0 \\ 0 & 0 & 1 \end{bmatrix} \begin{bmatrix} 1 & 2 & 3 & 4 \\ 5 & 6 & 7 & 8 \\ 9 & 10 & 11 & 12 \end{bmatrix} = \begin{bmatrix} 10 & 12 & 14 & 16 \\ 5 & 6 & 7 & 8 \\ 9 & 10 & 11 & 12 \end{bmatrix}.$$

Note that rows 2 and 3 are unchanged, but row 1 is now the sum of rows 1 and 3. In the notation we have used for row operations, this product gives R1 + R3 → R1. Next we add row 1 to row 3 (R1 + R3 → R3) and multiply.

$$\begin{bmatrix} 1 & 0 & 0 \\ 0 & 1 & 0 \\ 1 & 0 & 1 \end{bmatrix} \begin{bmatrix} 1 & 2 & 3 & 4 \\ 5 & 6 & 7 & 8 \\ 9 & 10 & 11 & 12 \end{bmatrix} = \begin{bmatrix} 1 & 2 & 3 & 4 \\ 5 & 6 & 7 & 8 \\ 10 & 12 & 14 & 16 \end{bmatrix}$$

Row 1 has been added to row 3 and the result stored in row 3.

Let's give one more example. Modify $I$ by the row operation 2R2 + R1 → R1,

$$\text{giving } I_M = \begin{bmatrix} 1 & 2 & 0 \\ 0 & 1 & 0 \\ 0 & 0 & 1 \end{bmatrix}.$$

$$\text{Now } I_M A = \begin{bmatrix} 1 & 2 & 0 \\ 0 & 1 & 0 \\ 0 & 0 & 1 \end{bmatrix} \begin{bmatrix} 1 & 2 & 3 & 4 \\ 5 & 6 & 7 & 8 \\ 9 & 10 & 11 & 12 \end{bmatrix} = \begin{bmatrix} 11 & 14 & 17 & 20 \\ 5 & 6 & 7 & 8 \\ 9 & 10 & 11 & 12 \end{bmatrix}.$$

The result is the row operation $2R2 + R1 \rightarrow R1$ on $A$.

We summarize these examples with a general property they represent.

---

**Row Operations Using Matrix Multiplication**

If a matrix $A$ has $n$ rows and $I$ is the $n \times n$ identity matrix, then modify $I$ by a row operation like $5R2 + R4 \rightarrow R4$, giving matrix $I_M$. The product $I_M A$ performs the same row operation on $A$.

---

## 2.5 EXERCISES

Access end-of-section exercises online at **www.webassign.net**  **WebAssign**

---

## Using Your TI Graphing Calculator

### Matrix Multiplication

The multiplication of matrices is straightforward. When [A] and [B] contain the matrices

$$[A] = \begin{bmatrix} 2 & 1 & -3 \\ 1 & -1 & 1 \\ 4 & 2 & 5 \end{bmatrix} \quad [B] = \begin{bmatrix} 1 & 2 & 1 & 1 \\ 3 & -1 & 4 & 2 \\ 2 & 5 & 0 & 3 \end{bmatrix}$$

then their product is obtained by [A]$\boxed{\times}$[B]$\boxed{\text{ENTER}}$.

Here is the matrix [A] $\times$ [B]:

```
[A]*[B]
[[-1 -12 6  -5]
 [0  8   -3  2]
 [20 31  12 23]]
```

Use $\boxed{\wedge}$ to obtain powers of matrices. For example, compute [A]³ with [A]$\boxed{\wedge}$$\boxed{3}$$\boxed{\text{ENTER}}$:

```
[A]^3
[[-99 -42 -84]
 [18  3   -6]
 [132 48  -3 ]]
```

*Note*: If you attempt to obtain the matrix product [A] $\times$ [B] when the number of columns of [A] does not equal the number of rows of [B], you will get the following error message: **ERR:DIM MISMATCH**.

*(continued)*

## Exercises

**1.** $[A] = \begin{bmatrix} 2 & -4 \\ 3 & 7 \end{bmatrix}$ and $[B] = \begin{bmatrix} 5 & 1 \\ 2 & -2 \end{bmatrix}$. Find $[A]\,[B]$.

**2.** $[A] = \begin{bmatrix} 2 & 1 & 3 \\ 4 & 6 & -2 \\ 5 & 9 & 1 \end{bmatrix}$ and $[B] = \begin{bmatrix} 1 & 0 & 2 \\ 6 & -2 & 1 \\ 3 & 1 & 1 \end{bmatrix}$. Find $[A]\,[B]$.

**3.** $[A] = \begin{bmatrix} 8 & -2 & 1 & 4 \\ 3 & 0 & -1 & 5 \end{bmatrix}$ and $[B] = \begin{bmatrix} 1 & 3 \\ 2 & 0 \\ 1 & -1 \\ 2 & -4 \end{bmatrix}$. Find $[A]\,[B]$.

**4.** $[A] = \begin{bmatrix} 3 & 1 \\ 1 & 2 \end{bmatrix}$. Find $[A]^2$ and $[A]^3$.

**5.** $[A] = \begin{bmatrix} 1 & 0 & 1 \\ 2 & 1 & 2 \\ 1 & 3 & 1 \end{bmatrix}$. Find $[A]^2$, $[A]^3$, and $[A]^4$.

 ## Using Excel

Excel has a command, MMULT, that performs the multiplication of two matrices. To illustrate, let's use the matrices

$$A = \begin{bmatrix} 1 & 3 & 2 \\ 4 & 1 & 5 \end{bmatrix} \qquad B = \begin{bmatrix} 2 & 1 & 1 & 3 \\ 5 & 0 & 2 & 4 \\ 3 & 2 & 2 & -1 \end{bmatrix}$$

Because $A$ has 2 rows and $B$ has 4 columns, the product $AB$ has 2 rows and 4 columns.

Enter matrix $A$ in cells A2:C3 and matrix $B$ in E2:H4. We will put $AB$ in cells A6:D7.

To calculate $AB$, select the cells A6:D7 and type =MMULT(A2:C3,E2:H4). Notice that this has the form =MMULT(Location of matrix $A$, Location of matrix $B$). The next step differs from the usual press ENTER. To activate matrix multiplication, simultaneously press the CTRL + SHIFT + ENTER keys. Then the product shows in cells A6:D7.

|   | A | B | C | D | E | F | G | H |
|---|---|---|---|---|---|---|---|---|
| 1 | Matrix A in A2:C3 | | | | Matrix B in E2:H4 | | | |
| 2 | 1 | 3 | 2 | | 2 | 1 | 1 | 3 |
| 3 | 4 | 1 | 5 | | 5 | 0 | 2 | 4 |
| 4 | | | | | 3 | 2 | 2 | -1 |
| 5 | Matrix AB in A6:D7 | | | | | | | |
| 6 | 23 | 5 | 11 | 13 | | | | |
| 7 | 28 | 14 | 16 | 11 | | | | |

## Exercises

Calculate $AB$ in the following exercises:

**1.** $A = \begin{bmatrix} 2 & 1 & 1 \\ 3 & 4 & 2 \end{bmatrix}$ $B = \begin{bmatrix} 4 & 5 \\ 1 & 2 \\ 3 & 3 \end{bmatrix}$

**2.** $A = \begin{bmatrix} -1 & 1 & 2 \\ 5 & 0 & 3 \end{bmatrix}$ $B = \begin{bmatrix} 3 & 1 \\ -2 & 5 \\ 4 & 3 \end{bmatrix}$

**3.** $A = \begin{bmatrix} 1 & 5 & 1 \\ 2 & 3 & 2 \end{bmatrix}$ $B = \begin{bmatrix} 4 & 0 & 1 \\ -2 & 1 & 3 \\ 3 & 2 & 1 \end{bmatrix}$

**4.** $A = \begin{bmatrix} 1 & 2 & 1 \\ 2 & 1 & 2 \\ 3 & 3 & 1 \end{bmatrix}$ $B = \begin{bmatrix} 4 & 5 & 4 \\ 3 & 0 & 3 \\ 1 & 2 & 3 \end{bmatrix}$

---

## 2.6 THE INVERSE OF A MATRIX

- Inverse of a Square Matrix
- Matrix Equations
- Using $A^{-1}$ to Solve a System

### Inverse of a Square Matrix

We can extend another number fact to matrices. The simple multiplication facts

$$2 \times \frac{1}{2} = 1$$
$$\frac{3}{4} \times \frac{4}{3} = 1$$
$$1.25 \times 0.8 = 1$$

have a common property. Each of the numbers $2, \frac{3}{4}$, and $1.25$ can be multiplied by another number to obtain 1. In general, for any real number $a$, except zero, there is a number $b$ such that $a \times b = 1$. We call $b$ the **inverse** of $a$. The standard notation for the inverse of $a$ is $a^{-1}$.

**Example 1**

$$3^{-1} = \frac{1}{3} \quad 2^{-1} = 0.5 \quad \left(\frac{5}{8}\right)^{-1} = \frac{8}{5}$$
$$0.4^{-1} = 2.5 \quad 625^{-1} = 0.0016$$

A similar property exists in terms of matrix multiplication. For example,

$$\begin{bmatrix} 1 & 1 \\ 1 & 2 \end{bmatrix}\begin{bmatrix} 2 & -1 \\ -1 & 1 \end{bmatrix} = \begin{bmatrix} 1 & 0 \\ 0 & 1 \end{bmatrix}$$

We can restate this equation as $AA^{-1} = I$, where

$$A = \begin{bmatrix} 1 & 1 \\ 1 & 2 \end{bmatrix} \quad \text{and} \quad A^{-1} = \begin{bmatrix} 2 & -1 \\ -1 & 1 \end{bmatrix}$$

We call $A^{-1}$ the inverse of the matrix $A$.

> **DEFINITION**
> **Inverse of a Matrix A**
>
> If $A$ and $B$ are square matrices such that $AB = BA = I$, then $B$ is the **inverse matrix** of $A$. The inverse of $A$ is denoted $A^{-1}$. If $B$ is found so that $AB = I$, then a theorem from linear algebra states that $BA = I$, so it is sufficient to just check $AB = I$.

Only square matrices have inverses. You can use the definition of an inverse matrix to check for an inverse.

**Example 2**    (a) For the two matrices

$$A = \begin{bmatrix} 2 & 5 & 4 \\ 1 & 4 & 3 \\ 1 & -3 & -2 \end{bmatrix} \quad \text{and} \quad B = \begin{bmatrix} -1 & 2 & 1 \\ -5 & 8 & 2 \\ 7 & -11 & -3 \end{bmatrix}$$

determine whether $B$ is the inverse of $A$.

(b) For the two matrices

$$A = \begin{bmatrix} 4 & 7 \\ 2 & 1 \end{bmatrix} \quad \text{and} \quad B = \begin{bmatrix} -\frac{1}{10} & \frac{7}{10} \\ \frac{1}{5} & -\frac{2}{5} \end{bmatrix}$$

determine whether $B = A^{-1}$.

(c) Determine whether $B$ is the inverse of $A$ for

$$A = \begin{bmatrix} 0 & 1 & 0 \\ 1 & 1 & 0 \\ 0 & 1 & 1 \end{bmatrix} \quad \text{and} \quad B = \begin{bmatrix} -1 & 1 & 0 \\ 1 & 0 & 1 \\ -1 & 1 & 0 \end{bmatrix}$$

**Solution**

In each case it suffices to compute $AB$. If $AB = I$, then $B$ is the inverse of $A$. If $AB \neq I$, then $B$ is not the inverse of $A$.

(a) $AB = \begin{bmatrix} 2 & 5 & 4 \\ 1 & 4 & 3 \\ 1 & -3 & -2 \end{bmatrix} \begin{bmatrix} -1 & 2 & 1 \\ -5 & 8 & 2 \\ 7 & -11 & -3 \end{bmatrix}$

$= \begin{bmatrix} -2 - 25 + 28 & 4 + 40 - 44 & 2 + 10 - 12 \\ -1 - 20 + 21 & 2 + 32 - 33 & 1 + 8 - 9 \\ -1 + 15 - 14 & 2 - 24 + 22 & 1 - 6 + 6 \end{bmatrix}$

$= \begin{bmatrix} 1 & 0 & 0 \\ 0 & 1 & 0 \\ 0 & 0 & 1 \end{bmatrix} = I$

so $B$ is the inverse of $A$.

(b) $AB = \begin{bmatrix} 4 & 7 \\ 2 & 1 \end{bmatrix} \begin{bmatrix} -\frac{1}{10} & \frac{7}{10} \\ \frac{1}{5} & -\frac{2}{5} \end{bmatrix}$

$= \begin{bmatrix} -\frac{4}{10} + \frac{7}{5} & \frac{28}{10} - \frac{14}{5} \\ -\frac{2}{10} + \frac{1}{5} & \frac{14}{10} - \frac{2}{5} \end{bmatrix} = \begin{bmatrix} 1 & 0 \\ 0 & 1 \end{bmatrix}$

so $B = A^{-1}$.

**(c)** $AB = \begin{bmatrix} 0 & 1 & 0 \\ 1 & 1 & 0 \\ 0 & 1 & 1 \end{bmatrix} \begin{bmatrix} -1 & 1 & 0 \\ 1 & 0 & 1 \\ -1 & 1 & 0 \end{bmatrix} = \begin{bmatrix} 1 & 0 & 1 \\ 0 & 1 & 1 \\ 0 & 1 & 1 \end{bmatrix} \ne I$

so $B$ is not the inverse of $A$.

In general, a matrix $A$ has an inverse if there is a matrix $A^{-1}$ that fulfills the conditions that $AA^{-1} = A^{-1}A = I$. Not all matrices have inverses. In fact, a matrix must be square in order to have an inverse, and some square matrices have no inverse. We now come to the problem of deciding if a square matrix has an inverse. If it does, how do we find it? Let's approach this problem with a simple $2 \times 2$ example.

**Example 3**    If we have the square matrix

$$A = \begin{bmatrix} 2 & 1 \\ 3 & 2 \end{bmatrix}$$

find its inverse, if possible.

**Solution**
We want to find a $2 \times 2$ matrix $A^{-1}$ such that $AA^{-1} = I$. Because we don't know the entries in $A^{-1}$, let's enter variables, $x_1, x_2, y_1$, and $y_2$ and attempt to find their values. Write

$$A^{-1} = \begin{bmatrix} x_1 & y_1 \\ x_2 & y_2 \end{bmatrix}$$

The condition $AA^{-1} = I$ can now be written

$$AA^{-1} = \begin{bmatrix} 2 & 1 \\ 3 & 2 \end{bmatrix} \begin{bmatrix} x_1 & y_1 \\ x_2 & y_2 \end{bmatrix} = \begin{bmatrix} 1 & 0 \\ 0 & 1 \end{bmatrix}$$

We want to find values of $x_1, x_2, y_1$, and $y_2$ so that the product on the left equals the identity matrix on the right. First, form the product $AA^{-1}$. We get

$$\overset{AA^{-1}}{\begin{bmatrix} (2x_1 + x_2) & (2y_1 + y_2) \\ (3x_1 + 2x_2) & (3y_1 + 2y_2) \end{bmatrix}} = \overset{I}{\begin{bmatrix} 1 & 0 \\ 0 & 1 \end{bmatrix}}$$

Recall that two matrices are equal only when they have equal entries in corresponding positions. So the matrix equality gives us the equations

$$\begin{array}{cc} 2x_1 + x_2 = 1 & \\ & \text{and} \\ 3x_1 + 2x_2 = 0 & \end{array} \qquad \begin{array}{c} 2y_1 + y_2 = 0 \\ \\ 3y_1 + 2y_2 = 1 \end{array}$$

Note that we have one system of two equations with variables $x_1$ and $x_2$ :

**1.** $\begin{array}{l} 2x_1 + x_2 = 1 \\ 3x_1 + 2x_2 = 0 \end{array}$    with augmented matrix    $\begin{bmatrix} 2 & 1 & | & 1 \\ 3 & 2 & | & 0 \end{bmatrix}$

and a system with variables $y_1$ and $y_2$ :

**2.** $\begin{array}{l} 2y_1 + y_2 = 0 \\ 3y_1 + 2y_2 = 1 \end{array}$    with augmented matrix    $\begin{bmatrix} 2 & 1 & | & 0 \\ 3 & 2 & | & 1 \end{bmatrix}$

The solution to system 1 gives $x_1 = 2$, $x_2 = -3$. The solution to system 2 gives $y_1 = -1, y_2 = 2$, so the inverse of

$$A = \begin{bmatrix} 2 & 1 \\ 3 & 2 \end{bmatrix} \quad \text{is} \quad A^{-1} = \begin{bmatrix} 2 & -1 \\ -3 & 2 \end{bmatrix}$$

We check our results by computing $AA^{-1}$ and $A^{-1}A$:

$$AA^{-1} = \begin{bmatrix} 2 & 1 \\ 3 & 2 \end{bmatrix}\begin{bmatrix} 2 & -1 \\ -3 & 2 \end{bmatrix} = \begin{bmatrix} 1 & 0 \\ 0 & 1 \end{bmatrix}$$

$$A^{-1}A = \begin{bmatrix} 2 & -1 \\ -3 & 2 \end{bmatrix}\begin{bmatrix} 2 & 1 \\ 3 & 2 \end{bmatrix} = \begin{bmatrix} 1 & 0 \\ 0 & 1 \end{bmatrix}$$

It checks. (*Note:* It suffices to check just one of these.)

Look at the two systems we just solved. The two systems have precisely the same coefficients; they differ only in the constant terms. The left-hand portions of the augmented matrices are exactly the same. In fact, each is the matrix A.

This means that when we solve each of the two systems using the Gauss-Jordan Method, we use precisely the same row operations. Thus, we can solve both systems using one matrix. Here's how: Combine the two augmented matrices into one using the common coefficient portion on the left, and list both columns from the right sides. This gives the matrix

$$\left[\begin{array}{cc|cc} 2 & 1 & 1 & 0 \\ 3 & 2 & 0 & 1 \end{array}\right]$$

Note that the left portion of the matrix is $A$ and the right portion is the identity matrix.

Now proceed in the same way you do to solve a system of equations with an augmented matrix; that is, use row operations to reduce the left-hand portion to the identity matrix. This gives the following sequence:

$$\left[\begin{array}{cc|cc} 2 & 1 & 1 & 0 \\ 3 & 2 & 0 & 1 \end{array}\right] \qquad \tfrac{1}{2}\text{R1} \to \text{R1}$$

$$\left[\begin{array}{cc|cc} 1 & \tfrac{1}{2} & \tfrac{1}{2} & 0 \\ 3 & 2 & 0 & 1 \end{array}\right] \qquad -3\text{R1} + \text{R2} \to \text{R2}$$

$$\left[\begin{array}{cc|cc} 1 & \tfrac{1}{2} & \tfrac{1}{2} & 0 \\ 0 & \tfrac{1}{2} & -\tfrac{3}{2} & 1 \end{array}\right] \qquad -\text{R2} + \text{R1} \to \text{R1}$$

$$\left[\begin{array}{cc|cc} 1 & 0 & 2 & -1 \\ 0 & \tfrac{1}{2} & -\tfrac{3}{2} & 1 \end{array}\right] \qquad 2\text{R2} \to \text{R2}$$

$$\left[\begin{array}{cc|cc} 1 & 0 & 2 & -1 \\ 0 & 1 & -3 & 2 \end{array}\right]$$

The final matrix has the identity matrix formed by the first two columns. The third column gives the solution to the first system, and the fourth column gives the solution to the second system. Note that the last two columns form $A^{-1}$. This is no accident; one may find the inverse of a square matrix in this manner.

**Method to Find the Inverse of a Square Matrix**

1. To find the inverse of a matrix $A$, form an augmented matrix $[A \mid I]$ by writing down the matrix $A$ and then writing the identity matrix to the right of $A$.

2. Perform a sequence of row operations that reduces the $A$ portion of this matrix to reduced echelon form.

3. If the $A$ portion of the reduced echelon form is the identity matrix, then the matrix found in the $I$ portion is $A^{-1}$.

4. If the reduced echelon form produces a row in the $A$ portion that is all zeros, then $A$ has no inverse.

Now use this method to find the inverse of a matrix.

**Example 4**   Find the inverse of the matrix

$$A = \begin{bmatrix} 1 & 3 & 2 \\ 2 & 4 & 2 \\ 1 & 2 & -1 \end{bmatrix}$$

**Solution**

First, set up the augmented matrix $[A \mid I]$:

$$\left[\begin{array}{ccc|ccc} 1 & 3 & 2 & 1 & 0 & 0 \\ 2 & 4 & 2 & 0 & 1 & 0 \\ 1 & 2 & -1 & 0 & 0 & 1 \end{array}\right] \quad \begin{array}{l} -2R1 + R2 \rightarrow R2 \\ -R1 + R3 \rightarrow R3 \end{array}$$

Next, use row operations indicated above to get zeros in column 1:

$$\left[\begin{array}{ccc|ccc} 1 & 3 & 2 & 1 & 0 & 0 \\ 0 & -2 & -2 & -2 & 1 & 0 \\ 0 & -1 & -3 & -1 & 0 & 1 \end{array}\right] \quad -\tfrac{1}{2}R2 \rightarrow R2$$

Now divide the entries in row 2 by $-2$:

$$\left[\begin{array}{ccc|ccc} 1 & 3 & 2 & 1 & 0 & 0 \\ 0 & 1 & 1 & 1 & -\tfrac{1}{2} & 0 \\ 0 & -1 & -3 & -1 & 0 & 1 \end{array}\right] \quad \begin{array}{l} -3R2 + R1 \rightarrow R1 \\[4pt] R2 + R3 \rightarrow R3 \end{array}$$

Next, get zeros in the second column:

$$\left[\begin{array}{ccc|ccc} 1 & 0 & -1 & -2 & \tfrac{3}{2} & 0 \\ 0 & 1 & 1 & 1 & -\tfrac{1}{2} & 0 \\ 0 & 0 & -2 & 0 & -\tfrac{1}{2} & 1 \end{array}\right] \quad -\tfrac{1}{2}R3 \rightarrow R3$$

Now divide the entries in row 3 by $-2$:

$$\left[\begin{array}{ccc|ccc} 1 & 0 & -1 & -2 & \tfrac{3}{2} & 0 \\ 0 & 1 & 1 & 1 & -\tfrac{1}{2} & 0 \\ 0 & 0 & 1 & 0 & \tfrac{1}{4} & -\tfrac{1}{2} \end{array}\right] \quad \begin{array}{l} R3 + R1 \rightarrow R1 \\[4pt] -R3 + R2 \rightarrow R2 \end{array}$$

Finally, use the indicated operations to get zeros in the third column:

$$\left[\begin{array}{ccc|ccc} 1 & 0 & 0 & -2 & \frac{7}{4} & -\frac{1}{2} \\ 0 & 1 & 0 & 1 & -\frac{3}{4} & \frac{1}{2} \\ 0 & 0 & 1 & 0 & \frac{1}{4} & -\frac{1}{2} \end{array}\right]$$

When the left-hand portion of the augmented matrix reduces to the identity matrix, $A^{-1}$ comes from the right-hand portion:

$$A^{-1} = \left[\begin{array}{ccc} -2 & \frac{7}{4} & -\frac{1}{2} \\ 1 & -\frac{3}{4} & \frac{1}{2} \\ 0 & \frac{1}{4} & -\frac{1}{2} \end{array}\right]$$

Now look at a case in which the matrix has no inverse.

**Example 5** Find the inverse of

$$A = \begin{bmatrix} 1 & 3 \\ 3 & 9 \end{bmatrix}$$

**Solution**
Adjoin $I$ to $A$ to obtain

$$\left[\begin{array}{cc|cc} 1 & 3 & 1 & 0 \\ 3 & 9 & 0 & 1 \end{array}\right]$$

Now reduce this matrix using row operations:

$$\left[\begin{array}{cc|cc} 1 & 3 & 1 & 0 \\ 3 & 9 & 0 & 1 \end{array}\right] \qquad -3R1 + R2 \rightarrow R2$$

$$\left[\begin{array}{cc|cc} 1 & 3 & 1 & 0 \\ 0 & 0 & -3 & 1 \end{array}\right]$$

The bottom row of the matrix represents two equations $0 = -3$ and $0 = 1$. Both of these are impossible, so in our attempt to find $A^{-1}$ we reached an inconsistency. Whenever we reach an inconsistency in trying to solve a system of equations, we conclude that there is no solution. Therefore, in this case $A$ has no inverse.

In general, when we use an augmented matrix $[A \mid I]$ to find the inverse of $A$ and reach a step where a row of the $A$ portion is all zeros, then $A$ has no inverse.

## Matrix Equations

We can write systems of equations using matrices and solve some systems using matrix inverses.
The **matrix equation**

$$\begin{bmatrix} 5 & 3 & -4 & 12 \\ 8 & -21 & 7 & -19 \\ 2 & 1 & -15 & 1 \end{bmatrix} \begin{bmatrix} x_1 \\ x_2 \\ x_3 \\ x_4 \end{bmatrix} = \begin{bmatrix} 7 \\ 16 \\ -22 \end{bmatrix}$$

becomes the following when the multiplication on the left is performed:

$$\begin{bmatrix} 5x_1 + 3x_2 - 4x_3 + 12x_4 \\ 8x_1 - 21x_2 + 7x_3 - 19x_4 \\ 2x_1 + x_2 - 15x_3 + x_4 \end{bmatrix} = \begin{bmatrix} 7 \\ 16 \\ -22 \end{bmatrix}$$

These matrices are equal only when corresponding components are equal; that is,

$$\begin{aligned} 5x_1 + 3x_2 - 4x_3 + 12x_4 &= 7 \\ 8x_1 - 21x_2 + 7x_3 - 19x_4 &= 16 \\ 2x_1 + x_2 - 15x_3 + x_4 &= -22 \end{aligned}$$

In general, we can write a system of equations in the compact matrix form

$$AX = B$$

where $A$ is a matrix formed from the coefficients of the variables

$$A = \begin{bmatrix} 5 & 3 & -4 & 12 \\ 8 & -21 & 7 & -19 \\ 2 & 1 & -15 & 1 \end{bmatrix}$$

$X$ is a column matrix formed by listing the variables

$$X = \begin{bmatrix} x_1 \\ x_2 \\ x_3 \\ x_4 \end{bmatrix}$$

and $B$ is the column matrix formed from the constants in the system

$$B = \begin{bmatrix} 7 \\ 16 \\ -22 \end{bmatrix}$$

**Example 6**  Here is a system of equations.

$$\begin{aligned} 4x_1 + 7x_2 - 2x_3 &= 5 \\ 3x_1 - x_2 + 7x_3 &= 8 \\ x_1 + 2x_2 - x_3 &= 9 \end{aligned}$$

We can use matrices to represent this system in the following ways:

The coefficient matrix of this system is

$$\begin{bmatrix} 4 & 7 & -2 \\ 3 & -1 & 7 \\ 1 & 2 & -1 \end{bmatrix}$$

and the augmented matrix that represents the system is

$$\left[\begin{array}{ccc|c} 4 & 7 & -2 & 5 \\ 3 & -1 & 7 & 8 \\ 1 & 2 & -1 & 9 \end{array}\right]$$

The system of equations can also be written in the matrix form, $AX = B$, as

$$\begin{bmatrix} 4 & 7 & -2 \\ 3 & -1 & 7 \\ 1 & 2 & -1 \end{bmatrix} \begin{bmatrix} x_1 \\ x_2 \\ x_3 \end{bmatrix} = \begin{bmatrix} 5 \\ 8 \\ 9 \end{bmatrix}$$

## Using $A^{-1}$ to Solve a System

Now we can illustrate the use of the inverse in solving a system of equations when the matrix of coefficients has an inverse. Sometimes it helps to be able to solve a system by using the inverse matrix. One such situation occurs when a number of systems need to be solved, and all have the same coefficients; that is, the constant terms change, but the coefficients don't. Here is a simple example.

A doctor treats patients who need adequate calcium and iron in their diet. The doctor has found that two foods, A and B, provide these. Each unit of food A has 0.5 milligram (mg) iron and 25 mg calcium. Each unit of food B has 0.3 mg iron and 7 mg calcium. Let $x$ = number of units of food A eaten by the patient; let $y$ = number of units of food B eaten by the patient. Then $0.5x + 0.3y$ gives the total milligrams of iron consumed by the patient and $25x + 7y$ gives the total milligrams of calcium. Suppose the doctor wants patient Jones to get 6 mg iron and 60 mg calcium. The amount of each food to be consumed is the solution to

$$\begin{aligned} 0.5x + 0.3y &= 6 \\ 25x + 7y &= 60 \end{aligned}$$

If patient Smith requires 7 mg iron and 80 mg calcium, the amount of food required is found in the solution of the system

$$\begin{aligned} 0.5x + 0.3y &= 7 \\ 25x + 7y &= 80 \end{aligned}$$

These two systems have the same coefficients; they differ only in the constant terms.

The inverse of the coefficient matrix

$$A = \begin{bmatrix} 0.5 & 0.3 \\ 25 & 7 \end{bmatrix}$$

may be used to avoid going through the Gauss-Jordan elimination process with each patient.

Here's how $A^{-1}$ may be used to solve a system. Let $AX = B$ be a system for which $A$ actually has an inverse. When both sides of $AX = B$ are multiplied by $A^{-1}$, the equation reduces to

$$\begin{aligned} A^{-1}AX &= A^{-1}B \\ IX &= A^{-1}B \\ X &= A^{-1}B \end{aligned}$$

The product $A^{-1}B$ gives the solution. The solution to such a system exists, and it is unique.

**Example 7**   Use an inverse matrix to solve the system of equations:

$$\begin{aligned} x_1 + 3x_2 + 2x_3 &= 3 \\ 2x_1 + 4x_2 + 2x_3 &= 8 \\ x_1 + 2x_2 - x_3 &= 10 \end{aligned}$$

**Solution**

First, write the system in matrix form, $AX = B$:

$$\begin{bmatrix} 1 & 3 & 2 \\ 2 & 4 & 2 \\ 1 & 2 & -1 \end{bmatrix} \begin{bmatrix} x_1 \\ x_2 \\ x_3 \end{bmatrix} = \begin{bmatrix} 3 \\ 8 \\ 10 \end{bmatrix}$$

In matrix form the solution is

$$\begin{bmatrix} x_1 \\ x_2 \\ x_3 \end{bmatrix} = \begin{bmatrix} 1 & 3 & 2 \\ 2 & 4 & 2 \\ 1 & 2 & -1 \end{bmatrix}^{-1} \begin{bmatrix} 3 \\ 8 \\ 10 \end{bmatrix}$$

The inverse was found in Example 4. Substitute it and obtain

$$\begin{bmatrix} x_1 \\ x_2 \\ x_3 \end{bmatrix} = \begin{bmatrix} -2 & \frac{7}{4} & -\frac{1}{2} \\ 1 & -\frac{3}{4} & \frac{1}{2} \\ 0 & \frac{1}{4} & -\frac{1}{2} \end{bmatrix} \begin{bmatrix} 3 \\ 8 \\ 10 \end{bmatrix} = \begin{bmatrix} 3 \\ 2 \\ -3 \end{bmatrix}$$

The system has the unique solution $x_1 = 3$, $x_2 = 2$, $x_3 = -3$. (Check this solution in each of the original equations.)

**Example 8** Solve the systems

$$AX = B$$

where

$$A = \begin{bmatrix} 1 & 2 \\ 4 & 3 \end{bmatrix} \quad \text{and} \quad X = \begin{bmatrix} x \\ y \end{bmatrix}$$

using

$$B = \begin{bmatrix} 6 \\ 3 \end{bmatrix}, \quad \begin{bmatrix} 10 \\ 15 \end{bmatrix}, \quad \text{and} \quad \begin{bmatrix} 2 \\ 11 \end{bmatrix}$$

**Solution**

First find $A^{-1}$. Adjoin the identity matrix of $A$:

$$\begin{bmatrix} 1 & 2 & | & 1 & 0 \\ 4 & 3 & | & 0 & 1 \end{bmatrix}$$

This reduces to

$$\begin{bmatrix} 1 & 0 & | & -\frac{3}{5} & \frac{2}{5} \\ 0 & 1 & | & \frac{4}{5} & -\frac{1}{5} \end{bmatrix}$$

so the inverse of $A$ is

$$\begin{bmatrix} -\frac{3}{5} & \frac{2}{5} \\ \frac{4}{5} & -\frac{1}{5} \end{bmatrix}$$

**NOTE**

Using the inverse of the coefficient matrix may not be the most efficient way to solve a *single* system of equations. However, some applications require the solution of several systems of equations in which *all* the systems have the same *coefficient matrix*. Using the inverse of the coefficient matrix can be more efficient in this situation.

Using the inverse of the coefficient matrix to solve a single system of equations may be the most efficient way when solving using a computer or graphing calculator.

For $B = \begin{bmatrix} 6 \\ 3 \end{bmatrix}$, the solution is

$$\begin{bmatrix} x \\ y \end{bmatrix} = \begin{bmatrix} -\frac{3}{5} & \frac{2}{5} \\ \frac{4}{5} & -\frac{1}{5} \end{bmatrix} \begin{bmatrix} 6 \\ 3 \end{bmatrix} = \begin{bmatrix} -\frac{12}{5} \\ \frac{21}{5} \end{bmatrix}$$

so $x = -\frac{12}{5}$, $y = \frac{21}{5}$ is the solution.

For $B = \begin{bmatrix} 10 \\ 15 \end{bmatrix}$,

$$\begin{bmatrix} x \\ y \end{bmatrix} = \begin{bmatrix} -\frac{3}{5} & \frac{2}{5} \\ \frac{4}{5} & -\frac{1}{5} \end{bmatrix} \begin{bmatrix} 10 \\ 15 \end{bmatrix} = \begin{bmatrix} 0 \\ 5 \end{bmatrix}$$

For $B = \begin{bmatrix} 2 \\ 11 \end{bmatrix}$,

$$\begin{bmatrix} x \\ y \end{bmatrix} = \begin{bmatrix} -\frac{3}{5} & \frac{2}{5} \\ \frac{4}{5} & -\frac{1}{5} \end{bmatrix} \begin{bmatrix} 2 \\ 11 \end{bmatrix} = \begin{bmatrix} \frac{16}{5} \\ -\frac{3}{5} \end{bmatrix}$$

Use the matrix solution $X = A^{-1}B$ to work the next example.

**Example 9**    Let's return to the earlier example where a doctor prescribed foods containing calcium and iron. Let

$$x = \text{the number of units of food A}$$
$$y = \text{the number of units of food B}$$

where A contains 0.5 mg iron and 25 mg calcium and B contains 0.3 mg iron and 7 mg calcium per unit.

**(a)** Find the amount of each food for patient Jones, who needs 1.3 mg iron and 49 mg calcium.

**(b)** Find the amount of each food for patient Smith, who needs 2.6 mg iron and 106 mg calcium.

**Solution**

**(a)** We need the solution to

$$0.5x + 0.3y = 1.3 \quad \text{(amount of iron)}$$
$$25x + 7y = 49 \quad \text{(amount of calcium)}$$

In matrix form this is

$$\begin{bmatrix} 0.5 & 0.3 \\ 25 & 7 \end{bmatrix} \begin{bmatrix} x \\ y \end{bmatrix} = \begin{bmatrix} 1.3 \\ 49 \end{bmatrix}$$

The inverse of

$$\begin{bmatrix} 0.5 & 0.3 \\ 25 & 7 \end{bmatrix} \quad \text{is} \quad \begin{bmatrix} -1.75 & 0.075 \\ 6.25 & -0.125 \end{bmatrix}$$

The solution to the system is

$$\begin{bmatrix} x \\ y \end{bmatrix} = \begin{bmatrix} -1.75 & 0.075 \\ 6.25 & -0.125 \end{bmatrix}\begin{bmatrix} 1.3 \\ 49 \end{bmatrix} = \begin{bmatrix} -1.75(1.3) + 0.075(49) \\ 6.25(1.3) - 0.125(49) \end{bmatrix}$$

$$= \begin{bmatrix} 1.4 \\ 2.0 \end{bmatrix}$$

so 1.4 units of food A and 2.0 units of food B are required.

**(b)** In this case, the solution is

$$\begin{bmatrix} x \\ y \end{bmatrix} = \begin{bmatrix} -1.75 & 0.075 \\ 6.25 & -0.125 \end{bmatrix}\begin{bmatrix} 2.6 \\ 106.0 \end{bmatrix} = \begin{bmatrix} 3.4 \\ 3.0 \end{bmatrix}$$

# 2.6 EXERCISES

Access end-of-section exercises online at **www.webassign.net**    WebAssign

## Using Your TI Graphing Calculator

### The Inverse of a Matrix

The inverse of a square matrix can be obtained by using the $\boxed{x^{-1}}$ key. For example, to find the inverse of a matrix stored in [A] where

$$[A] = \begin{bmatrix} 3 & 2 & 1 \\ 0 & 4 & 1 \\ 1 & 2 & 1 \end{bmatrix}$$

use [A]$\boxed{x^{-1}}$ $\boxed{\text{ENTER}}$, and the screen will show

```
[A]-1
  [[.5   0   -.5 ]
   [.25  .5  -.75]
   [-1   -1   3  ]]
■
```

Some matrices such as

$$[A] = \begin{bmatrix} 3 & 2 & 2 \\ 0 & 4 & 1 \\ 1 & 2 & 1 \end{bmatrix}$$

have no inverse. In this case, an error message is given:

```
ERR:SINGULAR MAT
1:Quit
2:Goto
```

The term "singular" indicates that the matrix has no inverse.

*(continued)*

## Exercises

Find the inverse of each of the following.

**1.** $\begin{bmatrix} 2 & -1 & 3 \\ 3 & 1 & 2 \\ 4 & 1 & 4 \end{bmatrix}$

**2.** $\begin{bmatrix} 1 & 0 & 1 \\ 0 & 1 & 1 \\ 1 & 1 & 0 \end{bmatrix}$

**3.** $\begin{bmatrix} 4 & -2 & 1 \\ 1 & -4 & -1 \\ 3 & 2 & 2 \end{bmatrix}$

## Using Excel

Excel has the **MINVERSE** command that calculates the inverse of a square matrix. To find the inverse of

$$A = \begin{bmatrix} 1 & 2 & 1 \\ 2 & 4 & 1 \\ 1 & 3 & 2 \end{bmatrix}$$ enter the matrix in cells A2:C4. Next, select the cells E2:G4 for the location of the inverse of

$A$, type =MINVERSE(A2:C4) and simultaneously press **CTRL + SHIFT + ENTER**. The inverse of $A$ then appears in E2:G4.

|   | A | B | C | D | E | F | G |
|---|---|---|---|---|---|---|---|
| 1 | Matrix A in A2:C4 | | | | Invers of A in E2:G4 | | |
| 2 | 1 | 2 | 1 | | 5 | -1 | -2 |
| 3 | 2 | 4 | 1 | | -3 | 1 | 1 |
| 4 | 1 | 3 | 2 | | 2 | -1 | 0 |
| 5 | | | | | | | |
| 6 | If a matrix has no inverse, you will get a matrix like this | | | | | | |
| 7 | #NUM! | #NUM! | #NUM! | | | | |
| 8 | #NUM! | #NUM! | #NUM! | | | | |
| 9 | #NUM! | #NUM! | #NUM! | | | | |

## Exercises

Find the inverse of $A$ in the following exercises.

**1.** $A = \begin{bmatrix} -2 & 6 & 3 \\ 7 & -3 & 1 \\ 9 & 2 & 5 \end{bmatrix}$

**2.** $A = \begin{bmatrix} 1 & 2 & 1 \\ 2 & 5 & 1 \\ 1 & 3 & 2 \end{bmatrix}$

**3.** $A = \begin{bmatrix} -0.4 & 6 & 3.75 \\ 1.4 & -3 & 1.25 \\ 1.8 & 2 & 6.25 \end{bmatrix}$

**4.** $A = \begin{bmatrix} 2 & -1 & 5 & 5 \\ 3 & 5 & 1 & -1 \\ 1 & 3 & 6 & -2 \\ 2 & 2 & -1 & 1 \end{bmatrix}$

**5.** $A = \begin{bmatrix} -2 & 3 & 4 \\ 7 & -3 & 1 \\ 1 & 2 & 5 \end{bmatrix}$

## 2.7  LEONTIEF INPUT–OUTPUT MODEL IN ECONOMICS

- The Leontief Economic Model

The economic health of our country affects each of us in some way. You want a good job upon graduation. The availability of a good job depends on the ability of an economic system to deal with problems that arise. Some problems are challenging indeed. How do we control inflation? How do we avoid a depression? How will a change in interest rates affect my options in buying a house or a car?

A better understanding of the interrelationships between prices, production, interest rates, consumer demand, and the like could improve our ability to deal with economic problems. Matrix theory has been successful in describing mathematical models used to analyze how industries depend on one another in an economic system. Wassily Leontief of Harvard University pioneered work in this area with a massive analysis of the U.S. economic system. As a result of the mathematical models he developed, he received the Nobel Prize in Economics in 1973. Since that time, we have seen the applications of the **Leontief input–output model** mushroom. The model is widely used to study the economic structure of businesses, corporations, and political units like cities, states, and countries. In practice, a large number of variables are required to describe an economic situation. Thus, the problems are quite complicated, so they can best be handled with computers using matrix techniques.

This section introduces the concept of the Leontief economic model, using relatively simple examples to suggest its use in more realistic situations.

### The Leontief Economic Model

In the input–output model, we have an economic system consisting of a number of industries that both produce goods and use goods. Some of the goods produced are used in the industrial processes themselves, and some goods are available to outsiders, the consumers.

To illustrate the input–output model, imagine a simple economy with just two industries: electricity and steel. These industries exist to produce electricity and steel for the consumers. However, both production processes themselves use electricity and steel. The electricity industry uses steel in the generating equipment and uses electricity to light the plant and to heat and cool the buildings. The steel industry uses electricity to run some of its equipment, and that equipment in turn contains steel components.

In the economic model, the quantities in the input–output matrix can be approximate measures of the goods, such as weight, volume, or value. We will describe the quantities in terms of dollar values. We are interested in the quantities of each product needed to provide for the consumers (their demand) and to provide for that consumed internally.

**Example 1**  The amount of electricity and steel consumed by the electric company and the steel company in producing their own products depends on the amount they produce. For example, an electric company may find that whatever the value of electricity produced, 15% of that goes to pay for the electricity used internally and 5% goes to pay for the steel used in production. Thus, if the electric company produces $200,000 worth of electricity, then $30,000 worth (15%) of electricity and $10,000 worth (5%) of steel are consumed by the electric company. Now let's state this concept in a more general form and bring in the cost of producing steel as well.

Let $x$ be the value of electricity produced and $y$ the value of steel produced. The cost of producing electricity includes the following: Of the value of the electricity produced, $x$, 15% of it, $0.15x$, pays for the electricity, and 5%, $0.05x$, pays for the steel consumed internally. Similarly, the cost of producing steel includes the following: Of the value of the steel produced, $y$, 40% of it, $0.40y$, pays for the electricity, and 10% of it, $0.10y$, pays for the steel consumed internally. We can express this information as

$$\text{Total electricity consumed internally} = 0.15x + 0.40y$$
$$\text{Total steel consumed internally} = 0.05x + 0.10y$$

Now note that this can also be expressed as

$$\begin{bmatrix} 0.15 & 0.40 \\ 0.05 & 0.10 \end{bmatrix} \begin{bmatrix} x \\ y \end{bmatrix}$$

The coefficient matrix used here is called the **input–output matrix** of the production model. The *column headings identify the output,* the amount produced, by each industry. The *row headings identify the input,* the amount used in production, by each industry.

$$\text{Input} \begin{pmatrix} \text{Amount used} \\ \text{in production} \end{pmatrix} \quad \begin{array}{cc} & \text{Output} \\ & \text{(Amount produced by)} \\ & \text{Electricity} \quad\; \text{Steel} \end{array}$$

$$\begin{array}{c} \text{Electricity} \\ \text{Steel} \end{array} \begin{bmatrix} 0.15 & 0.40 \\ 0.05 & 0.10 \end{bmatrix} = A$$

There is one row for each industry. The row labeled electricity gives the value of the electricity ($0.15) needed to produce $1 worth of electricity and the value of the electricity ($0.40) needed to produce $1 worth of steel. The second row shows the value of the steel ($0.05) needed to produce $1 worth of electricity and the value of the steel ($0.10) needed to produce $1 worth of steel.

We have shown how the internal consumption of two industries can be represented with matrices. The input–output matrix is the central element of the Leontief economic model. We proceed to show how we can use matrices to answer other questions about an economy. But first, be sure you understand the makeup of the input–output matrix. The entries give the fraction of goods produced by one industry that are used in producing 1 unit of goods in another industry. The row heading of the input–output matrix identifies the industry that provides goods to the industry identified by the column heading.

The electricity–steel input–output matrix

$$\begin{array}{c} \\ \text{E} \\ \text{S} \end{array} \begin{array}{cc} \text{E} & \text{S} \\ \begin{bmatrix} 0.15 & 0.40 \\ 0.05 & 0.10 \end{bmatrix} \end{array}$$

should be interpreted in the following way.

| | User | |
|---|---|---|
| **Supplier** | **Electricity** | **Steel** |
| Electricity | *Electric* industry provides $0.15 worth of electricity to the **electric** industry to produce $1.00 worth of electricity. | *Electric* industry provides $0.40 worth of electricity to the **steel** industry to produce $1.00 worth of steel. |
| Steel | *Steel* industry provides $0.05 worth of steel to the **electric** industry to produce $1.00 worth of electricity. | *Steel* industry provides $0.10 worth of steel to the **steel** industry to produce $1.00 worth of steel. |

**NOTE**

Don't make the error of mixing up the rows and columns of the input–output matrix. Remember that a row in the matrix represents the amounts of *one* material or good used by *all* industries.

You might be wondering why this information is presented in matrix form. It makes it easier to answer questions such as these:

1. The production capacity of industry is $9 million worth of electricity and $7 million worth of steel. How much of each is consumed internally by the production processes?

2. The consumers want $6 million worth of electricity and $8 million worth of steel for their use. How much of each should be produced to satisfy their demands and also to provide for the amounts consumed internally?

Before we learn how to answer these two questions, let's make some observations that will help set up the problems.

First, let $x$ = the dollar value of electricity produced and $y$ = the dollar value of steel produced. These values include that used internally for production and that available to the consumers. Then the total amounts consumed internally are

$$\text{Electricity consumed internally} = 0.15x + 0.40y$$
$$\text{Steel consumed internally} = 0.05x + 0.10y$$

which we expressed in matrix form as

$$\begin{bmatrix} \text{electricity consumed internally} \\ \text{steel consumed internally} \end{bmatrix} = \begin{bmatrix} 0.15 & 0.40 \\ 0.05 & 0.10 \end{bmatrix} \begin{bmatrix} x \\ y \end{bmatrix} = A\begin{bmatrix} x \\ y \end{bmatrix}$$

If production capacities are $9 million worth of electricity ($x = 9$) and $7 million worth of steel ($y = 7$), the amount consumed internally is

$$\begin{bmatrix} 0.15 & 0.40 \\ 0.05 & 0.10 \end{bmatrix} \begin{bmatrix} 9 \\ 7 \end{bmatrix} = \begin{bmatrix} 4.15 \\ 1.15 \end{bmatrix}$$

$4.15 million worth of electricity and $1.15 million worth of steel.

Another fact relates the amount of electricity and steel produced to that available to the consumer:

$$[\text{amount produced}] = \begin{bmatrix} \text{amount consumed} \\ \text{internally} \end{bmatrix} + \begin{bmatrix} \text{amount available} \\ \text{to consumer} \end{bmatrix}$$

We call these matrices **output**, **internal demand**, and **consumer demand** matrices, respectively.

In the case in which $9 million worth of electricity and $7 million worth of steel were produced with $4.15 and $1.15 million consumed internally, we have

$$\underset{\text{Output}}{\begin{bmatrix} 9 \\ 7 \end{bmatrix}} - \underset{\substack{\text{Internal} \\ \text{Demand}}}{\begin{bmatrix} 4.15 \\ 1.15 \end{bmatrix}} = \underset{\text{Consumer Demand}}{\begin{bmatrix} \text{electricity available to consumer} \\ \text{steel available to consumer} \end{bmatrix}}$$

We get $4.85 million and $5.85 million worth of electricity and steel available to the consumers.

If we call the output matrix $X$, the consumer demand matrix $D$, and the input–output matrix $A$, then the internal demand matrix is $AX$ and $X - AX$ is the quantity of goods available for consumers, and so $X - AX = D$ expresses the relation between output, internal demand, and consumer demand.

We can look at the production problem from another perspective, the problem of determining the production needed to provide a known consumer demand. For example, the question, "What total output is necessary to supply consumers with $6 million worth of electricity and $8 million worth of steel?" asks for the output $X$ when consumer demand $D$ is given. Using the same input–output matrix, we want to find $x$ and $y$ (output) so that

$$X - AX = D$$

$$\begin{bmatrix} x \\ y \end{bmatrix} - \begin{bmatrix} 0.15 & 0.40 \\ 0.05 & 0.10 \end{bmatrix} \begin{bmatrix} x \\ y \end{bmatrix} = \begin{bmatrix} 6 \\ 8 \end{bmatrix}$$

Note that the variables $x$ and $y$ appear in two matrices.

Let's use the equation $X - AX = D$ to apply some matrix algebra to find the matrix $X$. You need to solve for the matrix $X$ in

$$X - AX = D$$

This is equivalent to solving for $X$ in the following:

$$X - AX = D$$
$$IX - AX = D$$
$$(I - A)X = D$$
$$X = (I - A)^{-1}D$$

The last equation is the most helpful. To find the total production $X$ that meets the final demand $D$ and also provides the quantities needed to carry out the internal production processes, find the inverse of the matrix $I - A$ (as you did in Section 2.6) and multiply it by the matrix $D$. For example, using

$$A = \begin{bmatrix} 0.15 & 0.40 \\ 0.05 & 0.10 \end{bmatrix}$$

$$I - A = \begin{bmatrix} 0.85 & -0.40 \\ -0.05 & 0.90 \end{bmatrix}$$

and

$$(I - A)^{-1} = \begin{bmatrix} 1.208 & 0.537 \\ 0.0671 & 1.141 \end{bmatrix}$$

where the entries in $(I - A)^{-1}$ are rounded. For the demand matrix,

$$D = \begin{bmatrix} 6 \\ 8 \end{bmatrix}$$

$$X = \begin{bmatrix} 1.208 & 0.537 \\ 0.0671 & 1.141 \end{bmatrix} \begin{bmatrix} 6 \\ 8 \end{bmatrix} = \begin{bmatrix} 11.544 \\ 9.531 \end{bmatrix}$$

So, $11.544 million worth of electricity and $9.531 million worth of steel must be produced to provide $6 million worth of electricity and $8 million worth of steel to the consumers and to provide for the electricity and steel used internally in production.

**Leontief Input–Output Model**

The matrix equation for the Leontief input–output model that relates total production to the internal demands of the industries and to consumer demand is given by

$$X - AX = D$$

or the equivalent,

$$(I - A)X = D$$

where $A$ is the input–output matrix giving information on internal demands, $D$ represents consumer demands, and $X$ represents the total goods produced.

The solution to $(I - A)X = D$ is

$$X = (I - A)^{-1}D \quad [\text{provided } (I - A)^{-1} \text{ exists}]$$

**Example 2**   An input–output matrix for electricity and steel is

$$A = \begin{bmatrix} 0.25 & 0.20 \\ 0.50 & 0.20 \end{bmatrix}$$

**(a)** If the production capacity of electricity is $15 million and the production capacity for steel is $20 million, how much of each is consumed internally for capacity production?

**(b)** How much electricity and steel must be produced to have $5 million worth of electricity and $8 million worth of steel available for consumer use?

**Solution**

**(a)** We are given

$$A = \begin{bmatrix} 0.25 & 0.20 \\ 0.50 & 0.20 \end{bmatrix} \quad \text{and} \quad X = \begin{bmatrix} 15 \\ 20 \end{bmatrix}$$

We want to find $AX$:

$$AX = \begin{bmatrix} 0.25 & 0.20 \\ 0.50 & 0.20 \end{bmatrix}\begin{bmatrix} 15 \\ 20 \end{bmatrix} = \begin{bmatrix} 7.75 \\ 11.50 \end{bmatrix}$$

So, $7.75 million worth of electricity and $11.50 million worth of steel are consumed internally.

**(b)** We are given

$$A = \begin{bmatrix} 0.25 & 0.20 \\ 0.50 & 0.20 \end{bmatrix} \quad \text{and} \quad D = \begin{bmatrix} 5 \\ 8 \end{bmatrix}$$

and we need to solve for $X$ in $(I - A)^{-1}D = X$ or in $(I - A)X = D$. We use the latter this time:

$$I - A = \begin{bmatrix} 0.75 & -0.20 \\ -0.50 & 0.80 \end{bmatrix}$$

Then the augmented matrix for $(I - A)X = D$ is

$$\left[\begin{array}{cc|c} 0.75 & -0.20 & 5 \\ -0.50 & 0.80 & 8 \end{array}\right]$$

which reduces to

$$\left[\begin{array}{cc|c} 1 & 0 & 11.2 \\ 0 & 1 & 17.0 \end{array}\right] \qquad \text{(Check it.)}$$

The two industries must produce $11.2 million worth of electricity and $17.0 million worth of steel to have $5 million worth of electricity and $8 million worth of steel available to the consumers.

**Example 3**    Fantasy Island has an economy of three industries with the input–output matrix, $A$. Help the Industrial Planning Commission by computing the output levels of each industry to meet the demands of the consumers and the other industries for each of the two demand levels given. The units of $D$ are millions of dollars.

$$A = \begin{bmatrix} 0.3 & 0.3 & 0.2 \\ 0.4 & 0.4 & 0 \\ 0 & 0 & 0.2 \end{bmatrix} \quad D = \begin{bmatrix} 6 \\ 9 \\ 12 \end{bmatrix}, \begin{bmatrix} 12 \\ 15 \\ 18 \end{bmatrix}$$

**Solution**

We need to find the values of $X$ that correspond to each $D$. That comes from the solution of

$$X = (I - A)^{-1}D$$

For the given matrix $A$,

$$I - A = \begin{bmatrix} 1 & 0 & 0 \\ 0 & 1 & 0 \\ 0 & 0 & 1 \end{bmatrix} - \begin{bmatrix} 0.3 & 0.3 & 0.2 \\ 0.4 & 0.4 & 0 \\ 0 & 0 & 0.2 \end{bmatrix} = \begin{bmatrix} 0.7 & -0.3 & -0.2 \\ -0.4 & 0.6 & 0 \\ 0 & 0 & 0.8 \end{bmatrix}$$

We can find $(I - A)^{-1}$ by using Gauss-Jordan elimination:

$$(I - A)^{-1} = \begin{bmatrix} 2 & 1 & \frac{1}{2} \\ \frac{4}{3} & \frac{7}{3} & \frac{1}{3} \\ 0 & 0 & \frac{5}{4} \end{bmatrix}$$

We can obtain $X = (I - A)^{-1}D$ with one matrix multiplication by forming a matrix of two columns using the two $D$ matrices as columns.

$$X = \begin{bmatrix} 2 & 1 & \frac{1}{2} \\ \frac{4}{3} & \frac{7}{3} & \frac{1}{3} \\ 0 & 0 & \frac{5}{4} \end{bmatrix} \begin{bmatrix} 6 & 12 \\ 9 & 15 \\ 12 & 18 \end{bmatrix} = \begin{bmatrix} 27 & 48 \\ 33 & 57 \\ 15 & 22.5 \end{bmatrix}$$

$$\begin{array}{ccc} (I - A)^{-1} & \text{Values} & \text{Corresponding} \\ & \text{of } D & \text{Outputs} \end{array}$$

The output levels required to meet the demands $\begin{bmatrix} 6 \\ 9 \\ 12 \end{bmatrix}$ and $\begin{bmatrix} 12 \\ 15 \\ 18 \end{bmatrix}$ are $\begin{bmatrix} 27 \\ 33 \\ 15 \end{bmatrix}$

and $\begin{bmatrix} 48 \\ 57 \\ 22.5 \end{bmatrix}$, respectively, with the units being millions of dollars.

**Example 4**    Hubbs, Inc., has three divisions: grain, lumber, and energy. An analysis of their operations reveals the following information. For each $1.00 worth of grain produced, they use $0.10 worth of grain, $0.20 worth of lumber, and $0.50 worth of energy. For each $1.00 worth of lumber produced, they use $0.10 worth of grain, $0.15 worth of lumber, and $0.40 worth of energy. For each $1.00 worth of energy produced, they use $0.05 worth of grain, $0.35 worth of lumber, and $0.15 worth of energy.

**(a)** How much energy is used in the production of $750,000 worth of lumber?

**(b)** Which division uses the largest amount of energy per unit of goods produced? The least?

**(c)** Set up the input–output matrix.

**(d)** Find the cost of producing $1.00 worth of lumber.

**(e)** Find the internal demand if the production levels are $640,000 worth of grain, $800,000 worth of lumber, and $980,000 worth of energy.

**(f)** Find the total production required in order to have the following available to consumers: $300,000 worth of grain, $500,000 worth of lumber, and $800,000 worth of energy.

### Solution

**(a)** Since $0.40 worth of energy is required to produce $1.00 worth of lumber, $0.40(750,000) = $300,000 worth of energy is required.

**(b)** The largest amount of energy per unit is $0.50 worth of energy used to produce $1.00 worth of grain. The least amount of energy per unit is $0.15 worth of energy used to produce $1.00 worth of energy.

**(c)** Let  $x_1$ = total value of grain produced
$x_2$ = total value of lumber produced
$x_3$ = total value of energy produced

The value of grain used internally is

$$0.10x_1 + 0.10x_2 + 0.05x_3$$

The value of lumber used internally is

$$0.20x_1 + 0.15x_2 + 0.35x_3$$

The value of energy used internally is

$$0.50x_1 + 0.40x_2 + 0.15x_3$$

so the input–output matrix is

$$
\begin{array}{c}
 \\
 \\
\text{Supplier}
\end{array}
\begin{array}{c}
 \\
\text{G} \\
\text{L} \\
\text{E}
\end{array}
\overset{\displaystyle \text{User}}{
\overset{\text{G} \quad \text{L} \quad \text{E}}{
\begin{bmatrix}
0.10 & 0.10 & 0.05 \\
0.20 & 0.15 & 0.35 \\
0.50 & 0.40 & 0.15
\end{bmatrix}}}
$$

**(d)** The total cost of producing $1.00 worth of lumber is found by totaling the entries in column 2 of the input–output matrix, because those entries represent the value of G, L, and E to produce $1.00 worth of lumber. Total cost = $0.65.

**(e)** The internal demand is given by

$$
\begin{bmatrix}
0.10 & 0.10 & 0.05 \\
0.20 & 0.15 & 0.35 \\
0.50 & 0.40 & 0.15
\end{bmatrix}
\begin{bmatrix}
640,000 \\
800,000 \\
980,000
\end{bmatrix}
=
\begin{bmatrix}
193,000 \\
591,000 \\
787,000
\end{bmatrix}
$$

The internal consumption of the system is $193,000 worth of grain, $591,000 worth of lumber, and $787,000 worth of energy to produce a total of $640,000 worth of grain, $800,000 worth of lumber, and $980,000 worth of energy.

**(f)** In this case,

$$D = \begin{bmatrix} 300,000 \\ 500,000 \\ 800,000 \end{bmatrix}$$

so we need to solve $X = (I - A)^{-1}D$.

$$\left( \begin{bmatrix} 1 & 0 & 0 \\ 0 & 1 & 0 \\ 0 & 0 & 1 \end{bmatrix} - \begin{bmatrix} 0.10 & 0.10 & 0.05 \\ 0.20 & 0.15 & 0.35 \\ 0.50 & 0.40 & 0.15 \end{bmatrix} \right)^{-1} \begin{bmatrix} 300,000 \\ 500,000 \\ 800,000 \end{bmatrix} = \begin{bmatrix} x \\ y \\ z \end{bmatrix}$$

$$\begin{bmatrix} 0.90 & -0.10 & -0.05 \\ -0.20 & 0.85 & -0.35 \\ -0.50 & -0.40 & 0.85 \end{bmatrix}^{-1} \begin{bmatrix} 300,000 \\ 500,000 \\ 800,000 \end{bmatrix} = \begin{bmatrix} x \\ y \\ z \end{bmatrix}$$

With entries rounded to three decimal places, $(I - A)^{-1}$ is used.

$$\begin{bmatrix} 1.254 & 0.226 & 0.167 \\ 0.743 & 1.593 & 0.700 \\ 1.087 & 0.883 & 1.604 \end{bmatrix} \begin{bmatrix} 300,000 \\ 500,000 \\ 800,000 \end{bmatrix} = \begin{bmatrix} 622,800 \\ 1,579,400 \\ 2,050,800 \end{bmatrix}$$

Hubbs, Inc., needs to produce $622,800 worth of grain, $1,579,400 worth of lumber, and $2,050,800 worth of energy to meet the consumer demand specified.

———•

Today the concept of a world economy has become a reality. In 1973, the United Nations commissioned an input–output model of the world economy. The aim of the model was to transform the vast collection of economic facts that describe the world economy into an organized system from which economic projections could and have been made. In the model, the world is divided into 15 distinct geographic regions, each one described by an individual input–output matrix. The regions are then linked by a larger matrix that is used in an input–output model. Overall, more than 200 variables enter into the model, and the computations are of course done on a computer. By feeding in projected values for certain variables, researchers use the model to create scenarios of future world economic possibilities.

We need to understand that this model is not a crystal ball that shows exactly what the future holds. It gives an indication of situations that can develop *if trends continue unchanged.* We can change the trends and thereby alter future conditions. For example, the model predicted energy problems of major negative consequences by 2025. Conservation, more efficient automobile design, and recycling have helped change the energy use patterns that were in place in 1973. Although energy problems have not vanished, they are different.

Mathematical models can provide an "early warning" of what might be and allow policymakers to make adjustments to avoid or soften potential problems.

## 2.7 EXERCISES

Access end-of-section exercises online at **www.webassign.net**   ENHANCED WebAssign

## Using Your TI Graphing Calculator

Matrix multiplication can assist in the computation of input–output models. Let's illustrate with the input–output matrix

$$A = \begin{bmatrix} 0.3 & 0.3 & 0.2 \\ 0.4 & 0.4 & 0 \\ 0 & 0 & 0.2 \end{bmatrix} \quad \text{stored in } [A]$$

### Internal Demand

To compute the internal demand given the input–output matrix $[A]$ and the output matrix $[B] = \begin{bmatrix} 28 \\ 31 \\ 25 \end{bmatrix}$, calculate the internal demand with

$$[A]\,[B]\ \boxed{\text{ENTER}}\ \text{which gives}\ \begin{bmatrix} 22.7 \\ 23.6 \\ 5.0 \end{bmatrix}$$

### Find Output Given Demand

Given the input–output matrix A and $[D] = \begin{bmatrix} 50 & 36 \\ 44 & 53 \\ 35 & 28 \end{bmatrix}$, find the production matrix $X$ with $(I - [A])^{-1}[D] = X$.

Store the $3 \times 3$ identity matrix in $[C]$ and find $X$ by

$$([C] - [A])\,\boxed{x^{-1}}\,[D]\ \text{which gives}\ \begin{bmatrix} 161.5 & 139 \\ 181.0 & 181 \\ 43.75 & 35 \end{bmatrix}$$

## Using Excel

Let's illustrate how to use Excel in input–output calculations. We use the input–output matrix

$$A = \begin{bmatrix} 0.3 & 0.3 & 0.2 \\ 0.4 & 0.4 & 0 \\ 0 & 0 & 0.2 \end{bmatrix}.$$

### Internal Demand

To find the internal demand for the output matrix $X = \begin{bmatrix} 28 \\ 31 \\ 25 \end{bmatrix}$, do the following: Enter $A$ in the cells A2:C4 and $X$ in

*(continued)*

cells E2:E4. Since $AX$ will be a column matrix with three rows, select cells G2:G4 and type =MMULT(A2:C4, E2:E4); then press CTRL+SHIFT+ENTER, which gives the internal demand matrix

$$\begin{bmatrix} 22.7 \\ 23.6 \\ 5.0 \end{bmatrix}$$

## Find Output Given Demand

When given the demand matrix $D = \begin{bmatrix} 50 & 36 \\ 44 & 53 \\ 35 & 28 \end{bmatrix}$, we find the output matrix $(I - [A])^{-1}[D] = X$ as follows.

- Enter $A$ in cells A2:C4.

- Enter $I = \begin{bmatrix} 1 & 0 & 0 \\ 0 & 1 & 0 \\ 0 & 0 & 1 \end{bmatrix}$ in cells A6:C8.

- Enter $D$ in cells E6:F8.
- Next compute $I - A$ and place the result in A10:C12. Type =A6-A2 in A10 and drag A10 to the rest of the cells in A10:C12, which gives

$$\begin{bmatrix} 0.7 & -0.3 & -0.2 \\ -0.4 & 0.6 & 0 \\ 0 & 0 & 0.8 \end{bmatrix}$$

- Find $(I - A)^{-1}$ and store it in cells A14:C16 by selecting the cells A14:C16 and then typing

  =MINVERSE(A10:C12), which gives

$$\begin{bmatrix} 2 & 1 & 0.5 \\ 1.333 & 2.333 & 0.333 \\ 0 & 0 & 1.25 \end{bmatrix}$$

- Multiply $(I - A)^{-1} D$ and store the result in A18:B20 by selecting the cells A18:B20 and typing

  =MMULT(A14:C16,E6:F8); then press CTRL+SHIFT+ENTER. This gives the result

$$\begin{bmatrix} 161.5 & 139 \\ 181.0 & 181 \\ 43.75 & 35 \end{bmatrix}$$

## 2.8    LINEAR REGRESSION

Business, industry, and governments are interested in answers to such questions as, "How many cell phones will be needed in the next three years?" "What will be the population growth of Boulder, Colorado, over the next five years?" "Will the Midway School District need to build an additional elementary school within three years?" "How much will the number of 18-wheelers on I-35 increase over the next decade?"

Answers to these kinds of questions are important because it may take years to build the manufacturing plants, schools, or highways. In some cases, information from past years can indicate a trend from which reasonable estimates of future growth can be obtained. For example, the problem of air quality has been addressed by cities in recent years. The cities of Los Angeles and Long Beach, California have made significant progress, as shown by the table, which gives the number of days during the year that the Air Quality Index exceeded 100. As the index exceeds 100, the environment becomes unhealthy for sensitive groups.

| Year | Days |
|------|------|
| 1994 | 139 |
| 1995 | 113 |
| 1996 | 94 |
| 1997 | 60 |
| 1998 | 56 |
| 1999 | 27 |

When we plot this information using the year (1 for 1994, 2 for 1995, and so on) for the $x$-value and the number of days for the $y$-value, we get the graph of Figure 2–10.

We call a graph of points such as this a **scatter plot**. Notice that the points have a general, approximate linear downward trend, even though the points do not lie on a line. Even when the pattern of data points is not exactly linear, it may be useful to approximate their trend with a line so that we can estimate future behavior.

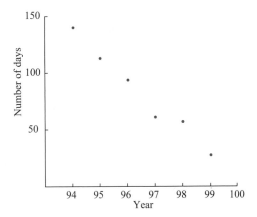

**FIGURE 2–10**

There may be cases when we believe the variables are related in a linear manner, but the data deviate from a line because (1) the data collected may not be accurate or (2) the assumption of a linear relationship is not valid. In either case, it may be useful to find the line that best approximates the trend and use it to obtain additional information and make predictions.

Figure 2–11 (on the next page) shows a line that approximates the trend of the points in Figure 2–10.

Although the line seems to be a reasonable representation of the general trend, we would like to know if this is the *best* approximation of the trend. Mathematicians use the **least squares line**, also called the **regression line**, for the line that best fits the data. In Figure 2–11, the line shown is the least squares, or regression, line $y = -21.9x + 158$.

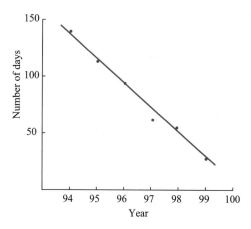

**FIGURE 2–11**

You have not been told how to find that equation. Before doing so, let's look at the idea behind the least squares procedure.

In Figure 2–12, we look at a simple case where we have drawn a line in the general direction of the trend of the scatter plot of the four points $P_1$, $P_2$, $P_3$, and $P_4$. For each of the points, we have indicated the vertical distance from each point to the line and labeled the distances $d_1$, $d_2$, $d_3$, and $d_4$. The distances to points above the line will be positive, and the distances to the points below the line will be negative. The basic idea of the least squares procedure seeks to find a line that somehow minimizes the entirety of these distances. To do so, the method finds the line that gives the smallest possible sum of the *squares* of the $d$'s—that is, the line $y = mx + b$ that makes

$$d_1{}^2 + d_2{}^2 + d_3{}^2 + d_4{}^2$$

the least value possible.

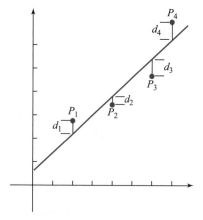

**FIGURE 2–12**

The following example shows the method.

**Example 1**   We find the values of $m$ and $b$ of the regression line $y = mx + b$ from a system of two equations in the variables $m$ and $b$. Let's illustrate the procedure using the points (2, 5), (3, 7), (5, 9), and (6, 11). The scatter plot is shown in Figure 2–13.

The system takes the form

$$Am + Bb = C$$
$$Dm + Eb = F$$

where

$A$ = the sum of the squares of the $x$-coordinates; in this case,
    $2^2 + 3^2 + 5^2 + 6^2 = 74$.

$B$ = the sum of the $x$-coordinates of the given points; in
    this case, $2 + 3 + 5 + 6 = 16$.

$C$ = the sum of the products of the $x$- and $y$-coordinates of the given points; in this
    case, $(2 \times 5) + (3 \times 7) + (5 \times 9) + (6 \times 11) = 142$.

$D = B$, the sum of the $x$-coordinates, 16.

$E$ = the number of given points, in this case, four points.

$F$ = the sum of the $y$-coordinates; in this case, $5 + 7 + 9 + 11 = 32$.

Thus, the solution to the system

$$74m + 16b = 142$$
$$16m + 4b = 32$$

gives the coefficients of the regression line that best fits the four given points. The solution is $m = 1.4$ and $b = 2.4$ (be sure you can find the solution), and the linear regression line is $y = 1.4x + 2.4$ (see Figure 2–14).

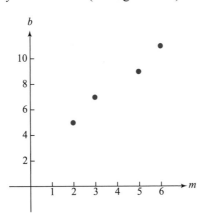

**FIGURE 2–13**

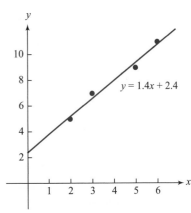

**FIGURE 2–14**

**Least Squares Line**   The linear function $y = mx + b$ is the **least squares line** for the points $(x_1, y_1)$, $(x_2, y_2), \dots, (x_n, y_n)$ when $m$ and $b$ are solutions of the system of equations.

$$(x_1^2 + x_2^2 + \cdots + x_n^2)m + (x_1 + x_2 + \cdots + x_n)b = (x_1 y_1 + x_2 y_2 + \cdots + x_n y_n)$$
$$(x_1 + x_2 + \cdots + x_n)m + \qquad\qquad nb = (y_1 + y_2 + \cdots + y_n)$$

We now show you two ways to organize the data in Example 1 that makes it easier to keep track of the computations needed to find the coefficients of the system.

**Method I.** Form a table like the following.

| | $x$ | $y$ | $x^2$ | $xy$ |
|---|---|---|---|---|
| | 2 | 5 | 4 | 10 |
| | 3 | 7 | 9 | 21 |
| | 5 | 9 | 25 | 45 |
| | 6 | 11 | 36 | 66 |
| Sum | 16 | 32 | 74 | 142 |

This gives the system

$$74m + 16b = 142$$
$$16m + 4b = 32$$

**Method II.** We find the augmented matrix $A$ of the system of equations, with a matrix product. For this example, it is

$$A = \begin{bmatrix} 2 & 3 & 5 & 6 \\ 1 & 1 & 1 & 1 \end{bmatrix} \begin{bmatrix} 2 & 1 & 5 \\ 3 & 1 & 7 \\ 5 & 1 & 9 \\ 6 & 1 & 11 \end{bmatrix} = \begin{bmatrix} 74 & 16 & 142 \\ 16 & 4 & 32 \end{bmatrix}$$

We state the general case as follows: Given the points $(x_1, y_1), (x_2, y_2), \ldots, (x_n, y_n)$, the augmented matrix $M$ of the system

$$Am + Bb = C$$
$$Dm + Eb = F$$

whose solution gives the least squares line of best fit for the given points is the product

$$M = \begin{bmatrix} x_1 & x_2 \ldots x_n \\ 1 & 1 \ldots 1 \end{bmatrix} \begin{bmatrix} x_1 & 1 & y_1 \\ x_2 & 1 & y_2 \\ \vdots & \vdots & \vdots \\ x_n & 1 & y_n \end{bmatrix}$$

This gives a useful method for a calculator with matrix operations.

Let's use the data on the number of cell phones in use in the United States and work an example.

**Example 2** A wireless industry survey estimates the number of cell-phone subscribers in the United States for 1998–2003 as follows.

| Year | Cell-phone subscribers (in millions) |
|---|---|
| 1998 | 69.2 |
| 1999 | 86.0 |
| 2000 | 109.5 |
| 2001 | 128.4 |
| 2002 | 140.8 |
| 2003 | 158.7 |

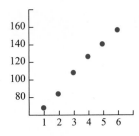

**FIGURE 2–15**

Find the scatter plot and the least squares line $y = mx + b$, where $x$ is the year ($x = 1$ for 1998, $x = 2$ for 1999, etc.) and $y$ is the number of subscribers.

**Solution**

We will find the coefficients of the system of equations using both methods. Figure 2–15 shows the scatter plot.

   **Method I.**

| $x$ | $y$ | $x^2$ | $xy$ |
|---|---|---|---|
| 1 | 69.2 | 1 | 69.2 |
| 2 | 86.0 | 4 | 172.0 |
| 3 | 109.5 | 9 | 328.5 |
| 4 | 128.4 | 16 | 513.6 |
| 5 | 140.8 | 25 | 704.0 |
| 6 | 158.7 | 36 | 952.2 |
| Sum   21 | 692.6 | 91 | 2739.5 |

The system of equations is

$$91m + 21b = 2739.5$$
$$21m + 6b = 692.6$$

Multiplying the first equation by 2 and the second equation by 7, we obtain the system

$$182m + 42b = 5479.0$$
$$\underline{147m + 42b = 4848.2} \quad \text{(subtract)}$$
$$35m = 630.8$$
$$m = 18.02$$
$$b = \frac{692.6 - 21(18.02)}{6} = \frac{314.18}{6} = 52.36$$

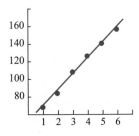

**FIGURE 2–16**

The least squares line is $y = 18.02x + 52.36$. Its graph is shown on the scatter plot in Figure 2–16.
   **Method II.**   Using matrices, we can write the augmented matrix of the least squares system as

$$A = \begin{bmatrix} 1 & 2 & 3 & 4 & 5 & 6 \\ 1 & 1 & 1 & 1 & 1 & 1 \end{bmatrix} \begin{bmatrix} 1 & 1 & 69.2 \\ 2 & 1 & 86.0 \\ 3 & 1 & 109.5 \\ 4 & 1 & 128.4 \\ 5 & 1 & 140.8 \\ 6 & 1 & 158.7 \end{bmatrix} = \begin{bmatrix} 91 & 21 & 2739.5 \\ 21 & 6 & 692.6 \end{bmatrix}$$

Using this augmented matrix, we obtain the same line as in Method I:

$$y = 18.02x + 52.36.$$

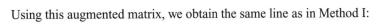

## **2.8**   EXERCISES

Access end-of-section exercises online at **www.webassign.net**   <sup>ENHANCED</sup> Web**Assign**

# Using Your TI Graphing Calculator

## Regression Lines

A TI graphing calculator can be used to find the equation of the regression line. We illustrate with the points $(2, 5)$, $(4, 6)$, $(6, 7)$, and $(7, 9)$.

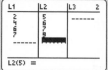

Enter the points in the lists L1 and L2 with the $x$-coordinates in L1 and the $y$-coordinates in L2.

To obtain the regression coefficients, use

$$\boxed{\text{STAT}} \ <\text{CALC}> \ <4:\text{LinReg(ax + b)}> \ \boxed{\text{L1}}, \boxed{\text{L2}} \ \boxed{\text{ENTER}}$$

This sequence of commands will give the following screens:

The last screen indicates that the least squares line is $y = 0.7288x + 3.288$.

## Exercises

1. Find the least squares line for the points $(15, 22)$, $(17, 25)$, and $(18, 27)$.
2. Find the least squares line for the points $(21, 44)$, $(24, 40)$, $(26, 38)$, and $(30, 32)$.
3. Find the least squares line for the points $(3.2, 5.7)$, $(4.1, 6.3)$, $(5.3, 6.7)$, and $(6.0, 7.2)$.

## Scatter Plot

The scatter plot of a set of points may be obtained by the following steps.
1. **Enter data**. Enter the points with the $x$-coordinates in the list L1 and the corresponding $y$-coordinates in L2.
2. **Set the horizontal and vertical scales**. Set Xmin, Xmax, Xscl, Ymin, Ymax, and Yscl using $\boxed{\text{WINDOW}}$ in the same way it is used to set the screen for graphing functions.
3. **Define the scatter plot**. Press $\boxed{\text{STAT PLOT}}$ $<1:\text{PLOT1}>$ $\boxed{\text{ENTER}}$

   You will see a screen similar to the following.

Turn on plot — Select scatter plot

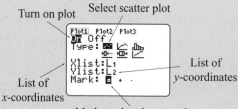

List of $x$-coordinates — List of $y$-coordinates — Mark used to denote points

On that screen select, as shown in the figure, $<$ON$>$, scatter plot for Type, L1 for the list of $x$-coordinates, L2 for the list of $y$-coordinates, and the kind of mark that will show the location of the points. Press $\boxed{\text{ENTER}}$ after each selection.

**4. Display the Scatter plot.**

Press GRAPH.

## Example

Show the scatter plot of the points (3, 5), (4, 6), (1, 2), (5, 5).

Set the Window to (0, 10) for the $x$ and $y$-ranges.

Enter the points in the L1 and L2 lists.

Graph the scatter plot.

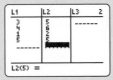
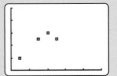

 **Using Excel**

## To Draw and Find a Linear Regression Line

We can use features of Excel to obtain a scatter plot of points that are given, find the equation of the regression line, and graph the line. We illustrate with the points (1, 3), (2, 5), (4, 9), (5, 8).

- For the given points, enter the $x$-values in cells A2:A5 and the corresponding $y$-values in cells B2:B5.
- Select the cells A2:B5
- Click on the **Insert** tab. Move to the **Charts** group. Select **Scatter** with the scatter sub-type.

- A graph will be inserted in the worksheet, which is now in the **Chart Tools** environment.

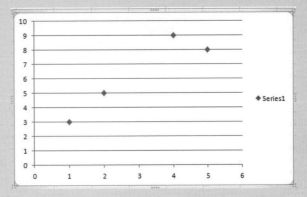

*(continued)*

- Now enter some names and labels. Click on the **Layout** tab under Chart Tools. Select **Chart Title** and the **Above Chart** option. In **Chart Title**, enter the name of the line, such as *Regression*.

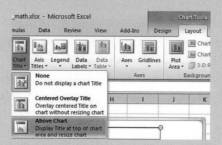

- Under **Axis Titles**, enter a horizontal axis title for what $x$ represents, such as "Number of ___." For a vertical axis title, identify what $y$ represents.
- The "Series 1" legend may be removed by clicking on it and deleting it.

  Now, we are ready to add the regression line.

- Right click on one of the data points in the graph.
- Click on **Add Trendline**.
- In the dialog box, click on Linear and check the box to **Display equation on chart**.

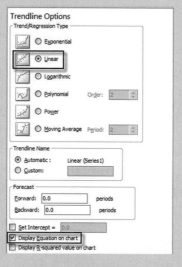

- Click **Close**.

  The line and its equation, $y = 1.4x + 2.05$, are shown on the chart.

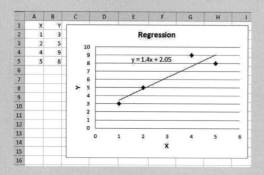

## To Find Regression Line with Systems of Equations

- Make headings for $x, y, x^2$ and $xy$ in cells B1:E1.
- For the given points, enter the $x$-values in cells B2:B5 and the corresponding $y$-values in cells C2:C5.
- In cell D2, type the formula $=$**B2^2**, for $x^2$. Copy this formula to D3:D5.
- In cell E2, type the formula $=$**B2*C2**, for $xy$. Copy this formula to E3:E5.
- Type the headings **Sum** in A6. In B6, enter the formula $=$**SUM(B2:B5)**. Copy this formula across to C6:E6.

| | A | B | C | D | E |
|---|---|---|---|---|---|
| 1 | | $x$ | $y$ | $x^2$ | $xy$ |
| 2 | | 1 | 3 | =B2^2 | =B2*C2 |
| 3 | | 2 | 5 | =B3^2 | =B3*C3 |
| 4 | | 4 | 9 | =B4^2 | =B4*C4 |
| 5 | | 5 | 8 | =B5^2 | =B5*C5 |
| 6 | Sum | =SUM(B2:B5) | =SUM(C2:C5) | =SUM(D2:D5) | =SUM(E2:E5) |

- Next, set up the matrix for the system of equations using the sums. In A9, type $=$**D6**. In B9 and A10, type $=$**B6**. Type 4 in B10, since $n = 4$, the number of data points.
- The right-hand side is typed in as follows: in C9, type $=$**E6** and in C10, type $=$**C6**. The result is shown, with formulas and without.

| | A | B | C |
|---|---|---|---|
| 9 | =D6 | =B6 | =E6 |
| 10 | =B6 | 4 | =C6 |

| | A | B | C |
|---|---|---|---|
| 9 | 46 | 12 | 89 |
| 10 | 12 | 4 | 25 |

- Calculate the inverse of the matrix in A9:B10. Select cells E9:F10. Type $=$**MINVERSE(A9:B10)** and press CTRL-SHIFT-ENTER, all at the same time.

| | A | B | C | D | E | F |
|---|---|---|---|---|---|---|
| 8 | | | | | Inverse | |
| 9 | 46 | 12 | 89 | | 0.1 | -0.3 |
| 10 | 12 | 4 | 25 | | -0.3 | 1.15 |

- Calculate $m$ and $b$ by multiplying the matrix inverse in E9:F10 by the right-hand side of the equation, in cells C9:C10. Select the cells B12:B13, type $=$**MMULT(E9:F10,C9:C10)**, and press CTRL-SHIFT-ENTER, all at the same time.

| | A | B | C |
|---|---|---|---|
| 11 | | | |
| 12 | m | 1.4 | |
| 13 | b | 2.05 | |

- From the values in B12:B13, the equation of the regression line is $y = 1.4x + 2.05$.

## Exercises

Find the least squares regression line for the points given.
1. $(2, 4), (4, 5), (5, 6), (8, 11)$.
2. $(10, 13), (13, 11), (14, 10), (17, 6), (20, 3)$.
3. $(31, 6), (33, 9), (37, 13), (40, 15)$.

## IMPORTANT TERMS

**2.1**
System of Equations
Solution of a System
Substitution Method
Elimination Method
Equivalent System
Inconsistent System
Many Solutions to a System
Parameter
Parametric Form of a Solution
Supply and Demand
Equilibrium Price

**2.2**
Matrix
Elements of the Matrix
Row Matrix
Column Matrix
Coefficient Matrix
Augmented Matrix

Row Operations
Equivalent Augmented Matrices
Back Substitution
Gauss-Jordan Method
Diagonal Locations
Pivoting

**2.3**
Reduced Echelon Form
No Solution
Many Solutions
Unique Solution

**2.4**
Size of Matrix
Square Matrix
Equal Matrices
Addition of Matrices
Scalar Multiplication

**2.5**
Dot Product
Matrix Multiplication
Identity Matrix

**2.6**
Main Diagonal of Matrix
Inverse Matrix
Matrix Equation

**2.7**
Leontief Input–Output Model
Input–Output Matrix
Output Matrix
Internal Demand Matrix
Consumer Demand Matrix

**2.8**
Linear Regression
Scatter Plot
Least Squares Line

## IMPORTANT CONCEPTS

**Elimination Method**    Solving a system of equations by:

   1. Multiplying one or both equations by an appropriate constant so that a variable in one equation has coefficient $c$ and the coefficient of the same variable in the other equation has coefficient $-c$.
   2. Adding the equations to eliminate the variable.
   3. Replacing one of the equations by the resulting equation.

**Inconsistent System**    A system of equations that has no solutions.

**Supply and Demand Analysis**    An estimate of the price at which consumer demand of an item equals the supply.

**Equilibrium Price**    The price at which consumer demand equals supply.

**Matrix**    A rectangular array of numbers.

**Matrix Row operations**
   1. Interchange two rows.
   2. Multiply or divide a row by a nonzero constant.
   3. Multiply a row by a constant and add or subtract it from another row.

**Equivalent Matrices**    One matrix is obtained from the other using row operations.

**Gauss-Jordan Method**    Solve a system of equations using row operations on the augmented matrix. (See Section 2.2)

| | |
|---|---|
| **Number of Solutions** | The number of solutions of a system of equations can be determined from the reduced augmented matrix: **No solution** occurs when a row has all zeroes to the left of the vertical line and a nonzero number to the right. **Unique solution** occurs when the number of nonzero rows equals the number of variables. **Infinite number of solutions** occurs when the number of nonzero rows is less than the number of variables. |
| **Matrix Addition** | Add corresponding elements of the two matrices. The matrices must be the same size. |
| **Scalar Multiplication** | Multiply a matrix by a constant by multiplying each element of the matrix by the constant. |
| **Matrix Multiplication** | The matrix product $AB$ is possible if the number of columns of $A$ equal the number of rows of $B$. (See section 2.5) |
| **Inverse of a Matrix** | $B$ is the inverse of matrix of $A$ if $AB = BA = I$ (See section 2.6) |
| **Least Squares Line** | A line that models a set of nonlinear points. (See section 2.8) |

## 2 REVIEW EXERCISES

**Solve the systems in Exercises 1 and 2 by substitution.**

**1.** $3x + 2y = 5$
$\quad 2x + 4y = 9$

**2.** $x + 5y = 2$
$\quad 3x - 7y = 12$

**Solve the systems in Exercises 3 through 6 by elimination.**

**3.** $5x - y = 34$
$\quad 2x + 3y = 0$

**4.** $x + 3y - 2z = -15$
$\quad 4x - 3y + 5z = 50$
$\quad 3x + 2y - 2z = -4$

**5.** $x - 2y + 3z = 3$
$\quad 4x + 7y - 6z = 6$
$\quad -2x + 4y + 12z = 0$

**6.** $2x - 3y + z = -10$
$\quad 3x - 2y + 4z = -5$
$\quad x + y + 3z = 5$

**Solve the systems in Exercises 7 through 16 by the Gauss-Jordan Method.**

**7.** $2x_1 - 4x_2 - 14x_3 = 50$
$\quad x_1 - x_2 - 5x_3 = 17$
$\quad 2x_1 - 4x_2 - 17x_3 = 65$

**8.** $3x_1 + 2x_2 = 3$
$\quad 6x_1 - 6x_2 = 1$

**9.** $x - y = 3$
$\quad 4x + 3y = 5$
$\quad 6x + y = 9$

**10.** $x + y - z = 0$
$\quad 2x - 3y + 3z = 10$
$\quad 5x - 5y + 5z = 20$

**11.** $x + z = 0$
$\quad 2x - y + z = -1$
$\quad x - y = -1$

**12.** $x_1 + 2x_2 - x_3 + 3x_4 = 3$
$\quad x_1 + 3x_2 + x_3 - x_4 = 0$
$\quad 2x_1 + x_2 - 6x_3 + 2x_4 = -11$
$\quad 2x_1 - 2x_2 + x_3 = 9$

**13.** $x_1 + 2x_2 - x_3 + 3x_4 = 3$
$\quad x_1 + 3x_2 + x_3 - x_4 = 0$
$\quad 2x_1 + x_2 - 6x_3 + 2x_4 = -11$
$\quad 3x_1 + 7x_2 - x_3 + 5x_4 = 6$

**14.** $x_1 + 2x_2 - x_3 + 3x_4 = 3$
$\quad x_1 + 3x_2 + x_3 - x_4 = 0$
$\quad 2x_1 + x_2 - 6x_3 + 2x_4 = -11$
$\quad 3x_1 + 4x_2 - 5x_3 + x_4 = 7$

**15.** $2x_1 + 3x_2 - 5x_3 = 8$
$\quad 6x_1 - 3x_2 + x_3 = 16$

**16.** $x - 2y = 12$
$\quad 3x + 4y = 16$
$\quad x + 8y = -8$

**17.** Find the value of $x$ that makes the matrices equal.

$$\begin{bmatrix} 4 & 3 \\ 3x + 2 & 6 \end{bmatrix} = \begin{bmatrix} 4 & 3 \\ 5 - x & 6 \end{bmatrix}$$

**Perform the indicated matrix operations in Exercises 18 through 25, if possible.**

**18.** $-3\begin{bmatrix} 1 & 4 \\ -2 & 7 \end{bmatrix}$

**19.** $-1\begin{bmatrix} 3 & 2 \\ -6 & -7 \end{bmatrix}$

**20.** $\begin{bmatrix} 1 & 5 \\ -2 & 6 \end{bmatrix} + \begin{bmatrix} 3 & 1 \\ 0 & -4 \end{bmatrix}$

**21.** $\begin{bmatrix} 3 & 2 \\ 6 & -4 \\ 1 & 1 \end{bmatrix} + \begin{bmatrix} 8 & -5 \\ 1 & 3 \\ 2 & -1 \end{bmatrix}$

**22.** $\begin{bmatrix} 2 & 1 & 5 \\ 3 & 0 & 2 \end{bmatrix} + \begin{bmatrix} 1 & 1 \\ -2 & 4 \\ 3 & 1 \end{bmatrix}$

**23.** $[3 \quad 1 \quad -2]\begin{bmatrix} 4 \\ 1 \\ 5 \end{bmatrix}$

**24.** $\begin{bmatrix} 1 & 0 & 2 \\ 3 & 1 & 1 \end{bmatrix} \begin{bmatrix} 6 & 4 & -2 \\ 3 & 5 & -3 \\ -1 & 0 & 1 \end{bmatrix}$

**25.** $\begin{bmatrix} 5 & 9 & 1 \\ 6 & -2 & 4 \end{bmatrix} \begin{bmatrix} 3 & 5 \\ -7 & 2 \end{bmatrix}$

**Find the inverse, when possible, of the matrices in Exercises 26 through 30.**

**26.** $\begin{bmatrix} 5 & -7 \\ -3 & 4 \end{bmatrix}$     **27.** $\begin{bmatrix} 8 & 6 \\ 7 & 5 \end{bmatrix}$

**28.** $\begin{bmatrix} 5 & -2 \\ -10 & 4 \end{bmatrix}$     **29.** $\begin{bmatrix} 1 & 0 & 3 \\ 2 & -5 & 4 \\ 1 & -2 & 2 \end{bmatrix}$

**30.** $\begin{bmatrix} 1 & 1 & 2 \\ 0 & 1 & -4 \\ 3 & 2 & 10 \end{bmatrix}$

**31.** Write the augmented matrix of the system

$$\begin{aligned} 6x_1 + 4x_2 - 5x_3 &= 10 \\ 3x_1 - 2x_2 \phantom{- 5x_3} &= 12 \\ x_1 + x_2 - 4x_3 &= -2 \end{aligned}$$

**Find the reduced echelon form of the matrices in Exercises 32 through 34.**

**32.** $\begin{bmatrix} 1 & 3 & 2 & 1 \\ 2 & 4 & -2 & 6 \\ 3 & 1 & 4 & -3 \end{bmatrix}$     **33.** $\begin{bmatrix} 2 & 4 & 6 & -2 \\ 3 & 1 & 0 & 5 \\ -2 & 1 & 3 & -11 \end{bmatrix}$

**34.** $\begin{bmatrix} 3 & -1 & 2 \\ 1 & 4 & -1 \\ 4 & 3 & 1 \\ 1 & -9 & 4 \end{bmatrix}$

**35.** LaShawn scored 19 times in a basketball game for a total of 36 points. Her number of two-pointers was twice the number of three-pointers plus the number of free throws. How many of each kind of score did she make?

**36.** Determine the equilibrium solutions of the following. The demand equation is given first and the supply equation second.

**(a)** $y = -4x + 241$
$\phantom{(a)\ }y = \phantom{-}3x - 158$

**(b)** $y = -7x + 1544$
$\phantom{(b)\ }y = \phantom{-}5x - 832$

**37.** One year an investor earned $2750 from a $50,000 investment in bonds and stocks. She earned 4% from bonds and 6.5% from stocks. How much was invested in each?

**38.** A firm makes three mixes of nuts that contain peanuts, cashews, and almonds in the following proportions.

Mix I:   6 pounds peanuts, 3 pounds cashews, and 1 pound almonds

Mix II:   5 pounds peanuts, 2 pounds cashews, and 2 pounds almonds

Mix III:   8 pounds peanuts, 3 pounds cashews, and 3 pounds almonds

The firm has 183 pounds of peanuts, 78 pounds of cashews, and 58 pounds of almonds in stock. How much of each mix can it make?

**39.** An investor owns High Tech and Big Burger stock. On Monday, the closing prices were $38 for High Tech and $16 for Big Burger, and the value of the portfolio was $5648. On Friday, the closing prices were $40$\frac{1}{2}$ for High Tech and $15$\frac{3}{4}$ for Big Burger and the value of the portfolio was $5931. How many shares of each stock does the investor own?

**40.** An investor has $25,000 to invest in two funds. One fund pays 6.5%, and the other pays 8.1%. How much should be invested in each so that the yield on the total investment is 7.5%?

**41.** A toy company makes its best-selling doll at plants A and B. At plant A, the unit cost is $3.60, and the fixed cost is $1260. At plant B, the unit cost is $3.30, and the fixed cost is $2637. The company wants the two plants to produce a combined total of 900 dolls, but the combined costs of both plants must be within the budget of $7056. How many dolls should be produced at each plant?

**42.** Tank A has 150 gallons of water, and tank B has 60 gallons. Water is added to tank A at the rate of 2.5 gallons per minute and to tank B at the rate of 3.3 gallons per minute.

**(a)** How long will it take for the two tanks to contain the same amount of water?

**(b)** How much water will they each contain at that time?

**43.** The population of Laverne is growing linearly. Six years ago, the population was 4600. Today it is 5400. What is the expected population 15 years from now?

**44.** There are 150,000 registered Democrats, Republicans, and Independents in a county. The number of Independents is 20% of the total number of Democrats and Republicans. In an election, 40% of the Democrats voted, 50% of the Republicans, and 70% of the Independents. The votes totaled 72,000. How many Democrats, Republicans, and Independents are registered?

**45.** Find the regression line for the points $(0, 4)$, $(2, 9)$, $(4, 13)$, and $(6, 18)$.

**46.** Find the regression line for the points $(3, 18)$, $(5, 15)$, $(8, 12)$, $(9, 11)$, and $(12, 15)$. Find the value of $y$ on the regression line when $x = 15$.

**47.** The Austin Avenue Bakery devoted all their production to their famous Christmas fruitcake. They made 1.5- and 2.5-pound fruitcakes that sold for $18 and $26, respectively. The day's production was sent to the storeroom to be processed for customers worldwide. The accounting office reported that $8400 worth of fruitcakes was produced and the storeroom reported 756 pounds added to the inventory. When receiving the information, the inventory clerk moaned, "They didn't tell me how many of each size. I need the quantities in the inventory records before I can go home."

Help the clerk by finding the number of 1.5- and 2.5-pound fruitcakes.

# LINEAR PROGRAMMING

The claim "We're number one!" is proclaimed in many ways and places, sometimes with good reason and sometimes not. We have the World Series in baseball, the Super Bowl in football, the Olympic Games, and numerous championship competitions that attempt to determine the best, the fastest, and the strongest.

Corporations spend millions of dollars in TV advertising to convince the public that their drink, snack, automobile, or toothpaste is the best. Spectators don't hesitate to debate, often with great emotion, which team or athlete is number one.

You may ask what this has to do with a mathematics course. It has to do with using mathematics to help corporations and businesses compete in the marketplace. To compete successfully, a company would like to be the best in performance and in its products. Answers should be sought to important questions such as the following: "How can we achieve the most efficient production?" "How can we obtain maximum profit?" "What personnel and equipment are needed to complete the project on time and hold costs to a minimum?" Although not the same as the question of who is number one, these are questions of determining the best strategy or procedure that should be achieved. Individuals and corporations look for ways to improve performance to gain a competitive edge. A small reduction in cost per item can mean big savings for a large corporation. The workers who give that little extra often move ahead of their fellow workers. Thus, optimizing performance interests individuals and corporations. We call the solutions to such problems **optimal solutions**. Some of these problems can be solved with a mathematical procedure known as *linear programming*.

One of those receiving a great deal of credit for the development of linear programming is George B. Dantzig, a mathematician who worked on logistic planning problems for the U.S. Air Force during World War II. He noticed that many problems that he and his colleagues worked on had similar characteristics and could be put in a form that we now call linear programming. Dantzig realized that problems in game theory and economic models might have an important underlying mathematical structure. Since that time, linear programming has evolved, and applications abound in industry, government, and business.

Today, linear programming helps determine the best diets, the most efficient production scheduling, the least waste of materials used in manufacturing, and the most economical transportation of goods. The term **linear programming** refers to a precise procedure that will solve certain types of optimization problems that involve linear conditions.

The linear programming problems encountered in business and industry are generally more complex than the examples and exercises in this chapter. The business and industry problems generally have many variables and the data are not described in small integers. Compared to industry and business, this chapter has nice, neat solutions and linear conditions so that we can focus on an understanding of the principles and methods of linear programming while giving you a glimpse of the important role of linear programming.

## 3.1   LINEAR INEQUALITIES IN TWO VARIABLES

A linear programming problem uses **linear inequalities** to help define the problem, so let's do a brief study of linear inequalities, their solutions, and their graphs. Such a study will aid us in setting up and solving linear programming problems. Let's look at an example subject to restrictions expressed by linear inequalities.

The recommended minimum daily requirement of vitamin $B_6$ for adults is 2.0 mg (milligram). A deficiency of vitamin $B_6$ in men may increase their cholesterol level and lead to a thickening and degeneration in the walls of their arteries. Many meats, breads, and vegetables contain no vitamin $B_6$. Fruits usually contain this vitamin. For example, one small banana contains 0.45 mg of vitamin $B_6$, and 1 ounce of grapes contains 0.02 mg of vitamin $B_6$. What quantities of bananas and grapes should an adult consume in order to meet or exceed the minimum requirements?

Mathematically, this question is equivalent to asking for solutions of the linear inequality

$$0.45x + 0.02y \geq 2.0$$

where $x =$ the number of bananas eaten, and $y =$ the number of ounces of grapes eaten.

Let's see how we solve inequalities like this. If a point is selected—say, (3, 5)—and its coordinates are substituted into the inequality for $x$ and $y$, we obtain

$$0.45(3) + 0.02(5) = 1.35 + 0.10$$
$$= 1.45 \quad \text{which is not greater than 2.0}$$

Because (3, 5) makes $0.45x + 0.02y > 2.0$ false, (3, 5) is *not* a solution. However, the point (4, 12) is a solution because

$$0.45(4) + 0.02(12) = 1.80 + 0.24$$
$$= 2.04$$

Thus, $x = 4$ and $y = 12$ make the inequality true. Actually, an infinite number of points make the statement true, as we shall see.

When you select an arbitrary point and substitute its coordinates for $x$ and $y$ in

$$0.45x + 0.02y$$

then we can expect one of three different outcomes:

$$0.45x + 0.02y = 2.0$$
$$0.45x + 0.02y < 2.0$$

or

$$0.45x + 0.02y > 2.0$$

You should recognize the first of these, $0.45x + 0.02y = 2.0$, as the equation of a straight line. Any point that makes this statement true lies on that line. This line holds the key to finding the solution to the original inequality. A line divides the plane into two parts. The areas on either side of the line are called **half planes**. (See Figure 3–1.)

We find it useful that all the points that satisfy

$$0.45x + 0.02y < 2.0$$

lie on one side of the line $0.45x + 0.02y = 2.0$ and all the points that satisfy

$$0.45x + 0.02y > 2.0$$

lie on the other side. We find the solution when we determine which half plane satisfies $0.45x + 0.02y > 2.0$.

There is a simple way to determine the correct half plane. Just pick a point and substitute its coordinates for $x$ and $y$. For example, the point (10, 10) lies in the half plane above the line. As $0.45(10) + 0.02(10) = 4.5 + 0.2$, the point (10, 10) satisfies the inequality $0.45x + 0.02y > 2.0$. Consequently, *all* points above the line satisfy the same inequality. The **graph of a linear inequality** is the half plane above the line. Indicate this by shading that half plane. (See Figure 3–2.) The line itself is not a part of the solution, so show the line as dotted.

The next two examples illustrate the kinds of inequalities that might be encountered in a linear programming problem.

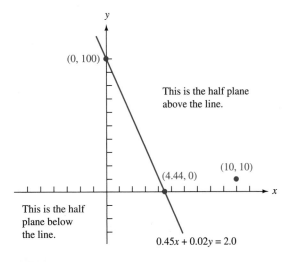

**FIGURE 3–1** The line $0.45x + 0.02y = 2.0$ divides the plane into two half planes.

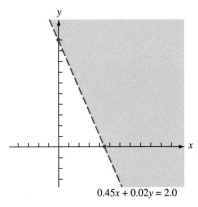

**FIGURE 3–2** The graph of $0.45x + 0.02y > 2.0$.

**Example 1**    (a)  Solve $2x + 3y \leq 12$.

(b)  Solve $2x + 3y > 12$.

**Solution**

First, replace the inequality by equals and obtain

$$2x + 3y = 12$$

This line divides the plane into two half planes (see Figure 3–3), the part where

$$2x + 3y > 12$$

and the part where

$$2x + 3y < 12$$

(a)  To decide which part satisfies

$$2x + 3y < 12$$

select a point and substitute its coordinates for $x$ and $y$. For example, the point $(0, 0)$ lies in the half plane below the line. Since $2(0) + 3(0) = 0 < 12$, the point $(0, 0)$ satisfies the inequality $2x + 3y < 12$. Consequently, *all* points below the line satisfy the same inequality. The original graph of the inequality consists of the points on the line and the points in the half plane below the line. Indicate this by shading that half plane. (See Figure 3–4.) The line $2x + 3y = 12$ is drawn as a solid line, indicating that it is a part of the solution.

If you select the point $(5, 6)$, a point above the line, you find that

$$2(5) + 3(6) = 28 > 12$$

so the half plane above the line does not represent $2x + 3y < 12$. This is another way to conclude that the solution lies in the half plane below the line.

(b)  In part (a), we found that $(0, 0)$ satisfies $2x + 3y < 12$. Thus, all points in the other half plane, the one opposite $(0, 0)$, satisfy $2x + 3y > 12$. The solution does not include the points on the line because the " $=$ " is not a part of the inequality. In such a case, draw a dotted line to indicate that it is not a part of the solution. (See Figure 3–5.)

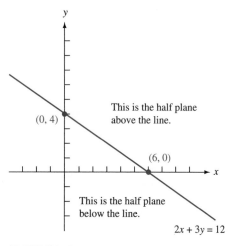

**FIGURE 3–3**    The line $2x + 3y=12$ divides the plane into two half planes.

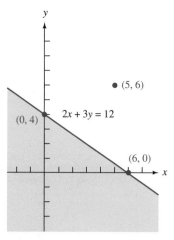

**FIGURE 3–4**    The graph of $2x + 3y \leq 12$.

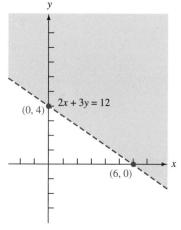

**FIGURE 3–5**    The graph of $2x + 3y > 12$.

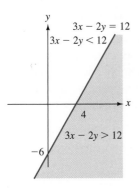

**FIGURE 3–6**

You should not assume that the solution to an inequality

$$ax + by > 0$$

will *always* lie above the line. For example, the line $3x - 2y = 12$ divides the plane in two parts. (See Figure 3–6.) The point $(0, 0)$ satisfies $3x - 2y < 12$, so the half plane above the line represents $3x - 2y < 12$ and the half plane below the line represents $3x - 2y > 12$.

**Example 2**   Graph the inequality $x \le 5$.

**Solution**
We use the line $x = 5$ as a boundary and determine the correct half plane by checking the point $(0, 0)$. Because its coordinates satisfy $x \le 5$, it lies in the half plane forming the solution. (See Figure 3–7.)

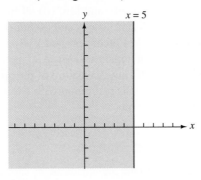

**FIGURE 3–7**   The graph of $x \le 5$.

**Procedure for Graphing a Linear Inequality**

Graph the inequality of the form $ax + by \le c$. (The procedure also applies if the inequality symbols are $<$, $>$, or $\ge$.)

1. Graph the line $ax + by = c$. This line forms the boundary between two half planes.
2. Select a point from one half plane that is not on the line. The point $(0, 0)$ is usually a good choice when it is not on the line. If $(0, 0)$ lies on the line, use a point that is not on the line.
3. Substitute the coordinates of the point for $x$ and $y$ in the inequality.
   (a) If the selected point satisfies the inequality, then shade the half plane where the point lies. These points lie on the graph.
   (b) If the selected point does not satisfy the inequality, shade the half plane opposite the point.
   (c) If the inequality symbol is $<$ or $>$, use a dotted line for the graph of $ax + by = c$. This indicates that the points on the line are *not* a part of the graph.
   (d) If the inequality symbol is $\le$ or $\ge$, use a solid line for the graph of $ax + by = c$. This indicates that the line *is* a part of the graph.

Use this procedure to graph the inequality defined in the next example.

**Example 3**    An automobile assembly plant has an assembly line that produces the Hatchback Special and the Sportster. Each Hatchback requires 2.5 hours of assembly line time, and each Sportster requires 3.5 hours. The assembly line has a maximum operating time of 140 hours per week. Graph the number of cars of each type that can be produced in one week.

**Solution**

Let $x$ be the number of Hatchback Specials and $y$ be the number of Sportsters produced per week. The total amount of assembly line time required is

$$2.5x + 3.5y$$

Because the total assembly line time is restricted to 140 hours, we need to graph the solution of the inequality

$$2.5x + 3.5y \leq 140$$

Graph the line $2.5x + 3.5y = 140$. Use a solid line. Because the point $(0, 0)$ is a solution to the system, it lies in the half plane of the graph. Figure 3–8 shows the graph of the inequality.

Note that the points $(20, 20)$, $(10, 30)$, and $(40, 10)$ all lie in the region of solutions. This tells us that the combination 20 Hatchbacks and 20 Sportsters could be produced, or 10 Hatchbacks and 30 Sportsters, or 40 Hatchbacks and 10 Sportsters. The point $(-10, 20)$ also lies in the region of solutions, but it makes no sense to say that negative 10 Hatchbacks and 20 Sportsters could be produced. Thus, the nature of the problem requires that neither $x$ nor $y$ can be negative. We can state this restriction with $x \geq 0$ and $y \geq 0$. Therefore, the inequalities that describe the problem are

$$2.5x + 3.5y \leq 140$$
$$x \geq 0, y \geq 0$$

These conditions restrict the graph to the first quadrant, as shown in Figure 3–9. When you set up a problem, you should be careful to include restrictions like this, even though they may not be stated explicitly.

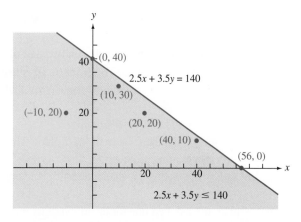

**FIGURE 3–8**    The graph of $2.5x + 3.5y \leq 140$.

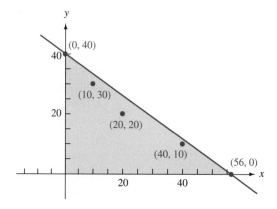

**FIGURE 3–9**    The graph of $2.5x + 3.5y \leq 140$, $x \geq 0, y \geq 0$.

This example illustrates how we can use methods of graphing an inequality to find the region that satisfies an inequality, but the context dictates that common sense be used in interpreting the answer. For example, the point (10.3, 5.2) is in the region of solutions, but it doesn't make sense to talk about producing 10.3 Hatchbacks and 5.2 Sportsters. Once the region of solutions is found, in this instance only points with whole number coordinates are reasonable.

If the previous problem had referred to quantities for which a fractional measure makes sense, such as acres of corn and wheat instead of numbers of cars, then fractional values in the solution would be appropriate.

---

## 3.1    EXERCISES

Access end-of-section exercises online at **www.webassign.net**      **WebAssign**

---

## 3.2    SOLUTIONS OF SYSTEMS OF INEQUALITIES: A GEOMETRIC PICTURE

- Feasible Region
- Boundaries and Corners
- No Feasible Solution

- Bounded and Unbounded Feasible Solutions
- Graphing a System of Inequalities

### Feasible Region

Linear programming problems are described by **systems of linear inequalities** rather than systems of linear equations. The solution to a linear programming problem depends on the ability to solve and graph such systems.

Because a linear inequality determines a region, we find it helpful to identify such a region; that is, we want to **graph the solution set of a system of inequalities**. Let's look at an example to illustrate.

---

**Example 1**   Graph the region determined by the following system of inequalities.

$$2x + y \leq 10$$
$$x + 3y \leq 12$$

**Solution**

We want to find and graph all points that make both inequalities true. We do so in the following manner.

1. We find the graph of $2x + y \leq 10$.
   (a) Graph the line $2x + y = 10$ (Figure 3–10a). Recall that this line divides the plane into two half planes, one of which is included in the solution of the inequality.

**(b)** Select a test point not on the line, say $(0, 0)$. If the test point satisfies $2x + y = 10$, it lies on the line, and you need to select another one.

**(c)** Substitute $x = 0$ and $y = 0$ into $2x + y \leq 10$, that is $2(0) + 0 \leq 10$. As this is true, the point $(0, 0)$ lies in the half plane of the solution. The arrows attached to the graph of $2x + y = 10$ indicate the solution half plane (Figure 3–10b).

**2.** Next, graph $x + 3y \leq 12$.

**(a)** Graph the line $x + 3y = 12$.

**(b)** Select a test point, say, $(3, 4)$.

**(c)** Substitute $x = 3$, $y = 4$ into $x + 3y \leq 12$; that is, $3 + 3(4) \leq 12$. As this makes $x + 3y \leq 12$ false, the point $(3, 4)$ is not in the solution half plane. The other half plane is the correct one. The arrows attached to $x + 3y = 12$ indicate the correct half plane (Figure 3–10c).

**3.** The region determined by the system of inequalities is the region of points that satisfy both inequalities. These points lie in the region where the two half planes overlap and is indicated by the shaded region in Figure 3–10d. This region of intersection is the **solution set of the system of linear inequalities**, and we call it the **feasible region**.

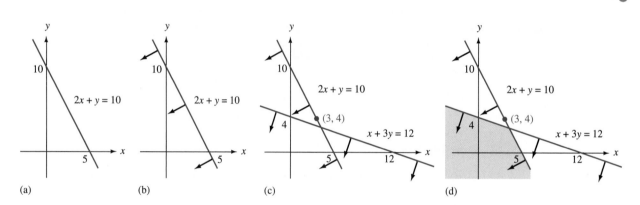

**FIGURE 3–10**    Steps to finding the solution of a system of inequalities.

Linear programming problems generally have a **nonnegative condition** on the variables, which states that some or all of the variables can never be negative because the quantities they measure (number of items, weight of materials) can never be negative. This restriction is used in determining a feasible region.

## Boundaries and Corners

**Example 2**    Graph the solutions (feasible region) to the following system

$$x + y \leq 4$$
$$-3x + 2y \leq 3$$
$$x \geq 0$$

**Solution**

The lines $x + y = 4$, $-3x + 2y = 3$, and $x = 0$ determine **boundaries of the solution set**. The half plane of the solution to each inequality is indicated with arrows (Figure 3–11a). The intersection of these half planes forms the feasible region (Figure 3–11b). We call points $A$ and $B$ in Figure 3–11b **corners of the feasible region** because they are points in the feasible region where boundaries intersect. Even though $C$ is a point of intersection of boundaries $x = 0$ and $x + y = 4$, it is not a corner because it lies outside the feasible region.

You will learn that the corners of the feasible region determine the optimal solution to a linear programming problem. Thus, finding the corners becomes a critical step. We find corners by solving pairs of simultaneous equations, using equations of lines forming the boundary.

To find corner $A$, the point of intersection of lines $x = 0$ and $-3x + 2y = 3$, we solve the system

$$x = 0$$
$$-3x + 2y = 3$$

and obtain $(0, \frac{3}{2})$ for corner $A$.

We find the point of intersection of $x + y = 4$ and $-3x + 2y = 3$ to obtain corner $B$. The solution of the system

$$x + \phantom{2}y = 4$$
$$-3x + 2y = 3$$

gives $(1, 3)$ for corner $B$.

The point $C$ $(0, 4)$ is the solution of the system

$$x = 0$$
$$x + y = 4$$

and so it gives the intersection of two boundary lines. However, the point $(0, 4)$ does not satisfy the inequality $-3x + 2y \leq 3$, so $(0, 4)$ lies outside the feasible region.

This illustrates that you cannot pick two boundary lines arbitrarily and expect their intersection to determine a corner point. You need to determine whether the point lies in the feasible region.

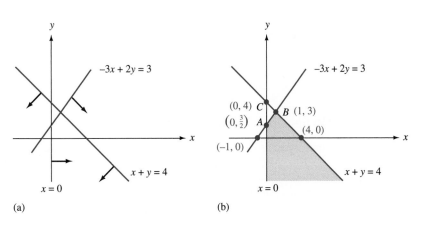

**FIGURE 3–11**    (a) Half planes determined by $x + y \leq 4$, $-3x + 2y \leq 3$, and $x \geq 0$. (b) Solution to the system $x + y \leq 4$, $-3x + 2y \leq 3$, $x \geq 0$.

Example 2 contained the nonnegative restriction, $x \geq 0$. Both conditions $x \geq 0$ and $y \geq 0$ generally enter into linear programming problems because the variables usually represent quantities such as the number or price of Mp3 players. The other inequalities represent restrictions associated with these quantities that are imposed by equipment capacity, safety regulations, cost constraints, availability of materials, and so on.

The next example shows how the feasible region of Example 2 changes when both of the nonnegative conditions are included.

**Example 3** Sketch the feasible region (solution set) determined by the system

$$
\begin{aligned}
x + y &\leq 4 \\
-3x + 2y &\leq 3 \\
x &\geq 0, y \geq 0
\end{aligned}
$$

**Solution**

The feasible region of this system is bounded by the lines

$$
\begin{aligned}
x + y &= 4 \\
-3x + 2y &= 3 \\
x &= 0, y = 0
\end{aligned}
$$

Note that the nonnegative conditions restrict the feasible region to the first quadrant of the plane. As in Example 2, the points $(0, \frac{3}{2})$ and $(1, 3)$ are corners, but the condition $y \geq 0$ introduces the corners $(0, 0)$ and $(4, 0)$ (see Figure 3–12).

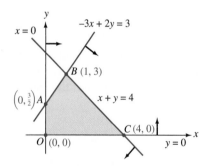

**FIGURE 3–12** Corners of the feasible region: $(0, \frac{3}{2})$ $B(1, 3)$, $C(4, 0)$, $O(0, 0)$.

## No Feasible Solution

Some systems of inequalities have no solution set, as the following example illustrates.

**Example 4** Find the solution set (feasible region) of the system

$$
\begin{aligned}
5x + 7y &\geq 35 \\
3x + 4y &\leq 12 \\
x &\geq 0, y \geq 0
\end{aligned}
$$

### Solution

The inequalities $x \geq 0$ and $y \geq 0$ force the solutions to be in the first quadrant. The test point $(0, 0)$ shows that the points that satisfy $5x + 7y > 35$ lie above the line $5x + 7y = 35$ and the points that satisfy $3x + 4y < 12$ lie below the line $3x + 4y = 12$. As shown in Figure 3–13, these two regions do not intersect in the first quadrant. The system then has no solution. We sometimes say that there is **no feasible solution**.

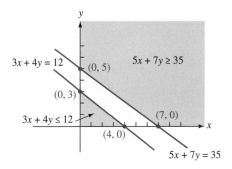

**FIGURE 3–13**   This system has no feasible solution.

## Bounded and Unbounded Feasible Solutions

A system of inequalities such as

$$x + 2y \leq 4$$
$$3x + 2y \leq 6$$
$$x \geq 0, y \geq 0$$

determines a feasible region, as shown in Figure 3–14. We say this system of inequalities has **bounded feasible solutions**, because the feasible region can be enclosed in a region where all the points are a finite distance apart.

On the other hand, a system of inequalities such as

$$x + 3y \geq 9$$
$$5x + 2y \geq 10$$
$$x \geq 0, y \geq 0$$

with a feasible region as shown in Figure 3–15, has **unbounded feasible solutions**, because some of the points in the feasible region are infinitely far apart.

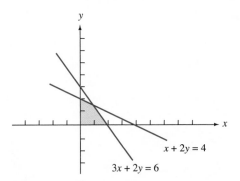

**FIGURE 3–14**   A system of inequalities with a bounded feasible region.

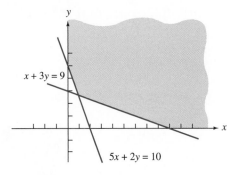

**FIGURE 3–15**   A system of inequalities with an unbounded feasible region.

**Example 5**   A theater wishes to book a musical group that requires a guarantee of $7000. Ticket prices are $10 for students and $15 for adults, and the theater's maximum capacity is 550 seats. State the inequalities that represent this information.

**Solution**

Let $x$ equal the number of student tickets and let $y$ equal the number of adult tickets. Then the total receipts is represented by $10x + 15y$, and the inequality $10x + 15y \geq 7000$ states that total receipts must meet or exceed \$7000. The inequality $x + y \leq 550$ states that seating is limited to 550 people. Because a negative number of tickets makes no sense, the restrictions $x \geq 0$ and $y \geq 0$ are also needed.

**Procedure to Graph a System of Inequalities**

1. Replace each inequality symbol with an equals sign to obtain a linear equation.

2. Graph each line. Use a solid line if it is a part of the solution. Use a dotted line if it is not a part of the solution. The line is a part of the solution when $\leq$ or $\geq$ is used. The line is not a part of the solution when $<$ or $>$ is used.

3. Select a test point not on the line.

4. If the test point satisfies the original inequality, it lies in the correct half plane. If it does not satisfy the inequality, the other half plane is the correct one.

5. Shade the correct half plane.

6. When the above steps are completed for each inequality, determine where the shaded half planes overlap. This region is the graph of the system of inequalities.

---

## 3.2    EXERCISES

Access end-of-section exercises online at **www.webassign.net**     WebAssign

## Using Your TI Graphing Calculator

### Finding Corner Points of a Feasible Region

Use the TI graphing calculator to find the corners of the feasible region defined by the following constraints:

$$y + \frac{x}{2} \leq 15 \qquad y + \frac{4}{5}x \leq \frac{84}{5} \qquad y + \frac{5}{4}x \leq 24 \qquad x \geq 0, y \geq 0$$

We show the feasible region and the corners that we seek to find on the calculator:

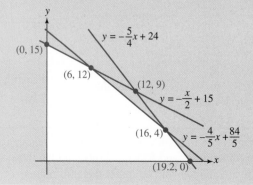

First, we graph the equations

$$y = -\frac{x}{2} + 15 \qquad y = -\frac{4}{5}x + \frac{84}{5} \qquad y = -\frac{5}{4}x + 24$$

using (0, 20) for both the x-range and y-range. We use the **intersect** command to find the corners. *Note:* Write $y = -\frac{5}{4}x + 24$ as $y = (-5/4)x + 24$ (include the parentheses).

With the graph of the three lines on the screen, press the CALC key ( 2nd TRACE ) and select <5 : intersect> from the menu.

```
CALCULATE
1:value
2:zero
3:minimum
4:maximum
5∎intersect
6:dy/dx
7:∫f(x)dx
```

Press ENTER and the following screen will appear:

Location of cursor

Note that the cursor lies at the intersection of two lines. Because the cursor lies on the line selected, we need to move it to determine which line is selected. Use the < or > key to move the cursor along a line. We move it and from the second screen we see the line selected.

Line selected

Now press ENTER , and we are ready to select the second line. Use the ∧ or ∨ key to move to another line.

Line selected

Press ENTER and obtain the following screen:

Press ENTER again and the point of intersection and its coordinates, (6, 12), are displayed:

Point of intersection (6, 12)

To find other corners, follow the same procedure and select two lines that intersect at another corner. Other intersections are

    and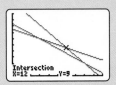

Notice the last intersection is not a corner because it is outside the feasible region.

*(continued)*

## Exercises

Find the corners of the feasible regions.

**1.** $1.5x + y \le 14$
  $0.8x + y \le 8.4$
  $x + 3y \le 21$
  $x \ge 0, y \ge 0$

**2.** $0.5x + y \le 13$
  $x + y \le 15$
  $2.5x + y \le 33$
  $x \ge 0, y \ge 0$

**3.** $5x + 11y \le 149$
  $x + 5y \le 55$
  $4x + 6y \le 108$
  $x \ge 0, y \ge 0$

# Using Excel

We will show how to graph equations of boundary lines and how to find corners of a feasible region. We illustrate using the constraints

$$y + 3x \le 30$$
$$y + \tfrac{7}{4}x \le 20$$
$$y + \tfrac{1}{2}x \le 15$$
$$x \ge 0, y \ge 0$$

To graph the boundary lines, we write the constraints as equations in the following form:

$$y = -3x + 30$$
$$y = -\tfrac{7}{4}x + 20$$
$$y = -\tfrac{1}{2}x + 15$$
$$y = 0, x = 0$$

## To Set Up the Spreadsheet

We use line 1 of the spreadsheet to identify the contents of the columns. In columns A through D, enter $x, y = -3x + 30, y = -7x/4 + 20$, and $y = -0.5x + 15$ to identify the contents of those columns.

In cells A2 through A7, enter the numbers 0, 2, 4, 6, 8, 10. These values of $x$ will be used to plot points on the graph of each line. When you use zero for a value of $x$, the graph will show the $y$-intercept of the boundary lines. Next enter the formulas:

In B2, enter =-3*A2+30
In C2, enter =-7*A2/4+20
In D2, enter =-.5*A2+15

Then drag each formula down to line 7. The values of $y$ corresponding to each value of $x$ appear for each function. This gives points used in plotting the graph.

|   | A | B | C | D |
|---|---|---|---|---|
| 1 | x | y=-3*x+30 | y=-7*x/4+20 | y=-.5*x+15 |
| 2 | 0 | 30 | 20 | 15 |
| 3 | 2 | 24 | 16.5 | 14 |
| 4 | 4 | 28 | 13 | 13 |
| 5 | 6 | 12 | 9.5 | 12 |
| 6 | 8 | 6 | 6 | 11 |
| 7 | 10 | 0 | 2.5 | 10 |

## To Show the Graphs

- Select the cells A2:D7.
- Click on the **Insert** tab. Move to the **Charts** group. Select **Scatter** with the smoothed lines subtype.
- A graph will be inserted in the worksheet, which is now in the **Chart Tools** environment.
- Now enter some names and labels. Click on the **Layout** tab under Chart Tools. Select **Chart Title** and the **Above Chart** option. In **Chart Title**, you may enter a name, such as *Sec 3.2 Example*. You may also identify the *x*- and *y*-axes by selecting **Axis Title** under the **Layout** tab.

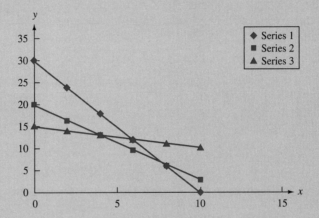

The Series 1, 2, and 3 legends shown refer to the three functions in the order they were entered in columns B, C, and D, respectively.

From the graph we see that the corners of the feasible region are $(0, 0)$, $(0, 15)$, $(10, 0)$, as well as the point where Series 2 and 3 intersect and the point where Series 1 and 2 intersect. Here's how we find those points of intersection.

## Points of Intersection

The point of intersection of Series 1 and Series 2 is the point where $-3x + 30 = -\frac{7}{4}x + 20$, or equivalently, $(-3x + 30) - (-\frac{7}{4}x + 20) = 0$. We use the second form and **Goal Seek** to find the point of intersection.

- In cell E2, enter =C2-D2, and in F2, enter =B2-C2. We will use E2 to find the intersection of Series 2 and 3, and we will use F2 to find the intersection of Series 1 and 2.
- Select cell E2.
- Select **Goal Seek** under the What-If Analysis option in the Data tab. In the menu that appears, fill the blanks with E2, 0, A2 as shown. The zero indicates the desired difference in *y*-values of the two lines, and A2 tells the location of the variable *x*.

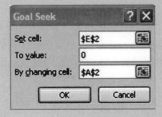

*(continued)*

- Click **OK**.

  You will see that A2 is 4, and C2 and D2 are both 13. Thus, the point of intersection of the two lines is (4, 13).

- Select cell F2.

- Select **Goal Seek** under the What-If Analysis option in the Data tab. As you did for E2, fill the blanks in the menu that appears with F2, 0, A2.

- Click **OK**. Cell A2 shows $x = 8$ and cells B2 and C2 show $y = 6$. Thus, the two lines intersect at the point (8, 6).

  We have found the other two corners of the feasible region, so we can list all corners: (0, 0), (0, 15), (4, 13), (8, 6), and (10, 0). You can use **Goal Seek** to find the x-intercept of a line, say $y = -\frac{7}{4}x + 20$, by entering C2, 0, A2 in the blanks of the Goal Seek menu that appears. The value of $x$ in A2 is the x-intercept.

## Exercises

1.  Graph the following linear functions and find their point of intersection.
    $y = -2x + 19$
    $3x + 4y = 41$ (Put this in the form $y = (-3x + 41)/4$)

2.  Graph the following linear functions and find their point of intersection.
    $y = 2.5x - 4.5$
    $6x + 5y = -4$

3.  Graph the following linear functions and find their points of intersection.
    $2x + y = 32$
    $2x - 5y = -40$
    $2x + 3y = 40$

4.  Graph the following linear functions and find their points of intersection.
    $x + y = 14$
    $x + 3y = 32$
    $2x + 3y = 40$

5.  Graph the following linear functions and find their point of intersection.
    $4.2x + 2.2y = 24.4$
    $3.8x + 5.3y = 44.25$

---

## 3.3    LINEAR PROGRAMMING: A GEOMETRIC APPROACH

- Constraints and Objective Function
- Geometric Solution
- Why Optimal Values Occur at Corners
- Unbounded Feasible Region
- Unusual Linear Programming Situations

### Constraints and Objective Function

Managers in business and industry often make decisions in an effort to maximize or minimize some quantity. For example, a plant manager wants to minimize overtime pay for production workers, a store manager makes an effort to maximize revenue, and a stockbroker tries to maximize the return on investments. Most of these decisions are complicated by restrictions that limit choices. The plant manager might not be able to eliminate all overtime and still meet the contract deadline. A store might be swamped by customers because of its low prices, but the prices might be so low that the store loses money.

A successful manager tries to make the right decision or to choose the best decision from several possible decisions. We will look at some simple examples where linear programming helps a manager make those decisions. First is an example that illustrates the form of a linear programming problem.

**Example 1**  An appliance store manager plans to offer a special on washers and dryers. The storeroom capacity is limited to 50 items. Each washer requires 2 hours to unpack and set up, and each dryer requires 1 hour. The manager has 80 hours of employee time available for unpacking and setup. Washers sell for $300 each, and dryers sell for $200 each. How many of each should the manager order to obtain the maximum revenue?

**Solution**
Convert the given information to mathematical statements:

$$\text{Let} \quad x = \text{the number of washers}$$
$$y = \text{the number of dryers}$$

The total number to be placed in the storeroom is

$$x + y$$

and the total setup time is

$$2x + 1y$$

Because the manager has space for only 50 items and setup time of 80 hours, we have the restrictions

$$x + y \leq 50$$
$$2x + y \leq 80$$

As $x$ and $y$ cannot be negative, we also have

$$x \geq 0 \quad \text{and} \quad y \geq 0$$

Because washers sell for $300 and dryers sell for $200, we want to find values of $x$ and $y$ that maximize the total revenue

$$z = 300x + 200y$$

Here is the problem stated in concise form:
    Maximize $z$, where

$$z = 300x + 200y$$

is subject to

$$x + y \leq 50$$
$$2x + y \leq 80$$
$$x \geq 0, y \geq 0$$

We now discuss how you can solve a linear programming problem. In the preceding example, the inequalities

$$x + y \leq 50 \quad \text{and} \quad 2x + y \leq 80$$

impose restrictions on the problem, and we call them **constraints**. The restrictions $x \geq 0$, $y \geq 0$ are **nonnegative conditions**. We call the function $z = 300x + 200y$ the **objective function**. Find the values of $x$ and $y$ that satisfy the system of constraints (inequalities) by the methods from the last section. You should obtain the feasible region as shown in Figure 3–16a. Corner $A$ is (30, 20).

Each point in the feasible region determines a value for the objective function. We want to find the point in the feasible region that maximizes the objective function. The point (10, 10) gives $z = 300x + 200y$ the value $z = 5000$, whereas the point (20, 20) gives the value $z = 10,000$. Because the feasible region contains an infinite number of points, we will not easily find the maximum value of $z$ by a haphazard trial-and-error process.

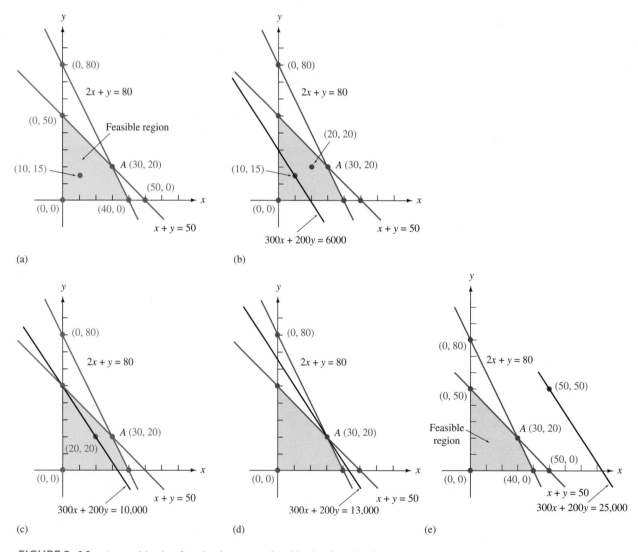

(a)

(b)

(c)

(d)

(e)

**FIGURE 3–16** As an objective function increases, the objective function line moves toward a corner. In (e) the objective function moves out of the feasible region and out of consideration.

**DEFINITION**
**Constraints, Nonnegative Condition**

A linear inequality of the form

$$a_1 x + a_2 y \leq b$$

or

$$a_1 x + a_2 y \geq b$$

is called a **constraint** of a linear programming problem. The restrictions

$$x \geq 0 \quad \text{and} \quad y \geq 0$$

are **nonnegative conditions**.

## Geometric Solution

We now state a basic theorem that makes it easier to find a maximum or minimum value of an objective function. The term **optimal** refers to either maximum or minimum.

**THEOREM**
**Optimal Values**

Given a linear **objective function** subject to linear inequality constraints, if the objective function has an **optimal value** (maximum or minimum), it must occur at a corner of the feasible region.

The feasible region in Example 1 has corners $(0, 0)$, $(0, 50)$, $(30, 20)$, and $(40, 0)$ (Figure 3–16a). The value of the objective function in each case is $z = 0$ at $(0, 0)$, $z = 10,000$ at $(0, 50)$, $z = 13,000$ at $(30, 20)$, and $z = 12,000$ at $(40, 0)$. Therefore, by the theorem, the maximum value of $z$ within the feasible region is 13,000, and the minimum value is 0.

**Graphical Solution to Linear Programming Problem**

1. Use each constraint (linear inequality) in turn to sketch the boundary of the feasible region.
2. Determine the corners of the feasible region by solving pairs of linear equations (equations obtained from the constraints).
3. Evaluate the objective function at each corner.
4. The largest (smallest) value of the objective function at corners yields the desired maximum (minimum).

## Why Optimal Values Occur at Corners

Let's look at Example 1 in order to understand why optimal values of the objective function occur at corners. The point $(10, 15)$ lies in the feasible region. (See Figure 3–16a.) The value of the objective function $z = 300x + 200y$ at that point is

$$z = 300(10) + 200(15) = 6000$$

This value of $z$ can be obtained at other points in the feasible region; in fact, $z = 6000$ will occur at any point in the feasible region that lies on the line $300x + 200y = 6000$. (See Figure 3–16b.) Furthermore, for any point $(x, y)$ in the feasible region, one of the following will occur:

$$300x + 200y = 6000$$
$$300x + 200y < 6000$$

or

$$300x + 200y > 6000$$

We have already observed that equality holds for points on the line $300x + 200y = 6000$. The point $(0, 0)$ satisfies $300x + 200y < 6000$, so all points below the line do also. From graphing inequalities we then know that the points in the half plane *above* the line $300x + 200y = 6000$ must satisfy

$$300x + 200y > 6000$$

From Figure 3–16b, we observe that the point $(20, 20)$ lies in the feasible region and lies above $300x + 200y = 6000$, so that point will yield a larger value of the objective function. It yields $z = 300(20) + 200(20) = 10,000$. Again, points above the line $300x + 200y = 10,000$ exist in the feasible region, so there are points in the feasible region that make the objective function greater than 10,000. (See Figure 3–16c.) When, if ever, will we reach the maximum value of the objective function $z = 300x + 200y$? As we move the line representing the objective function farther away from the origin, we obtain larger values of the objective function.

As shown in Figure 3–16e, we can move the line completely out of the feasible region, where points on the line cannot be considered. So, where do we stop moving the line away from the origin so that it still intersects the feasible region and gives the largest possible value of the objective function?

Look at Figures 3–16b and 3–16c. As we moved the line through $(10, 15)$ away from the origin to the point $(20, 20)$, the value of the objective function increased from 6000 to 10,000. If we continue to move it farther away, we will get larger values of the objective function. Observe in Figure 3–16d that the farthest we can move it and still intersect the feasible region is at the point $(30, 20)$. At $(30, 20)$, $z = 300(30) + 200(20) = 13,000$. Thus, the line $300x + 200y = 13,000$ divides the plane into half planes with all points on the line, giving

$$300x + 200y = 13,000$$

all points above the line, giving

$$300x + 200y > 13,000$$

and all points below the line, giving

$$300x + 200y < 13,000$$

Because all points above the line are outside the feasible region, we cannot consider them. The only point in the feasible region lying on the line is $(30, 20)$. Thus, $(30, 20)$ gives the value 13,000 to the objective function, and all other points in the feasible region give smaller values. The maximum value of the objective function indeed occurs at the corner point $(30, 20)$.

In this example, the feasible region is **bounded**, because it can be enclosed in a finite rectangle. For a bounded feasible region, the objective function will have both a maximum and a minimum.

In some cases the constraints lead to an inconsistency, so there are no points in the feasible region—it is **empty** (see Example 4 of Section 3.2).

| **THEOREM** | When the feasible region is not empty and is bounded, the objective function has both a maximum and a minimum value, which must occur at corners. |
| :--- | :--- |
| **Bounded Feasible Region** | |

Example 1 and the discussion that follows show a setup of the problem, how it is solved, and why an optimal solution occurs at a corner. With an understanding of those ideas, we can solve a problem like Example 1 in a more concise manner, as shown in the next example.

**Example 2** Find the maximum value of the objective function

$$z = 10x + 15y$$

subject to the constraints

$$x + 4y \le 360$$
$$2x + y \le 300$$
$$x \ge 0, y \ge 0$$

**Solution**
Graph the feasible region of the system of inequalities (Figure 3–17). The corners of the feasible region are (0, 90), (0, 0), (150, 0), and (120, 60). The point (120, 60) is found by solving the system

$$x + 4y = 360$$
$$2x + y = 300$$

Find the value of $z$ at each corner point.

| Corner | $z = 10x + 15y$ |
|--------|-----------------|
| (0, 90) | $10(0) + 15(90) = 1350$ |
| (0, 0) | $10(0) + 15(0) = 0$ |
| (150, 0) | $10(150) + 15(0) = 1500$ |
| (120, 60) | $10(120) + 15(60) = 2100$ |

The maximum value of $z$ is 2100 and occurs at the corner (120, 60).

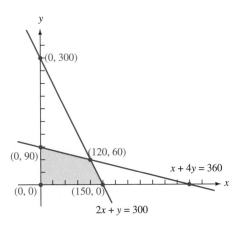

**FIGURE 3–17** The maximum value of $z = 10x + 15y$ occurs at the corner (120, 60).

## Unbounded Feasible Region

In applications of linear programming, we generally expect all coefficients in the objective function will be positive. In such cases, the value of the objective function tends to increase for points farther from the origin and tends to decrease at points closer to the origin.

To find the optimal values in such a case, we observe that for a bounded feasible region the maximum value of the objective function occurs at the last corner touched as we move away from the origin. The minimum value occurs at the last corner touched as we move toward the origin. When the feasible region is unbounded there may be no maximum value for $z$. We illustrate with the following example of an unbounded feasible region with no maximum solution.

**Example 3**    Find the maximum value of $z = 5x + 10y$ subject to the constraints

$$3x + 2y \geq 60$$
$$x + 4y \geq 40$$
$$x \geq 0, y \geq 0$$

**Solution**

The corner points of the feasible region are $(0, 30)$, $(16, 6)$, and $(40, 0)$. (See Figure 3–18.) The points on the line $5x + 10y = 275$ shown give 275 for the objective value, $z$. Points above this line give larger values of the objective function and points below the line give smaller values. Observe that we can move the objective function line farther and farther away from the origin without limit. Therefore, the objective function has no maximum value and we say there is no optimal maximum solution.

  Doesn't it appear that as the objective function line is moved toward the origin that it last touches the feasible region at the corner $(16, 6)$? In fact, when we evaluate the objective function at each corner point we obtain $z = 300$ at $(0, 30)$; $z = 140$ at $(16, 6)$; and $z = 200$ at $(40, 0)$. Thus, the minimum value occurs at $(16, 6)$.

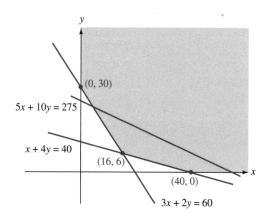

**FIGURE 3–18**    $z = 5x + 10y$ has no maximum value, but the minimum value occurs at $(16, 6)$ where $z = 140$.

**Example 4**    Find the maximum and minimum values of

$$z = 4x + 6y$$

subject to the constraints

$$5x + 3y \geq 15$$
$$x + 2y \leq 20$$
$$7x + 9y \leq 105$$
$$x \geq 0, y \geq 0$$

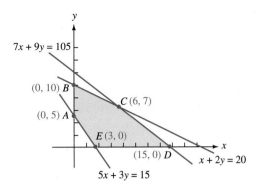

**FIGURE 3-19** The maximum value of $z$ occurs at the corner $(6, 7)$, and the minimum value of $z$ occurs at the corner $(3, 0)$.

**Solution**

The feasible region of this system and its corners are shown in Figure 3–19. Compute the value of $z$ at each corner to determine the maximum and minimum values of $z$.

| Corner | Value of $z = 4x + 6y$ |
|--------|------------------------|
| $(0, 5)$ | $4(0) + 6(5) = 30$ |
| $(0, 10)$ | $4(0) + 6(10) = 60$ |
| $(6, 7)$ | $4(6) + 6(7) = 66$ |
| $(15, 0)$ | $4(15) + 6(0) = 60$ |
| $(3, 0)$ | $4(3) + 6(0) = 12$ |

The maximum value of $z$ is 66 and occurs at the corner $(6, 7)$. The minimum value of $z$ is 12 and occurs at the corner $(3, 0)$.

We now look at two simple applications.

**Example 5** Lisa's grandfather, "Daddy Bill," survived triple bypass surgery and was placed on a strict diet. His wife, "Mama Lou," monitored his food carefully. One Tuesday, she served him soup and a sandwich that contained the amounts of fat, sodium, and protein shown in the table.

| | Amounts per Serving | | |
|---|---|---|---|
| | **Fat (g)** | **Sodium (mg)** | **Protein (g)** |
| Sandwich | 11 | 275 | 17 |
| Soup | 2 | 875 | 4 |

The lunch is to provide no more than 15 grams of fat and 700 milligrams of sodium. How many servings of soup and sandwich should Daddy Bill have for lunch in order to maximize his protein intake?

### Solution

Two restrictions apply to this problem: the amount of fat intake and the amount of sodium intake. At the same time, the amount of protein is to be maximized. The solution to the problem involves the following steps:

1. Identify the variables to be used.
2. State the restrictions and the quantity to be maximized with mathematical statements.
3. Sketch the graph of given conditions.
4. Use the graph to find the solution.

The quantities that can be varied are the number of servings of soup and sandwiches.

$$\text{Let} \quad x = \text{number of servings of sandwiches}$$
$$y = \text{number of servings of soup}$$

The fat in $x$ servings of sandwiches is $11x$ and the fat in $y$ servings of soup is $2y$, so the total fat intake is

$$11x + 2y$$

The inequality

$$11x + 2y \leq 15$$

states that the fat intake should not exceed 15 g.
The total sodium intake is

$$275x + 875y$$

and the inequality

$$275x + 875y \leq 700$$

states that the sodium intake is limited to 700 mg.
The total protein intake is

$$17x + 4y$$

We call this quantity to be maximized or minimized the objective function, in this case, $z = 17x + 4y$. Because $x$ and $y$ represent numbers of servings, they can never be negative, so $x \geq 0$ and $y \geq 0$.

We can now state this linear programming problem as:

Maximize $z = 17x + 4y$, subject to the constraints

$$11x + \phantom{0}2y \leq \phantom{0}15$$
$$275x + 875y \leq 700$$
$$x \geq 0, y \geq 0$$

We want to find the solutions of this system of inequalities that maximize $z = 17x + 4y$. A graph of the system and its feasible region is shown in Figure 3–20.

We find the corner points of the feasible region at the points where pairs of lines intersect. We find them from the following system.

| Corner $A$ | Corner $B$ | Corner $C$ | Corner $D$ |
|---|---|---|---|
| $275x + 875y = 700$ | $11x + \phantom{0}2y = \phantom{0}15$ | $11x + 2y = 15$ | $x = 0$ |
| $x \phantom{275x + 875y} = \phantom{00}0$ | $275x + 875y = 700$ | $y = \phantom{0}0$ | $y = 0$ |

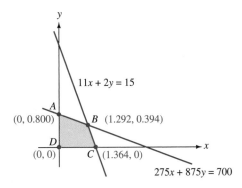

**FIGURE 3–20**

These systems give the corners (shown to three decimals)

$$A: (0, 0.800) \qquad B: (1.292, 0.394) \qquad C: (1.364, 0) \qquad D: (0, 0)$$

The maximum value of $z = 17x + 4y$ is attained at one of these corners. Examine each one:

At $(0, 0.800)$: $\qquad z = 17(0) + 4(0.800) = 3.200$

At $(1.292, 0.394)$: $\qquad z = 17(1.292) + 4(0.394) = 23.540$

At $(1.364, 0)$: $\qquad z = 17(1.364) + 4(0) = 23.188$

At $(0, 0)$: $\qquad z = 0$

The maximum value of $z = 17x + 4y$ occurs when $x = 1.292$ and $y = 0.394$. This indicates that approximately 1.3 servings of sandwiches and 0.4 serving of soup will yield the greatest quantity of protein while keeping the fat and sodium intake at acceptable levels.

**Example 6** Mom's Old-Fashioned Casseroles produces a luncheon casserole that consists of 50% carbohydrates, 30% protein, and 20% fat. The dinner casserole consists of 75% carbohydrates, 20% protein, and 5% fat. The luncheon casserole costs $2.00 per pound and the dinner casserole costs $2.50 per pound. How much of each type of casserole should be used to provide at least 3 pounds of carbohydrates, 1.50 pounds of protein, and 0.50 pound of fat at a minimum cost?

**Solution**
Let $x =$ the number of pounds of the luncheon casserole and $y =$ the number of pounds of dinner casserole. We wish to minimize the cost, $z = 2x + 2.50y$. The constraints are

$$0.50x + 0.75y \geq 3 \qquad \text{(Total carbohydrates)}$$
$$0.30x + 0.20y \geq 1.50 \qquad \text{(Total protein)}$$
$$0.20x + 0.05y \geq 0.50 \qquad \text{(Total fat)}$$
$$x \geq 0, y \geq 0$$

Figure 3–21 shows the feasible region with corners $(0, 10)$, $(1, 6)$, $(4.2, 1.2)$, and $(6, 0)$. The values of $z$ at each of these corners are $z = 25$ at $(0, 10)$; $z = 17$ at $(1, 6)$; $z = 11.4$ at $(4.2, 1.2)$; and $z = 12$ at $(6, 0)$. Then the minimum cost is $11.40 when 4.2 pounds of the luncheon casserole and 1.2 pounds of the dinner casserole are used.

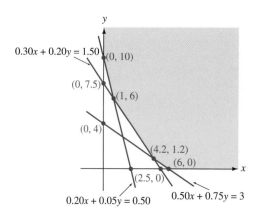

$$0.30x + 0.20y = 1.50$$

(0, 10)

(0, 7.5)

(1, 6)

(0, 4)

(4.2, 1.2)

(6, 0)

(2.5, 0)

$$0.20x + 0.05y = 0.50$$

$$0.50x + 0.75y = 3$$

**FIGURE 3–21**

A common optimization problem is called the **transportation problem**. It seeks the minimum cost of shipping goods from several sources to several destinations. The next example shows a simple case.

**Example 7**    The Garden Center has two stores, one in Clifton and one in Marlin. They sell a popular garden tractor that is supplied to them from Supplier A and Supplier B. The shipping costs (in dollars) per tractor from suppliers A and B to the Clifton and Marlin stores are the following:

|  | Store | |
| --- | --- | --- |
| **Supplier** | **Clifton** | **Marlin** |
| A | 70 | 55 |
| B | 85 | 60 |

One spring, the Garden Center ordered 40 tractors for Clifton and 50 for Marlin. Supplier A had 45 tractors in stock, and Supplier B had 70 tractors in stock. Find the number of tractors that should be shipped from each supplier to each store to minimize shipping costs.

**Solution**

Let    $x$ = number of tractors shipped from Supplier A to Clifton
       $y$ = number of tractors shipped from Supplier A to Marlin

The balance of the tractors to each store must come from Supplier B, so

$40 - x$ = number of tractors shipped from Supplier B to Clifton
$50 - y$ = number of tractors shipped from Supplier B to Marlin

We now summarize the information with the following table:

|  | Store | | Available |
|---|---|---|---|
|  | **Clifton** | **Marlin** |  |
| Supplier A |  |  |  |
| Cost (each) | $70 | $55 |  |
| Number | $x$ | $y$ | 45 |
| Supplier B |  |  |  |
| Cost (each) | $85 | $60 |  |
| Number | $40 - x$ | $50 - y$ | 70 |
| Number needed | 40 | 50 |  |

From this we find the total shipping cost,

$$C = 70x + 55y + 85(40 - x) + 60(50 - y)$$
$$= -15x - 5y + 6400$$

The constraints are

$$x + \quad y \le 45 \qquad \text{(available from A)}$$

$$40 - x + 50 - y \le 70 \qquad \text{(available from B)}$$
$$x \ge 0, y \ge 0, 40 - x \ge 0, 50 - y \ge 0$$

(No negative quantities are delivered.)
    These constraints simplify to

$$x + y \le 45$$
$$x + y \ge 20$$
$$x \ge 0, y \ge 0, x \le 40, y \le 50$$

This problem is a linear programming problem that requires that we minimize

$$C = 6400 - 15x - 5y$$

subject to

$$x + y \le 45$$
$$x + y \ge 20$$
$$x \ge 0, y \ge 0, x \le 40, y \le 50$$

Figure 3–22 shows the feasible region and the corner points.
    We now find the value of $C = 6400 - 15x - 5y$ at each corner:

| Corner | $C = 6400 - 15x - 5y$ |
|---|---|
| $(20, 0)$ | 6100 |
| $(0, 20)$ | 6300 |
| $(0, 45)$ | 6175 |
| $(40, 5)$ | 5775 |
| $(40, 0)$ | 5800 |

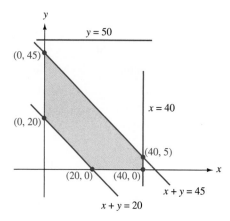

**FIGURE 3-22**

The minimum cost is \$5775 and occurs when $x = 40$, $y = 5$. Thus, Supplier A should ship 40 tractors to Clifton and 5 to Marlin. Supplier B should ship none to Clifton and 45 to Marlin.

## Unusual Linear Programming Situations

**Multiple Optimal Solutions.**   In the preceding example, just one corner point gave an optimal solution. It is possible for more than one optimal solution to exist. In other cases, there may be no solution at all. First, look at an example with multiple solutions.

**Example 8**   Find the maximum value of

$$z = 12x + 9y$$

subject to the constraints

$$4x + 3y \leq 36$$
$$8x + 3y \leq 48$$
$$x \geq 0, y \geq 0$$

**Solution**
Figure 3–23 shows the feasible region with corners $(0, 0)$, $(6, 0)$, $(3, 8)$, and $(0, 12)$. The value of $z$ at each corner point is

At $(0, 0)$:     $z = 0$

At $(6, 0)$:     $z = 72$

At $(3, 8)$:     $z = 108$

At $(0, 12)$:    $z = 108$

In this case, the maximum value of $z = 12x + 9y$ occurs at two corners, $(3, 8)$ and $(0, 12)$.

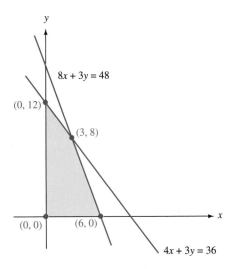

**FIGURE 3–23**   All points on the line segment between (0, 12) and (3, 8) give optimal solutions.

Actually, the value of $z = 12x + 9y$ is 108 for every point on the line segment between the points (3, 8) and (0, 12). For example, the points (1.5, 10) and (0.75, 11) lie on the line and

For (1.5, 10):          $z = 12(1.5) + 9(10) = 108$

For (0.75, 11):        $z = 12(0.75) + 9(11) = 108$

Generally, when two different points yield a maximum for the objective function, we say the problem has **multiple solutions**. (See Figure 3–23.)

---

**Multiple Solutions**    When two corners of a feasible region yield maximum solutions to a linear programming problem, then all points on the line segment joining those corners also yield maximum values. Note that the two corners are adjacent.

---

Multiple solutions allow a number of choices of $x$ and $y$ that yield the same optimal value for the objective functions. For example, if $x$ and $y$ represent production quantities, then management can achieve optimal production with a variety of production levels. If problems should occur with one production line, management might adjust production on the other line and still achieve optimal production. When just one optimal solution exists, management has no flexibility, having just one choice of $x$ and $y$ by which management can achieve its objective.

You might wonder how to tell whether there are multiple solutions. It turns out that this happens when a boundary line has the same slope as the objective function.

**Unbounded Feasible Region.**   The constraints of a linear programming problem might define an unbounded feasible region for which the objective function has no maximum value. In such a case, the problem has no solution. Here is an illustration.

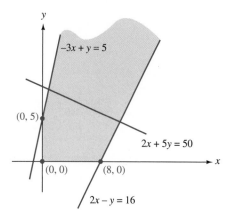

**FIGURE 3–24**

**Example 9**   Find the maximum value of the objective function $z = 2x + 5y$ subject to the constraints

$$2x - y \leq 16$$
$$-3x + y \leq \phantom{0}5$$
$$x \geq 0, y \geq 0$$

**Solution**

Figure 3–24 shows the feasible region of this system. Observe that the feasible region extends upward indefinitely. Suppose someone claims that the maximum value of the objective function is 50. This is equivalent to stating that

$$2x + 5y = 50$$

for points in the feasible region that lie on that line and that no other values of $x$ and $y$ in the feasible region give larger values. You recognize this as the equation of a straight line. Figure 3–24 shows that this line crosses the feasible region. Next, observe that the test point $(0, 0)$ does *not* satisfy the inequality $2x + 5y \geq 50$, so the points in the half plane above the line *must* satisfy it. Thus, every point in the feasible region that lies above the line $2x + 5y = 50$ will give a larger value of the objective function. If we substitute larger values—say, 100, 5000, and so on—instead of 50 in the equation, we essentially determine lines that are parallel to $2x + 5y = 50$ but lie farther away from the origin. In each case, points in the feasible region that lie above the line give an even larger value of $2x + 5y$. Because the feasible region extends upward indefinitely, we can never find a largest value.

We point out that the objective function $z = 2x + 5y$ does have a *minimum* value at $(0, 0)$. So, an unbounded feasible region does not rule out an optimal solution. It depends on the region and the kind of optimal solution sought.

**No Optimal Solution Because There Is No Feasible Region.**   The system of inequalities

$$5x + 7y \geq 35$$
$$3x + 4y \leq 12$$
$$x \geq 0, y \leq \phantom{0}0$$

is a system of inequalities that has no solution. (See Example 4 of Section 3.2.) Whenever the constraints of a linear programming problem do not define a feasible region, there can be no solution.

## 3.3 EXERCISES

Access end-of-section exercises online at **www.webassign.net**  Web**Assign**

### Using Your TI Graphing Calculator

#### Evaluate the Objective Function

An objective function can be evaluated for a series of corners using TI graphing calculator LISTs. You may use up to six lists with names L1, L2, . . . , L6 and these names appear on the keyboard above the keys for 1, 2, . . . , 6.

You may need to clear lists for use. For example, you can clear lists $L_1$ and $L_2$ with

$$\boxed{\text{STAT}} \ \text{<EDIT>} \ \text{< 4:CLRLST >} \ \boxed{\text{ENTER}} \ \boxed{\text{L1}} \ \boxed{,} \ \boxed{\text{L2}} \ \boxed{\text{ENTER}}$$

To illustrate the use of lists to evaluate an objective function, we set $z = 3x_1 + 7x_2 + 4x_3$ as the objective function and use the corner points $(0, 0, 0)$, $(3, 8, 2)$, $(5, 4, 7)$, $(6, 3, 5)$, and $(10, 6, 4)$. We will enter the values for $x_1$ in $L_1$, the values of $x_2$ in $L_2$, and the values of $x_3$ in $L_3$.

Obtain the LIST screen with

$$\boxed{\text{STAT}} \ \text{<EDIT>} \ \boxed{\text{ENTER}}$$

and you will see

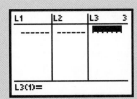

Enter the first value of $x_1$ in $L_1$, press $\boxed{\text{ENTER}}$, and enter the next value of $x_1$ until all are entered. Press $\boxed{>}$ to move to $L_2$ and enter the values of $x_2$. Then move the cursor to $L_3$ and enter the values of $x_3$. To exit LIST, press $\boxed{\text{QUIT}}$ by using $\boxed{\text{2nd}}$ $\boxed{\text{MODE}}$.

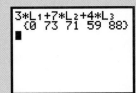

in order to evaluate the objective function, key in

$$\boxed{3} \ \boxed{\times} \ \boxed{\text{L1}} \ \boxed{+} \ \boxed{7} \ \boxed{\times} \ \boxed{\text{L2}} \ \boxed{+} \ \boxed{4} \ \boxed{\times} \ \boxed{\text{L3}} \ \boxed{\text{ENTER}}$$

and you will see $\{0 \quad 73 \quad 71 \quad 59 \quad 88\}$, which are the values of $z$ listed in the same order as the corner points.

```
3*L₁+7*L₂+4*L₃
   {0 73 71 59 88}
■
```

*(continued)*

## Exercises

Find the values of the objective function for each given corner of the feasible region.

1. $z = 45x + 32y$

   Corners: $(0, 0)$, $(0, 22)$, $(15, 18)$, and $(23, 0)$

2. $z = 48x + 63y$

   Corners: $(0, 0)$, $(0, 18)$, $(19, 15)$, and $(31, 0)$

3. $z = 5x + 11y$

   Corners: $(0, 0)$, $(0, 7)$, $(3, 9)$, $(5, 6)$, and $(4, 0)$

4. $z = 110x_1 + 230x_2 + 98x_3$

   Corners: $(0, 5, 9)$, $(8, 0, 6)$, $(14, 3, 4)$, and $(20, 5, 0)$

5. $z = 9x_1 + 15x_2 + 7x_3$

   Corners: $(0, 5, 6)$, $(3, 1, 7)$, $(5, 4, 6)$, and $(9, 8, 0)$

 # Using Excel

## Evaluate an Objective Function

We will show how to evaluate an objective function at a series of corner points of a feasible region. We illustrate by using $z = 3x_1 + 7x_2 + 4x_3$ for the objective function and the corner points $(0, 0, 0)$, $(3, 8, 2)$, $(5, 4, 7)$, $(6, 3, 5)$, and $(10, 6, 4)$.

Enter the corner points in A2:C6 with the values of $x_1$ in A, the values of $x_2$ in B, and the values of $x_3$ in C. In D2 enter $=3*A2 + 7*B2 + 4*C2$.

|   | A | B | C | D | E |
|---|---|---|---|---|---|
| 1 | x1 | x2 | x3 | z | |
| 2 | 0 | 0 | 0 | =3*A2+7*B2+4*C2 | |
| 3 | 3 | 8 | 2 | | |
| 4 | 5 | 4 | 7 | | |
| 5 | 6 | 3 | 5 | | |
| 6 | 10 | 6 | 4 | | |

Next, drag D2 down to D6 to obtain the values of the objective function listed in D2:D6.

|   | A | B | C | D |
|---|---|---|---|---|
| 1 | x1 | x2 | x3 | z |
| 2 | 0 | 0 | 0 | 0 |
| 3 | 3 | 8 | 2 | 73 |
| 4 | 5 | 4 | 7 | 71 |
| 5 | 6 | 3 | 5 | 59 |
| 6 | 10 | 6 | 4 | 88 |

## Exercises

Find the values of the objective function for each given corner of the feasible region.

1. $z = 45x + 32y$
   Corners: $(0, 0)$, $(0, 22)$, $(15, 18)$, and $(23, 0)$
2. $z = 48x + 63y$
   Corners: $(0, 0)$, $(0, 18)$, $(19, 15)$, and $(31, 0)$
3. $z = 5x + 11y$
   Corners: $(0, 0)$, $(0, 7)$, $(3, 9)$, $(5, 6)$, and $(4, 0)$
4. $z = 110x_1 + 230x_2 + 98x_3$
   Corners: $(0, 5, 9)$, $(8, 0, 6)$, $(14, 3, 4)$, and $(20, 5, 0)$
5. $z = 9x_1 + 15x_2 + 7x_3$
   Corners: $(0, 5, 6)$, $(3, 1, 7)$, $(5, 4, 6)$, and $(9, 8, 0)$

## 3.4 APPLICATIONS

The geometric method of solving linear programming problems works reasonably well when there are only two variables and a few constraints. In general, linear programming problems will likely require dozens of variables, not just two. In those cases, the geometric method becomes impractical. We address the method of solving problems with more than two variables in Chapter 4. The method applies to problems with a few or a very large number of variables and constraints. The procedure has the advantage of being a well-defined routine, so problems can be solved with pencil and paper or with computer programs.

Do not be misled by the fact that a solution can be obtained by a routine. A correct analysis and description of the problem must occur before applying the method. You may expect an erroneous constraint to yield an erroneous solution. Defining the problem and identifying the constraints require the work of a human mind.

In this section, you will concentrate on analyzing and setting up linear programming problems with more than two variables. When you have learned to correctly set up a problem, the methods of Chapter 4 can be used in a meaningful way.

To prepare you for solving larger systems, we now look at some examples of linear programming problems with more than two variables. We remind you that we do not yet have procedures to solve these problems.

The applications discussed are quite simple with nice inequalities compared to real-life applications encountered in business and industry. For this reason, we want you to focus on the principles and methods of linear programming that are required when faced with more complex problems.

**Example 1**    Cox Department Store plans a major advertising campaign for its going-out-of-business sale. They plan to advertise in the newspaper and on radio and TV. To provide a balance among the three types of media, they will place no more than ten ads in the newspaper, at

least 15% of the ads will be on TV, and no more than 40% will be on radio. The budget for this campaign is $30,000. These are the costs and expected audience exposure for the three types of media:

|  | **Radio** | **TV** | **Newspaper** |
|---|---|---|---|
| Cost per ad | $250 | $2200 | $700 |
| Audience per ad | 15,000 | 95,000 | 45,000 |

How many ads should be run in each type of media to maximize audience exposure? Find the constraints and objective function that describe this problem.

### Solution

The first task in setting up a problem is to identify the variables. The term *variable* suggests something that can vary. In this case, which items are fixed, and which can assume different values? The cost per ad is fixed, as well as the audience per each type of ad. The number of ads, and consequently the total audience exposure, can vary. In fact, the question "How many ads ... ?" suggests that we want variables to represent the number of ads. Because total audience exposure can also vary, we would expect that to be a variable. However, the number of ads determines total audience exposure. Because exposure depends on ads, we call total audience exposure a **dependent variable**. We call the number of ads in this case **independent variables**.

The first step in setting up a problem is to determine the independent variables—in this case, three variables:

$$x_1 = \text{number of ads on the radio}$$
$$x_2 = \text{number of ads on TV}$$
$$x_3 = \text{number of ads in the newspaper}$$

The dependent variable, audience exposure, is to be maximized, so we let $z =$ total audience exposure and write it in terms of the number of ads to obtain the objective function:

$$z = 15{,}000x_1 + 95{,}000x_2 + 45{,}000x_3$$

The constraints are inequalities that describe the restrictions imposed. One restriction requires that total cost must not exceed $30,000, so we have the cost constraint

$$250x_1 + 2200x_2 + 700x_3 \le 30{,}000$$

Other constraints follow:

$$x_3 \le 10 \qquad \text{(no more than 10 ads are in the paper)}$$
$$x_2 \ge 0.15(x_1 + x_2 + x_3) \qquad \text{(at least 15\% of the ads are on TV)}$$
$$x_1 \le 0.40\,(x_1 + x_2 + x_3) \qquad \text{(no more than 40\% of the ads are on radio)}$$

No variables are negative because we could not place a negative number of ads. The last two constraints can be written

$$0.15x_1 - 0.85x_2 + 0.15x_3 \le 0$$
$$-0.60x_1 + 0.40x_2 + 0.40x_3 \ge 0$$

As a linear programming problem, we can state it in the following way:

Maximize $z = 15{,}000x_1 + 95{,}000x_2 + 45{,}000x_3$, subject to

$$0.15x_1 - 0.85x_2 + 0.15x_3 \leq 0$$
$$-0.60x_1 + 0.40x_2 + 0.40x_3 \geq 0$$
$$x_1 \geq 0,\, x_2 \geq 0,\, 10 \geq x_3 \geq 0$$

The next example illustrates how some scheduling problems can be modeled as a linear programming problem.

**Example 2**

Adventure Time offers one-week summer vacations in the Rocky Mountains during the month of August. The package includes round-trip transportation and a week's accommodations at Rocky Mountain Lodge. The Lodge gives a discount to Adventure Time if they rent two- or three-week blocks of condos, giving a rent of $1000 per condo for a two-week period and $1300 per condo for a three-week period.

Adventure Time expects to need the following number of condos:

|  | Number of condos needed |
|---|---|
| First week | 30 |
| Second week | 42 |
| Third week | 21 |
| Fourth week | 32 |

How many condos should Adventure Time rent for two weeks, and how many should be rented for three weeks, to meet the needed number and to minimize Adventure Time's rental costs? Set this up as a linear programming problem.

**Solution**

First, we need to determine the possible ways to schedule two- and three-week blocks in August.

A glance at a "calendar" (Figure 3–25) shows three possible two-week periods and two possible three-week periods.

| | Two-week periods | | Three-week periods | | Number of condos needed |
|---|---|---|---|---|---|
| Week 1 | $x_1$ | | | | 30 |
| Week 2 | | $x_2$ | $x_4$ | | 42 |
| Week 3 | | | $x_3$ | $x_5$ | 21 |
| Week 4 | | | | | 32 |

**FIGURE 3–25** Possible ways to schedule two-week and three-week blocks in August.

We label them $x_1, x_2, x_3, x_4$, and $x_5$ as shown. The $x_i$ will also represent the number of condos rented during that period.

Note that the 30 condos needed for the first week will come from the $x_1$ and $x_4$ groups, and the 42 needed for the second week will come from the $x_1, x_2, x_4$, and $x_5$ groups. We want to minimize rent, that is,

$$\text{Minimize} \quad z = 1000x_1 + 1000x_2 + 1000x_3 + 1300x_4 + 1300x_5$$

We can summarize the relationship between the number of condos needed and the five time periods as follows.

$$
\begin{aligned}
x_1 + \quad\quad\quad x_4 \quad\quad &\geq 30 \\
x_1 + x_2 + \quad\quad x_4 + x_5 &\geq 42 \\
x_2 + x_3 + x_4 + x_5 &\geq 21 \\
x_3 + \quad\quad x_5 &\geq 32 \\
\end{aligned}
$$
$$x_1 \geq 0, x_2 \geq 0, x_3 \geq 0, x_4 \geq 0, x_5 \geq 0$$

We now turn to an example of a transportation problem.

**Example 3**   Spokes and Things has two bicycle stores, one in Calvert and one in Hico. They sell a popular model of a mountain bike that they obtain from two suppliers, A and B. In preparation for the upcoming biking season, they order a minimum of 80 bikes for the Calvert store and a minimum of 65 bikes for the Hico store. At that time, Supplier A has 62 bikes, and Supplier B has 90 bikes. The shipping cost from supplier to store is given in this chart:

**Shipping Cost per Bike**

| Supplier | Store Calvert | Hico |
|---|---|---|
| A | $ 8 | $6 |
| B | $10 | $7 |

Find the number each supplier ships to each store to minimize shipping costs.

**Solution**

We summarize the given information in the following table:

| | Store Calvert | Hico | No. Available |
|---|---|---|---|
| Supplier A | | | |
| Cost (each) | $8 | $6 | |
| Number supplied | $x_1$ | $x_2$ | 62 |
| Supplier B | | | |
| Cost (each) | $10 | $7 | |
| Number | $x_3$ | $x_4$ | 90 |
| Number needed | 80 or more | 65 or more | |

Let   $x_1$ = number shipped from Supplier A to Calvert

$x_2$ = number shipped from Supplier A to Hico

$x_3$ = number shipped from Supplier B to Calvert

$x_4$ = number shipped from Supplier B to Hico

Notice that we cannot use $80 - x_1$ for the number shipped from supplier B to Calvert (as we did in Example 7 of Section 3.3) because the exact number shipped to Calvert is not specified; it is 80 *or more.*

We now state the linear programming problem.

Minimize   $z = 8x_1 + 6x_2 + 10x_3 + 7x_4$, subject to

$$x_1 + x_2 \leq 62 \text{ (available from A)}$$
$$x_3 + x_4 \leq 90 \text{ (available from B)}$$
$$x_1 + x_3 \geq 80 \text{ (needed at Calvert)}$$
$$x_2 + x_4 \geq 65 \text{ (needed at Hico)}$$
$$x_1 \geq 0, x_2 \geq 0, x_3 \geq 0, x_4 \geq 0 \quad \text{(no negative amounts are ordered)}$$

---

## **3.4**  EXERCISES

Access end-of-section exercises online at **www.webassign.net**    Web**Assign**

---

**3**

## IMPORTANT TERMS

**3.1**
Linear Inequality
Half Plane
Graph of a Linear Inequality

**3.2**
System of Linear Inequalities
Solutions to a System of Linear
   Inequalities
Graph of a System of Linear
   Inequalities

Feasible Solution
Feasible Region
Bounded Feasible Region
Unbounded Feasible Region
Boundary of a Feasible Region
Corners of a Feasible Region
No Feasible Solution

**3.3**
Constraints
Nonnegative Conditions

Objective Function
Maximize Objective Function
Minimize Objective Function
Optimal Solution
Multiple Optimal Solutions
Bounded Feasible Region
Unbounded Feasible Region

**3.4**
Dependent Variable
Independent Variable

# IMPORTANT CONCEPTS

| | |
|---|---|
| **Linear Inequality** | An expression obtained by replacing the "=" sign in a linear equation with an inequality symbol. |
| **Solution of an Inequality** | The half plane that satisfies an inequality. |
| **Feasible Region** | The set of points, the solution, that satisfy a system of inequalities. |
| **Boundaries of a Feasible Region** | The lines that determine the feasible region of a system of inequalities. |
| **Corners of a Feasible Region** | The points in the feasible region where boundary lines intersect. |
| **Bounded Feasible Region** | All points in a feasible region are a finite distance apart. |
| **Unbounded Feasible Region** | Some points in a feasible region are an infinite distance apart. |
| **Linear Programming Problem** | Find the maximum or minimum value of a function (Objective Function) restricted by linear inequalities (Constraints). |
| **Multiple Solutions** | All points on a line segment joining two corner points yield optimal solutions. |

# REVIEW EXERCISES

**1.** Graph the solution to the following inequalities.

   (a) $5x + 7y < 70$      (b) $2x - 3y > 18$

   (c) $x + 9y \le 21$      (d) $-2x + 12y \ge 26$

   (e) $y \ge -6$          (f) $x \le 3$

**2.** Graph the following systems of inequalities.

   (a) $2x + y \le 4$      (b) $x + y \le 5$

       $x + 3y < 9$            $x - y > 3$

                             $x \ge 1, y \le 3$

**Find the feasible region and corner points of the systems in Exercises 3 through 7.**

**3.** $x - 3y \ge 6$      **4.** $5x + 2y \le 50$

    $x - y \le 4$            $x + 4y \le 28$

       $y \ge -5$                $x \ge 0$

**5.** $-3x + 4y \le 20$      **6.** $3x + 10y \le 150$

      $x + y \ge -2$           $2x + y \le 32$

    $8x + y \le 40$            $x \le 14$

         $y \ge 0$              $x \ge 0, y \ge 0$

**7.**   $x - 2y \le 0$

    $-2x + y \le 2$

    $x \le 2, y \le 2$

**8.** Maximize $z = x + 2y$, subject to

$$x + y \le 8$$
$$x \quad\quad \le 5$$
$$x \ge 0, y \ge 0$$

**9.** Maximize $z = 5x + 4y$, subject to

$$3x + 2y \le 12$$
$$x + y \le 5$$
$$x \ge 0, y \ge 0$$

**10.** Find the maximum and minimum values of $z = 2x + 5y$, subject to

$$2x + y \le 9$$
$$4x + 3y \ge 23$$
$$x \ge 0, y \ge 0$$

**11.** **(a)** Find the minimum value of $z = 5x + 4y$, subject to

$$3x + 2y \ge 18$$
$$x + 2y \ge 10$$
$$5x + 6y \ge 46$$
$$x \ge 0, y \ge 0$$

    **(b)** Find the minimum value of

$$z = 10x + 12y$$

    subject to the constraints of part (a).

**12.** Maximize $z = x + 5y$, subject to

$$x + y \le 10$$
$$2x + y \ge 10$$
$$x + 2y \ge 10$$
$$x \ge 0, y \ge 0$$

**13.** Maximize $z = 4x + 7y$, subject to

$$2x + y \le 90$$
$$x + 2y \le 80$$
$$x + y \le 50$$
$$x \ge 0, y \ge 0$$

**14.** Minimize $z = 7x + 3y$, subject to

$$x + 2y \ge 16$$
$$3x + 2y \ge 32$$
$$5x + 2y \ge 40$$
$$x \ge 0, y \ge 0$$

**15.** An assembly plant has two production lines. Line A can produce 65 items per hour, and line B can produce 105 per hour. The loading dock can ship a maximum of 700 items per eight-hour day.

**(a)** Express the information with an inequality.

**(b)** Graph the inequality.

**16.** A building supplies truck has a load capacity of 25,000 pounds. A delivery requires at least 21 pallets of brick weighing 950 pounds each and at least 15 pallets of roofing material weighing 700 pounds each. Express these restrictions with a system of inequalities.

**17.** The Ivy Twin Theater has a seating capacity of 275 seats. Adult tickets sell for $7.50 each, and children's tickets sell for $4.00 each. The theater must have ticket sales of at least $1100 to break even for the night. Write these constraints as a system of inequalities.

**18.** A tailor makes suits and dresses. A suit requires 1 yard of polyester and 4 yards of wool. Each dress requires 2 yards of polyester and 2 yards of wool. The tailor has a supply of 80 yards of polyester and 150 yards of wool. What restrictions does this place on the number of suits and dresses she can make?

**19.** The Hoover Steel Mill produces two grades of stainless steel, which is sold in 100-pound bars. The standard grade is 90% steel and 10% chromium by weight, and the premium grade is 80% steel and 20% chromium. The company has 80,000 pounds of steel and 12,000 pounds of chromium on hand.

If the price per bar is $90 for the standard grade and $100 for the premium grade, how much of each grade should it produce to maximize revenue?

**20.** The Nut Factory produces a mixture of peanuts and cashews. It guarantees that at least one third of the total weight is cashews. A retailer wants 1200 pounds or more of the mixture. The peanuts cost the Nut Factory $0.75 per pound, and the cashews cost $1.40 per pound. Find the amount of each kind of nut the company should use to minimize the cost

**(a)** if 600 pounds of peanuts are available.

**(b)** if 900 pounds of peanuts are available.

**21.** A massive inoculation program is initiated in an area devastated by floods. Doctors and nurses form teams of the following sizes: A-teams are composed of one doctor and three nurses, B-teams are composed of one doctor and two nurses, C-teams are composed of one doctor and one nurse. An A-team can inoculate an estimated 175 people per hour, a B-team can inoculate an estimated 110 people per hour, and a C-team can inoculate an estimated 85 people per hour. There are 75 doctors and 200 nurses available. How many teams of each type should be formed to maximize the number of inoculations per hour? Set up the constraints and objective function. Do not solve.

**22.** A school cafeteria serves three foods for lunch: A, B, and C. There is pressure on the cafeteria director to reduce lunch costs. Help the director by finding the quantities of each food that will minimize costs and still maintain the desired nutritional level. The three foods have the following nutritional characteristics:

| | Foods | | |
|---|---|---|---|
| **Per unit** | **A** | **B** | **C** |
| Protein (g) | 15 | 10 | 23 |
| Carbohydrates (g) | 20 | 30 | 11 |
| Calories | 500 | 400 | 200 |
| Cost ($) | 1.40 | 1.65 | 1.95 |

A lunch must contain at least 80 grams of protein, 95 grams of carbohydrates, and 800 calories. How many units of each food should be served to minimize cost? Set up the problem. Do not solve.

**23.** Southside Landscapes proposes to plant the flower beds for the historic Earle House gardens. The company uses four patterns. Pattern I uses 40 tulips, 25 daffodils, and 6 boxwood. Pattern II uses 25 tulips, 50 daffodils, and 4 boxwood. Pattern III uses 30 tulips, 40 daffodils, and 8 boxwood. Pattern IV uses 45 tulips, 45 daffodils, and 2 boxwood. The profit for each pattern is $48 for Pattern I, $45 for Pattern II, $55 for Pattern III, and $65 for Pattern IV. Southside Landscape has 1250 tulips, 1600 daffodils, and 195 boxwood available. How many of each pattern should be used to maximize profit? Set up the problem. Do not solve.

# 4
# LINEAR PROGRAMMING: THE SIMPLEX METHOD

Chapter 3 introduced you to the basic ideas of linear programming. Perhaps you noticed that most of the examples and problems involved two variables. In practice, linear programming problems involve dozens of variables. The graphical method is not practical in problems having more than two variables. Fortunately, we have a procedure for solving linear programming problems involving several variables, thanks to the mathematician George B. Dantzig. Dantzig and his colleagues observed and defined the class of linear programming problems and originally had no efficient method for solving them. This changed in the mid-1940s when Dantzig invented a procedure that became the workhorse in solving linear programming problems. He called it the **simplex method,** a procedure for examining corners of a feasible region in an intelligent manner that speeds the process of finding the optimal solution.

Basically, the simplex method moves along a boundary from one vertex to another, seeking the vertex that gives the optimal solution. When the problem involves many variables, many boundaries may leave from the current vertex, and the best route may be difficult to ascertain. Many mathematicians, including Dantzig, thought there should be a better search technique. A great deal of research focused on developing a better algorithm. Then, in 1984, a breakthrough came when 28-year-old Narendra Karmarkar announced an improved method for large-scale problems (those involving thousands of variables). His method is considerably more complicated than the simplex method, so we study only the simplex method.

We emphasize that the examples and exercises in this chapter do not reflect the complexity of real life linear programming problems. We use relatively simple examples and exercises because the primary purpose is for you to set up linear programming problems, to understand the procedures, and get a glimpse of the usefulness of linear programming without the burden of complex computations. Contrary to what one expects in business and industry applications, we generally use integers and rational numbers for the data and solutions with fairly simple numbers.

## 4.1   SETTING UP THE SIMPLEX METHOD

- Standard Maximum
- The Geometric Form of the Problem
- Slack Variables
- Simplex Tableau

We now study an algebraic technique that applies to any number of variables and enables us to solve larger linear programming problems. It has the added advantage of being well suited to a computer, thereby making it possible to avoid tedious pencil-and-paper solutions. We call this technique the **simplex method**.

In practice, a linear programming application of any consequence does not lend itself to a pencil-and-paper solution. Thus, you will use a computer if you are asked to solve a real-world problem. You may wonder why this chapter contains mainly pencil-and-paper exercises. We do so because a correct solution to an application requires an understanding of the procedure used so that you can set up the analysis correctly. It helps if you understand the purpose and the restrictions of the procedure. Consequently, this chapter contains relatively simple linear programming exercises to give you experience in working through the details of the procedure, as well as exercises in which you will learn to analyze and set up linear programming problems.

In Chapter 3 you learned that optimal solutions of a linear programming problem occur at corners of a feasible region. It may be prohibitively time consuming to find and check every corner of a problem that has dozens of constraints. The simplex method restricts the search to a relatively small number of corners and provides a sequence of corners that yield a more efficient way to reach an optimal solution.

Basically, the linear programming method involves modifying the constraints so that one has a system of linear equations and then finding selected solutions of the system. Remember how we solved a system of linear equations using an augmented matrix and reducing it with row operations? The simplex method follows a similar procedure. We introduce the simplex method in several steps, and we refer to the graphical method to illustrate the steps involved.

### Standard Maximum

We will refer to the following linear programming problem several times as we develop the concepts.

**Illustrative Example**   Maximize the objective function

$$z = 4x_1 + 12x_2$$

subject to the constraints

$$3x_1 + x_2 \le 180$$
$$x_1 + 2x_2 \le 100$$
$$-2x_1 + 2x_2 \le 40 \qquad \qquad (1)$$

and the nonnegative conditions

$$x_1 \ge 0, x_2 \ge 0$$

Note that we now use the notation $x_1$ and $x_2$ for the variables instead of $x$ and $y$. This notation allows us to use several variables without running out of letters for variables.

In this section, we deal only with **standard maximum** linear programming problems. They are problems like the illustrative example that have the following properties.

Standard Maximum Problem

1. The objective function is to be maximized.
2. Each constraint is written using the $\leq$ inequality (excluding the nonnegative conditions).
3. The constants in the constraints to the right of $\leq$ are never negative (180, 100, and 40 in the example).
4. The variables are restricted to nonnegative values (nonnegative conditions).

## The Geometric Form of the Problem

We now show the graph of the feasible region of the illustrative example to help us follow the steps of the simplex method. The problem is as follows:

Maximize $z = 4x_1 + 12x_2$, subject to

$$\begin{aligned} 3x_1 + x_2 &\leq 180 \\ x_1 + 2x_2 &\leq 100 \\ -2x_1 + 2x_2 &\leq 40 \\ x_1 \geq 0, x_2 &\geq 0 \end{aligned}$$

Recall that the boundary to the feasible region is formed by the lines

$$\begin{aligned} 3x_1 + x_2 &= 180 \\ x_1 + 2x_2 &= 100 \\ -2x_1 + 2x_2 &= 40 \\ x_1 = 0, x_2 &= 0 \end{aligned}$$

The feasible region is shown in Figure 4–1.

In the geometric approach, we look at the corners of the feasible region for optimal values. A corner point occurs at the intersection of a pair of boundary lines such as (52, 24) and (60, 0). Not all pairs of boundary lines intersect at a corner point. Note that the point (40, 60) is not a corner point; it lies outside the feasible region.

The equation of the boundary lines can give us only points on the boundaries. To use equations to represent points in the interior of the feasible region, we must introduce some new variables called *slack variables*. Slack variables allow us to view the problem as a system of equations.

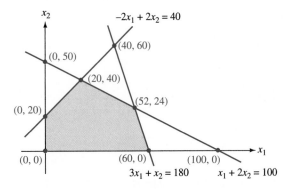

**FIGURE 4–1** The feasible region.

## Slack Variables

The first step in the simplex method converts the constraints to linear equations. To do this, we must introduce additional variables called **slack variables**. We introduce one slack variable for each constraint. The first constraint, $3x_1 + x_2 \leq 180$, is true for pairs of numbers such as $x_1 = 10$ and $x_2 = 20$ because $3(10) + 20 \leq 180$. Observe that $3x_1 + x_2 + 130 = 180$ when $x_1 = 10$ and $x_2 = 20$. When $x_1 = 30$ and $x_2 = 50$, the constraint $3x_1 + x_2 \leq 180$ is true, as is the equation $3x_1 + x_2 + 40 = 180$. In general, when a pair of values for $x_1$ and $x_2$ make the statement $3x_1 + x_2 \leq 180$ true, there will also be a value of $s_1$ ($s_1$ depends on the values of $x_1$ and $x_2$) that makes the statement $3x_1 + x_2 + s_1 = 180$ true. In each case, $s_1$ is not negative. We call $s_1$ a *slack variable* because it takes up the slack between $3x_1 + x_2$ and 180. Whenever $s_1 = 0$, then $x_1$ and $x_2$ must be coordinates of a point on the boundary line $3x_1 + x_2 = 180$.

The values $x_1 = 10$ and $x_2 = 20$ also make the constraint $x_1 + 2x_2 \leq 100$ true, and also make the equation $x_1 + 2x_2 + 50 = 100$ true. When we used these same values for $x_1$ and $x_2$ in the constraint $3x_1 + x_2 \leq 180$, we used the value 130 to take up the slack. We point this out to illustrate that whenever values of $x_1$ and $x_2$ make two constraints true, then *different* values of the slack variable may be required to take up the slack. Thus, we need one nonnegative slack variable for each constraint.

Figure 4–2 illustrates the slack variables' relationship to the feasible region determined by

$$3x_1 + x_2 \leq 180$$
$$x_1 + 2x_2 \leq 100$$
$$x_1 \geq 0, x_2 \geq 0$$

that is,

$$3x_1 + x_2 + s_1 = 180$$
$$x_1 + 2x_2 + s_2 = 100$$
$$x_1 \geq 0, x_2 \geq 0$$

when the constraints are written with slack variables.

Note that points that lie on the boundary $3x_1 + x_2 + s_1 = 180$ yield $s_1 = 0$ [like (40, 60) and (52, 24)], and points that lie on $x_1 + 2x_2 + s_2 = 100$ yield $s_2 = 0$ [like (10, 45) and (52, 24)]. A point that lies in the interior of the feasible region, such as (10, 20), yields positive values for both $s_1$ and $s_2$. The point (40, 60), which lies outside the feasible region, yields a negative value for $s_2$. In general, a point that lies outside the feasible region will yield one or more negative slack variables.

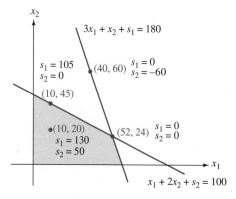

**FIGURE 4–2** Values of slack variables at selected points.

| | |
|---|---|
| **Criteria for the Feasible Region** | A point $(x_1, x_2, \ldots, x_n)$ lies in the feasible region of a linear programming problem if and only if all variables $x_1, x_2, \ldots, x_n$ and all slack variables $s_1, s_2, \ldots, s_k$ are *nonnegative*. |

**Example 1**  With the addition of slack variables, the constraints (1) in the illustrative example become

$$\begin{aligned}
3x_1 + x_2 + s_1 &= 180 \\
x_1 + 2x_2 + s_2 &= 100 \\
-2x_1 + 2x_2 + s_3 &= 40 \\
x_1 \geq 0,\ x_2 \geq 0,\ s_1 \geq 0,\ s_2 \geq 0,\ s_3 &\geq 0
\end{aligned} \qquad (2)$$

The last line gives the nonnegative conditions that apply to the variables.

You might wonder how this system of equations will help. Here's how. When nonnegative values of $x_1$, $x_2$, and $s_1$ are chosen to satisfy $3x_1 + x_2 + s_1 = 180$, then the point obtained from the values of $x_1$ and $x_2$ will satisfy the constraint $3x_1 + x_2 \leq 180$. Furthermore, when $x_1$ and $x_2$ come from an interior point such as $(20, 30)$, for example, then the nonnegative values of the slack variables ($s_1 = 90$, $s_2 = 20$, and $s_3 = 20$, in this case) satisfy the system of equations (2).

In general we determine the points in the feasible region from the solutions to a system of equations like (2), provided that we use only those solutions in which *all* the variables ($x_1$, $x_2$, $s_1$, $s_2$, $s_3$ in this case) are nonnegative.

**Example 2**  Write the following constraints as a system of equations using slack variables:

$$\begin{aligned}
5x_1 + 3x_2 + 17x_3 &\leq 140 \\
7x_1 + 2x_2 + 4x_3 &\leq 256 \\
3x_1 + 9x_2 + 11x_3 &\leq 540 \\
2x_1 + 16x_2 + 8x_3 &\leq 99
\end{aligned}$$

**Solution**
Introduce a slack variable for each equation:

$$\begin{aligned}
5x_1 + 3x_2 + 17x_3 + s_1 &= 140 \\
7x_1 + 2x_2 + 4x_3 + s_2 &= 256 \\
3x_1 + 9x_2 + 11x_3 + s_3 &= 540 \\
2x_1 + 16x_2 + 8x_3 + s_4 &= 99
\end{aligned}$$

The objective function needs to be included in the system of equations because we want to find the value of $z$ that comes from the solution of the above system. Its form needs to be modified by writing all terms on the left-hand side. For example,

$$z = 3x_1 + 5x_2$$

is modified to

$$z - 3x_1 - 5x_2 = 0$$

which we write as

$$-3x_1 - 5x_2 + z = 0$$

because we want to write the $x$'s first, as we do in the constraints.

The objective function

$$z = 6x_1 + 7x_2 + 15x_3 + 2x_4$$

is modified to

$$z - 6x_1 - 7x_2 - 15x_3 - 2x_4 = 0$$

and then to

$$-6x_1 - 7x_2 - 15x_3 - 2x_4 + z = 0$$

**Example 3**    Include the objective function

$$z = 20x_1 + 35x_2 + 40x_3$$

with the constraints of Example 2 and write them as a system of equations.

**Solution**

$$
\begin{aligned}
5x_1 + \ \ 3x_2 + 17x_3 + s_1 &= 140 \\
7x_1 + \ \ 2x_2 + \ \ 4x_3 + s_2 &= 256 \\
3x_1 + \ \ 9x_2 + 11x_3 + s_3 &= 540 \\
2x_1 + 16x_2 + \ \ 8x_3 + s_4 &= \ \ 99 \\
-20x_1 - 35x_2 - 40x_3 + z &= \ \ \ \ 0
\end{aligned}
$$

**Example 4**    Write the following as a system of equations. Maximize $z = 4x_1 + 12x_2$, subject to

$$
\begin{aligned}
3x_1 + \ \ x_2 &\le 180 \\
x_1 + 2x_2 &\le 100 \\
-2x_1 + 2x_2 &\le \ \ 40 \\
x_1 \ge 0, x_2 &\ge 0
\end{aligned}
$$

**Solution**

$$
\begin{aligned}
3x_1 + \ \ x_2 + s_1 \ \ \ \ \ \ \ \ \ \ \ \ \ \ \ \ &= 180 \\
x_1 + 2x_2 \ \ \ \ \ \ + s_2 \ \ \ \ \ \ \ \ &= 100 \\
-2x_1 + 2x_2 \ \ \ \ \ \ \ \ \ \ \ + s_3 \ \ \ \ &= \ \ 40 \\
-4x_1 - 12x_2 \ \ \ \ \ \ \ \ \ \ \ \ \ \ \ + z &= \ \ \ \ 0
\end{aligned}
$$

Let's pause to make some observations about the systems of equations that we obtain.

1. One slack variable is introduced for each constraint (four in Example 3 and three in Example 4).

2. The total number of variables in the system is the number of original variables (number of $x$'s) plus the number of constraints (number of slack variables) plus one for $z$. (The total is $3 + 4 + 1 = 8$ in Example 3 and $2 + 3 + 1 = 6$ in Example 4.)

**3.** The introduction of slack variables always results in fewer equations than variables. Recall from the summary in Chapter 2 that this situation generally yields an infinite number of solutions. In fact, one variable for each equation can be written in terms of some other variables.

Look back at the system of equations in Example 4. One general form of the solution occurs when we solve for $s_1$, $s_2$, $s_3$, and $z$ in terms of $x_1$ and $x_2$.

$$
\begin{aligned}
s_1 &= 180 - 3x_1 - x_2 \\
s_2 &= 100 - x_1 - 2x_2 \\
s_3 &= 40 + 2x_1 - 2x_2 \\
z &= 4x_1 + 12x_2
\end{aligned}
$$

You find a particular solution when you substitute values for $x_1$ and $x_2$. The first step of the simplex method always uses zero for that substitution because it gives corner points. So when $x_1 = 0$ and $x_2 = 0$, we obtain $s_1 = 180$, $s_2 = 100$, $s_3 = 40$, and $z = 0$.

The preceding discussion enables us to make the first step in solving a linear programming problem with the simplex method, setting up the simplex tableau.

## Simplex Tableau

The simplex method uses matrices and row operations on matrices to determine an optimal solution. Let's use the problem from Example 4 to set up the matrix that we call the **simplex tableau**.

**Example 5**    Set up the simplex tableau that represents the following problem.
Maximize $z = 4x_1 + 12x_2$, subject to

$$
\begin{aligned}
3x_1 + x_2 &\le 180 \\
x_1 + 2x_2 &\le 100 \\
-2x_1 + 2x_2 &\le 40 \\
x_1 \ge 0, x_2 &\ge 0
\end{aligned}
$$

**Solution**
First, write the problem as a system of equations using slack variables:

$$
\begin{aligned}
3x_1 + x_2 + s_1 &= 180 \\
x_1 + 2x_2 + s_2 &= 100 \\
-2x_1 + 2x_2 + s_3 &= 40 \\
-4x_1 - 12x_2 + z &= 0
\end{aligned}
$$

Next, form the augmented matrix of this system:

$$
\begin{array}{cccccc}
x_1 & x_2 & s_1 & s_2 & s_3 & z \\
\end{array}
$$

$$
\left[
\begin{array}{cccccc|c}
3 & 1 & 1 & 0 & 0 & 0 & 180 \\
1 & 2 & 0 & 1 & 0 & 0 & 100 \\
-2 & 2 & 0 & 0 & 1 & 0 & 40 \\
\hline
-4 & -12 & 0 & 0 & 0 & 1 & 0
\end{array}
\right]
$$

This is the **initial** simplex tableau.

The line drawn above the bottom row emphasizes that the bottom row is the objective function.

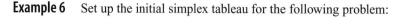

**Example 6**  Set up the initial simplex tableau for the following problem:

Garden Tools, Inc., manufactures three items: hoes, rakes, and shovels. It takes 3 minutes of labor to produce each hoe, 5 minutes of labor to produce each rake, and 4 minutes of labor to produce each shovel. Each hoe costs $2.50 to produce, each rake costs $3.15, and each shovel costs $3.35. The profit is $3.20 per hoe, $3.35 per rake, and $4.10 per shovel. If the company has 108,000 minutes of labor and $6800 in operating funds available per week, how many of each item should it produce to maximize profit?

**Solution**
If you are puzzled about which quantities to call variables, ask yourself what quantities are mentioned, which quantities are given and therefore fixed, which quantities are mentioned and no values given, and, finally, how are the quantities related?

The problem gives information about three items: hoes, rakes, and shovels. The minutes of labor and the cost to produce each item are given. The profit for each item is also given. The available minutes of labor and operating funds are known. These last two quantities are related to the number of items produced, but those numbers are not given. In fact, you are asked to find the number of each item. Generally, the statement of a problem identifies variables by asking something like "How many . . ."; in this example, you are asked to find how many of each item should be produced. Thus, the variables represent the number of hoes, rakes, and shovels. Those are the quantities that can vary, so we let

$$x_1 = \text{number of hoes}$$
$$x_2 = \text{number of rakes}$$
$$x_3 = \text{number of shovels}$$

It often helps to summarize the given information in a chart before writing the constraints. In this problem, there are three variables and three types of information: labor time, cost, and profit. Look at the information and observe that it can be summarized as follows:

|  | $x_1$ (Hoes) | $x_2$ (Rakes) | $x_3$ (Shovels) | Maximum available |
|---|---|---|---|---|
| Labor (minutes) | 3 | 5 | 4 | 108,000 |
| Cost | $2.50 | $3.15 | $3.35 | $6800 |
| Profit | $3.20 | $3.35 | $4.10 | |

The information on labor requirements, operating costs, and profit can be written as follows:

$$3x_1 + 5x_2 + 4x_3 \leq 108{,}000 \quad \text{(Labor)}$$
$$2.50x_1 + 3.15x_2 + 3.35x_3 \leq 6{,}800 \quad \text{(Operating cost)}$$
$$\text{Maximize:} \quad 3.20x_1 + 3.35x_2 + 4.10x_3 = z \quad \text{(Profit)}$$

We introduce slack variables into the constraints to form equations and rewrite the objective function (profit) as follows:

$$3x_1 + 5x_2 + 4x_3 + s_1 = 108{,}000$$
$$2.50x_1 + 3.15x_2 + 3.35x_3 + s_2 = 6{,}800$$
$$-3.20x_1 - 3.35x_2 - 4.10x_3 + z = 0$$

The simplex tableau is

$$
\begin{array}{cccccc}
x_1 & x_2 & x_3 & s_1 & s_2 & z \\
\left[\begin{array}{cccccc|c}
3 & 5 & 4 & 1 & 0 & 0 & 108{,}000 \\
2.50 & 3.15 & 3.35 & 0 & 1 & 0 & 6{,}800 \\
-3.20 & -3.35 & -4.10 & 0 & 0 & 1 & 0
\end{array}\right]
\end{array}
$$

---

# **4.1**      EXERCISES

Access end-of-section exercises online at **www.webassign.net**

---

# **4.2**      THE SIMPLEX METHOD

- System of Equations: Many Solutions
- Basic Solution

- Pivot Column, Row, and Element
- Final Tableau

## System of Equations: Many Solutions

In Section 4.1, you learned how to convert a linear programming problem to a system of equations using slack variables and then putting the system into a matrix form called a simplex tableau. The simplex method uses the simplex tableau to find the optimal solution. By the nature of a linear programming problem, the simplex method involves a system of equations in which the number of variables exceeds the number of equations. Such a system generally has an infinite number of solutions where some of the variables can be chosen arbitrarily. We will see that the simplex method chooses the variables in a way that gives corner points of the feasible region. Recall that when the solutions of a system are represented by an augmented matrix such as

$$
\begin{array}{cccc}
x_1 & x_2 & x_3 & x_4 \\
\left[\begin{array}{cccc|c}
1 & 0 & 0 & 2 & 3 \\
0 & 1 & 0 & -1 & 5 \\
0 & 0 & 1 & 4 & 7
\end{array}\right]
\end{array}
$$

then the solutions can be written as

$$x_1 = 3 - 2x_4$$
$$x_2 = 5 + x_4$$
$$x_3 = 7 - 4x_4$$

where any value may be assigned to $x_4$, thereby determining values of $x_1$, $x_2$, and $x_3$. We solve for $x_1$, $x_2$, and $x_3$ in terms of $x_4$ because $x_1$ occurs in only one row. (All entries in the $x_1$ column are zero except for one entry.) Similarly, $x_2$ and $x_3$ occur only once. The variable $x_4$ occurs in all rows, so it appears in each equation of the solution. Observe that the entries in the $x_1$, $x_2$, and $x_3$ columns form a column from an identity matrix (call these **unit columns**). The $x_4$ column does not look like a column from an identity matrix. In general, we solve for variables corresponding to unit columns in terms of variables whose columns are not unit columns. For example, in the augmented matrix

$$\begin{array}{ccccc} x_1 & x_2 & x_3 & x_4 & x_5 \end{array}$$
$$\left[\begin{array}{ccccc|c} 3 & 0 & -1 & 0 & 1 & 4 \\ 1 & 1 & 0 & 0 & 0 & -6 \\ 2 & 0 & 5 & 1 & 0 & 10 \end{array}\right]$$

the unit columns are the $x_2$, $x_4$, and $x_5$ columns, so we solve for $x_2$, $x_4$, and $x_5$ in terms of $x_1$ and $x_3$, giving

$$\begin{array}{lll} x_2 = -6 - x_1 & \text{(from row 2)} \\ x_4 = 10 - 2x_1 - 5x_3 & \text{(from row 3)} \\ x_5 = 4 - 3x_1 + x_3 & \text{(from row 1)} \end{array}$$

Obtain specific solutions by assigning arbitrary values to $x_1$ and $x_3$.

We will see how the simplex method selects certain solutions to a system by assigning zeros to the arbitrary variables.

## Basic Solution

The simplex method finds a sequence of selected solutions to a system of equations. We make the selections so that we find the optimal solution in a relatively small number of steps.

Let's look at the illustrative example in Section 4.1 again. It is the following: Maximize the objective function $z = 4x_1 + 12x_2$, subject to the constraints

$$\begin{array}{rr} 3x_1 + x_2 \leq 180 \\ x_1 + 2x_2 \leq 100 \\ -2x_1 + 2x_2 \leq 40 \\ x_1 \geq 0, x_2 \geq 0 \end{array} \qquad (1)$$

We introduce slack variables to obtain the following. Maximize $z = 4x_1 + 12x_2$, subject to

$$\begin{array}{rrrr} 3x_1 + x_2 + s_1 & = 180 \\ x_1 + 2x_2 & + s_2 & = 100 \\ -2x_1 + 2x_2 & + s_3 = 40 \end{array} \qquad (2)$$

where $x_1$, $x_2$, $s_1$, $s_2$, and $s_3$ are all nonnegative.

The simplex tableau is

$$\begin{array}{cccccc} x_1 & x_2 & s_1 & s_2 & s_3 & z \end{array}$$
$$\left[\begin{array}{cccccc|c} 3 & 1 & 1 & 0 & 0 & 0 & 180 \\ 1 & 2 & 0 & 1 & 0 & 0 & 100 \\ -2 & 2 & 0 & 0 & 1 & 0 & 40 \\ \hline -4 & -12 & 0 & 0 & 0 & 1 & 0 \end{array}\right]$$

Note that unit columns occur in the $s_1$, $s_2$, and $s_3$ columns. Because the feasible region is determined by the constraints, for now we will ignore the bottom row (objective function) and the $z$ column. Thus we can solve for $s_1$, $s_2$, and $s_3$ as

$$s_1 = 180 - 3x_1 - x_2$$
$$s_2 = 100 - x_1 - 2x_2$$
$$s_3 = 40 + 2x_1 - 2x_2$$

We can substitute any values for $x_1$ and $x_2$ to obtain $s_1$, $s_2$, and $s_3$. Those five numbers will form a solution to the system of equations. However, only those nonnegative values of $x_1$ and $x_2$ that also yield nonnegative values for $s_1$, $s_2$, and $s_3$ give **feasible solutions**. The simplest choices are $x_1 = 0$ and $x_2 = 0$. This gives $s_1 = 180$, $s_2 = 100$, and $s_3 = 40$. Solutions like this, *where the arbitrary variables are set to zero,* are called **basic solutions**. The number of variables set to zero in this case was two, which happens to be the number of $x$'s involved. In general, the number of $x$'s determines the number of arbitrary variables set to zero.

We chose zero for the values of $x_1$ and $x_2$, not just for simplicity, but also because that gives a corner point $(0, 0)$ of the feasible region.

| **DEFINITION** **Basic Solution** | If a linear programming problem has $k$ $x$'s in the constraints, then a **basic solution** is obtained by setting $k$ variables (except $z$) to zero and solving for the others. |
|---|---|

Actually, in the above example we can find a basic solution by setting any two variables to zero and solving for the others.

If we set $s_1 = 0$ and $s_3 = 0$, we can obtain another basic solution by solving for the other variables from the simplex tableau. If $s_1 = 0$ and $s_3 = 0$, the system (2) reduces to

$$3x_1 + x_2 = 180$$
$$x_1 + 2x_2 + s_2 = 100$$
$$-2x_1 + 2x_2 = 40$$
$$-4x_1 - 12x_2 + z = 0$$

We have four equations in four unknowns, $x_1$, $x_2$, $s_2$, and $z$, to solve. This system has the solution (we omit the details)

$$x_1 = 40, \quad x_2 = 60, \quad s_2 = -60, \quad z = 880$$

Note that $s_2$ is *negative*. This violates the nonnegative conditions on the $x$'s and slack variables. Although the solution

$$x_1 = 40, \quad x_2 = 60, \quad s_1 = 0, \quad s_2 = -60, \quad s_3 = 0, \quad z = 880$$

is a basic solution, it is not feasible.

If you look at Figure 4–1, you will see that two boundary lines intersect at the point $(40, 60)$, but it lies outside the feasible region. Properly carried out, the simplex method finds only basic *feasible* solutions that are corner points of the feasible region.

| **DEFINITION** **Basic Feasible Solution** | Call the number of $x$ variables in a linear programming problem $k$. A **basic feasible solution** of the system of equations is a solution with $k$ variables (except $z$) set to zero and with none of the slack variables or $x$'s negative. |
|---|---|

We call the variables set to zero **nonbasic variables**. We call the others **basic variables**.

With this background, let's proceed to solve this example by finding the appropriate basic feasible solution.

Solve by the simplex method:

Maximize $z = 4x_1 + 12x_2$, subject to

$$3x_1 + x_2 \leq 180$$
$$x_1 + 2x_2 \leq 100$$
$$-2x_1 + 2x_2 \leq 40$$
$$x_1 \geq 0, x_2 \geq 0$$

**Step 1.** Begin with the *initial tableau:*

|       |       |       |       |       |       |       | Basic variables |
|-------|-------|-------|-------|-------|-------|-------|-----------------|
| $x_1$ | $x_2$ | $s_1$ | $s_2$ | $s_3$ | $z$ |       |                 |
| 3     | 1     | 1     | 0     | 0     | 0     | 180   | $s_1$           |
| 1     | 2     | 0     | 1     | 0     | 0     | 100   | $s_2$           |
| $-2$  | 2     | 0     | 0     | 1     | 0     | 40    | $s_3$           |
| $-4$  | $-12$ | 0     | 0     | 0     | 1     | 0     |                 |

We find the first basic feasible solutions. Note that $s_1$, $s_2$, $s_3$ form unit columns, so we can solve for them in terms of the $x$'s. Now we can set the $x$'s to zero to obtain $s_1 = 180$, $s_2 = 100$, and $s_3 = 40$. (You can read these from the tableau.) Because we find this basic feasible solution first, we call it the **initial basic feasible solution**. It always occurs at the origin of the feasible region.

In this case, the unit columns give $s_1$, $s_2$, and $s_3$ as the basic variables. The nonbasic variables are $x_1$ and $x_2$.

---

**Basic and Nonbasic Variables—Summary**

A linear programming problem with $n$ constraints using $k$ variables is converted into a system of equations by including a slack variable for each constraint.

- A system of $n$ constraints in $k$ variables converts to a system of $n$ equations in $n + k$ variables.

- The $n + k$ variables are classified as either *basic* or *nonbasic* variables.

- There is a basic variable for each equation, giving $n$ basic variables. The other $k$ variables are nonbasic. A unit column determines a basic variable.

- The nonunit columns determine the nonbasic variables.

- The values of the basic variables are found by setting all nonbasic variables to zero in each of the equations and then solving for the basic variables.

## Pivot Column, Row, and Element

**Step 2.** Next we want to modify the tableau so that the new tableau has a basic feasible solution that increases the value of $z$. This step requires the selection of a **pivot element** from the tableau as follows:

**(a)** To select the column containing the pivot element, do the following. Select the *most negative* entry from the *bottom* row:

$$
\begin{array}{cccccc}
x_1 & x_2 & s_1 & s_2 & s_3 & z \\
\end{array}
$$

$$
\left[\begin{array}{cccccc|c}
3 & 1 & 1 & 0 & 0 & 0 & 180 \\
1 & 2 & 0 & 1 & 0 & 0 & 100 \\
-2 & 2 & 0 & 0 & 1 & 0 & 40 \\
\hline
-4 & \boxed{-12} & 0 & 0 & 0 & 1 & 0
\end{array}\right]
$$

└──── Most negative entry gives pivot column.

This selects the **pivot column** containing the pivot element. The pivot element itself is an entry in this column *above* the line. We must now determine which row contains the pivot element.

**(b)** To select the row, called the **pivot row**, containing the pivot element, do the following. Divide each constant above the line in the last column by the corresponding entries in the pivot column. The ratios are written to the right of the tableau. In this example, all ratios are positive. However, negative or zero ratios can occur. The case of a zero ratio is discussed in the summary of the simplex method.

  **(i)** If a negative ratio occurs, do not use that row for the pivot row.

  **(ii)** Select the *smallest nonnegative ratio* for which the pivot element is positive (20 in this case). The row containing this ratio is the pivot row.

Now look at the ratios on the right of the tableau.

$$
\begin{array}{cccccc}
x_1 & x_2 & s_1 & s_2 & s_3 & z \\
\end{array}
$$

$$
\left.\begin{array}{cccccc|c}
3 & 1 & 1 & 0 & 0 & 0 & 180 \\
1 & 2 & 0 & 1 & 0 & 0 & 100 \\
-2 & ② & 0 & 0 & 1 & 0 & 40 \\
-4 & -12 & 0 & 0 & 0 & 1 & 0
\end{array}\right]
\begin{array}{l}
\frac{180}{1} = 180 \\
\frac{100}{2} = 50 \\
\frac{40}{2} = 20 \\
\\
\end{array}
$$

Pivot row → (third row)
Pivot element

**NOTE**

The **pivot element** will always lie above the line drawn above the bottom row.

Because all ratios are nonnegative, the smallest, 20, determines the pivot row. The positive entry 2 in the pivot row and pivot column is the **pivot element**.

**Step 3.** Move to the next basic feasible solution. We call this process **pivoting** on the pivot element. In this case we pivot on 2 in row 3, column 2.

To pivot on 2, use row operations to modify the tableau so that the pivot element becomes a 1 and the rest of the pivot column contains zeros. (You recognize that this is part of the Gauss-Jordan Method for solving systems.)

Multiply each entry in the third row (pivot row) by $\frac{1}{2}$ so that the pivot entry becomes 1. The third row becomes

$$[\,-1 \quad 1 \quad 0 \quad 0 \quad \tfrac{1}{2} \quad 0 \quad 20\,]$$

giving the tableau

$$
\begin{array}{cccccc}
x_1 & x_2 & s_1 & s_2 & s_3 & z \\
\end{array}
$$

$$
\left[
\begin{array}{cccccc|c}
3 & \textcircled{1} & 1 & 0 & 0 & 0 & 180 \\
1 & \textcircled{2} & 0 & 1 & 0 & 0 & 100 \\
-1 & 1 & 0 & 0 & \frac{1}{2} & 0 & 20 \\
\hline
-4 & \boxed{-12} & 0 & 0 & 0 & 1 & 0
\end{array}
\right]
$$

We now need zeros in the circled locations of the pivot column.

We replace each row where a zero is needed by multiplying row 3 by a constant and adding it to the row to be replaced. Each time, use the constant that gives a zero in the pivot column. This is accomplished as follows:

Replace row 1 with (row 1 − row 3): i.e., R1 − R3 → R1

$$= [\,4 \quad 0 \quad 1 \quad 0 \quad -\tfrac{1}{2} \quad 0 \quad 160\,]$$

Replace row 2 with (row 2 + (−2)row 3): i.e., R2 − 2R3 → R2

$$= [\,3 \quad 0 \quad 0 \quad 1 \quad -1 \quad 0 \quad 60]$$

Replace row 4 with (row 4 + (12)row 3): i.e., R4 + 12R3 → R4

$$= [\,-16 \quad 0 \quad 0 \quad 0 \quad 6 \quad 1 \quad 240]$$

This gives the tableau

$$
\begin{array}{ccccccc}
 & x_1 & x_2 & s_1 & s_2 & s_3 & z & & \text{Basic variables}
\end{array}
$$

$$
\left[
\begin{array}{cccccc|c}
4 & 0 & 1 & 0 & -\frac{1}{2} & 0 & 160 \\
3 & 0 & 0 & 1 & -1 & 0 & 60 \\
-1 & 1 & 0 & 0 & \frac{1}{2} & 0 & 20 \\
\hline
-16 & 0 & 0 & 0 & 6 & 1 & 240
\end{array}
\right]
\begin{array}{c}
s_1 \\
s_2 \\
x_2 \\
\\
\end{array}
$$

To determine the basic feasible solution from this tableau, observe that the columns under $x_2$, $s_1$, and $s_2$ form unit columns. These variables are the *basic variables*, the ones we solve for. The other two variables, $x_1$ and $s_3$, are *nonbasic variables*, the ones we set to zero.

The basic feasible solution from this tableau is

$$x_1 = 0, \qquad x_2 = 20, \qquad s_1 = 160, \qquad s_2 = 60, \qquad s_3 = 0, \qquad z = 240$$

The initial solution gave $z = 0$, and this solution gives $z = 240$, so we do indeed have a larger value of the objective function.

We need to know when we have reached the optimal solution, the maximum value of $z$. We can tell when the maximum has been achieved from the simplex tableau.

**Step 4.** Is $z$ maximum?

If the last row contains any negative coefficients, $z$ is not maximum. Because $-16$ is a coefficient from the last row, 240 is not the maximum value of $z$, so we proceed to move to another basic feasible solution.

## Final Tableau

**Step 5.** Find another basic feasible solution.

Proceed as in Steps 2 and 3 with the most recent tableau:

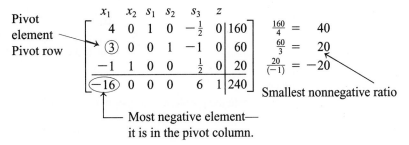

Pivot
element
Pivot row

$$\begin{bmatrix} x_1 & x_2 & s_1 & s_2 & s_3 & z & \\ 4 & 0 & 1 & 0 & -\frac{1}{2} & 0 & 160 \\ ③ & 0 & 0 & 1 & -1 & 0 & 60 \\ -1 & 1 & 0 & 0 & \frac{1}{2} & 0 & 20 \\ \boxed{-16} & 0 & 0 & 0 & 6 & 1 & 240 \end{bmatrix}$$

$\frac{160}{4} = 40$

$\frac{60}{3} = 20$

$\frac{20}{(-1)} = -20$ — Smallest nonnegative ratio

Most negative element— it is in the pivot column.

The smallest nonnegative ratio determines the pivot row, row 2 in this case, and 3 is the pivot element.

We now use row operations to obtain a 1 in the pivot element position and zeros in the rest of the pivot column. We do so by dividing each entry in row 2 by 3 to obtain a new row 2 as shown in the tableau below. We complete the pivot with the indicated row operations:

$$\begin{bmatrix} x_1 & x_2 & s_1 & s_2 & s_3 & z & \\ 4 & 0 & 1 & 0 & -\frac{1}{2} & 0 & 160 \\ 1 & 0 & 0 & \frac{1}{3} & -\frac{1}{3} & 0 & 20 \\ -1 & 1 & 0 & 0 & \frac{1}{2} & 0 & 20 \\ -16 & 0 & 0 & 0 & 6 & 1 & 240 \end{bmatrix}$$

$-4\text{R2} + \text{R1} \rightarrow \text{R1}$

$\text{R2} + \text{R3} \rightarrow \text{R3}$

$16\text{R2} + \text{R4} \rightarrow \text{R4}$

giving the tableau

Basic variables

$$\begin{bmatrix} x_1 & x_2 & s_1 & s_2 & s_3 & z & \\ 0 & 0 & 1 & -\frac{4}{3} & \frac{5}{6} & 0 & 80 \\ 1 & 0 & 0 & \frac{1}{3} & -\frac{1}{3} & 0 & 20 \\ 0 & 1 & 0 & \frac{1}{3} & \frac{1}{6} & 0 & 40 \\ 0 & 0 & 0 & \frac{16}{3} & \frac{2}{3} & 1 & 560 \end{bmatrix}$$

$s_1$

$x_1$

$x_2$

The basic variables are $x_1, x_2$, and $s_1$, because these columns form unit columns. Find the basic feasible solution by setting $s_2$ and $s_3 = 0$ and solving for the others. The solution is

$$x_1 = 20, \qquad x_2 = 40, \qquad s_1 = 80, \qquad s_2 = 0, \qquad s_3 = 0, \qquad z = 560$$

with $x_1, x_2$, and $s_1$ the basic variables and $s_2$ and $s_3$ nonbasic.

The value of $z$ can be increased by going to another tableau only when a negative number appears in the bottom row. The last tableau, the **final tableau**, has no negative numbers in the last row, so $z = 560$ is the maximum value of $z$ and occurs at the point (20, 40).

If you review the steps of the example just completed (and observe the steps in the examples that follow), you will see that the basic variables of the initial tableau consist of the slack variables. Each pivot replaces one of the basic variables with another, nonbasic, variable. The ultimate goal of the simplex method is to determine the nonnegative $x$'s that yield the maximum value of the objective function. The sequence of pivots begins with only slack variables as basic, and it brings in a nonbasic variable with each pivot to replace a basic variable until an optimal solution is reached.

**Example 1**   Use the simplex method to maximize $z = 2x_1 + 3x_2 + 2x_3$, subject to

$$2x_1 + x_2 + 2x_3 \leq 13$$
$$x_1 + x_2 - 3x_3 \leq 8$$
$$x_1 \geq 0, x_2 \geq 0, x_3 \geq 0$$

**Solution**

We first write the problem as a system of equations:

$$
\begin{aligned}
2x_1 + x_2 + 2x_3 + s_1 \quad\quad\quad\quad &= 13 \\
x_1 + x_2 - 3x_3 \quad + s_2 \quad\quad &= 8 \\
-2x_1 - 3x_2 - 2x_3 \quad\quad\quad + z &= 0
\end{aligned}
$$

The initial simplex tableau is

$$
\begin{array}{c}
\phantom{x} \\
\begin{array}{cccccc}
x_1 & x_2 & x_3 & s_1 & s_2 & z
\end{array} \\
\left[\begin{array}{cccccc|c}
2 & 1 & 2 & 1 & 0 & 0 & 13 \\
1 & 1 & -3 & 0 & 1 & 0 & 8 \\
\hline
-2 & -3 & -2 & 0 & 0 & 1 & 0
\end{array}\right]
\end{array}
\begin{array}{c}
\text{Basic} \\
\text{variables} \\
s_1 \\
s_2 \\
\phantom{x}
\end{array}
$$

Because we have three $x$'s, all basic solutions will have three variables set to zero. From the initial tableau, the initial basic feasible solution is

$$x_1 = 0, \quad x_2 = 0, \quad x_3 = 0, \quad s_1 = 13, \quad s_2 = 8, \quad z = 0$$

With negative entries in the last row, the solution is not optimal.

Find the pivot element:

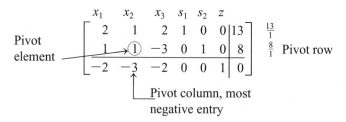

Because all ratios are nonnegative, the smallest, 8, determines the pivot row. Use row 2 to pivot on 1 in row 2, column 2 to find the next tableau. Use the following row operations to convert column 2 to a unit column.

$$\begin{bmatrix} 2 & 1 & 2 & 1 & 0 & 0 & 13 \\ 1 & 1 & -3 & 0 & 1 & 0 & 8 \\ -2 & -3 & -2 & 0 & 0 & 1 & 0 \end{bmatrix} \quad \begin{array}{l} -R2 + R1 \to R1 \\ \\ 3R2 + R3 \to R3 \end{array}$$

giving

|  | $x_1$ | $x_2$ | $x_3$ | $s_1$ | $s_2$ | $z$ |  | Basic variables |
|---|---|---|---|---|---|---|---|---|
|  | 1 | 0 | 5 | 1 | −1 | 0 | 5 | $s_1$ |
|  | 1 | 1 | −3 | 0 | 1 | 0 | 8 | $x_2$ |
|  | 1 | 0 | −11 | 0 | 3 | 1 | 24 |  |

Again, the solution is not optimal because a negative entry, $-11$, occurs in the last row. Find the new pivot element:

$$\begin{array}{c} \text{Pivot} \\ \text{element} \end{array} \quad \begin{array}{ccccccc} x_1 & x_2 & x_3 & s_1 & s_2 & z & \\ \begin{bmatrix} 1 & 0 & \circledS & 1 & -1 & 0 & 5 \\ 1 & 1 & -3 & 0 & 1 & 0 & 8 \\ 1 & 0 & -11 & 0 & 3 & 1 & 24 \end{bmatrix} \end{array} \quad \begin{array}{l} \frac{5}{5} = 1 \text{ (smallest nonnegative ratio)} \\ \frac{8}{-3} = -2.67 \end{array}$$

Pivot column

This tableau has no zero ratios, so choose the smallest positive, 1, which gives row 1 as the pivot row. Now use row 1 to pivot on 5 in row 1, column 3. First divide row 1 by 5 to obtain the tableau below. Then perform the indicated row operations.

$$\begin{array}{ccccccc} x_1 & x_2 & x_3 & s_1 & s_2 & z & \\ \begin{bmatrix} \frac{1}{5} & 0 & 1 & \frac{1}{5} & -\frac{1}{5} & 0 & 1 \\ 1 & 1 & -3 & 0 & 1 & 0 & 8 \\ 1 & 0 & -11 & 0 & 3 & 1 & 24 \end{bmatrix} \end{array} \quad \begin{array}{l} \\ 3R1 + R2 \to R2 \\ 11R1 + R3 \to R3 \end{array}$$

giving the tableau

$$\begin{array}{cccccc} x_1 & x_2 & x_3 & s_1 & s_2 & z \\ \begin{bmatrix} \frac{1}{5} & 0 & 1 & \frac{1}{5} & -\frac{1}{5} & 0 & 1 \\ \frac{8}{5} & 1 & 0 & \frac{3}{5} & \frac{2}{5} & 0 & 11 \\ \frac{16}{5} & 0 & 0 & \frac{11}{5} & \frac{4}{5} & 1 & 35 \end{bmatrix} \end{array}$$

The basic feasible solution from the tableau is

$$x_1 = 0, \qquad x_2 = 11, \qquad x_3 = 1, \qquad s_1 = 0, \qquad s_2 = 0, \qquad z = 35$$

With no negative entries in the last row, $z = 35$ is a maximum.

*Note:* An explanation of the rationale for the choices of the pivot row and pivot column is found in Section 4.6.

**Summary of the Simplex Method**

**Standard Maximization Problem**

1. Convert the problem to a system of equations:

   (a) Convert each inequality to an equation by adding a slack variable.

   (b) Write the objective function

   $$z = ax_1 + bx_2 + \cdots + kx_n$$

   as

   $$-ax_1 - bx_2 - \cdots - kx_n + z = 0$$

2. Form the initial simplex tableau from the equations.

3. Locate the pivot element of the tableau:

   (a) Locate the most negative entry in the bottom row. This determines the pivot column. In case of a tie for most negative, choose either.

   (b) Divide each entry in the last column (above the line) by the corresponding entry in the pivot column. Choose the smallest nonnegative ratio for which the pivot entry is positive. It determines the pivot row. In case of a tie for pivot row, choose either.

   (c) The element where the pivot column and pivot row intersect is the pivot element.

4. Modify the simplex tableau by using row operations to obtain a new basic feasible solution.

   (a) Divide each entry in the pivot row by the pivot element to obtain a 1 in the pivot position.

   (b) Use row operations on the pivot row to obtain zeros in the other entries of the pivot column.

5. Determine whether $z$ has reached its maximum.

   (a) If there is a negative entry in the last row of the tableau, $z$ is not maximum. Repeat the process in steps 3 and 4.

   (b) If the bottom row contains no negative entries, $z$ is maximum and the solution is available from the final tableau.

6. Determine the solution from the final tableau.

   (a) Set $k$ variables to zero, where $k$ is the number of $x$'s used in the constraints. These are the nonbasic variables. They correspond to the columns that contain more than one nonzero entry.

   (b) Determine the values of the basic variables. These basic variables correspond to unit columns.

   (c) State the maximum value and the values of the original variables that give the maximum value.

7. Perhaps you have noticed that the $z$-column in the simplex tableau never changes in the pivoting operations. Thus, you may omit that column. If you do, remember that the number in the last row of the last column is the current value of $z$.

**NOTE**

There may be occasions when the ratio is zero. If the ratio giving zero has a *positive* divisor, then choose the row where this occurs as the pivot row. If all zero ratios have negative divisors, then choose the smallest *positive* ratio.

The next example illustrates a case where you have a tie for the choice of pivot column.

**Example 2**    Maximize $z = 9x_1 + 5x_2 + 9x_3$, subject to

$$6x_1 + x_2 + 4x_3 \leq 72$$
$$3x_1 + 4x_2 + 2x_3 \leq 30$$
$$x_1 \geq 0, x_2 \geq 0, x_3 \geq 0$$

**Solution**

Form the system of equations

$$6x_1 + x_2 + 4x_3 + s_1 \qquad\qquad = 72$$
$$3x_1 + 4x_2 + 2x_3 \qquad + s_2 \qquad = 30$$
$$-9x_1 - 5x_2 - 9x_3 \qquad\qquad + z = 0$$

From this system, we write the initial tableau:

$$
\begin{array}{cccccc}
x_1 & x_2 & x_3 & s_1 & s_2 & z
\end{array}
\qquad
\begin{array}{c}
\text{Basic} \\
\text{variables}
\end{array}
$$

$$
\left[
\begin{array}{cccccc|c}
6 & 1 & 4 & 1 & 0 & 0 & 72 \\
3 & 4 & 2 & 0 & 1 & 0 & 30 \\
\hline
-9 & -5 & -9 & 0 & 0 & 1 & 0
\end{array}
\right]
\begin{array}{c}
s_1 \\
s_2 \\
\\
\end{array}
$$

A tie in the last row for the most negative entry $(-9)$ tells us that we have two choices for the pivot column. We use the first one.

$$
\begin{array}{cccccc}
x_1 & x_2 & x_3 & s_1 & s_2 & z
\end{array}
$$

$$
\begin{array}{c}
\text{Pivot} \\
\text{element}
\end{array}
\longrightarrow
\left[
\begin{array}{cccccc|c}
6 & 1 & 4 & 1 & 0 & 0 & 72 \\
\boxed{3} & 4 & 2 & 0 & 1 & 0 & 30 \\
\hline
-9 & -5 & -9 & 0 & 0 & 1 & 0
\end{array}
\right]
\begin{array}{c}
\frac{72}{6} = 12 \\
\frac{30}{3} = 10 \\
\\
\end{array}
$$

We give the sequence of tableaux to find the optimal solution but leave out some of the details. Be sure that you follow each step.

$$
\begin{array}{c}
-6R2 + R1 \rightarrow R1 \\
\\
9R2 + R3 \rightarrow R3
\end{array}
\left[
\begin{array}{cccccc|c}
6 & 1 & 4 & 1 & 0 & 0 & 72 \\
1 & \frac{4}{3} & \frac{2}{3} & 0 & \frac{1}{3} & 0 & 10 \\
-9 & -5 & -9 & 0 & 0 & 1 & 0
\end{array}
\right]
$$

$$
\begin{array}{c}
\\
\text{Basic} \\
\text{variables}
\end{array}
$$

$$
\frac{3}{2}R2 \rightarrow R2
\left[
\begin{array}{cccccc|c}
0 & -7 & 0 & 1 & -2 & 0 & 12 \\
1 & \frac{4}{3} & \boxed{\frac{2}{3}} & 0 & \frac{1}{3} & 0 & 10 \\
0 & 7 & \boxed{-3} & 0 & 3 & 1 & 90
\end{array}
\right]
\begin{array}{c}
s_1 \\
x_1 \\
\\
\end{array}
$$

Not a maximum yet

New pivot element

$$
3R2 + R3 \rightarrow R3
\left[
\begin{array}{cccccc|c}
0 & -7 & 0 & 1 & -2 & 0 & 12 \\
\frac{3}{2} & 2 & 1 & 0 & \frac{1}{2} & 0 & 15 \\
0 & 7 & -3 & 0 & 3 & 1 & 90
\end{array}
\right]
$$

Basic
variables

$$\begin{bmatrix} 0 & -7 & 0 & 1 & -2 & 0 & | & 12 \\ \frac{3}{2} & 2 & 1 & 0 & \frac{1}{2} & 0 & | & 15 \\ \frac{9}{2} & 13 & 0 & 0 & \frac{9}{2} & 1 & | & 135 \end{bmatrix} \begin{array}{l} s_1 \\ x_3 \\ \end{array}$$

This is the final tableau with the solution

$$x_1 = 0, \qquad x_2 = 0, \qquad x_3 = 15, \qquad s_1 = 12, \qquad s_2 = 0, \qquad z = 135$$

We note that if we had chosen the 2 in row 2, column 3 as the pivot element in the first step, we have the sequence:

$$\begin{bmatrix} 6 & 1 & 4 & 1 & 0 & 0 & | & 72 \\ 3 & 4 & 2 & 0 & 1 & 0 & | & 30 \\ -9 & -5 & -9 & 0 & 0 & 1 & | & 0 \end{bmatrix} \qquad \frac{1}{2}R2 \rightarrow R2$$

$$\begin{bmatrix} 6 & 1 & 4 & 1 & 0 & 0 & | & 72 \\ \frac{3}{2} & 2 & 1 & 0 & \frac{1}{2} & 0 & | & 15 \\ -9 & -5 & -9 & 0 & 0 & 1 & | & 0 \end{bmatrix} \begin{array}{l} -4R2 + R1 \rightarrow R1 \\ \\ 9R2 + R3 \rightarrow R3 \end{array}$$

$$\begin{bmatrix} 0 & -7 & 0 & 1 & -2 & 0 & | & 12 \\ \frac{3}{2} & 2 & 1 & 0 & \frac{1}{2} & 0 & | & 15 \\ \frac{9}{2} & 13 & 0 & 0 & \frac{9}{2} & 1 & | & 135 \end{bmatrix}$$

This gives the same final tableau in fewer steps, but you could not have predicted that this solution would be shorter.

In case of a tie for the most negative entry in the last row, each one will lead to the same answer. However, one might lead to the answer in fewer steps. You have no way of knowing the shorter choice.

Now let's look at an example that has a tie for pivot row.

**Example 3**    Maximize $z = 3x_1 + 8x_2$, subject to

$$\begin{aligned} x_1 + 2x_2 &\leq 80 \\ 4x_1 + x_2 &\leq 68 \\ 5x_1 + 3x_2 &\leq 120 \\ x_1 \geq 0, x_2 &\geq 0 \end{aligned}$$

**Solution**
The initial tableau is

$$\begin{bmatrix} 1 & 2 & 1 & 0 & 0 & 0 & | & 80 \\ 4 & 1 & 0 & 1 & 0 & 0 & | & 68 \\ 5 & 3 & 0 & 0 & 1 & 0 & | & 120 \\ -3 & -8 & 0 & 0 & 0 & 1 & | & 0 \end{bmatrix} \begin{array}{l} \frac{80}{2} = 40 \\ \frac{68}{1} = 68 \\ \frac{120}{3} = 40 \\ \\ \end{array}$$

         ↑
         └── Pivot column

Since we have two ratios of 40 each, either row 1 or row 3 may be selected as the pivot row. If we select row 1, then 2 is the pivot element, and our sequence of tableaux is the following. You should work the row operations so that you see that each tableau is correct.

$$
\begin{bmatrix}
\frac{1}{2} & 1 & \frac{1}{2} & 0 & 0 & 0 & 40 \\
4 & 1 & 0 & 1 & 0 & 0 & 68 \\
5 & 3 & 0 & 0 & 1 & 0 & 120 \\
-3 & -8 & 0 & 0 & 0 & 1 & 0
\end{bmatrix}
\quad
\begin{aligned}
& -R1 + R2 \to R2 \\
& -3R1 + R3 \to R3 \\
& 8R1 + R4 \to R4
\end{aligned}
$$

$$
\begin{bmatrix}
\frac{1}{2} & 1 & \frac{1}{2} & 0 & 0 & 0 & 40 \\
\frac{7}{2} & 0 & -\frac{1}{2} & 1 & 0 & 0 & 28 \\
\frac{7}{2} & 0 & -\frac{3}{2} & 0 & 1 & 0 & 0 \\
1 & 0 & 4 & 0 & 0 & 1 & 320
\end{bmatrix}
\quad
\begin{aligned}
& \text{Basic} \\
& \text{variables} \\
& x_2 \\
& s_2 \\
& s_3 \\
& \\
\end{aligned}
$$

This final tableau gives the optimal solution:

$$
x_1 = 0, \quad x_2 = 40, \quad s_1 = 0, \quad s_2 = 28, \quad s_3 = 0, \quad z = 320
$$

We now give an example that arises less frequently, but it is mentioned in the note to the summary of the simplex method. It is possible for a situation to arise where a ratio used to determine the pivot row is zero. Here is an example where that occurs.

**Example 4**   Maximize $z = 6x + 5y$, subject to

$$
\begin{aligned}
3x + y &\le 30 \\
3x + 4y &\le 48 \\
4x + y &\le 40 \\
x \ge 0, y &\ge 0
\end{aligned}
$$

**Solution**
The initial tableau is

$$
\begin{bmatrix}
3 & 1 & 1 & 0 & 0 & 0 & 30 \\
3 & 4 & 0 & 1 & 0 & 0 & 48 \\
4 & 1 & 0 & 0 & 1 & 0 & 40 \\
-6 & -5 & 0 & 0 & 0 & 1 & 0
\end{bmatrix}
\quad
\begin{aligned}
& \tfrac{30}{3} = 10 \\
& \tfrac{48}{3} = 16 \\
& \tfrac{40}{4} = 10 \\
& \\
\end{aligned}
$$

The pivot column is column 1, and the pivot row is either row 1 or row 3 because the smallest ratio, 10, occurs twice. Let's use row 1 as the pivot row and see what happens.

First, we divide each element in row 1 by 3 and then use the row operations indicated below:

$$
\begin{bmatrix}
1 & \frac{1}{3} & \frac{1}{3} & 0 & 0 & 0 & 10 \\
3 & 4 & 0 & 1 & 0 & 0 & 48 \\
4 & 1 & 0 & 0 & 1 & 0 & 40 \\
-6 & -5 & 0 & 0 & 0 & 1 & 0
\end{bmatrix}
\quad
\begin{aligned}
& -3R1 + R2 \to R2 \\
& -4R1 + R3 \to R3 \\
& 6R1 + R4 \to R4
\end{aligned}
$$

This yields the tableau

$$\begin{bmatrix} 1 & \frac{1}{3} & \frac{1}{3} & 0 & 0 & 0 & 10 \\ 0 & 3 & -1 & 1 & 0 & 0 & 18 \\ 0 & -\frac{1}{3} & -\frac{4}{3} & 0 & 1 & 0 & 0 \\ 0 & -3 & 2 & 0 & 0 & 1 & 60 \end{bmatrix}$$

The solution is not optimal, so we use column 2 as the next pivot column. We check the ratios and obtain

$$\frac{10}{\frac{1}{3}} = 30, \qquad \frac{18}{3} = 6, \qquad \frac{0}{-\frac{1}{3}} = 0$$

The smallest nonnegative ratio is zero; however, a negative entry in the pivot column rules out this as the pivot row (page 210). We use the smallest positive ratio, 6, to determine the pivot row. So we pivot on 3 in row 2, column 2. To do so, we replace row 2 by first dividing each element in row 2 by 3 and then using the indicated row operations,

$$\begin{bmatrix} 1 & \frac{1}{3} & \frac{1}{3} & 0 & 0 & 0 & 10 \\ 0 & 1 & -\frac{1}{3} & \frac{1}{3} & 0 & 0 & 6 \\ 0 & -\frac{1}{3} & -\frac{4}{3} & 0 & 1 & 0 & 0 \\ 0 & -3 & 2 & 0 & 0 & 1 & 60 \end{bmatrix} \quad \begin{array}{l} -\frac{1}{3}R2 + R1 \rightarrow R1 \\ \\ \frac{1}{3}R2 + R3 \rightarrow R3 \\ 3R2 + R4 \rightarrow R4 \end{array}$$

to obtain the tableau

$$\begin{bmatrix} 1 & 0 & \frac{4}{9} & -\frac{1}{9} & 0 & 0 & 8 \\ 0 & 1 & -\frac{1}{3} & \frac{1}{3} & 0 & 0 & 6 \\ 0 & 0 & -\frac{13}{9} & \frac{1}{9} & 1 & 0 & 2 \\ 0 & 0 & 1 & 1 & 0 & 1 & 78 \end{bmatrix}$$

This solution is optimal with maximum $z = 78$ at $(8, 6)$.

Now let's go back to the initial tableau and observe what happens if we choose the other of two possible pivot rows. The initial tableau is

$$\begin{bmatrix} 3 & 1 & 1 & 0 & 0 & 0 & 30 \\ 3 & 4 & 0 & 1 & 0 & 0 & 48 \\ 4 & 1 & 0 & 0 & 1 & 0 & 40 \\ -6 & -5 & 0 & 0 & 0 & 1 & 0 \end{bmatrix}$$

The first time we used row 1 as the pivot row. Now we use row 3 as the pivot row. To pivot on 4 in row 3, column 1, we first divide each element in row 3 by 4 and then use the row operations indicated below:

$$\begin{bmatrix} 3 & 1 & 1 & 0 & 0 & 0 & 30 \\ 3 & 4 & 0 & 1 & 0 & 0 & 48 \\ 1 & \frac{1}{4} & 0 & 0 & \frac{1}{4} & 0 & 10 \\ -6 & -5 & 0 & 0 & 0 & 1 & 0 \end{bmatrix} \quad \begin{array}{l} -3R3 + R1 \rightarrow R1 \\ -3R3 + R2 \rightarrow R2 \\ \\ 6R3 + R4 \rightarrow R4 \end{array}$$

We obtain

$$\begin{bmatrix} 0 & \frac{1}{4} & 1 & 0 & -\frac{3}{4} & 0 & 0 \\ 0 & \frac{13}{4} & 0 & 1 & -\frac{3}{4} & 0 & 18 \\ 1 & \frac{1}{4} & 0 & 0 & \frac{1}{4} & 0 & 10 \\ 0 & -\frac{7}{2} & 0 & 0 & \frac{3}{2} & 1 & 60 \end{bmatrix}$$

For the next pivot, use column 2 as the pivot column. From it, we have the ratios

$$\frac{0}{\frac{1}{4}} = 0, \quad \frac{18}{\frac{13}{4}} = 5.54, \quad \frac{10}{\frac{1}{4}} = 40$$

In this case, the smallest nonnegative ratio is zero and the corresponding entry in the pivot column is positive so the summary (page 210) indicates that this determines the pivot row. We pivot on $\frac{1}{4}$ in row 1, column 2, by first replacing row 1 via multiplication of row 1 by 4. Then use the row operations indicated below.

$$\left[\begin{array}{cccccc|c} 0 & 1 & 4 & 0 & -3 & 0 & 0 \\ 0 & \frac{13}{4} & 0 & 1 & -\frac{3}{4} & 0 & 18 \\ 1 & \frac{1}{4} & 0 & 0 & \frac{1}{4} & 0 & 10 \\ 0 & -\frac{7}{2} & 0 & 0 & \frac{3}{2} & 1 & 60 \end{array}\right] \quad \begin{array}{l} \\ -\frac{13}{4}\,\text{R1} + \text{R2} \rightarrow \text{R2} \\ -\frac{1}{4}\,\text{R1} + \text{R3} \rightarrow \text{R3} \\ \frac{7}{2}\,\text{R1} + \text{R4} \rightarrow \text{R4} \end{array}$$

We obtain

$$\left[\begin{array}{cccccc|c} 0 & 1 & 4 & 0 & -3 & 0 & 0 \\ 0 & 0 & -13 & 1 & 9 & 0 & 18 \\ 1 & 0 & -1 & 0 & 1 & 0 & 10 \\ 0 & 0 & 14 & 0 & -9 & 1 & 60 \end{array}\right]$$

which is not optimal. The ratios

$$\frac{0}{-3} = 0, \quad \frac{18}{9} = 2, \quad \frac{10}{1} = 10$$

have zero for the smallest nonnegative ratio, but the pivot column entry for that row is negative, so we do not use row 1 for the pivot row. The smallest positive ratio, 2, then indicates row 2 is the pivot row. To pivot on 9 in row 2, column 5, we first replace row 2 by dividing each entry in row 2 by 9. Then we use the row operations indicated below,

$$\left[\begin{array}{cccccc|c} 0 & 1 & 4 & 0 & -3 & 0 & 0 \\ 0 & 0 & -\frac{13}{9} & \frac{1}{9} & 1 & 0 & 2 \\ 1 & 0 & -1 & 0 & 1 & 0 & 10 \\ 0 & 0 & 14 & 0 & -9 & 1 & 60 \end{array}\right] \quad \begin{array}{l} 3\text{R2} + \text{R1} \rightarrow \text{R1} \\ \\ -\text{R2} + \text{R3} \rightarrow \text{R3} \\ 9\text{R2} + \text{R4} \rightarrow \text{R4} \end{array}$$

to obtain

$$\left[\begin{array}{cccccc|c} 0 & 1 & -\frac{1}{3} & \frac{1}{3} & 0 & 0 & 6 \\ 0 & 0 & -\frac{13}{9} & \frac{1}{9} & 1 & 0 & 2 \\ 1 & 0 & \frac{4}{9} & -\frac{1}{9} & 0 & 0 & 8 \\ 0 & 0 & 1 & 1 & 0 & 1 & 78 \end{array}\right]$$

which is optimal with maximum $z = 78$ at $(8, 6)$, the same result as before.

*To review:* When we had two choices for a pivot row, either one led to the same solution. However, one sequence required more steps.

Now we want to show you what happens had we not used the row with a zero for the pivot row. Look back at the step where the tableau was

$$\begin{bmatrix} 0 & \frac{1}{4} & 1 & 0 & -\frac{3}{4} & 0 & 0 \\ 0 & \frac{13}{4} & 0 & 1 & -\frac{3}{4} & 0 & 18 \\ 1 & \frac{1}{4} & 0 & 0 & \frac{1}{4} & 0 & 10 \\ 0 & -\frac{7}{2} & 0 & 0 & \frac{3}{2} & 1 & 60 \end{bmatrix}$$

with ratios

$$\frac{0}{\frac{1}{4}} = 0, \qquad \frac{18}{\frac{13}{4}} = 5.54, \qquad \frac{10}{\frac{1}{4}} = 40$$

and we used the row with ratio zero for the pivot row.

If we use the smallest *positive* ratio, 5.54, and pivot on $\frac{13}{4}$ in row 2 column 2, we would obtain the tableau

$$\begin{bmatrix} 0 & 0 & 1 & -\frac{1}{13} & -\frac{9}{13} & 0 & -\frac{18}{13} \\ 0 & 1 & 0 & \frac{4}{13} & -\frac{3}{13} & 0 & \frac{72}{13} \\ 1 & 0 & 0 & -\frac{1}{13} & \frac{4}{13} & 0 & \frac{112}{13} \\ 0 & 0 & 0 & \frac{14}{13} & \frac{9}{13} & 1 & \frac{1032}{13} \end{bmatrix}$$

The negative entry in the last column, $-\frac{18}{13}$, is not a valid entry; it indicates that we are not in the feasible region. The pivot on $\frac{13}{4}$ took us outside the feasible region. Thus, if a zero ratio ever occurs, you must be careful to pivot correctly, or an invalid situation may arise.

---

## 4.2    EXERCISES

Access end-of-section exercises online at **www.webassign.net**

---

## Using Your TI Graphing Calculator

### A Program for the Simplex Method

A TI graphing calculator program for the simplex method, SMPLX, allows you to enter the location of the pivot element, and it computes the next tableau. It also computes the ratios used to determine the next pivot row. The initial tableau is stored in [A]. Here is the SMPLX program.

: [A] → [B]
: dim ([A]) → L1
: Lbl 7
: Disp "PIVOT COL"
: Input J
: 1 → L
: Lbl 1
: If abs([B](L,J)) < 10 ^-7
: Goto 2
: [B](L,L1(2))/[B](L,J) → P
: round(P,2) → P
: Disp P
: Lbl 3
: L + 1→ L
: If L < L1(1)
: Goto 1
: Pause
: Goto 4
: Lbl 2
: Disp "ZERO DIV"

: Goto 3
: Lbl 4
: Disp "PIVOT ROW"
: Input I
: *Row(1/[B](I,J),[B],I) → [B]
: 1 → K
: Lbl 5
: If K = I
: Goto 6
: *Row + (-[B](K,J),[B],I,K) → [B]
: Lbl 6
: 1 + K → K
: If K ≤ L1(1)
: Goto 5
: Pause [B] ▷ Frac
: round([B],2) → [C]
: Pause [C]
: Goto 7
: End

(*Note:* The step after Disp "ZERO DIV" is at the top of the second column.)

Note that this program contains the steps "Pause [B] ▷ Frac" and "round([B],2) → [C]". The first displays the tableau in fractional form and the second in decimal form. To display more than two decimals, change 2 to the desired number of decimals.

We use the SMPLX program to solve the following problem: Maximize $z = 5x_1 + 3x_2$, subject to

$$x_1 + x_2 \leq 240$$
$$2x_1 + x_2 \leq 300$$
$$x_1 \geq 0, x_2 \geq 0$$

First, enter the initial tableau in matrix [A]:

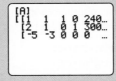

(*Note:* The *z*-column has been omitted throughout, because it does not change when pivoting and would only take up space.)

To initiate the program, select  PRGM  <SMPLX>  ENTER , and you will see the screens

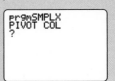

(*continued*)

Enter the first pivot column, column 1 and ENTER, to obtain the screens

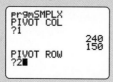

The ratios 240 and 150 indicate the second row is the pivot row. Press ENTER and enter 2.

```
prgmSMPLX
PIVOT COL
?1
                240
                150
PIVOT ROW
?2▮
```

Press ENTER to obtain the next tableau.

## Using Excel

### Using Solver

Excel has a program, **Solver**, that does the computations for a linear programming problem. We use the following example to illustrate how it is used.

Maximize $z = 5x_1 + 8x_2 + 7x_3$ subject to

$$3x_1 + 2x_2 + x_3 \leq 660$$
$$x_1 + 4x_2 + 2x_3 \leq 740$$
$$2x_1 + x_2 + 3x_3 \leq 853$$
$$x_1 \geq 0, x_2 \geq 0, x_3 \geq 0$$

Here's how we set up the problem.

We use the cells A1:C1 for the values of the variables $x_1$, $x_2$, and $x_3$. **Solver** will adjust the values of the variables until the objective function reaches the optimal value. We begin by entering $x_1 = 0$ in A1, $x_2 = 0$ in B1, and $x_3 = 0$ in C1. Next we enter the formulas for the left side of the constraints, the nonnegative conditions and the objective function in column D, and the number on the right side in column E.

|   | A | B | C | D | E | F |
|---|---|---|---|---|---|---|
| 1 | 0 | 0 | 0 | =3*A1+2*B1+C1 | 660 | Constraint 1 |
| 2 |   |   |   | =A1+4*B1+2*C1 | 740 | Constraint 2 |
| 3 |   |   |   | =2*A1+B1+3*C1 | 853 | Constraint 3 |
| 4 |   |   |   | =A1 | 0 | Nonnegative condition |
| 5 |   |   |   | =B1 | 0 | Nonnegative condition |
| 6 |   |   |   | =C1 | 0 | Nonnegative condition |
| 7 |   |   |   | =5*A1+8*B1+7*C1 |   |   |

To start up **Solver**, click on the **Data** tab. Move to the **Analysis** group and click on **Solver**.

- Place the cursor in the box next to **Set Target Cell**, or **Set Objective**, depending on your version of Excel.
- Select cell **D7**, since this is the location of the formula for the objective function.
- Check the button to **maximize**, since this is a maximization problem.

Place the cursor in the box under **By changing cells:** and then select cells A1:C1 (the cells containing the variables).

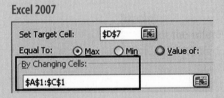 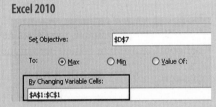

You are now ready to enter the constraints.

Move the cursor to the box under **Subject to the Constraints:** and click on the button shown as **Add** to obtain

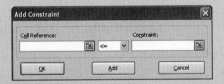

Make sure the cursor is in the box under **Cell Reference:** and select cell D1; then move the cursor to the box under **Constraint:** and select the cell E1.

Click the **Add** button. Repeat the preceding step using cells D2 and E2, then D3 and E3. Click **OK**.

You have entered the constraints. Note that the constraints have the symbol <= between **Cell Reference:** and **Constraint:** (this represents ≤).

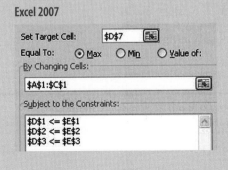 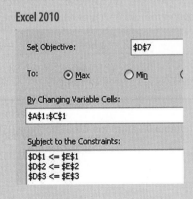

*(continued)*

Click on the **Add** button to enter the nonnegative conditions. Again place the cursor in the **Cell Reference:** box and select cell D4. Then select E4 for the **Constraint:** box. Since this is a nonnegative condition, we need the $\geq$ symbol; get it by selecting $>=$. Then click **Add.**

In the same way, enter the nonnegative constraints using D5, E5 and D6, E6. You will then have the screen

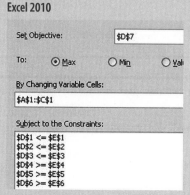

**Excel 2007**

Set Target Cell: $D$7
Equal To: ⦿ Max ◯ Min ◯ Value of:
By Changing Cells:
$A$1:$C$1
Subject to the Constraints:
$D$1 <= $E$1
$D$2 <= $E$2
$D$3 <= $E$3
$D$4 >= $E$4
$D$5 >= $E$5
$D$6 >= $E$6

**Excel 2010**

Set Objective: $D$7
To: ⦿ Max ◯ Min ◯ Val
By Changing Variable Cells:
$A$1:$C$1
Subject to the Constraints:
$D$1 <= $E$1
$D$2 <= $E$2
$D$3 <= $E$3
$D$4 >= $E$4
$D$5 >= $E$5
$D$6 >= $E$6

Now set the options for **Solver**, depending on your version.

**Excel 2007**

Click into the Options box, and make sure that the **Assume Linear Model** checkbox is checked, as in the following figure.

Leave all other options as is, and click **OK.**

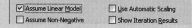

**Excel 2010**

Make sure the Simplex LP method is selected in the dialog box.

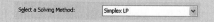

Click on the **Solve** button to obtain the solution.

|   | A | B | C | D | E | F |
|---|---|---|---|---|---|---|
| 1 | 116 | 63 | 186 | 660 | 660 | Constraint 1 |
| 2 |   |   |   | 740 | 740 | Constraint 2 |
| 3 |   |   |   | 853 | 853 | Constraint 3 |
| 4 |   |   |   | 116 | 0 | Nonnegative condition |
| 5 |   |   |   | 63 | 0 | Nonnegative condition |
| 6 |   |   |   | 186 | 0 | Nonnegative condition |
| 7 |   |   |   | 2386 |   |   |

The maximum value of $z = 2386$ at $(116, 63, 186)$.

# 4.3  THE STANDARD MINIMUM PROBLEM: DUALITY

- Standard Minimum Problem
- Dual Problem
- Solve the Minimization Problem
- Using the Dual Problem

## Standard Minimum Problem

We have used the simplex method to solve standard maximum problems. However, a variety of optimization problems that are not standard maximum need to be solved. We consider one form in this section, the **standard minimum**, which can be solved by a procedure called the **dual method**.

The standard minimum problem can be solved with other techniques, to be discussed in Section 4.4. Those techniques also provide a means of solving other problems that are not standard. We include the dual method because it is a commonly used one.

The dual method converts a standard minimum problem to a standard maximum problem. Before we show you this method, we define a standard minimum problem.

**DEFINITION**
**Standard Minimum**

A linear programming problem is **standard minimum** if

1. The objective function is to be minimized.
2. All the inequalities are $\geq$.
3. The constants to the right of the inequalities are nonnegative.
4. The variables are restricted to nonnegative values (nonnegative conditions).

## Dual Problem

We can solve the standard minimum problem by converting it to a *dual* maximum problem. Let's look at an example to describe how to set up the **dual problem**.

**Example 1**    A doctor specifies that a patient's diet contain certain minimum amounts of iron and calcium, but calories are to be held to a minimum.

Two foods, A and B, are used in a meal, and the amounts of iron, calcium, and calories are given in the following table:

|  | Amount provided by one unit of | | Amount required |
|---|---|---|---|
|  | A | B |  |
| Iron | 4 | 1 | 12 or more |
| Calcium | 2 | 3 | 10 or more |
| Calories | 90 | 120 |  |

We convert this information into a linear programming form as follows:

$$\text{Let} \quad x_1 = \text{the number of units of A}$$
$$x_2 = \text{the number of units of B}$$

The iron requirement is $4x_1 + x_2 \geq 12$, the calcium requirement is $2x_1 + 3x_2 \geq 10$, and the calorie count is $90x_1 + 120x_2$, where $x_1 \geq 0$ and $x_2 \geq 0$.

The problem is
Minimize $z = 90x_1 + 120x_2$, subject to

$$4x_1 + x_2 \geq 12$$
$$2x_1 + 3x_2 \geq 10$$
$$x_1 \geq 0, x_2 \geq 0$$

**CAUTION**

This is *not* a simplex tableau, because it does not contain slack variables and the objective function has not been rewritten. We wrote the matrix in this form because it is used to obtain a maximization problem first.

**Solution**
To form the dual problem, we first write the minimum problem in a matrix form, using an augmented matrix of the constraints and objective function. For this example, the matrix takes the following form. Note its similarity to the augmented matrix of a system of equations.

$$\begin{bmatrix} 4 & 1 & 12 \\ 2 & 3 & 10 \\ \hline 90 & 120 & 1 \end{bmatrix}$$

Next, we obtain a new matrix by taking each *row* of

$$A = \begin{bmatrix} 4 & 1 & 12 \\ 2 & 3 & 10 \\ \hline 90 & 120 & 1 \end{bmatrix}$$

and making it the *column* of the new matrix. (The new matrix is called the **transpose of the matrix**.) The new matrix is

$$B = \begin{bmatrix} 4 & 2 & 90 \\ 1 & 3 & 120 \\ \hline 12 & 10 & 1 \end{bmatrix}$$

From the new matrix, $B$, we set up a standard maximum problem. To do so, we introduce new variables, $y_1$ and $y_2$, because they play a different role from the original ones. We use the rows above the line to form constraints for the maximum problem. The row below the line forms the new objective function. Because we want this matrix to give a standard maximum problem, all inequalities are $\leq$. We write the new constraints and objective functions next to their rows:

$$\begin{array}{cc} y_1 & y_2 \\ \end{array}$$
$$\begin{bmatrix} 4 & 2 & 90 \\ 1 & 3 & 120 \\ \hline 12 & 10 & 1 \end{bmatrix}$$ 
New constraint: $4y_1 + 2y_2 \leq 90$
New constraint: $y_1 + 3y_2 \leq 120$
New objective function: $w = 12y_1 + 10y_2$

This gives the dual problem:
Maximize $w = 12y_1 + 10y_2$, subject to

$$4y_1 + 2y_2 \leq 90$$
$$y_1 + 3y_2 \leq 120$$
$$y_1 \geq 0, y_2 \geq 0$$

**Example 2**   Set up the dual problem to the following standard minimum problem.

Minimize $z = 30x_1 + 40x_2 + 50x_3$, subject to

$$10x_1 + 14x_2 + 5x_3 \geq 220$$
$$5x_1 + 3x_2 + 9x_3 \geq 340$$
$$x_1 \geq 0, x_2 \geq 0, x_3 \geq 0$$

**Solution**

Form the augmented matrix of the problem with the objective function written in the last row:

$$A = \begin{array}{c} \phantom{A=} \begin{array}{ccc} x_1 & x_2 & x_3 \end{array} \\ \left[\begin{array}{ccc|c} 10 & 14 & 5 & 220 \\ 5 & 3 & 9 & 340 \\ \hline 30 & 40 & 50 & 1 \end{array}\right] \end{array}$$

Form the transpose of $A$:

$$B = \begin{array}{c} \phantom{B=} \begin{array}{cc} y_1 & y_2 \end{array} \\ \left[\begin{array}{cc|c} 10 & 5 & 30 \\ 14 & 3 & 40 \\ 5 & 9 & 50 \\ \hline 220 & 340 & 1 \end{array}\right] \end{array}$$

Set up the dual problem from this matrix using $\leq$ on all constraints:

Maximize $w = 220y_1 + 340y_2$, subject to

$$10y_1 + 5y_2 \leq 30$$
$$14y_1 + 3y_2 \leq 40$$
$$5y_1 + 9y_2 \leq 50$$
$$y_1 \geq 0, y_2 \geq 0$$

---

**Set Up the Dual Problem of a Standard Minimum Problem**

1. Start with a standard minimum problem.
2. Write the augmented matrix, $A$, of the minimum problem. Write the objective function in the last row.
3. Write the transpose of the matrix $A$ to obtain matrix $B$. Each row of $A$ becomes the corresponding column of $B$.
4. Form a constraint for the dual problem from each row of $B$ (except the last) using the new variables and $\leq$.
5. Form the objective function of the dual problem from the last row of $B$. It is to be maximized.

## Solve the Minimization Problem Using the Dual Problem

The theory relating a standard minimum problem to its dual problem is beyond the level of this course. We state the relationship between the solution of a minimum problem and its dual problem with a fundamental theorem, which we will use.

| **THEOREM** | A standard minimum problem has a solution if and only if its dual problem has a solution. If a solution exists, the standard minimum problem and its dual problem *have the same* optimal value. |
| --- | --- |
| **Fundamental Theorem of Duality** | |

This theorem states that the maximum value of the dual problem objective function is the minimum value of the objective function for the minimum problem. To help see this, let's work through the diet example at the beginning of the section. The problem and its dual are as follows:

| **Standard minimum problem** | **Dual problem** |
| --- | --- |
| Minimize $z = 90x_1 + 120x_2$, subject to | Maximize $w = 12y_1 + 10y_2$, subject to |
| $4x_1 + x_2 \geq 12$ | $4y_1 + 2y_2 \leq 90$ |
| $2x_1 + 3x_2 \geq 10$ | $y_1 + 3y_2 \leq 120$ |
| $x_1 \geq 0, x_2 \geq 0$ | $y_1 \geq 0, y_2 \geq 0$ |

The procedure is straightforward: Solve the dual problem by the simplex method. We first write the dual problem as a system of equations using slack variables and obtain

$$
\begin{aligned}
4y_1 + 2y_2 + x_1 \qquad\qquad &= 90 \\
y_1 + 3y_2 \qquad + x_2 \qquad &= 120 \\
-12y_1 - 10y_2 \qquad\qquad + w &= 0
\end{aligned}
$$

Note that we use $x_1$ and $x_2$ for slack variables. We intend these to be the same as the variables in the original minimum problem, because it turns out that certain values of the slack variables of the dual problem give the desired values of the original variables in the minimum problem. Let's set up the simplex tableau and work through the solution.

The initial tableau is

$$
\begin{array}{ccccc}
y_1 & y_2 & x_1 & x_2 & w \\
\left[\begin{array}{ccccc|c}
4 & 2 & 1 & 0 & 0 & 90 \\
1 & 3 & 0 & 1 & 0 & 120 \\
\hline
-12 & -10 & 0 & 0 & 1 & 0
\end{array}\right]
\end{array}
$$

We now proceed to find the pivot element and perform row operations in the usual manner. You should fill in details that are omitted.

$$
\begin{array}{ccccc}
y_1 & y_2 & x_1 & x_2 & w \\
\left[\begin{array}{ccccc|c}
4 & 2 & 1 & 0 & 0 & 90 \\
1 & 3 & 0 & 1 & 0 & 120 \\
\hline
-12 & -10 & 0 & 0 & 1 & 0
\end{array}\right] & \tfrac{1}{4}R1 \to R1
\end{array}
$$

$$
\begin{array}{ccccc}
y_1 & y_2 & x_1 & x_2 & w \\
\left[\begin{array}{ccccc|c}
1 & \tfrac{1}{2} & \tfrac{1}{4} & 0 & 0 & \tfrac{90}{4} \\
1 & 3 & 0 & 1 & 0 & 120 \\
\hline
-12 & -10 & 0 & 0 & 1 & 0
\end{array}\right] & \begin{array}{l} \\ -R1 + R2 \to R2 \\ 12R1 + R3 \to R3 \end{array}
\end{array}
$$

$$
\begin{array}{ccccc}
y_1 & y_2 & x_1 & x_2 & w \\
\end{array}
$$

$$
\left[\begin{array}{ccccc|c}
1 & \frac{1}{2} & \frac{1}{4} & 0 & 0 & \frac{90}{4} \\
0 & \frac{5}{2} & -\frac{1}{4} & 1 & 0 & \frac{390}{4} \\
0 & -4 & 3 & 0 & 1 & 270
\end{array}\right]
\qquad \frac{2}{5}R2 \rightarrow R2
$$

$$
\begin{array}{ccccc}
y_1 & y_2 & x_1 & x_2 & w \\
\end{array}
$$

$$
\left[\begin{array}{ccccc|c}
1 & \frac{1}{2} & \frac{1}{4} & 0 & 0 & \frac{90}{4} \\
0 & 1 & -\frac{1}{10} & \frac{2}{5} & 0 & 39 \\
0 & -4 & 3 & 0 & 1 & 270
\end{array}\right]
\qquad \begin{array}{l} -\frac{1}{2}R2 + R1 \rightarrow R1 \\[10pt] 4R2 + R3 \rightarrow R3 \end{array}
$$

$$
\begin{array}{ccccc}
y_1 & y_2 & x_1 & x_2 & w \\
\end{array}
$$

$$
\left[\begin{array}{ccccc|c}
1 & 0 & \frac{6}{20} & -\frac{1}{5} & 0 & 3 \\
0 & 1 & -\frac{1}{10} & \frac{2}{5} & 0 & 39 \\
0 & 0 & \frac{26}{10} & \frac{8}{5} & 1 & 426
\end{array}\right]
$$

As no negative entries appear in the bottom row, the solution is optimal, and the maximum value is 426. The maximum value occurs when

$$y_1 = 3 \qquad \text{and} \qquad y_2 = 39$$

By the fundamental theory of duality, the *minimum* value of the original objective function, $z = 90x_1 + 120x_2$, is also 426. The values of $x_1$ and $x_2$ that yield this minimum value are found in the bottom row of the final tableau of the dual problem. That bottom row is

$$
\begin{array}{ccccc}
y_1 & y_2 & x_1 & x_2 & w \\
[0 & 0 & 2.6 & 1.6 & 1 \quad 426]
\end{array}
$$

The numbers under $x_1$ and $x_2$ are the values of $x_1$ and $x_2$ that give the optimal value of the original minimization problem. So the objective function $z = 90x_1 + 120x_2$ has the minimum value of 426 at $x_1 = 2.6$, $x_2 = 1.6$.

**NOTE**

To find the solution of a standard minimum problem, *look at the bottom row of the final tableau of the dual problem.*

**Example 3**  Solve the following minimization problem by the dual problem method:
Minimize $z = 8x_1 + 15x_2$, subject to

$$
\begin{aligned}
4x_1 + 5x_2 &\geq 80 \\
2x_1 + 5x_2 &\geq 60 \\
x_1 \geq 0, x_2 &\geq 0
\end{aligned}
$$

**Solution**
The augmented matrix of this problem is

$$
A = \left[\begin{array}{cc|c}
4 & 5 & 80 \\
2 & 5 & 60 \\
8 & 15 & 1
\end{array}\right]
$$

The transpose of $A$ is

$$
B = \left[\begin{array}{cc|c}
4 & 2 & 8 \\
5 & 5 & 15 \\
80 & 60 & 1
\end{array}\right]
$$

*B* represents the following maximization problem:

Maximize $w = 80y_1 + 60y_2$, subject to

$$4y_1 + 2y_2 \le 8$$
$$5y_1 + 5y_2 \le 15$$
$$y_1 \ge 0, y_2 \ge 0$$

The initial simplex tableau of this problem is

$$
\begin{array}{c c c c c}
y_1 & y_2 & x_1 & x_2 & w \\
\end{array}
$$
$$
\left[
\begin{array}{c c c c c|c}
4 & 2 & 1 & 0 & 0 & 8 \\
5 & 5 & 0 & 1 & 0 & 15 \\
\hline
-80 & -60 & 0 & 0 & 1 & 0
\end{array}
\right]
$$

Now proceed with the pivot and row operations to obtain the sequence of tableaux:

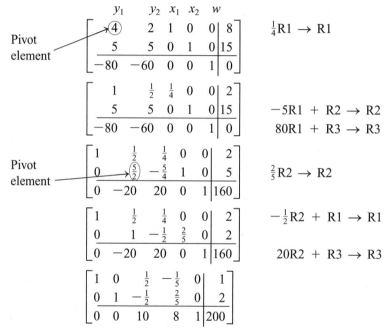

This is the final tableau of the dual problem. The last row gives the solution to the minimum problem:

$$x_1 = 10 \qquad x_2 = 8 \qquad z = 200$$

---

## 4.3 EXERCISES

Access end-of-section exercises online at **www.webassign.net**

ENHANCED
WebAssign

## 4.4 MIXED CONSTRAINTS

- Minimizing a Function
- Problems with $\geq$ Constraints
- Negative Constant in $\leq$ Constraints
- Examples and Applications

The simplex method has been used to solve standard maximum and standard minimum problems. Although these are important problems, other types of optimization problems arise.

In this section, we study more general problems. The constraints may be a mixture of $\leq$ and $\geq$, and we may wish to either maximize or minimize the objective function. Because such problems contain a mixture of $\leq$ and $\geq$, they are referred to as having **mixed constraints**.

### Minimizing a Function

A minimization problem may or may not be standard. In either case, we can make an adjustment that converts a minimization problem to a maximization problem whose solution enables us to find the solution to the minimization problem.

The adjustment is simple. If $z$ is the objective function to be minimized, then solve the maximization problem using $w = -z$ as the objective function. This works because if $k$ is the maximum value of $w$, then $-k$ is the minimum value of $z$. For example, when you multiply a set of numbers by $-1$, you reverse the order. The set of numbers $\{1, 5, 7, 16\}$ has 1 as the smallest number and 16 as the largest number. The set made of the negatives of these numbers is $\{-1, -5, -7, -16\}$. It has $-1$ as the *largest* number and $-16$ as the *smallest*.

Let's look at an example that illustrates the simplex solution of a minimization problem.

**Example 1**  Minimize $z = 2x_1 - 3x_2$, subject to

$$x_1 + 2x_2 \leq 10$$
$$2x_1 + x_2 \leq 11$$
$$x_1 \geq 0, x_2 \geq 0$$

**Solution**

Convert the objective function to $w = -2x_1 + 3x_2$. We now seek to maximize $w = -2x_1 + 3x_2$, subject to the original constraints.

The tableaux for the solution are as follows:

$$
\begin{array}{ccccc}
x_1 & x_2 & s_1 & s_2 & w \\
\end{array}
$$

$$
\left[\begin{array}{ccccc|c}
1 & 2 & 1 & 0 & 0 & 10 \\
2 & 1 & 0 & 1 & 0 & 11 \\
\hline
2 & -3 & 0 & 0 & 1 & 0
\end{array}\right] \quad \tfrac{1}{2}R1 \rightarrow R1
$$

$$
\left[\begin{array}{ccccc|c}
\tfrac{1}{2} & 1 & \tfrac{1}{2} & 0 & 0 & 5 \\
2 & 1 & 0 & 1 & 0 & 11 \\
\hline
2 & -3 & 0 & 0 & 1 & 0
\end{array}\right] \quad \begin{array}{l} -R1 + R2 \rightarrow R2 \\ 3R1 + R3 \rightarrow R3 \end{array}
$$

$$
\left[\begin{array}{ccccc|c}
\tfrac{1}{2} & 1 & \tfrac{1}{2} & 0 & 0 & 5 \\
\tfrac{3}{2} & 0 & -\tfrac{1}{2} & 1 & 0 & 6 \\
\hline
\tfrac{7}{2} & 0 & \tfrac{3}{2} & 0 & 1 & 15
\end{array}\right]
$$

The optimal solution (maximum) is $w = 15$ when $x_1 = 0$, $x_2 = 5$. The original problem then has as its optimal (minimum) solution $z = -15$ at $x_1 = 0$, $x_2 = 5$.

## Problems with $\geq$ Constraints

The simplex method assumes that all constraints of a maximum problem are of the form

$$a_1 x_1 + a_2 x_2 + \cdots + a_n x_n \leq b$$

Realistically, some of the constraints can be of the form

$$a_1 x_1 + a_2 x_2 + \cdots + a_n x_n \geq b$$

We can use the simplex method by modifying any constraints of this type in the following way.

We modify a $\geq$ constraint by multiplying through by $-1$, which reverses the sign, giving a $\leq$ constraint. This may introduce a negative constant in the new constraint, but we will describe how to handle that situation later.

**Modification of $\geq$ Constraints**    For the simplex method:

$$\begin{array}{ll} \text{Replace} & a_1 x_1 + a_2 x_2 + \cdots + a_n x_n \geq b \\ \text{with} & -a_1 x_1 - a_2 x_2 - \cdots - a_n x_n \leq -b \end{array}$$

The following example illustrates the modifications when $\geq$ constraints occur.

**Example 2**    Modify the following problem and set up the initial simplex tableau:
Maximize $z = 8x_1 + 2x_2 + 6x_3$, subject to

$$\begin{aligned} 6x_1 + 4x_2 + 5x_3 &\leq 68 \\ 4x_1 + 3x_2 + x_3 &\geq 32 \\ 2x_1 + 4x_2 + 3x_3 &\geq 36 \\ x_1 \geq 0, x_2 \geq 0, x_3 &\geq 0 \end{aligned}$$

**Solution**
Write all the constraints, other than the nonnegative conditions, as $\leq$ constraints. The problem then becomes
Maximize $z = 8x_1 + 2x_2 + 6x_3$, subject to

$$\begin{aligned} 6x_1 + 4x_2 + 5x_3 &\leq 68 \\ -4x_1 - 3x_2 - x_3 &\leq -32 \\ -2x_1 - 4x_2 - 3x_3 &\leq -36 \\ x_1 \geq 0, x_2 \geq 0, x_3 &\geq 0 \end{aligned}$$

The initial simplex tableau is

$$\begin{bmatrix} 6 & 4 & 5 & 1 & 0 & 0 & 0 & | & 68 \\ -4 & -3 & -1 & 0 & 1 & 0 & 0 & | & -32 \\ -2 & -4 & -3 & 0 & 0 & 1 & 0 & | & -36 \\ -8 & -2 & -6 & 0 & 0 & 0 & 1 & | & 0 \end{bmatrix}$$

In the last example, the last column contains negative constants. The standard simplex method assumes nonnegative entries in the last column, so we need to know how to proceed in such a situation.

## Negative Constant in $\leq$ Constraints

Let's use a simple example to illustrate the adjustments that enable us to deal with a negative constant.

**Example 3**   Maximize $z = 20x + 50y$, subject to

$$\begin{aligned} 3x + 4y &\leq 72 \\ -5x + 2y &\leq -16 \\ x \geq 0, y &\geq 0 \end{aligned}$$

**Solution**
The graph of the feasible region is shown in Figure 4–3. Notice that the corners of the feasible region are $(\frac{16}{5}, 0)$ (24, 0), and (8, 12).

If we set up the initial tableau for this problem in the usual way, we obtain

$$\begin{bmatrix} 3 & 4 & 1 & 0 & 0 & | & 72 \\ -5 & 2 & 0 & 1 & 0 & | & -16 \\ -20 & -50 & 0 & 0 & 1 & | & 0 \end{bmatrix}$$

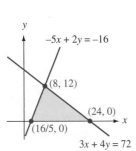

**FIGURE 4–3**   The feasible region of Example 3.

Generally, the initial basic solution in the simplex method occurs at the origin. That holds true in this case with $x_1 = 0$, $x_2 = 0$, $s_1 = 72$, and $s_2 = -16$. However, this solution is *not feasible* because $s_2$ is negative. Figure 4–3 shows that the origin is outside the feasible region. Because the basic solution is not feasible, we need to adjust the tableau so that the simplex procedure can be used. We do so by a sequence of pivots that move us from the origin to a corner point of the feasible region. Here's how.

In the initial tableau, select the row with the most negative constant in the *right-most* (last) column. In this case use row 2 because of $-16$. Next, select the pivot element. The *left-most* entry in the row is usually a good choice, but another negative entry may be chosen when, for example, it yields simpler coefficients. In this case, $-5$ is the choice. This entry becomes a pivot element. Divide the entries in the second row by $-5$ to obtain the following tableau. Then pivot on the 1 in row 2, column 1, using the indicated row operations.

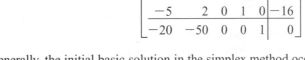

$$\begin{bmatrix} 3 & 4 & 1 & 0 & 0 & | & 72 \\ 1 & -\frac{2}{5} & 0 & -\frac{1}{5} & 0 & | & \frac{16}{5} \\ -20 & -50 & 0 & 0 & 1 & | & 0 \end{bmatrix} \quad \begin{matrix} -3R2 + R1 \rightarrow R1 \\ \\ 20R2 + R3 \rightarrow R3 \end{matrix}$$

After pivoting on 1 in row 2, column 1, we have the following tableau:

$$\begin{bmatrix} 0 & \frac{26}{5} & 1 & \frac{3}{5} & 0 & \frac{312}{5} \\ 1 & -\frac{2}{5} & 0 & -\frac{1}{5} & 0 & \frac{16}{5} \\ 0 & -58 & 0 & -4 & 1 & 64 \end{bmatrix}$$

This tableau yields the basic feasible solution $x = \frac{16}{5}$, $y = 0$, $s_1 = \frac{312}{5}$, and $s_2 = 0$. This solution takes us to the point $(\frac{16}{5}, 0)$, which is a corner point of the feasible region. Because the basic solution is feasible, we can apply the simplex method to this tableau in the usual manner. The next pivot element is $\frac{26}{5}$ in row 1. We multiply the first row by $\frac{5}{26}$ to obtain the following matrix. Then perform the indicated row operations.

$$\begin{bmatrix} 0 & 1 & \frac{5}{26} & \frac{3}{26} & 0 & 12 \\ 0 & -\frac{2}{5} & 0 & -\frac{1}{5} & 0 & \frac{16}{5} \\ 0 & -58 & 0 & -4 & 1 & 64 \end{bmatrix} \quad \begin{array}{l} \frac{2}{5}R1 + R2 \to R2 \\ 58R1 + R3 \to R3 \end{array}$$

We get the tableau

$$\begin{bmatrix} 0 & 1 & \frac{5}{26} & \frac{3}{26} & 0 & 12 \\ 1 & 0 & \frac{1}{13} & -\frac{2}{13} & 0 & 8 \\ 0 & 0 & \frac{145}{13} & \frac{35}{13} & 1 & 760 \end{bmatrix}$$

This tableau gives the optimal solution of $z = 760$ at $x = 8$, $y = 12$.

Look back over the solution to the problem and observe two phases that generally apply to problems of this type.

**Phase I.**   Set up the initial tableau. When a negative constant appears in the right-most column of the tableau, the basic solution is not feasible and we are outside the feasible region. We need to modify the tableau to give a feasible basic solution and thereby bring us into the feasible region. We do this with a pivot whose pivot row and pivot column are determined in the following manner:

1. Determine the pivot row by selecting the row with the *most negative* entry in its right-most column.

2. A negative entry in the pivot row determines the pivot column, and it becomes the pivot element.

Using this pivot will remove the negative entry in the pivot row; however, it might be necessary to pivot more than once to remove all negative entries in the last column.

**Phase II.**   When the modifications in Phase I produce a tableau with feasible basic solutions, then proceed with the usual simplex method.

Now look back at Figure 4–3 to see what happened in Phase I. The initial tableau gives the basic solution with $x = 0$, $y = 0$. In this case, this point $(0, 0)$ lies outside the feasible region. (That's why we got an infeasible solution.) When we pivot on $-5$, the next basic solution has $x = \frac{16}{5}$, $y = 0$ with both slack variables nonnegative. This solution is feasible and represents the corner $(\frac{16}{5}, 0)$ of the feasible region. Thus, Phase I moved from the origin to a corner of the feasible region.

When a feasible solution is reached in Phase I, we enter Phase II and follow the simplex procedure. In this example, the optimal solution is reached in one more pivot. Notice that the pivot takes us from the corner $(\frac{16}{5}, 0)$ to the corner $(8, 12)$, the optimal solution.

Look at another example to be sure that you understand the procedure followed in Phase I and Phase II.

**Example 4**  Maximize $z = 3x_1 + 8x_2 + 4x_3$, subject to

$$
\begin{aligned}
x_1 + x_2 + x_3 &\le 12 \\
2x_1 + 6x_2 + 3x_3 &\le 42 \\
x_1 - 2x_2 &\ge 6 \\
x_1 \ge 0, x_2 \ge 0, x_3 &\ge 0
\end{aligned}
$$

**Solution**
We need to replace the constraint

$$x_1 - 2x_2 \ge 6$$

with

$$-x_1 + 2x_2 \le -6$$

**Phase I.**  The initial simplex tableau is

$$
\left[
\begin{array}{ccccccc|c}
1 & 1 & 1 & 1 & 0 & 0 & 0 & 12 \\
2 & 6 & 3 & 0 & 1 & 0 & 0 & 42 \\
-1 & 2 & 0 & 0 & 0 & 1 & 0 & -6 \\
\hline
-3 & -8 & -4 & 0 & 0 & 0 & 1 & 0
\end{array}
\right]
\qquad
\begin{array}{l}
\text{R3} + \text{R1} \to \text{R1} \\[4pt]
\text{2R3} + \text{R2} \to \text{R2} \\[8pt]
\\
-\text{3R3} + \text{R4} \to \text{R4}
\end{array}
$$

Pivot on $-1$ in column 1, row 3, by using the operations indicated above. Then multiply row 3 of the result by $-1$, giving the tableau

$$
\left[
\begin{array}{ccccccc|c}
0 & 3 & 1 & 1 & 0 & 1 & 0 & 6 \\
0 & 10 & 3 & 0 & 1 & 2 & 0 & 30 \\
1 & -2 & 0 & 0 & 0 & -1 & 0 & 6 \\
\hline
0 & -14 & -4 & 0 & 0 & -3 & 1 & 18
\end{array}
\right]
$$

This tableau gives a feasible basic solution, $x_1 = 6, x_2 = 0, x_3 = 0, s_1 = 6, s_2 = 30, s_3 = 0$, so proceed to Phase II.

**Phase II.**  Pivot on 3 in column 2, row 1, by first dividing the entries of row 1 by 3 to obtain the tableau below. Complete the pivot with the indicated row operations.

$$
\left[
\begin{array}{ccccccc|c}
0 & 1 & \frac{1}{3} & \frac{1}{3} & 0 & \frac{1}{3} & 0 & 2 \\
0 & 10 & 3 & 0 & 1 & 2 & 0 & 30 \\
1 & -2 & 0 & 0 & 0 & -1 & 0 & 6 \\
\hline
0 & -14 & -4 & 0 & 0 & -3 & 1 & 18
\end{array}
\right]
\qquad
\begin{array}{l}
\\
-\text{10R1} + \text{R2} \to \text{R2} \\[4pt]
\text{2R1} + \text{R3} \to \text{R3} \\[4pt]
\text{14R1} + \text{R4} \to \text{R4}
\end{array}
$$

This yields the following tableau:

$$\begin{bmatrix} 0 & 1 & \frac{1}{3} & \frac{1}{3} & 0 & \frac{1}{3} & 0 & 2 \\ 0 & 0 & -\frac{1}{3} & -\frac{10}{3} & 1 & -\frac{4}{3} & 0 & 10 \\ 1 & 0 & \frac{2}{3} & \frac{2}{3} & 0 & -\frac{1}{3} & 0 & 10 \\ 0 & 0 & \frac{2}{3} & \frac{14}{3} & 0 & \frac{5}{3} & 1 & 46 \end{bmatrix}$$

The optimal solution is $z = 46$ at $(10, 2, 0)$.

## Examples and Applications

We now give some examples to illustrate problems with mixed constraints.

**Example 5**   Minimize $z = 8x_1 + 5x_2$, subject to

$$\begin{aligned} x_1 + x_2 &\leq 8 \\ 5x_1 + 3x_2 &\geq 21 \\ x_1 + 3x_2 &\geq 9 \\ x_1 \geq 0, x_2 &\geq 0 \end{aligned}$$

**Solution**

We need to make the following modifications to obtain the simplex tableau.

Change minimize $z = 8x_1 + 5x_2$ to maximize $w = -8x_1 - 5x_2$, change $5x_1 + 3x_2 \geq 21$ to $-5x_1 - 3x_2 \leq -21$, and change $x_1 + 3x_2 \geq 9$ to $-x_1 - 3x_2 \leq -9$. This converts the problem to

Maximize $w = -8x_1 - 5x_2$, subject to

$$\begin{aligned} x_1 + x_2 &\leq 8 \\ -5x_1 - 3x_2 &\leq -21 \\ -x_1 - 3x_2 &\leq -9 \\ x_1 \geq 0, x_2 &\geq 0 \end{aligned}$$

(See Figure 4–4.)

The initial simplex tableau is

$$\begin{bmatrix} 1 & 1 & 1 & 0 & 0 & 0 & 8 \\ -5 & -3 & 0 & 1 & 0 & 0 & -21 \\ -1 & -3 & 0 & 0 & 1 & 0 & -9 \\ 8 & 5 & 0 & 0 & 0 & 1 & 0 \end{bmatrix}$$

This has the basic solution

$$x_1 = 0, \quad x_2 = 0, \quad s_1 = 8, \quad s_2 = -21, \quad s_3 = -9$$

which is not feasible. [It gives the origin, $(0, 0)$, in Figure 4–4.] Because the solution is not feasible, we enter Phase I to modify the negative entries in the last column.

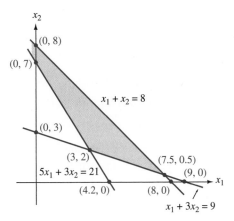

**FIGURE 4-4**

**Phase I.** Row 2 is the pivot row because of $-21$. Pivot on the left-most negative entry in row 2, the entry $-5$. We do so by first dividing each entry in row 2 by $-5$, giving the tableau below. We complete the pivot using the indicated operations.

$$
\begin{bmatrix}
1 & 1 & 1 & 0 & 0 & 0 & 8 \\
1 & \frac{3}{5} & 0 & -\frac{1}{5} & 0 & 0 & \frac{21}{5} \\
-1 & -3 & 0 & 0 & 1 & 0 & -9 \\
8 & 5 & 0 & 0 & 0 & 1 & 0
\end{bmatrix}
\quad
\begin{array}{l}
-R2 + R1 \to R1 \\
\\
R2 + R3 \to R3 \\
-8R3 + R4 \to R4
\end{array}
$$

This gives the next tableau

$$
\begin{bmatrix}
0 & \frac{2}{5} & 1 & \frac{1}{5} & 0 & 0 & \frac{19}{5} \\
1 & \frac{3}{5} & 0 & -\frac{1}{5} & 0 & 0 & \frac{21}{5} \\
0 & -\frac{12}{5} & 0 & -\frac{1}{5} & 1 & 0 & -\frac{24}{5} \\
0 & \frac{1}{5} & 0 & \frac{8}{5} & 0 & 1 & -\frac{168}{5}
\end{bmatrix}
$$

The negative $\frac{24}{5}$ in the last column of row 3 indicates we are not yet in the feasible region, so we pivot on $-\frac{12}{5}$ in row 3, column 2 and obtain the matrix

$$
\begin{bmatrix}
0 & 0 & 1 & \frac{1}{6} & \frac{1}{6} & 0 & 3 \\
1 & 0 & 0 & -\frac{1}{4} & \frac{1}{4} & 0 & 3 \\
0 & 1 & 0 & \frac{1}{12} & -\frac{5}{12} & 0 & 2 \\
0 & 0 & 0 & \frac{19}{12} & \frac{1}{12} & 1 & -34
\end{bmatrix}
$$

This tableau is feasible, and actually optimal, so we need not go to Phase II. This tableau yields the optimal solution $w = -34$ at $(3, 2)$, so the original problem has a minimum value of $z = 34$ at $(3, 2)$.

This illustrates that Phase I may take you to the optimal solution without using Phase II.

**Example 6**    An investment firm offers three types of investments to its clients. To help a client make a better-informed decision, each investment is assigned a risk factor. The risk factor and expected return of each investment are the following:

Investment A:        12% return per year, risk factor = 0.50

Investment B:        15% return per year, risk factor = 0.75

Investment C:         9% return per year, risk factor = 0.40

A client wishes to invest up to $50,000. He wants an annual return of at least $6300 and at least $10,000 invested in type C investments. How much should be invested in each type to minimize his total risk? (*Note:* If $20,000 is invested in A, that risk totals $0.50 \times 20,000 = 10,000$.)

**Solution**

Let $x_1$ = amount invested in A, $x_2$ = amount invested in B, and $x_3$ = amount invested in C. The total risk is to be minimized, so the objective function is

Minimize $z = 0.50x_1 + 0.75x_2 + 0.40x_3$.

The constraints are

$$
\begin{aligned}
x_1 + x_2 + x_3 &\le 50,000 \quad &\text{(total investment)} \\
0.12x_1 + 0.15x_2 + 0.09x_3 &\ge 6,300 \quad &\text{(total annual return)} \\
x_3 &\ge 10,000 \quad &\text{(at least \$10,000 in C)} \\
x_1 \ge 0, x_2 \ge 0, x_3 &\ge 0
\end{aligned}
$$

Modify the objective function to a maximum and the $\ge$ constraints to $\le$ and obtain

Maximize $w = -0.50x_1 - 0.75x_2 - 0.40x_3$, subject to

$$
\begin{aligned}
x_1 + x_2 + x_3 &\le 50,000 \\
-0.12x_1 - 0.15x_2 - 0.09x_3 &\le -6,300 \\
-x_3 &\le -10,000 \\
x_1 \ge 0, x_2 \ge 0, x_3 &\ge 0
\end{aligned}
$$

The sequence of tableaux that lead to the optimal solution follows:

**Phase I.**

$$
\left[
\begin{array}{ccccccc|c}
1 & 1 & 1 & 1 & 0 & 0 & 0 & 50,000 \\
-0.12 & -0.15 & -0.09 & 0 & 1 & 0 & 0 & -6,300 \\
0 & 0 & -1 & 0 & 0 & 1 & 0 & -10,000 \\
\hline
0.50 & 0.75 & 0.40 & 0 & 0 & 0 & 1 & 0
\end{array}
\right]
\quad
\begin{array}{l}
\text{R3 + R1} \rightarrow \text{R1} \\
-0.09\text{R3 + R2} \rightarrow \text{R2} \\
\\
0.40\text{R3 + R4} \rightarrow \text{R4}
\end{array}
$$

Pivot on $-1$ in row 3, column 3, by performing the row operations indicated above. After they are completed, multiply row 3 by $-1$ to obtain the following tableau.

$$
\left[
\begin{array}{ccccccc|c}
1 & 1 & 0 & 1 & 0 & 1 & 0 & 40,000 \\
-0.12 & -0.15 & 0 & 0 & 1 & -0.09 & 0 & -5,400 \\
0 & 0 & 1 & 0 & 0 & -1 & 0 & 10,000 \\
\hline
0.5 & 0.75 & 0 & 0 & 0 & \frac{2}{5} & 1 & -4,000
\end{array}
\right]
$$

Now pivot on $-0.15$ in row 2, column 2, by dividing each entry in row 2 by $-0.15$ to obtain the tableau below. Then perform the row operations indicated to complete the pivot.

$$\begin{bmatrix} 1 & 1 & 0 & 1 & 0 & 1 & 0 & 40{,}000 \\ 0.8 & 1 & 0 & 0 & -\frac{20}{3} & 0.6 & 0 & 36{,}000 \\ 0 & 0 & 1 & 0 & 0 & -1 & 0 & 10{,}000 \\ 0.5 & 0.75 & 0 & 0 & 0 & 0.4 & 1 & -4{,}000 \end{bmatrix} \quad \begin{array}{l} -R2 + R1 \rightarrow R1 \\ \\ \\ -0.75R2 + R4 \rightarrow R4 \end{array}$$

This gives the tableau

$$\begin{bmatrix} \frac{1}{5} & 0 & 0 & 1 & \frac{20}{3} & \frac{2}{5} & 0 & 4{,}000 \\ \frac{4}{5} & 1 & 0 & 0 & -\frac{20}{3} & \frac{3}{5} & 0 & 36{,}000 \\ 0 & 0 & 1 & 0 & 0 & -1 & 0 & 10{,}000 \\ -\frac{1}{10} & 0 & 0 & 0 & 5 & -\frac{1}{20} & 1 & -31{,}000 \end{bmatrix}$$

**Phase II.** Pivot on $\frac{1}{5}$ in row 1, column 1, by multiplying row 1 by 5 to obtain the following tableau. Then perform the indicated row operations to complete the pivot.

$$\begin{bmatrix} 1 & 0 & 0 & 5 & \frac{100}{3} & 2 & 0 & 20{,}000 \\ \frac{4}{5} & 1 & 0 & 0 & -\frac{20}{3} & \frac{3}{5} & 0 & 36{,}000 \\ 0 & 0 & 1 & 0 & 0 & -1 & 0 & 10{,}000 \\ -\frac{1}{10} & 0 & 0 & 0 & 5 & -\frac{1}{20} & 1 & -31{,}000 \end{bmatrix} \quad \begin{array}{l} \\ -\frac{4}{5}R1 + R2 \rightarrow R2 \\ \\ \frac{1}{10}R1 + R4 \rightarrow R4 \end{array}$$

The result is

$$\begin{bmatrix} 1 & 0 & 0 & 5 & \frac{100}{3} & 2 & 0 & 20{,}000 \\ 0 & 1 & 0 & -4 & -\frac{100}{3} & -1 & 0 & 20{,}000 \\ 0 & 0 & 1 & 0 & 0 & -1 & 0 & 10{,}000 \\ 0 & 0 & 0 & \frac{1}{2} & \frac{25}{3} & \frac{3}{20} & 1 & -29{,}000 \end{bmatrix}$$

This tableau gives the optimal solution maximum $w = -29{,}000$ when $x_1 = 20{,}000$, $x_2 = 20{,}000$, $x_3 = 10{,}000$. Therefore, the original problem has the optimal solution minimum $z = 29{,}000$ when $x_1 = 20{,}000$, $x_2 = 20{,}000$, $x_3 = 10{,}000$. The minimum risk occurs when \$20,000 is invested in A, \$20,000 in B, and \$10,000 in C.

**Example 7** A convenience store has to order three items, A, B, and C. The following table summarizes information about the items.

| Item | Cost | Selling price | Storage space required | Weight |
|------|------|------|------|------|
| A | \$10 | \$19 | 0.6 cu ft | 2 lb |
| B | \$12 | \$22 | 0.4 cu ft | 3 lb |
| C | \$ 8 | \$13 | 0.2 cu ft | 4 lb |

The purchasing agent must abide by the following guidelines:

> The order must provide at least 3400 items.
>
> The total cost of the order must not exceed \$36,000.
>
> The total storage space available is 1420 cubic feet.
>
> The total weight must not exceed 11,400 pounds.

How many of each item should be ordered to maximize profit?

### Solution

Let $x_1$ = number of items A, $x_2$ = number of items B, and $x_3$ = number of items C. The objective function and constraints described by the given information are the following:

Maximize $z = 9x_1 + 10x_2 + 5x_3$ (profit = selling price − cost), subject to

$$
\begin{array}{rrrll}
x_1 + & x_2 + & x_3 \geq & 3,400 & \text{(total number of items)} \\
10x_1 + & 12x_2 + & 8x_3 \leq & 36,000 & \text{(total cost)} \\
0.6x_1 + & 0.4x_2 + & 0.2x_3 \leq & 1,420 & \text{(storage space)} \\
2x_1 + & 3x_2 + & 4x_3 \leq & 11,400 & \text{(total weight)} \\
& x_1 \geq 0, x_2 \geq & 0, x_3 \geq 0 & & \text{(nonnegative conditions)}
\end{array}
$$

The initial tableau and subsequent tableaux that lead to the optimal solution are as follows. The initial tableau is

$$
\left[\begin{array}{cccccccc|c}
-1 & -1 & -1 & 1 & 0 & 0 & 0 & 0 & -3,400 \\
10 & 12 & 8 & 0 & 1 & 0 & 0 & 0 & 36,000 \\
0.6 & 0.4 & 0.2 & 0 & 0 & 1 & 0 & 0 & 1,420 \\
2 & 3 & 4 & 0 & 0 & 0 & 1 & 0 & 11,400 \\
\hline
-9 & -10 & -5 & 0 & 0 & 0 & 0 & 1 & 0
\end{array}\right]
\begin{array}{l}
\\
10R1 + R2 \to R2 \\
0.6R1 + R3 \to R3 \\
2R1 + R4 \to R4 \\
-9R1 + R5 \to R5
\end{array}
$$

**Phase I.** Because a negative number occurs in the last column, pivot on $-1$ in column 1 by using the operations indicated above. After those are completed, multiply row 1 by $-1$ to obtain the following tableau:

$$
\left[\begin{array}{cccccccc|c}
1 & 1 & 1 & -1 & 0 & 0 & 0 & 0 & 3,400 \\
0 & 2 & -2 & 10 & 1 & 0 & 0 & 0 & 2,000 \\
0 & -\frac{1}{5} & -\frac{2}{5} & \frac{3}{5} & 0 & 1 & 0 & 0 & -620 \\
0 & 1 & 2 & 2 & 0 & 0 & 1 & 0 & 4,600 \\
\hline
0 & -1 & 4 & -9 & 0 & 0 & 0 & 1 & 30,600
\end{array}\right]
$$

We still have a negative number in the last column, so we pivot on $-\frac{2}{5}$ in row 3, column 3 by first dividing row 3 by $-\frac{2}{5}$ to obtain the matrix below. Then perform the row operations indicated.

$$
\left[\begin{array}{cccccccc|c}
1 & 1 & 1 & -1 & 0 & 0 & 0 & 0 & 3,400 \\
0 & 2 & -2 & 10 & 1 & 0 & 0 & 0 & 2,000 \\
0 & \frac{1}{2} & 1 & -\frac{3}{2} & 0 & -\frac{5}{2} & 0 & 0 & 1,550 \\
0 & 1 & 2 & 2 & 0 & 0 & 1 & 0 & 4,600 \\
\hline
0 & -1 & 4 & -9 & 0 & 0 & 0 & 1 & 30,600
\end{array}\right]
\begin{array}{l}
-R3 + R1 \to R1 \\
2R3 + R2 \to R2 \\
\\
-2R3 + R4 \to R4 \\
-4R3 + R5 \to R5
\end{array}
$$

We obtain

$$
\begin{bmatrix}
1 & \frac{1}{2} & 0 & \frac{1}{2} & 0 & \frac{5}{2} & 0 & 0 & 1{,}850 \\
0 & 3 & 0 & 7 & 1 & -5 & 0 & 0 & 5{,}100 \\
0 & \frac{1}{2} & 1 & -\frac{3}{2} & 0 & -\frac{5}{2} & 0 & 0 & 1{,}550 \\
0 & 0 & 0 & 5 & 0 & 5 & 1 & 0 & 1{,}500 \\
\hline
0 & -3 & 0 & -3 & 0 & 10 & 0 & 1 & 24{,}400
\end{bmatrix}
$$

With no negative entries in the last column we are now in the feasible region. We use 3 in row 2, column 2 as the pivot element. Divide row 2 by 3 to obtain this tableau.

$$
\begin{bmatrix}
1 & \frac{1}{2} & 0 & \frac{1}{2} & 0 & \frac{5}{2} & 0 & 0 & 1{,}850 \\
0 & 1 & 0 & \frac{7}{3} & \frac{1}{3} & -\frac{5}{3} & 0 & 0 & 1{,}700 \\
0 & \frac{1}{2} & 1 & -\frac{3}{2} & 0 & -\frac{5}{2} & 0 & 0 & 1{,}550 \\
0 & 0 & 0 & 5 & 0 & 5 & 1 & 0 & 1{,}500 \\
\hline
0 & -3 & 0 & -3 & 0 & 10 & 0 & 1 & 24{,}400
\end{bmatrix}
\begin{array}{l}
-\frac{1}{2}\text{R2} + \text{R1} \rightarrow \text{R1} \\
\\
-\frac{1}{2}\text{R2} + \text{R3} \rightarrow \text{R3} \\
\\
3\text{R2} + \text{R5} \rightarrow \text{R5}
\end{array}
$$

Complete the pivot with the indicated row operations to obtain the final tableau.

$$
\begin{bmatrix}
1 & 0 & 0 & -\frac{2}{3} & -\frac{1}{6} & \frac{10}{3} & 0 & 0 & 1{,}000 \\
0 & 1 & 0 & \frac{7}{3} & \frac{1}{3} & -\frac{5}{3} & 0 & 0 & 1{,}700 \\
0 & 0 & 1 & -\frac{8}{3} & -\frac{1}{6} & -\frac{5}{3} & 0 & 0 & 700 \\
0 & 0 & 0 & 5 & 0 & 5 & 1 & 0 & 1{,}500 \\
\hline
0 & 0 & 0 & 4 & 1 & 5 & 0 & 1 & 29{,}500
\end{bmatrix}
$$

The optimal solution is maximum $z = 29{,}500$ when $x_1 = 1000$, $x_2 = 1700$, and $x_3 = 700$. The purchasing agent should order 1000 of item A, 1700 of item B, and 700 of item C.

---

**Summary of the Simplex Method for Problems with Mixed Constraints**

1. For minimization problems, maximize $w = -z$.

2. ($\geq$ constraint) For each constraint of the form

$$a_1 x_1 + a_2 x_2 + \cdots + a_n x_n \geq b$$

multiply the inequality by $-1$ to obtain

$$-a_1 x_1 - a_2 x_2 - \cdots - a_n x_n \leq -b$$

3. Form the initial simplex tableau.

4. If no negative entry appears in the last column of the initial tableau, proceed to Phase II; otherwise, proceed to Phase I.

5. (Phase I) If there is a negative entry in the last column, change it to a positive entry by pivoting in the following manner. (Ignore a negative entry in the objective function [last row] for this step.)

   (a) The pivot row is the row containing the most negative entry in the last column.

   (b) Select the left-most or most negative entry in the pivot row. This entry is the pivot element. These two choices may differ in the number of steps in the solution, but you don't know which ahead of time.

   (c) Reduce the pivot element to 1 and the other entries of the pivot column to 0 using row operations.

*(continued)*

**6.** Repeat the parts of step 5 as long as a negative entry occurs in the last column. When no negative entries remain in the last column (except possibly in the last row), proceed to Phase II.

**7.** (Phase II) The basic solution to the tableau is now feasible. Use the standard simplex procedure to obtain the optimal solution.

**8.** For a minimum problem, be sure to change the sign of the optimal value obtained.

## 4.4    EXERCISES

Access end-of-section exercises online at **www.webassign.net**

 **Using Excel**

We can use Excel's **Solver** program to solve minimization problems. We illustrate with the following example. (See Section 4.2 for details on using **Solver** for linear programming.)

Minimize $z = 8x_1 + 10x_2 + 2x_3$ subject to

$$10x_1 + 12x_2 + 5x_3 \geq 100$$
$$5x_1 + 7x_2 + 5x_3 \leq 75$$
$$10x_1 + 2x_2 + 10x_3 \leq 120$$
$$x_1 \geq 0, x_2 \geq 0, \quad x_3 \geq 0$$

We enter the problem in the spreadsheet as follows:

|   | A | B | C | D | E |
|---|---|---|---|---|---|
| 1 | 0 | 0 | 0 | =10*A1+12*B1+5*C1 | 100 |
| 2 |   |   |   | =5*A1+7*B1+5*C1 | 75 |
| 3 |   |   |   | =10*A1+2*B1+10*C1 | 120 |
| 4 |   |   |   | =A1 | 0 |
| 5 |   |   |   | =B1 | 0 |
| 6 |   |   |   | =C1 | 0 |
| 7 |   |   |   | =8*A1+10*B1+2*C1 |   |

This translates in **Solver** as:

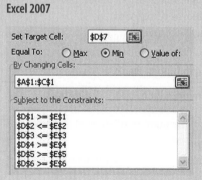

**Excel 2007**

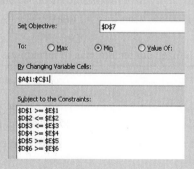

**Excel 2010**

Note that the Min choice has been selected. The solution follows:

| | A | B | C | D | E |
|---|---|---|---|---|---|
| 1 | 2.5 | 2.5 | 9 | 100.0 | 100 |
| 2 | | | | 75.0 | 75 |
| 3 | | | | 120.0 | 120 |
| 4 | | | | 2.5 | 0 |
| 5 | | | | 2.5 | 0 |
| 6 | | | | 9.0 | 0 |
| 7 | | | | 63.0 | |

The minimum $z = 63$ at $(2.5, 2.5, 9)$.

## Exercises

**1.** Maximize $z = 7x_1 + 7x_2 + 3x_3$ subject to
$$x_1 + 4x_2 + 3x_3 \leq 134$$
$$2x_1 + 10x_2 + 5x_3 \geq 280$$
$$5x_1 + x_2 + 3x_3 \leq 100$$
$$x_1 \geq 0, x_2 \geq 0, x_3 \geq 0$$

**2.** Maximize $z = 6x_1 + 5x_2 + 3x_3$ subject to
$$3x_1 + x_2 + 3x_3 \leq 9$$
$$2x_1 + 3x_2 + x_3 \leq 12$$
$$x_1 + x_2 - x_3 \geq 3$$
$$x_1 \geq 0, x_2 \geq 0, x_3 \geq 0$$

**3.** Maximize $z = 3x_1 + 4x_2$ subject to
$$5x_1 + 2x_2 \leq 10$$
$$x_1 + 2x_2 \leq 6$$
$$x_1 \geq 0, x_2 \geq 0$$

**4.** Maximize $z = 8x_1 + 4x_2$ subject to
$$3x_1 + 2x_2 \leq 48$$
$$2x_1 + 4x_2 \leq 64$$
$$4x_1 + 6x_2 \geq 84$$
$$x_1 \geq 0, x_2 \geq 0$$

## 4.5 MULTIPLE SOLUTIONS, UNBOUNDED SOLUTIONS, AND NO SOLUTIONS

- Multiple Solutions
- No Solutions
- Unbounded Solutions
- No Feasible Solution

Generally, a mathematics textbook introduces a new concept or method with examples and exercises that have nice, neat solutions.

Although well-behaved problems form a valid starting point, texts sometimes avoid problems that stray from these nice forms. In this chapter on the simplex method, we have studied the basic method and some variations, but the examples and problems generally have had unique solutions. That is not always the case. Reflect a moment on the fact that the simplex method converts a linear programming problem to a system of equations. We know that a system of equations can have a unique solution, no solution, or an infinity of solutions. Thus, with good reason, we should expect situations to arise when a linear program has no solutions or many solutions. Fortunately, the simplex method can signal when no solutions are possible or when multiple solutions exist.

### Multiple Solutions

We look at a problem with multiple optimal solutions.

**Example 1**    Maximize $z = 18x + 24y$, subject to

$$3x + 4y \leq 48$$
$$x + 2y \leq 22$$
$$3x + 2y \leq 42$$
$$x \geq 0, y \geq 0$$

**Solution**

Figure 4–5 shows the graph of the feasible region. Note that the corners are $(0, 0)$, $(0, 11)$, $(4, 9)$, $(12, 3)$, and $(14, 0)$, so the maximum value occurs at one or more of these corners.

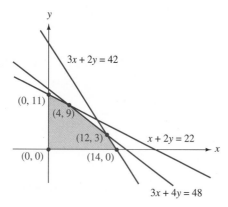

**FIGURE 4–5**

The initial tableau is

$$
\begin{array}{cccccc|c}
x & y & s_1 & s_2 & s_3 & z & \\
\hline
3 & 4 & 1 & 0 & 0 & 0 & 48 \\
1 & 2 & 0 & 1 & 0 & 0 & 22 \\
3 & 2 & 0 & 0 & 1 & 0 & 42 \\
-18 & -24 & 0 & 0 & 0 & 1 & 0
\end{array}
$$

The pivot element is 2 in row 2, column 2. We first divide the entries in row 2 by 2, which yields the following tableau. We then perform the row operations indicated there.

$$
\begin{array}{cccccc|c}
3 & 4 & 1 & 0 & 0 & 0 & 48 \\
\frac{1}{2} & 1 & 0 & \frac{1}{2} & 0 & 0 & 11 \\
3 & 2 & 0 & 0 & 1 & 0 & 42 \\
-18 & -24 & 0 & 0 & 0 & 1 & 0
\end{array}
\quad
\begin{array}{l}
-4R2 + R1 \rightarrow R1 \\
\\
-2R2 + R3 \rightarrow R3 \\
24R2 + R4 \rightarrow R4
\end{array}
$$

This gives the tableau

$$
\begin{array}{cccccc|c}
x & y & s_1 & s_2 & s_3 & z & \\
\hline
1 & 0 & 1 & -2 & 0 & 0 & 4 \\
\frac{1}{2} & 1 & 0 & \frac{1}{2} & 0 & 0 & 11 \\
2 & 0 & 0 & -1 & 1 & 0 & 20 \\
-6 & 0 & 0 & 12 & 0 & 1 & 264
\end{array}
\quad
\begin{array}{l}
\\
-\frac{1}{2}R1 + R2 \rightarrow R2 \\
-2R1 + R3 \rightarrow R3 \\
6R1 + R4 \rightarrow R4
\end{array}
$$

This is not optimal, so we pivot on 1 in row 1, column 1, using the operations indicated with the preceding tableau.

$$
\begin{array}{cccccc}
x & y & s_1 & s_2 & s_3 & z \\
\end{array}
$$
$$
\left[\begin{array}{cccccc|c}
1 & 0 & 1 & -2 & 0 & 0 & 4 \\
0 & 1 & -\frac{1}{2} & \frac{3}{2} & 0 & 0 & 9 \\
0 & 0 & -2 & 3 & 1 & 0 & 12 \\
\hline
0 & 0 & 6 & 0 & 0 & 1 & 288
\end{array}\right]
$$

This tableau gives an optimal solution $z = 288$ when $x = 4$, $y = 9$. Recall that we say that $x$, $y$, and $s_3$ are basic variables and $s_1$ and $s_2$ are nonbasic variables. Note that there is a zero in the bottom row of the $s_2$ column. This zero is the clue that there might be a different optimal solution.

To determine whether there is another optimal solution, use the $s_2$ column as the pivot column. The ratios of the entries in the last column are $-2$, $6$, and $4$. Because 4 is the small-est *nonnegative* ratio, row 3 is the pivot row, and 3 in row 3, column 4 is the pivot element. We pivot by first dividing each entry in row 3 by 3 to obtain the next tableau:

$$
\begin{array}{cccccc}
x & y & s_1 & s_2 & s_3 & z \\
\end{array}
$$
$$
\left[\begin{array}{cccccc|c}
1 & 0 & 1 & -2 & 0 & 0 & 4 \\
0 & 1 & -\frac{1}{2} & \frac{3}{2} & 0 & 0 & 9 \\
0 & 0 & -\frac{2}{3} & 1 & \frac{1}{3} & 0 & 4 \\
\hline
0 & 0 & 6 & 0 & 0 & 1 & 288
\end{array}\right]
\quad
\begin{array}{l}
2R3 + R1 \to R1 \\
-\frac{3}{2}R3 + R2 \to R2 \\
\\
\\
\end{array}
$$

We complete the pivot using the row operations above to give the following tableau:

$$
\begin{array}{cccccc}
x & y & s_1 & s_2 & s_3 & z \\
\end{array}
$$
$$
\left[\begin{array}{cccccc|c}
1 & 0 & -\frac{1}{3} & 0 & \frac{2}{3} & 0 & 12 \\
0 & 1 & \frac{1}{2} & 0 & -\frac{1}{2} & 0 & 3 \\
0 & 0 & -\frac{2}{3} & 1 & \frac{1}{3} & 0 & 4 \\
\hline
0 & 0 & 6 & 0 & 0 & 1 & 288
\end{array}\right]
$$

This tableau gives an optimal solution $z = 288$ when $x = 12$, $y = 3$. Thus, the same maxi-mum value of $z$, 288, occurs at another point. The optimal solution occurs at the corners $(4, 9)$ and $(12, 3)$ of the feasible region. Actually, *all* points on the line segment between $(4, 9)$ and $(12, 3)$ also yield the maximum value of $z = 288$.

---

**Multiple Solutions**   To determine whether a problem has more than one optimal solution:

1. Find an optimal solution by the usual simplex method.

2. Look at zeros in the bottom row of the final tableau. If a zero appears in the bottom row of a column for a *nonbasic* variable, there might be other optimal solutions.

3. To find another optimal solution, if any, use the column of a nonbasic variable with a zero at the bottom as the pivot column. Find the pivot row in the usual manner, and then pivot on the pivot element.

4. If this new tableau gives the same optimal value of $z$ at another point, then multiple solutions exist.

5. Given the two optimal solutions, all points on the line segment joining them are also optimal solutions.

## No Solutions

Two conditions give rise to no solution in a linear programming problem. In one case, an unbounded feasible region exists, so the objective function can be made arbitrarily large by selecting points farther away in the feasible region. In another case, some constraints are inconsistent, so no feasible region exists. We now find out how to recognize these situations from the simplex tableau.

## Unbounded Solutions

The following example illustrates a problem with an unbounded feasible region. In such a case, the objective function has no maximum value because it can be arbitrarily large.

**Example 2** Maximize $z = x_1 + 4x_2$, subject to

$$\begin{aligned}
x_1 - x_2 &\leq 3 \\
-4x_1 + x_2 &\leq 4 \\
x_1 \geq 0, x_2 &\geq 0
\end{aligned}$$

**Solution**

The graph of the feasible region is shown in Figure 4–6. This problem converts to the system

$$\begin{aligned}
x_1 - x_2 + s_1 &= 3 \\
-4x_1 + x_2 \quad\quad + s_2 &= 4 \\
-x_1 - 4x_2 \quad\quad\quad\quad + z &= 0
\end{aligned}$$

and to the initial simplex tableau

$$
\begin{array}{ccccc}
x_1 & x_2 & s_1 & s_2 & z \\
\end{array}
$$
$$
\left[
\begin{array}{ccccc|c}
1 & -1 & 1 & 0 & 0 & 3 \\
-4 & ① & 0 & 1 & 0 & 4 \\
-1 & -4 & 0 & 0 & 1 & 0
\end{array}
\right]
$$

Pivot element

Except for the pivot element, convert all entries in the pivot column to 0 to obtain the next tableau:

$$
\left[
\begin{array}{ccccc|c}
-3 & 0 & 1 & 1 & 0 & 7 \\
-4 & 1 & 0 & 1 & 0 & 4 \\
-17 & 0 & 0 & 4 & 1 & 16
\end{array}
\right]
$$

Because of the $-17$ in the last row, we know that $z$ is not maximal. When we check for the pivot row, we get the ratios $-7/3$ and $-1$. We cannot proceed with the simplex method because *all* ratios are negative. When this occurs, there is no maximum. Because the feasible region is unbounded (Figure 4–6), a maximum value of the objective function does not exist.

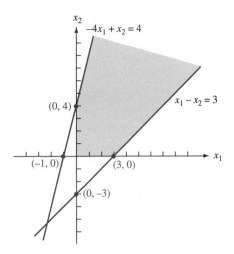

**FIGURE 4–6** Unbounded feasible region.

**Unbounded Solutions** When you arrive at a simplex tableau that has no positive entries in the pivot column, the feasible region is unbounded, and the objective function is unbounded. There is no maximum value.

## No Feasible Solution

The following example illustrates what happens in a simplex tableau when the problem has no feasible solution.

**Example 3** Maximize $z = 8x_1 + 24x_2$, subject to

$$\begin{aligned} x_1 + x_2 &\leq 10 \\ 2x_1 + 3x_2 &\geq 60 \\ x_1 \geq 0, x_2 &\geq 0 \end{aligned}$$

**Solution**

The graph of the constraints (Figure 4–7) shows that the half plane below $x_1 + x_2 = 10$ and the half plane above $2x_1 + 3x_2 = 60$ do not intersect in the first quadrant, so no feasible solution exists. Let's attempt to solve the problem with the simplex method and see what happens.

Here is the initial tableau:

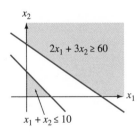

**FIGURE 4–7** No feasible region.

$$\begin{array}{ccccc} x_1 & x_2 & s_1 & s_2 & z \\ \left[\begin{array}{ccccc|c} 1 & 1 & 1 & 0 & 0 & 10 \\ -2 & -3 & 0 & 1 & 0 & -60 \\ -8 & -24 & 0 & 0 & 1 & 0 \end{array}\right] \end{array}$$

To remove the negative entry in the last column, we pivot on $-3$ in row 2, column 2 and obtain the next tableau:

$$
\begin{array}{ccccc}
x_1 & x_2 & s_1 & s_2 & z \\
\end{array}
$$

$$
\left[
\begin{array}{ccccc|c}
\frac{1}{3} & 0 & 1 & \frac{1}{3} & 0 & -10 \\
\frac{2}{3} & 1 & 0 & -\frac{1}{3} & 0 & 20 \\
\hline
8 & 0 & 0 & -8 & 1 & 480
\end{array}
\right]
$$

The first row indicates no feasible solution. Here's why. The first row represents the equation $\frac{1}{3}x_1 + s_1 + \frac{1}{3}s_2 = -10$. Because $x_1$, $s_1$, and $s_2$ cannot be negative, there are no values that can be used on the left-hand side that will give a negative number. Therefore, there is no feasible solution.

---

**No Feasible Solution**    When a simplex tableau has a negative entry in the last column and no other entries in that row are negative, then there is no feasible solution to the problem.

---

**Example 4**

A machine shop makes standard and heavy-duty gears. The process requires two steps. Step 1 takes 8 minutes for the standard gear and 10 minutes for the heavy-duty gear. Step 2 takes 6 minutes for the standard gear and 10 minutes for the heavy-duty gear. The company's labor contract requires that it use at least 200 labor-hours (12,000 minutes) per week on the step 1 equipment. The maintenance required on the step 2 machine restricts it to 140 hours per week or less (8400 minutes). The materials cost $15 for each standard gear and $22 for each heavy-duty gear. How many of each type of gear should be made each week to minimize material costs?

Show that this problem has no solution.

**Solution**

Let $x =$ number of standard gears and $y =$ number of heavy-duty gears. The problem is
Minimize $z = 15x + 22y$, subject to

$$
\begin{array}{rcl}
8x + 10y & \geq & 12,000 \\
6x + 10y & \leq & 8,400 \\
x \geq 0, y & \geq & 0
\end{array}
$$

To use the simplex method, we modify the problem to:
Maximize $w = -15x - 22y$, subject to

$$
\begin{array}{rcr}
-8x - 10y & \leq & -12,000 \\
6x + 10y & \leq & 8,400 \\
x \geq 0, y & \geq & 0
\end{array}
$$

The initial tableau is

$$
\left[
\begin{array}{ccccc|r}
-8 & -10 & 1 & 0 & 0 & -12,000 \\
6 & 10 & 0 & 1 & 0 & 8,400 \\
\hline
15 & 22 & 0 & 0 & 1 & 0
\end{array}
\right]
$$

Because the basic solution is not feasible, we must apply Phase I and pivot on $-10$ in row 1:

$$\begin{bmatrix} \frac{8}{10} & 1 & -\frac{1}{10} & 0 & 0 & 1,200 \\ 6 & 10 & 0 & 1 & 0 & 8,400 \\ 15 & 22 & 0 & 0 & 1 & 0 \end{bmatrix} \begin{matrix} \\ -10R1 + R2 \to R2 \\ -22R1 + R3 \to R3 \end{matrix}$$

$$\begin{bmatrix} \frac{8}{10} & 1 & -\frac{1}{10} & 0 & 0 & 1,200 \\ -2 & 0 & 1 & 1 & 0 & -3,600 \\ -\frac{13}{5} & 0 & \frac{11}{5} & 0 & 1 & -26,400 \end{bmatrix}$$

Now pivot on $-2$ in row 2:

$$\begin{bmatrix} 0 & 1 & \frac{3}{10} & \frac{2}{5} & 0 & -240 \\ 1 & 0 & -\frac{1}{2} & -\frac{1}{2} & 0 & 1,800 \\ 0 & 0 & \frac{9}{10} & -\frac{13}{10} & 1 & -21,720 \end{bmatrix}$$

Row 1 has a negative number in the constant column, and all coefficients to the left of the line are nonnegative. This indicates that there is no feasible region and therefore no solution.

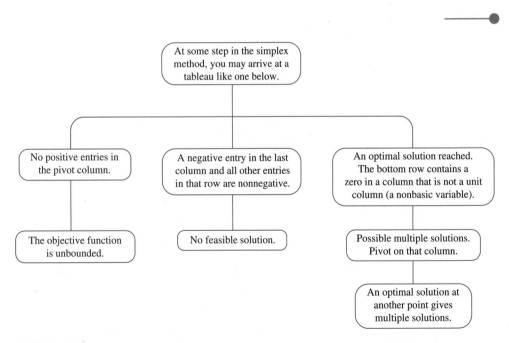

**FIGURE 4–8**   A summary of the indications that no feasible solution, an unbounded objective function, or multiple solutions are encountered in the simplex method.

## 4.5   EXERCISES

Access end-of-section exercises online at **www.webassign.net**

**4.6**    # WHAT'S HAPPENING IN THE SIMPLEX METHOD? (OPTIONAL)

The simplex method can be performed in a rather mechanical manner, thereby making it a procedure that can run on a computer. This makes it possible to solve linear programming problems with hundreds of constraints and variables. As much as we appreciate a machine handling the routine, tedious computation, the human mind sometimes becomes curious about what lies behind the computations. In this section, we look at the simplex process to better understand why the steps are performed. We outline the steps of the simplex method and explain why we perform these steps. We use Example 5 from Section 4.1 again.

Maximize $z = 4x_1 + 12x_2$, subject to

$$3x_1 + x_2 \leq 180$$
$$x_1 + 2x_2 \leq 100$$
$$-2x_1 + 2x_2 \leq 40$$
$$x_1 \geq 0, x_2 \geq 0$$

The graph of the feasible region and the lines forming its boundary is shown in Figure 4–9.

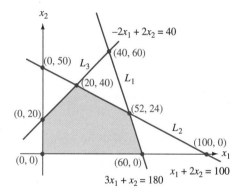

**FIGURE 4–9**

1. We convert the problem to a system of equations by adding a nonnegative slack variable to each inequality

$$3x_1 + x_2 + s_1 \qquad\qquad = 180$$
$$x_1 + 2x_2 \qquad + s_2 \qquad\qquad = 100$$
$$-2x_1 + 2x_2 \qquad\qquad + s_3 \qquad = 40$$
$$-4x_1 - 12x_2 \qquad\qquad\qquad + z = 0$$

where $x_1, x_2, s_1, s_2,$ and $s_3$ are all nonnegative.

2. The simplex method searches for solutions to this system of equations. Each simplex tableau gives a basic feasible solution. Recall that we set a variable to zero for each $x$ in the system to obtain a basic solution. A solution gives points where two of the boundary lines of the feasible region intersect.

Look at Figure 4–9, which illustrates this. Setting a slack variable to zero gives a boundary line. The boundary lines are

$$L_1: \quad 3x_1 + x_2 = 180 \quad \text{(where } s_1 = 0\text{)}$$
$$L_2: \quad x_1 + 2x_2 = 100 \quad \text{(where } s_2 = 0\text{)}$$
$$L_3: \quad -2x_1 + 2x_2 = 40 \quad \text{(where } s_3 = 0\text{)}$$

$$x_1\text{-axis:} \quad x_2 = 0$$
$$x_2\text{-axis:} \quad x_1 = 0$$

The corner points of the feasible region occur at intersections of boundary lines and are (0, 0), (0, 20), (20, 40), (52, 24), and (60, 0). Some boundary lines intersect outside the feasible region, such as (40, 60), and give no corner point (see Figure 4–9).

We have already seen that all slack variables are nonnegative for points in the feasible region and that the optimal solution occurs at a corner point.

The simplex method finds the corner at which the objective function is maximum in the following manner:

1. Begin at the origin, (0, 0).

2. Select a pivot column that tends to increase the value of $z$ the most.

3. At each step, move along a boundary line to an adjacent corner of the feasible region.

4. When $z$ can no longer be increased, the procedure stops.

Now let's see how these relate to the simplex tableaux. We refer to the tableaux in Section 4.2 that were obtained in finding the optimal solution.

1. The initial tableau is

$$\begin{bmatrix} 3 & 1 & 1 & 0 & 0 & 0 & | & 180 \\ 1 & 2 & 0 & 1 & 0 & 0 & | & 100 \\ -2 & 2 & 0 & 0 & 1 & 0 & | & 40 \\ \hline -4 & -12 & 0 & 0 & 0 & 1 & | & 0 \end{bmatrix}$$

which has the basic solution $x_1 = 0$, $x_2 = 0$, $s_1 = 180$, $s_2 = 100$, $s_3 = 40$, $z = 0$. This gives the point (0, 0), the origin.

2. The choice of the pivot element determines the solution in the next step, and it increases $z$ as much as possible for a unit increase in the variable.

Let's look at the initial tableau to illustrate this:

$$\begin{array}{cccccc} x_1 & x_2 & s_1 & s_2 & s_3 & z \\ \end{array}$$
$$\begin{bmatrix} 3 & 1 & 1 & 0 & 0 & 0 & | & 180 \\ 1 & 2 & 0 & 1 & 0 & 0 & | & 100 \\ -2 & 2 & 0 & 0 & 1 & 0 & | & 40 \\ \hline -4 & -12 & 0 & 0 & 0 & 1 & | & 0 \end{bmatrix}$$

Let's see how the choice of the pivot column increases $z$ as much as possible. The objective function in this problem is

$$z = 4x_1 + 12x_2$$

If values of $x_1$ and $x_2$ are given and if you are allowed to increase either one of them by a specified amount—say, 1—which one would you change to increase $z$ the most? The coefficients of $x_1$ and $x_2$ hold the key to your response. If $x_1$ is increased by 1, then the coefficient 4 causes $z$ to increase by 4. Similarly, an increase of 1 in $x_2$ causes $z$ to increase by 12. Thus, it appears that the greatest increase in $z$ is gained by increasing the variable with the largest positive coefficient, 12 in this case. In the tableau, $z = 4x_1 + 12x_2$ is written as $-4x_1 - 12x_2 + z = 0$. In this form, the choice of the *most negative* coefficient is equivalent to choosing the variable that seems to increase $z$ the most. So in the simplex method, the pivot column is chosen by the most negative entry in the bottom row because this gives our best guess of the variable that will tend to increase $z$ the most.

3. Let's see how the choice of the pivot row restricts basic solutions to corner points.

   In the initial tableau, the basic solution assumes that both $x_1$ and $x_2$ are zero. Then the $x_2$ column becomes the pivot column to obtain the next tableau because that column contains the most negative entry of the last row. Because this means that we want to increase $x_2$, $x_1$ remains zero. Using $x_1 = 0$, let's write each row of the initial tableau in equation form:

$$[3 \quad 1 \quad 1 \quad 0 \quad 0 \quad 0 \quad 180] \qquad \text{becomes}$$
$$x_2 + s_1 = 180$$

$$[1 \quad 2 \quad 0 \quad 1 \quad 0 \quad 0 \quad 100] \qquad \text{becomes}$$
$$2x_2 + s_2 = 100$$

and

$$[-2 \quad 2 \quad 0 \quad 0 \quad 1 \quad 0 \quad 40] \qquad \text{becomes}$$
$$2x_2 + s_3 = 40$$

(The first number in each row doesn't appear in the equation because it is the coefficient of $x_1$, which we are using as 0.)

We can write these three equations in the following form:

$$s_1 = 180 - x_2$$
$$s_2 = 100 - 2x_2$$
$$s_3 = 40 - 2x_2$$

Keep in mind that we want to increase $x_2$ to achieve the largest increase in $z$. The larger the increase in $x_2$, the more $z$ increases, but be careful: We must remain in the feasible region. Because $s_1$, $s_2$, and $s_3$ must not be negative, $x_2$ must be chosen to avoid making any one of them negative.

The equations

$$s_1 = 180 - x_2$$
$$s_2 = 100 - 2x_2$$
$$s_3 = 40 - 2x_2$$

and the nonnegative condition $s_1 \geq 0$, $s_2 \geq 0$, and $s_3 \geq 0$ indicate that

$$180 - x_2 \geq 0$$
$$100 - 2x_2 \geq 0$$
$$40 - 2x_2 \geq 0$$

must all be true. Solving each of these inequalities for $x_2$ gives

$$\frac{180}{1} \geq x_2, \quad \frac{100}{2} \geq x_2, \quad \frac{40}{2} \geq x_2.$$

In order for all three of the ratios to be larger than $x_2$, then $x_2$ must be equal to the smallest of these three ratios. Thus, for *all three* of $s_1$, $s_2$, and $s_3$ to be nonnegative, the smallest value of $x_2$, 20, must be used. The ratios

$$\frac{180}{1}, \quad \frac{100}{2}, \quad \frac{40}{2}$$

are exactly the ratios that we use in the simplex method to determine the pivot row. These ratios are also the $x_2$-coordinates where the boundary lines cross the $x_2$-axis. By the nature of the feasible region, the lowest point is the one in the feasible region. The selection of the smallest nonnegative ratio makes a basic solution a feasible basic solution; that is, a corner point is chosen.

4. Recall that the maximum value of $z$ occurs when the last row of the simplex tableau contains no negative entries. The final tableau of this problem was

$$
\begin{array}{ccccccc}
x_1 & x_2 & s_1 & s_2 & s_3 & z & \\
\left[\begin{array}{cccccc|c}
0 & 0 & 1 & -\frac{4}{3} & \frac{5}{6} & 0 & 80 \\
1 & 0 & 0 & \frac{1}{3} & -\frac{1}{3} & 0 & 20 \\
0 & 1 & 0 & \frac{1}{3} & \frac{1}{6} & 0 & 40 \\
\hline
0 & 0 & 0 & \frac{16}{3} & \frac{2}{3} & 1 & 560
\end{array}\right]
\end{array}
$$

This tableau tells us that $s_2$ and $s_3$ are set to zero (their columns are not unit columns) in the optimal solution; and as the last row contains no negative entries, we know that we cannot increase $z$ further. Here's why: Write the last row of this tableau in equation form. It is

$$\frac{16}{3}s_2 + \frac{2}{3}s_3 + z = 560$$

which can be written as

$$z = 560 - \frac{16}{3}s_2 - \frac{2}{3}s_3$$

This form tells us that if we use any positive number for $s_2$ or $s_3$, we will *subtract* something from 560, thereby making $z$ smaller. So we stop because another tableau will move us to another corner point, and $s_2$ or $s_3$ will become positive and therefore reduce $z$.

Let's compare this situation with the next to last tableau. Its last row was

$$[-16 \quad 0 \quad 0 \quad 0 \quad 6 \quad 1 \quad 240]$$

This row represents the equation

$$-16x_1 + 6s_3 + z = 240$$

which may be written

$$z = 240 + 16x_1 - 6s_3$$

If $x_1$ is increased from 0 to a positive number, then a positive quantity will be added to $z$, thereby increasing it. If you look back at this step in the solution, you will find that $x_1$ was increased from 0 to 10.

The value of $z$ can be increased as long as there is a negative entry in the last row. It can be increased no further when no entry is negative.

5. Reviewing the steps of this example, we observe that the sequence of simplex tableaux and the corner points determined are the following:

Initial tableau:

$$\begin{bmatrix} 3 & 1 & 1 & 0 & 0 & 0 & 180 \\ 1 & 2 & 0 & 1 & 0 & 0 & 100 \\ -2 & 2 & 0 & 0 & 1 & 0 & 40 \\ \hline -4 & -12 & 0 & 0 & 0 & 1 & 0 \end{bmatrix}$$

Corner $(0, 0)$ $z = 0$, slack $(180, 100, 40)$.

Tableau 1:

$$\begin{bmatrix} 4 & 0 & 1 & 0 & -\frac{1}{2} & 0 & 160 \\ 3 & 0 & 0 & 1 & -1 & 0 & 60 \\ -1 & 1 & 0 & 0 & \frac{1}{2} & 0 & 20 \\ \hline -16 & 0 & 0 & 0 & 6 & 1 & 240 \end{bmatrix}$$

Corner $(0, 20)$ $z = 240$, slack $(160, 60, 0)$.

Tableau 2:

$$\begin{bmatrix} 0 & 0 & 1 & -\frac{4}{3} & \frac{5}{6} & 0 & 80 \\ 1 & 0 & 0 & \frac{1}{3} & -\frac{1}{3} & 0 & 20 \\ 0 & 1 & 0 & \frac{1}{3} & \frac{1}{6} & 0 & 40 \\ \hline 0 & 0 & 0 & \frac{16}{3} & \frac{2}{3} & 1 & 560 \end{bmatrix}$$

Corner $(20, 40)$ $z = 560$, slack $(80, 0, 0)$.

Note that the initial corner point is the origin, $(0, 0)$. The first pivot moves to $(0, 20)$, which is a corner adjacent to $(0, 0)$. The second pivot moves to $(20, 40)$, which is a corner adjacent to $(0, 20)$. A pivot in the simplex method moves from the current corner to an adjacent corner (see Figure 4–10).

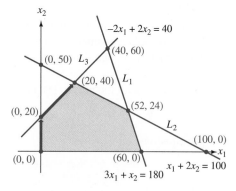

**FIGURE 4–10**   Sequence of corner points generated by the simplex tableaux.

On the basis of the preceding discussion, the simplex method can be summarized this way:

> The simplex method maximizes the objective function by computing it at selected corner points of the feasible region until the optimal solution is reached. The method begins at the origin and moves at each stage to an adjacent corner point determined by the variable that tends to yield the largest increase in $z$.

## 4.6 EXERCISES

Access end-of-section exercises online at **www.webassign.net**

## 4.7 SENSITIVITY ANALYSIS

- Changes in the Objective Function
- Resource Changes

In the linear programming examples and exercises we have used throughout this chapter, we have tacitly assumed the quantities, costs, and other data are accurately known. In practice, those quantities and costs are often estimates or subject to change. Thus, this question arises: "If some of the data change, how will that affect the optimal solution?" A study of the consequences of changes in data is called **sensitivity analysis**. Such an analysis can become complicated, but we look at two simpler situations: (a) when changes occur in the objective function, and (b) when changes occur in the resources available.

We use the following elementary example to illustrate the analysis.

**Example 1** Electronic Design makes two electronic control devices for an equipment manufacturer, the Primary Control and the Auxiliary Control. These controls require two types of circuit boards, the C-8 board and the H-2 board. The Primary Control requires 5 C-8 boards and 1 H-2 board. The Auxiliary Control requires 4 C-8 boards and 2 H-2 boards. The company expects to make a $16 profit on the Primary Control devices and a $24 profit on the Auxiliary Control devices. The company has 88 C-8 and 32 H-2 boards. How many of each control boards should be made to maximize profit?

**Solution**
Let's set up the constraints and objective function.

$$\text{Let } x = \text{number of Primary Control devices}$$
$$y = \text{number of Auxiliary Control devices}$$

Maximize $z = 16x + 24y$, subject to

$$5x + 4y \leq 88 \quad \text{(number of C-8 boards)}$$
$$x + 2y \leq 32 \quad \text{(number of H-2 boards)}$$
$$x \geq 0, y \geq 0$$

Figure 4–11 shows the corners of the feasible region.
The corners and the corresponding values of the objective function are the following:

| Corner | $z = 16x + 24y$ |
|---|---|
| (0, 0) | 0 |
| (0, 16) | 384 |
| (8, 12) | 416 |
| (17.6, 0) | 281.6 |

The maximum profit is $416, with 8 Primary Control and 12 Auxiliary Control devices made.

At its maximum, the objective function gives $16x + 24y = 416$, a line through (8, 12) as shown in Figure 4–12.

The fact that the line $16x + 24y = 416$ lies between the lines $5x + 4y = 88$ and $x + 2y = 32$ is significant. In fact, for the objective function to attain its maximum value at (8, 12), it *must* lie between the boundary lines that intersect at (8, 12).

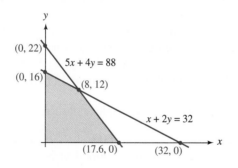

**FIGURE 4–11**

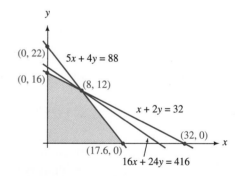

**FIGURE 4–12**

## Changes in the Objective Function

In Figure 4–13, we show three other lines, $L_1$, $L_2$, and $L_3$, that pass through (8, 12) but have different slopes from the objective function.

Because the maximum value of the objective function must occur at a corner point, for our example, it must be at (8, 12), (0, 16), or (17.6, 0).

In the case where the profit on a device changes, we could get a line with a slope similar to $L_1, L_2,$ or $L_3$. Recall that the objective function increases in value as it moves away from the origin. It reaches its maximum value at the last corner point it touches as it moves out. Note that the point (8, 12) is the last corner $L_1$ touches as it moves away from the origin, so (8, 12) is the point of maximum value for *any* line that lies between $5x + 4y = 88$ and $x + 2y = 32$. Also observe that $L_2$ can move farther away from the origin and still cross the feasible region. The corner (0, 16) is the last corner touched by $L_2$, so the maximum value of the $L_2$ objective function occurs at (0, 16).

Likewise, $L_3$ can move farther out until it touches the corner (17.6, 0), which is the point that makes the $L_3$ objective function maximum.

The discussion of Figure 4–13 can help to answer questions such as "What profit values can we have for each device so the point (8, 12) remains the point of maximum profit?" and "What profit values for each device will cause the point of maximum profit to move to (0, 16) or to (17.6, 0)?" The next three examples show how to answer these questions.

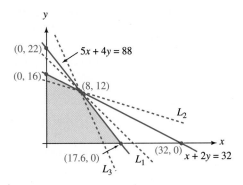

**FIGURE 4–13**

**Example 2**   The problem

Maximize $z = 16x + 24y$, subject to

$$5x + 4y \le 88 \quad \text{(number of C-8 boards)}$$
$$x + 2y \le 32 \quad \text{(number of H-2 boards)}$$
$$x \ge 0, y \ge 0$$

has the solution

$$\text{Maximum } z = 416 \text{ at } (8, 12)$$

How much change can occur in a device profit, such that the point (8, 12) still yields the maximum profit?

**Solution**
We use the slopes of the boundary lines $5x + 4y = 88$ and $x + 2y = 32$ and the slope of an arbitrary objective function, $z = Ax + By$, to find the answer. Because the objective function with slope $-A/B$ through (8, 12) must lie between the lines $5x + 4y = 88$ (slope $= -\frac{5}{4}$) and $x + 2y = 32$ (slope $= -\frac{1}{2}$), the slope of the objective function must lie between the slopes of the boundary lines; that is,

$$-\frac{5}{4} \le -\frac{A}{B} \le -\frac{1}{2}$$

which we can write as

$$\frac{1}{2} \le \frac{A}{B} \le \frac{5}{4}$$

Thus, the profits of the devices whose ratio lies between 0.5 and 1.25 will yield a maximum total profit at (8, 12). For example, device profits of $A = 18$ and $B = 24$ give $\frac{A}{B} = \frac{18}{24} = 0.75$, giving a total maximum profit at (8, 12).

**Example 3**    For the problem of Example 2, determine the device profits that will cause the maximum total profit to move to (0, 16).

**Solution**

For the objective function $z = Ax + By$ to be maximum at the point (0, 16), it must pass through (0, 16) and lie above $x + 2y = 32$, like the dotted line labeled $L_1$ in Figure 4–14.

Because the lines have negative slope and line $L_1$ is more nearly horizontal than $x + 2y = 32$, the slope of line $L_1$ is nearer to zero than the slope $-\frac{1}{2}$. We indicate this by

$$-\frac{1}{2} \le -\frac{A}{B} \le 0$$

and conclude that $z = Ax + By$ has its maximum value at (0, 16) when

$$0 \le \frac{A}{B} \le \frac{1}{2}, \text{ which is equivalent to } A \le \frac{1}{2}B$$

One such example is $z = 10x + 24y$.

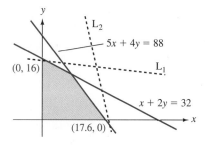

**FIGURE 4–14**

**Example 4**    For the problem of Example 2, determine the device profits that will cause the maximum total profit to move to (17.6, 0).

**Solution**

For the objective function $z = Ax + By$ to be maximum at the point (17.6, 0), it must pass through (17.6, 0) and lie above $5x + 4y = 88$, like the dotted line labeled $L_2$ in Figure 4–14. Because the slopes are negative and line $L_2$ is more nearly vertical than $5x + 4y = 88$, the slope of line $L_2$ is farther from zero than the slope $\frac{-5}{4}$. We indicate this by

$$-\frac{A}{B} \le -\frac{5}{4} \le 0$$

and conclude that $z = Ax + By$ has its maximum value at $(17.6, 0)$ when

$$0 \le \frac{5}{4} \le \frac{A}{B} \quad \text{or} \quad \frac{5}{4} \le \frac{A}{B}$$

which is equivalent to $A \ge 1.25B$. One such example is $z = 25x + 14y$.

Let's continue the analysis of Example 2 to see the effect a change in the objective function has on the point of maximum profit. Recall that the profit on each Primary Control is \$16 and the profit on each Auxiliary Control is \$24, so the objective function is $z = 16x + 24y$, and it attains its maximum value at $(8, 12)$.

**Example 5**    Suppose the manager of Electronic Design is confident that the profit of each Auxiliary Control is \$24, but the profit of each Primary Control may differ from \$16. How much can it differ and $(8, 12)$ remain the optimal point? We use the objective function $z = Ax + 24y$ and find the value of $A$ for which $(8, 12)$ gives the maximum value of $z$. As $B = 24$, we have

$$\frac{1}{2} \le \frac{A}{24} \le \frac{5}{4}$$

Multiply each term by 24 to obtain

$$12 \le A \le 30$$

The profit on each Primary Control device can vary from \$12 to \$30, and $(8, 12)$ remains the optimal point.

**Example 6**    Let's look at the situation when the profit for each Primary Control, \$16, is considered accurate, but the profit for each Auxiliary Control might differ from \$24. As $A = 16$, we have

$$\frac{1}{2} \le \frac{16}{B} \le \frac{5}{4}$$

We can take the reciprocal of each term and reverse the inequality signs to obtain

$$2 \ge \frac{B}{16} \ge \frac{4}{5}$$

> **NOTE**
>
> If you are in doubt about this, consider the numbers
> $$\frac{1}{2} < 3 < 5$$
> and observe that
> $$2 > \frac{1}{3} > \frac{1}{5}$$

Multiply each term in

$$2 \ge \frac{B}{16} \ge \frac{4}{5}$$

by 16 to obtain

$$32 \ge B \ge 12.8$$

In this case, the profit per Auxiliary Control can vary from \$12.80 to \$32 and $(8, 12)$ remains the optimal point.

## Resource Changes

A manager may determine the number of devices produced that yield maximum profit for the number of circuit boards available. That, however, can change. Additional circuit boards might become available, or a shipment may be delayed so that fewer circuit boards are available. We should not be surprised if the optimal solution changes as resources change, because corner points change when a boundary line is moved. Let's look at Example 1 and change the number of H-2 boards available from 32 to 38, with other data unchanged. The problem then becomes the next example.

**Example 7**   Maximize $z = 16x + 24y$, subject to

$$
\begin{aligned}
5x + 4y &\le 88 \\
x + 2y &\le 38 \\
x \ge 0, \; y &\ge 0
\end{aligned}
$$

**Solution**

Figure 4–15 shows the feasible region and corner points resulting from the change in available H-2 boards.

The corners are $(0, 0)$, $(0, 19)$, $(4, 17)$, and $(17.6, 0)$. The corresponding values of $z$ are the following:

| Corner | $z = 16x + 24y$ |
|--------|-----------------|
| $(0, 0)$ | 0 |
| $(0, 19)$ | 456 |
| $(4, 17)$ | 472 |
| $(17.6, 0)$ | 281.6 |

The maximum profit is $472 and occurs when 4 Primary Control and 17 Auxiliary Control devices are made. When 6 H-2 boards were added, profits increased by $56. We observe that the maximum value of $z$ still occurs at the corner formed by the intersection of the two constraints. This will not always happen. If the number of available H-2 boards increases enough – say, to 48 – we have the situation shown by Figure 4–16. The constraint lines intersect outside the first quadrant, so the constraint $x + 2y = 48$ can be ignored because the constraint $5x + 4y = 88$ completely determines the feasible region.

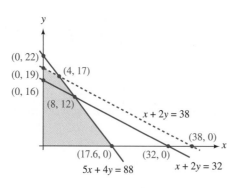

**FIGURE 4–15**

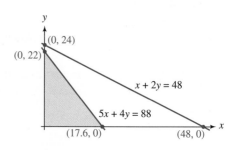

**FIGURE 4–16**

So we ask, "How does a change in the available number of H-2 boards affect maximum profit when the constraint lines still intersect in the first quadrant?"

We answer the question by looking at a more general form of the H-2 constraint in the next example.

**Example 8**   In the problem:

Maximize $z = 16x + 24y$, subject to

$$5x + 4y \leq 88 \quad \text{(number of C-8 boards)}$$
$$x + 2y \leq 32 \quad \text{(number of H-2 boards)}$$
$$x \geq 0, y \geq 0$$

How does a change in the available number of H-2 boards affect maximum profit when the constraint lines still intersect in the first quadrant?

**Solution**

Let $D$ = the change in the number of H-2 boards, where $D$ can be positive or negative. The problem becomes

Maximize $z = 16x + 24y$, subject to

$$5x + 4y \leq 88$$
$$x + 2y \leq 32 + D$$
$$x \geq 0, y \geq 0$$

Figure 4–17 shows the feasible region, where the $y$-intercept of $x + 2y = 32 + D$ is

$$\left(0, 16 + \frac{D}{2}\right)$$

and the $x$-intercept is $(32 + D, 0)$.

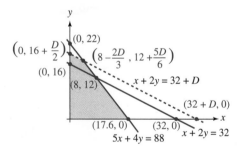

**FIGURE 4–17**

Solving the system

$$5x + 4y = 88$$
$$x + 2y = 32 + D$$

we find at the intersection of the two lines

$$x = 8 - \frac{2D}{3}$$

$$y = 12 + \frac{5D}{6}$$

and the value of $z$ at that point is

$$16\left(8 - \frac{2D}{3}\right) + 24\left(12 + \frac{5D}{6}\right) = 128 - \frac{32}{3}D + 288 + 20D$$

$$= 416 + \frac{28}{3}D$$

This tells us that for each unit change in the number of H-2 boards available, the number of Primary devices decreases by $\frac{2}{3}$, the number of Auxiliary devices increases by $\frac{5}{6}$, and profit increases by $\frac{28}{3} = 9.33$. However, we must be careful. For the constraint lines to intersect in the first quadrant, neither $x$ nor $y$ can be negative. We must have

$$x = 8 - \frac{2D}{3} \geq 0$$

which gives $12 \geq D$ and

$$y = 12 + \frac{5D}{6} \geq 0$$

which gives $D \geq -\frac{72}{5} = -14.4$. This indicates that the corner

$$\left(8 - \frac{2D}{3}, 12 + \frac{5D}{6}\right)$$

is the point where $z$ is maximum when $D$ is restricted to the interval $-14.4 \leq D \leq 12$.

---

## 4.7  EXERCISES

Access end-of-section exercises online at **www.webassign.net**

# IMPORTANT TERMS

**4.1**
Simplex Method
Standard Maximum Problem
Slack Variable
Simplex Tableau

**4.2**
Unit Column
Feasible Solution
Basic Solution
Basic Feasible Solution
Nonbasic Variable
Basic Variable
Initial Basic Feasible Solution

Pivot Element
Pivot Column
Pivot Row
Pivoting
Final Tableau

**4.3**
Standard Minimum Problem
Dual Problem
Transpose of a Matrix

**4.4**
Mixed Constraints
Phase I

Phase II
Equality Constraint

**4.5**
Multiple Solutions
No Solutions
Unbounded Solution
No Feasible Solution

**4.7**
Sensitivity Analysis

**4**

## IMPORTANT CONCEPTS

**Standard Maximum Problem**  The objective function is to be maximized subject to inequalities using the $\leq$ inequality and the constraints have nonnegative constants.

**Simplex Method**  Finds the optimal solution using row operations on a matrix representing the objective function and constraints. (See Section 4.2.)

**Basic Solution**  Obtain a basic solution for a problem with $k$ variables in the constraints by setting $k$ variables (except $z$) to zero and solving for the rest.

**Standard Minimum Problem**  The objective function is to be minimized, the constraints have the $\geq$ inequality and the constraints have nonnegative constants.

**Dual Problem**  A standard minimum problem is converted to a maximum problem by the dual method. (See Section 4.3.)

**Theorem of Duality**  A standard minimum problem has a solution if and only if its dual problem has a solution. The solution to the dual problem is the same as the solution to the minimum problem.

**Mixed Constraints**  The constraints have a mixture of $\leq$ and $\geq$ inequalities or a negative constant.

**Multiple Solutions**  A simplex tableau signals multiple solutions when a zero appears in the bottom row of a nonbasic variable's column (a non-unit column).

**Unbounded Solutions**  A simplex tableau signals unbounded solutions when a pivot column contains no positive entries.

**No Feasible Solution**  A simplex tableau signals no feasible solution when a row occurs with a negative entry in the last column and no other negative entries in the row.

**Sensitivity Analysis**  A study of the consequences of changes in data of a linear programming problem. (See Section 4.7.)

**4**

## REVIEW EXERCISES

**In Exercises 1 through 3, write the constraints as a system of equations using slack variables.**

1.  $6x_1 + 4x_2 + 3x_3 \leq 220$
    $x_1 + 5x_2 + x_3 \leq 162$
    $7x_1 + 2x_2 + 5x_3 \leq 139$

2.  $5x_1 + 3x_2 \leq 40$
    $7x_1 + 2x_2 \leq 19$
    $6x_1 + 5x_2 \leq 23$

3.  $6x_1 + 5x_2 + 3x_3 + 3x_4 \leq 89$
    $7x_1 + 4x_2 + 6x_3 + 2x_4 \leq 72$

**In Exercises 4 through 6, write the constraints and objective function as a system of equations.**

4.  Objective function: $z = 3x_1 + 7x_2$
    Constraints:

    $7x_1 + 5x_2 \leq 14$
    $3x_1 + 6x_2 \leq 25$
    $4x_1 + 3x_2 \leq 29$

5.  Objective function: $z = 20x_1 + 36x_2 + 19x_3$
    Constraints:

    $10x_1 + 12x_2 + 8x_3 \leq 24$
    $7x_1 + 13x_2 + 5x_3 \leq 35$

6. Objective function: $z = 5x_1 + 12x_2 + 8x_3 + 2x_4$
Constraints:

$$9x_1 + 7x_2 + x_3 + x_4 \leq 84$$
$$x_1 + 3x_2 + 5x_3 + x_4 \leq 76$$
$$2x_1 + x_2 + 6x_3 + 3x_4 \leq 59$$

**Write Exercises 7 through 10 as systems of equations.**

7. Maximize $z = 9x_1 + 2x_2$, subject to

$$3x_1 + 7x_2 \leq 14$$
$$9x_1 + 5x_2 \leq 18$$
$$x_1 - x_2 \leq 21$$
$$x_1 \geq 0, x_2 \geq 0$$

8. Maximize $z = x_1 + 5x_2 + 4x_3$, subject to

$$x_1 + x_2 + x_3 \leq 20$$
$$4x_1 + 5x_2 + x_3 \leq 48$$
$$2x_1 - 6x_2 + 5x_3 \leq 38$$
$$x_1 \geq 0, x_2 \geq 0, x_3 \geq 0$$

9. Maximize $z = 6x_1 + 8x_2 + 4x_3$, subject to

$$x_1 + x_2 + x_3 \leq 15$$

$$x_1 \geq 0, x_2 \geq 0, x_3 \geq 0$$

10. Maximize $z = 5x_1 + 5x_2$, subject to

$$5x_1 + 3x_2 \leq 15$$
$$2x_1 + 3x_2 \leq 12$$
$$x_1 \geq 0, x_2 \geq 0$$

11. Find the pivot element in each of the following tableaux.

(a)
$$\begin{bmatrix} 5 & 3 & 2 & 1 & 0 & 0 & 0 & 660 \\ 4 & 6 & 1 & 0 & 1 & 0 & 0 & 900 \\ 1 & 2 & 3 & 0 & 0 & 1 & 0 & 800 \\ -5 & -8 & -4 & 0 & 0 & 0 & 1 & 0 \end{bmatrix}$$

(b)
$$\begin{bmatrix} 1 & 4 & 0 & 3 & 0 & -2 & 0 & 60 \\ 0 & 6 & 1 & 5 & 0 & 4 & 0 & 60 \\ 0 & -3 & 0 & 1 & 1 & 2 & 0 & 60 \\ 0 & -1 & 0 & -2 & 0 & 3 & 1 & 48 \end{bmatrix}$$

12. Find the pivot element in each of the following tableaux.

(a)
$$\begin{bmatrix} 3 & 0 & 5 & 1 & 0 & 0 & 20 \\ 2 & 0 & -1 & 0 & 1 & 0 & 6 \\ 1 & 1 & 4 & 0 & 0 & 0 & 0 \\ 4 & 0 & -8 & 0 & 0 & 1 & 145 \end{bmatrix}$$

(b)
$$\begin{bmatrix} 6 & 1 & 0 & -5 & 0 & 2 & 0 & 0 \\ 4 & 0 & 1 & 2 & 0 & -4 & 0 & 10 \\ -2 & 0 & 0 & 7 & 1 & 3 & 0 & 21 \\ -1 & 0 & 0 & -9 & 0 & 6 & 1 & 572 \end{bmatrix}$$

13. Write the basic feasible solution for each of the following tableaux.

(a)
$$\begin{bmatrix} 6 & 0 & 10 & 1 & 0 & 42 \\ 5 & 1 & 8 & 0 & 0 & 80 \\ -2 & 0 & 5 & 0 & 1 & 98 \end{bmatrix}$$

(b)
$$\begin{bmatrix} 0 & 1 & 0 & 8 & 6 & 4 & 0 & 42 \\ 1 & 0 & 0 & -2 & 3 & 3 & 0 & 73 \\ 0 & 0 & 1 & 5 & -1 & 6 & 0 & 15 \\ 0 & 0 & 0 & 4 & 2 & 5 & 1 & 138 \end{bmatrix}$$

14. Write the initial simplex tableau for each of the following problems. Do not solve.

(a) Minimize $z = 4x_1 - 5x_2 + 3x_3$, subject to

$$9x_1 + 7x_2 + x_3 \leq 45$$
$$3x_1 + 2x_2 + 4x_3 \leq 39$$
$$x_1 + 5x_2 + 12x_3 \leq 50$$
$$x_1 \geq 0, x_2 \geq 0, x_3 \geq 0$$

(b) Maximize $z = 8x_1 + 13x_2$, subject to

$$9x_1 + 5x_2 \leq 45$$
$$6x_1 + 8x_2 \geq 48$$
$$x_1 \geq 0, x_2 \geq 0$$

15. Write the initial simplex tableau for each of the following problems. Do not solve.

(a) Maximize $z = 3x_1 + 5x_2 + 4x_3$, subject to

$$11x_1 + 5x_2 + 3x_3 \leq 142$$
$$3x_1 + 4x_2 + 7x_3 \geq 95$$
$$2x_1 + 15x_2 + x_3 \leq 124$$
$$x_1 \geq 0, x_2 \geq 0, x_3 \geq 0$$

(b) Minimize $z = 14x_1 + 22x_2$, subject to

$$7x_1 + 4x_2 \leq 28$$
$$x_1 + 3x_2 \geq 6$$
$$x_1 \geq 0, x_2 \geq 0$$

16. Write the initial simplex tableau for each of the following problems. Do not solve.

(a) Maximize $z = 24x_1 + 36x_2$, subject to

$$2x_1 + 9x_2 \leq 18$$
$$5x_1 + 7x_2 \leq 35$$
$$x_1 + 8x_2 \geq 8$$
$$x_1 \geq 0, x_2 \geq 0$$

**(b)** Minimize $z = 6x_1 + 11x_2$, subject to

$$5x_1 + 4x_2 \geq 20$$
$$3x_1 + 8x_2 \geq 24$$
$$x_1 + x_2 \leq 22$$
$$x_1 \geq 0, x_2 \geq 0$$

**17.** Write the initial simplex tableau for each of the following problems. Do not solve.

**(a)** Maximize $z = 5x_1 + 12x_2$, subject to

$$15x_1 + 8x_2 \geq 120$$
$$10x_1 + 12x_2 \leq 120$$
$$15x_1 + 5x_2 \geq 75$$
$$x_1 \geq 0, x_2 \geq 0$$

**(b)** Minimize $z = 3x_1 + 2x_2$, subject to

$$14x_1 + 9x_2 \leq 126$$
$$10x_1 + 11x_2 \geq 110$$
$$-5x_1 + x_2 \leq 9$$
$$x_1 \geq 0, x_2 \geq 0$$

**Solve Exercises 18 through 23.**

**18.** Maximize $z = 4x_1 + 5x_2$, subject to

$$x_1 + 3x_2 \leq 12$$
$$2x_1 + 4x_2 \leq 16$$
$$x_1 \geq 0, x_2 \geq 0$$

**19.** Maximize $z = 3x_1 + 5x_2 + 2x_3$, subject to

$$2x_1 + 4x_2 + 2x_3 \leq 34$$
$$3x_1 + 6x_2 + 4x_3 \leq 57$$
$$2x_1 + 5x_2 + x_3 \leq 30$$
$$x_1 \geq 0, x_2 \geq 0, x_3 \geq 0$$

**20.** Maximize $z = 3x_1 + 4x_2$, subject to

$$x_1 - 3x_2 \leq 6$$
$$x_1 + x_2 \leq 8$$
$$x_1 \geq 0, x_2 \geq 0$$

**21.** Maximize $z = 10x_1 + 15x_2$, subject to

$$-4x_1 + x_2 \leq 3$$
$$x_1 - 2x_2 \leq 12$$
$$x_1 \geq 0, x_2 \geq 0$$

**22.** Maximize $z = 3x_1 + 6x_2 + x_3$, subject to

$$4x_1 + 4x_2 + 8x_3 \leq 800$$
$$8x_1 + 6x_2 + 4x_3 \leq 1800$$
$$8x_1 + 4x_2 \leq 400$$
$$x_1 \geq 0, x_2 \geq 0, x_3 \geq 0$$

**23.** Maximize $z = x_1 + 3x_2 + x_3$, subject to

$$4x_1 + x_2 + x_3 \leq 372$$
$$x_1 + 8x_2 + 6x_3 \leq 1116$$
$$x_1 \geq 0, x_2 \geq 0, x_3 \geq 0$$

**24.** Find the values of the slack variables in the constraints

$$4x_1 + 3x_2 + 6x_3 + s_1 = 68$$
$$x_1 + 2x_2 + 5x_3 + s_2 = 90$$

for the points (3, 2, 1) and (5, 10, 3).

**25.** Which variable contributes the most to increasing $z$ in the following objective functions of a maximization problem?

**(a)** $z = 7x_1 + 2x_2 + 9x_3$

**(b)** $z = x_1 + 8x_2 + x_3 + 7x_4$

**26.** Given the constraint

$$7x_1 + 4x_2 + 17x_3 + s_1 = 56$$

when $x_2 = 0$, what is the largest possible value of $x_3$ so that $s_1$ is nonnegative?

**27.** Write the transpose of the matrices

$$\begin{bmatrix} 3 & 1 & -2 \\ 4 & 0 & 6 \\ 5 & 7 & 8 \end{bmatrix} \text{ and } \begin{bmatrix} 4 & 3 & 2 & 1 \\ -5 & 0 & 12 & 9 \end{bmatrix}$$

**Solve Exercises 28 through 36.**

**28.** Minimize $z = 3x_1 + 5x_2 + 4x_3$, subject to

$$3x_1 - 3x_2 + x_3 \leq 54$$
$$x_1 + x_2 + x_3 \geq 24$$
$$-3x_1 + 2x_3 \leq 15$$
$$x_1 \geq 0, x_2 \geq 0, x_3 \geq 0$$

**29.** Minimize $z = 6x_1 + 8x_2 + 16x_3$, subject to

$$2x_1 + x_2 \geq 6$$
$$x_2 + 2x_3 \geq 8$$
$$x_1 \geq 0, x_2 \geq 0, x_3 \geq 0$$

**30.** Minimize $z = 10x_1 + 20x_2 + 15x_3$, subject to

$$x_1 + x_2 + x_3 \geq 100$$
$$9x_1 - 4x_3 \leq 128$$
$$x_1 + 4x_2 \geq 48$$
$$x_1 \geq 0, x_2 \geq 0, x_3 \geq 0$$

**31.** Maximize $z = 6x_1 + 11x_2 + 8x_3$, subject to

$$2x_1 + 5x_2 + 4x_3 \leq 40$$
$$40x_1 + 45x_2 + 30x_3 \leq 430$$
$$6x_1 + 3x_2 + 4x_3 \geq 48$$
$$x_1 \geq 0, x_2 \geq 0, x_3 \geq 0$$

**32.** Maximize $z = 3x_1 + 4x_2 + x_3$, subject to

$$x_1 + x_2 + x_3 \leq 28$$
$$3x_1 + 4x_2 + 6x_3 \geq 60$$
$$2x_1 + x_2 \leq 30$$
$$x_1 \geq 0, x_2 \geq 0, x_3 \geq 0$$

**33.** Maximize $z = 2x_1 + 5x_2 + 3x_3$, subject to

$$x_1 + x_2 + x_3 \geq 6$$
$$2x_1 + x_2 + 3x_3 \leq 10$$
$$2x_2 - x_3 \leq 5$$
$$x_1 \geq 0, x_2 \geq 0, x_3 \geq 0$$

**34.** Minimize $z = 18x_1 + 24x_2$, subject to

$$3x_1 + 4x_2 \geq 48$$
$$x_1 + 2x_2 \leq 22$$
$$3x_1 + 2x_2 \leq 42$$
$$x_1 \geq 0, x_2 \geq 0$$

**35.** Maximize $z = 20x_1 + 32x_2$, subject to

$$x_1 - 3x_2 \leq 24$$
$$-5x_1 + 4x_2 \leq 20$$
$$x_1 \geq 0, x_2 \geq 0$$

**36.** Minimize $z = 8x_1 + 10x_2 + 25x_3$, subject to

$$x_1 + x_3 \geq 30$$
$$2x_1 + 4x_2 + 5x_3 \geq 70$$
$$2x_2 + x_3 \geq 27$$
$$x_1 \geq 0, x_2 \geq 0, x_3 \geq 0$$

**Solve Exercises 37 through 41.**

**37.** Minimize $z = 18x_1 + 36x_2$, subject to

$$3x_1 + 2x_2 \geq 24$$
$$5x_1 + 4x_2 \geq 46$$
$$4x_1 + 9x_2 \geq 60$$
$$x_1 \geq 0, x_2 \geq 0$$

**38.** Maximize $z = 5x_1 + 15x_2$, subject to

$$4x_1 + x_2 \leq 200$$
$$x_1 + 3x_2 \geq 120$$
$$x_1 \geq 0, x_2 \geq 0$$

**39.** Maximize $z = 5x_1 + 15x_2$, subject to

$$4x_1 + x_2 \leq 180$$
$$x_1 + 3x_2 \geq 120$$
$$-x_1 + 3x_2 \geq 150$$
$$x_1 \geq 0, x_2 \geq 0$$

**40.** Minimize $z = 3x_1 - 2x_2$, subject to

$$x_1 + 3x_2 \leq 30$$
$$3x_1 + x_2 \leq 21$$
$$x_1 \geq 0, x_2 \geq 0$$

**41.** Maximize $z = 2x_1 + x_2$, subject to

$$x_1 + 3x_2 \leq 9$$
$$x_1 - x_2 \leq -2$$
$$x_1 \geq 0, x_2 \geq 0$$

**42.** For each of the following minimization problems, set up the augmented matrix and the initial tableau for the dual problem. Do not solve.

**(a)** Minimize $z = 30x_1 + 17x_2$, subject to

$$4x_1 + 5x_2 \geq 52$$
$$7x_1 + 14x_2 \geq 39$$
$$x_1 \geq 0, x_2 \geq 0$$

**(b)** Minimize $z = 100x_1 + 225x_2 + 145x_3$, subject to

$$20x_1 + 35x_2 + 15x_3 \geq 130$$
$$40x_1 + 10x_2 + 6x_3 \geq 220$$
$$35x_1 + 22x_2 + 18x_3 \geq 176$$
$$x_1 \geq 0, x_2 \geq 0, x_3 \geq 0$$

**Set up the initial simplex tableau for Exercises 43 and 44. Do not solve.**

**43.** A company manufactures three items: hunting jackets, all-weather jackets, and ski jackets. It takes 3 hours of labor per dozen to produce hunting jackets, 2.5 hours per dozen for all-weather jackets, and 3.5 hours per dozen for ski jackets. The cost per dozen is $26 for hunting jackets, $20 for all-weather jackets, and $22 for ski jackets. The profit per dozen is $7.50 for hunting jackets, $9 for all-weather jackets, and $11 for ski jackets. The company has 3200 hours of labor and $18,000 in operating funds available. How many of each jacket should it produce to maximize profits?

**44.** A fertilizer company produces two kinds of fertilizers, lawn and tree. It has orders on hand that call for the production of at least 20,000 bags of lawn fertilizer and 5000 bags of tree fertilizer. Plant A can produce 1500 bags of lawn and 300 bags of tree fertilizer per day. Plant B can produce 750 bags of lawn and 250 bags of tree fertilizer per day. It costs $18,000 per day to operate plant A and $12,000 per day to operate plant B. How many days should the company operate each plant to minimize operating costs?

**45.** The constraints to a linear programming problem are

$$5x + 2y \leq 155$$
$$3x + 4y \leq 135$$
$$x \geq 0, y \geq 0$$

Find the possible values of $A$ so that the objective function $z = Ax + 8y$ has its maximum value at $(25, 15)$.

**46.** The constraints to a linear programming problem are

$$4x + 5y \leq 205$$
$$16x + 5y \leq 336$$
$$x \geq 0, y \geq 0$$

Find the possible values of $B$ so that the objective function $z = 10x + By$ has its maximum value at $(0, 41)$.

# MATHEMATICS OF FINANCE

Our modern economy depends on borrowed money. If you have a credit card or a student loan, you have firsthand experience with a loan. Borrowed money enables students to obtain an education or to own an automobile. Few families can own a home without borrowing money. Business depends on borrowed money for day-to-day operations and major expansions. Governments at all levels, schools, churches, and other institutions borrow money. Banks depend on loans for a major source of their income. Our economy would collapse if financial institutions quit making loans.

Here are some trivia questions for you: "What is the current national debt?" "When was the last time the federal government balanced its budget?" "When was the last time there was no national debt?"

"Rented money" describes "borrowed money" because a fee is paid for the use of money for a period of time. Just as you pay a rental fee for the use of an apartment for a semester, you pay a rental fee, or **interest** as it is called, for the use of money.

Even those few who may not borrow money may place money in a savings account or a certificate of deposit. In doing so, that person loans money to the bank, and the bank pays interest. In turn, the bank loans the money to an individual or business (at a higher rate, of course).

Some day in the distant future, you hope to retire with adequate funds to enjoy a reasonable quality of life. To do so, you, your employer, or both need to invest in a retirement plan that will pay you an **annuity** in your retirement years. An analysis of investments and annuities shows that the earlier you invest for retirement, the more likely you will have adequate retirement income.

The fee charged for the use of money works two ways. When you borrow money from a bank, you pay a fee for the use of the bank's money. When you buy a certificate of deposit from a bank, the bank pays you a fee for the use of your money. These fees help increase the value of your investment, they add to the cost of your student loan, and they help you to plan for future expenses. In this chapter you will learn how to compute the cost of borrowing money, monthly car payments, regular investments needed to pay cash for a future expense, and the impact of interest fees on the cost of a house.

We now look at some methods used to determine the fees charged for the use of money.

## 5.1   SIMPLE INTEREST

- Simple Interest
- Future Value
- Treasury Bills—Simple Discount

### Simple Interest

**Simple interest** is most often used for loans of shorter duration, in situations like a construction company that may be expecting payment in a week for work done, but it is payday and the workers are due their checks so the company borrows money for a week, or a store may obtain a 30-day loan to purchase inventory for a sale.

How does a simple interest loan work? First, let's mention some standard terms used when discussing simple interest. We call the money borrowed in a loan the **principal**. The number of dollars received by the borrower is the **present value**. In a simple interest loan, the principal and the present value are the same. The fee for a simple interest loan is usually expressed as a percentage of the principal and is called the **interest rate**. For an interest rate of 10% per year, each year the borrower pays 10% of the principal, the amount borrowed. For an interest rate of 1.5% per month, the borrower pays 1.5% of the principal for each month the money is borrowed.

You should be aware that the formulas in this chapter require that the interest rate be written in decimal form, not in percent. For example, 7.5% must be written as 0.075.

We denote the amount of the principal by $P$. For an interest rate of 10% per year, the interest paid is $0.10P$ for each year of the loan. Thus, for money borrowed for 3 years, the total interest paid is $3(0.10P)$.

This suggests the general formula that gives the total fee, interest, which is paid for a simple interest loan.

| | |
|---|---|
| **Simple Interest** | $$I = Prt$$ |

where

$P =$ principal (amount borrowed)
$r =$ interest rate per year (expressed in decimal form)
$t =$ time in years
$I =$ interest paid

An interest rate may be stated as 10% per year, 1% per month, or in terms of other time units. Sometimes no time units are specified. In such cases it is understood that the time unit is years. A statement that the interest rate is 12% should be interpreted as 12% *per year*. The time period will be stated when it is not annual. Simple interest is paid on the principal borrowed and is not paid on interest already earned.

**Example 1**    Compute the interest paid on a loan of $1400 at a 9% interest rate for 18 months.

**Solution**

> **CAUTION**
>
> The time units for $r$ and $t$ must be consistent, so months were converted to years.

$$P = \$1400$$
$$r = 9\% = 0.09 \text{ in decimal form}$$
$$t = 18 \text{ months } = \frac{18}{12} \text{ years} = 1.5 \text{ years}$$

$$I = 1400 \times 0.09 \times 1.5 = 189$$

So, the interest paid is $189.

**Example 2**    An individual borrows $300 for 6 months at 1% simple interest per month. How much interest is paid?

**Solution**
Note that the interest rate is given as 1% *per month*. We can still use the $I = Prt$ formula, provided that $r$ and $t$ are consistent in time units—in this case, months.

$$I = 300 \times 0.01 \times 6 = \$18$$

**Example 3**    Jose borrows money at 8% for 2 years. He paid $124 interest. How much did he borrow?

**Solution**
In this case, $I = 124$, $r = 0.08$, and $t = 2$, so

$$124 = P(0.08)(2) = 0.16P$$
$$P = \frac{124}{0.16} = 775$$

The loan was $775.

**Example 4**    Jane borrowed $950 for 15 months. The interest was $83.13. Find the interest rate.

**Solution**
We are given $P = 950$, $t = \frac{15}{12} = 1.25$ years, and $I = 83.13$, so

$$83.13 = 950r(1.25)$$
$$= 1187.5r$$
$$r = \frac{83.13}{1187.5} = 0.07$$

Thus, the interest rate was 7% per year.

## Future Value

A loan made at simple interest requires that the borrower pay back the sum borrowed (principal) plus the interest. We call this total, $P + I$, the **future value**, or **amount**, of the loan. Because $I = Prt$, we can write the future value as $A = P + Prt$.

---

**Amount, or Future Value, of a Loan**

$$A = P + I$$
$$= P + Prt$$
$$= P(1 + rt)$$

where

$P = $ principal, or present value
$r = $ annual interest rate
$t = $ time in years
$A = $ amount, or future value

---

**Example 5** Find the amount (future value) of a $2400 loan for 9 months at 11% interest rate.

**Solution**
We want to find $A$ in $A = P + I = P + Prt$. We know that $P = 2400$, $r = 0.11$, and $t = \frac{9}{12} = 0.75$ years, so

$$I = 2400(0.11)(0.75) = 198$$

and

$$A = 2400 + 198 = 2598$$

We can also use the formula $A = P(1 + rt)$ and compute $A$ as

$$A = 2400(1 + 0.11(0.75))$$
$$= 2400(1 + 0.0825)$$
$$= 2400(1.0825)$$
$$= 2598$$

The total of principal and interest is $2598.

Up to this point, the examples have dealt with the cost (interest) of a loan. In order for a person to obtain a loan, another party must provide the money to be borrowed. That party invests money and expects to receive income from the loan. The same formulas apply to the investor as to the borrower.

**Example 6** How much should you invest at 12% for 21 months to have $3000 at the end of the 21 months?

**Solution**
In this example, you are given the future value, $3000, and are asked to find the present value, $P$. We know that $r = 0.12$, $t = \frac{21}{12} = 1.75$ years, and $A = 3000$. Then,

$$3000 = P(1 + 0.12(1.75))$$
$$= P(1 + 0.21)$$
$$= 1.21P$$

So, $P = \frac{3000}{1.21} = \$2479.34$ (rounded to nearest cent).

**Example 7**  Your friend loaned some money. The debtor promises to pay him $550 in 4 months. (The future value is $550.) Your friend needs the money now, so you agree to pay him $525 for the note, and the debtor will pay you $550 in 4 months. What annual interest rate will you earn?

**Solution**
For the purposes of this problem, $A = 550$, $P = 525$, and $t = 4$ months $= \frac{1}{3}$ year. Using the formula $A = P + Prt$, we have

$$550 = 525 + 525r\left(\frac{1}{3}\right)$$
$$= 525 + 175r$$
$$25 = 175r$$
$$r = \frac{25}{175} = 0.1429$$

You will earn about 14.3% annual interest.

## Treasury Bills—Simple Discount

The federal government issues short-term securities called *treasury bills*. The bills do not specify a rate of interest. They are sold at weekly public auctions with financial institutions making competitive bids. For example, a bank may bid $978,300 for a 90-day $1 million treasury bill. At the end of 90 days, the bank receives $1 million, which covers the cost of the bill and interest earned on the bill. We call this a **simple discount** transaction.

The simple discount loan differs from the simple interest loan in that the interest is *deducted* from the principal and the borrower receives less than the principal. For example, if a person borrows $1000 at 6% for 1 year, the interest is $60, which is deducted from the $1000, and the borrower receives $940. When the loan is repaid, the borrower pays $1000. We use the terminology **simple discount note** for this type of loan. We use the term **discount** for interest deducted, we use **proceeds** for the amount received by the borrower, the term **discount rate** denotes the percentage (interest rate) used, and **maturity value** refers to the amount repaid. Here's how they are related:

**Simple Discount**

$$D = Mdt$$
$$PR = M - D$$
$$= M - Mdt$$
$$= M(1 - dt)$$

where

$$M = \text{maturity value (principal)}$$
$$d = \text{annual discount rate, written in decimal form}$$
$$t = \text{time in years}$$
$$D = \text{discount}$$
$$PR = \text{proceeds, the amount the borrower receives}$$

**Example 8**   Find the discount and the amount a borrower receives (proceeds) on a $1500 simple discount loan at 8% discount rate for 1.5 years.

**Solution**
In this case, $M = 1500$, $d = 0.08$, and $t = 1.5$ years. Then,

$$D = 1500(0.08)(1.5)$$
$$= 180$$
$$PR = 1500 - 180$$
$$= 1320$$

So, the bank keeps the discount, $180, and the borrower receives $1320.

**Example 9**   A bank wants to earn 7.5% simple discount interest on a 90-day $1 million treasury bill. How much should it bid?

**Solution**
We are given the maturity value, $M = 1,000,000$, the discount rate, $d = 0.075$, and the time, $t = \frac{90}{360}$. (Banks often use 360 days per year when computing daily interest.)
     We want to find the proceeds, $PR$. Putting our given values into the simple discount formula, we get

$$PR = 1,000,000\left[1 - (0.075)\frac{90}{360}\right]$$
$$= 1,000,000(1 - 0.01875)$$
$$= 1,000,000(0.98125)$$
$$= 981,250$$

So, the bank should bid $981,250.

**NOTE**

The practice of using 360 days for a year began before computers and calculators: dividing by 360 is easier than dividing by 365. It is also consistent with using 30 days for a month.

**Example 10**   A bank paid $983,000 for a 90-day $1 million treasury bill. Find the simple discount rate.

**Solution**
We are given $M = 1,000,000$, $PR = 983,000$, and $t = \frac{90}{360}$. We want to find $d$. We use the form $PR = M(1 - dt)$.

$$983,000 = 1,000,000\left[1 - d\left(\frac{90}{360}\right)\right]$$
$$= 1,000,000 - 250,000d$$
$$-17,000 = -250,000d$$
$$d = \frac{17,000}{250,000} = 0.068$$

So, the annual discount rate was 6.8%.

## 5.1     EXERCISES

Access end-of-section exercises online at **www.webassign.net**

## 5.2     COMPOUND INTEREST

- Compound Interest
- Present Value
- Doubling an Investment
- Effective Rate
- Zero-Coupon Bonds

### Compound Interest

When you deposit money into a savings account, the bank will pay for the use of your money. Normally, the bank pays interest at specified periods of time, such as every 3 months. Unless instructed otherwise, the bank credits your account with the interest, and for the next time period, the bank pays interest on the new total. We call this **compound interest**. Let's look at a simple example.

**Example 1**     You put $1000 into an account that pays 8% annual interest. The bank will compute interest and add it to your account at the end of each year. We call this **compounding interest annually**. Here's how your account builds up. We use the formula $A = P(1 + rt)$, where $r = 0.08$ and $t = 1$.

| End of year | Balance in your account |
|---|---|
| 0 (start) | $1000.00 |
| 1 | $1000(1.08) = 1080.00$ |
| 2 | $1080(1.08) = 1166.40 = 1000(1.08)^2$ |
| 3 | $1166.40(1.08) = 1259.71 = 1000(1.08)^3$ |
| 4 | $1259.71(1.08) = 1360.49 = 1000(1.08)^4$ |
| 5 | $1360.49(1.08) = 1469.33 = 1000(1.08)^5$ |

So, at the end of 5 years, the account has grown to $1469.33.

Note how the pattern of growth involves *powers* of 1.08. The amount at the end of 1 year is just 1.08 times the amount at the end of the preceding year. Thus, the amount at the end of 4 years is 1.08 times the amount at the end of 3 years, the amount at the end of 5 years is 1.08 times the amount at the end of 4 years; or for longer periods of time, the amount at the end of 17 years is 1.08 times the amount at the end of 16 years, and so on. A more helpful form gives the amount after, say, 5 years in terms of the *original* investment and 1.08. Note that the amount after 5 years is $1000(1.08)^5$.

This follows because multiplying 1000 by 1.08 to obtain the amount for the first year, multiplying the first year total by 1.08 to obtain the amount for the second year, and so on for 5 years is equivalent to multiplying the original investment by 1.08 five times. You may expect, quite correctly, that the amount after 10 years equals $1000(1.08)^{10}$.

In general, we have the following formula for interest compounded annually.

---

**Amount of Annual Compound Interest**

When $P$ dollars are invested at an annual interest rate $r$ and the interest is compounded annually, the amount $A$, or **future value**, at the end of $t$ years is

$$A = P(1 + r)^t$$

---

**Example 2**    $800 are invested at 6%, with interest compounded annually. Find the amount in the account at the end of 4 years.

**Solution**
Here, $P = 800$, $r = 0.06$, and $t = 4$, so

$$A = 800(1.06)^4 = 800(1.26248) = 1009.98$$

In the preceding examples, interest was compounded annually. Interest is often calculated and added to the principal at other regular intervals. The most common intervals are semi-annually, quarterly, monthly, and daily.

What do we mean by compounding semiannually, quarterly, and so on? It means that at the end of a fixed time period, interest is calculated and added to the account. Here is a summary of common compound interest intervals.

| **Interest Period** | **Length of Period** | **Frequency of Interest Payments** |
|---|---|---|
| Annually | 1 year | Once a year |
| Semiannually | 6 months | 2 times a year |
| Quarterly | 3 months | 4 times a year |
| Monthly | 1 month | 12 times a year |
| Daily | 1 day | 365 times a year |

*Note:* In Section 5.1, 360 days were used for a year. With the availability of calculators, 365 days is more accurate and just as easy to use.

We use the quarterly interest rate for interest compounded quarterly, the monthly interest rate for monthly compounding, and so on.

Because interest rates are usually stated as an annual rate, we need to convert to the appropriate **periodic interest rate**. We do this by dividing the annual rate by the number of interest periods in a year. An 8% annual rate becomes $8\%/4 = 2\%$ quarterly rate and a 6% annual rate becomes $6\%/12 = 0.5\%$ monthly rate.

Generally, we will use the letter $r$ to represent the annual rate and the letter $i$ to represent the periodic rate.

**Periodic Interest Rate**
Given the annual interest rate, $r$, with interest compounded $m$ times a year, the periodic interest rate, $i$, is

$$i = \frac{r}{m}$$

Let's repeat Example 1 but compound the interest quarterly.

**Example 3**
$1000 is invested at 8% annual rate with interest compounded quarterly. Find the amount in the account at the end of 5 years.

**Solution**
Because interest is compounded quarterly, it must be computed every three months and added to the principal. The quarterly interest rate is 2%, so the computations are the following:

| End of Quarter | Amount in Account |
|:---:|:---:|
| 0 | 1000.00 |
| 1 | $1000(1.02) = 1020.00$ |
| 2 | $1020(1.02) = 1040.40 = 1000(1.02)^2$ |
| 3 | $1040.40(1.02) = 1061.21 = 1000(1.02)^3$ |
| 4 | $1061.21(1.02) = 1082.43 = 1000(1.02)^4$ |

At this stage we have the amount at the end of 1 year. We will not continue for another 16 computations, but observe the pattern. At the end of 10 quarters, we correctly expect the amount to be $A = 1000(1.02)^{10}$. At the end of 5 years, 20 quarters, the amount is

$$A = 1000(1.02)^{20} = 1000(1.48595) = 1485.95$$

Note that compounding quarterly gives an amount of $1485.95, a larger amount than the amount $1469.33 obtained from compounding annually.

The general formula for finding the **amount** after a specified number of compound periods is the following.

**Compound Interest— Amount (Future Value)**

$$A = P(1 + i)^n$$

where

$r$ = annual interest rate
$m$ = number of times compounded per year
$i = r/m$ = the interest rate per period
$n$ = the number of periods, $n = mt$ where $t$ is the number of years
$A$ = amount (**future value**) at the end of $n$ compound periods
$P$ = principal (present value)

## NOTE

Two people may obtain different results of calculations in the examples and exercises. Calculators may use a different number of digits in computations. A person who rounds the calculations at each step will likely obtain different results from a person who uses all digits in a calculator at each step.

In this chapter, the examples and exercises are worked using all digits generated by the calculator. Some of the intermediate steps are shown with fewer than the 12 digits used by the calculator. When computing dollar amounts, the final answer is rounded to the nearest penny, although the intermediate steps are not rounded.

An example of the effect of rounding intermediate steps follows: If $1000 is invested at a 7% annual rate compounded monthly, the amount after 2 years is given by

$$1000\left(1 + \frac{0.07}{12}\right)^{24} \text{ where } \frac{0.07}{12} = 0.00583333\dots$$

with as many 3's as the calculator will contain. The person who rounds to 0.00583 will obtain $A = 1000(1.00583)^{24} = 1149.71$ (to the nearest penny). The person who uses all digits in the calculator will obtain $A = 1000(1.0058333\dots)^{24} = 1149.81$ (rounded to the nearest penny). The answers differ by $0.10.

---

**Example 4**  $800 is invested at 12% for 2 years. Find the amount at the end of 2 years for the interest compounded (a) annually, (b) semiannually, and (c) quarterly.

**Solution**
In this example, $P = 800$, $r = 0.12$, and $t = 2$ years.

**(a)** $m = 1$, so $i = 0.12$, $n = 2$:

$$A = 800(1 + 0.12)^2 = 800(1.12)^2 = 800(1.2544) = 1003.52$$

**(b)** $m = 2$, so $i = \dfrac{0.12}{2} = 0.06$, $n = 4$:

$$A = 800(1 + 0.06)^4 = 800(1.06)^4 = 800(1.26248) = 1009.98$$

**(c)** $m = 4$, so $i = \dfrac{0.12}{4} = 0.03$, $n = 8$:

$$A = 800(1 + 0.03)^8 = 800(1.03)^8 = 800(1.26677) = 1013.42$$

---

**Example 5**  $3000 is invested at 7.2%, compounded quarterly. Find the amount at the end of 5 years.

**Solution**

$$P = 3000, m = 4, r = 0.072, i = \frac{0.072}{4} = 0.018, \text{ and } n = 4 \times 5 = 20.$$

$$A = 3000(1.018)^{20} = 3000(1.428748) = 4286.24$$

The amount at the end of 5 years is $4286.24. Table 5–1 and Figure 5–1 show the growth by each quarter.

## TABLE 5–1

| Quarter | Interest | Amount | Quarter | Interest | Amount |
|---------|----------|---------|---------|----------|---------|
| 0 | 0 | 3000.00 | 11 | 64.54 | 3650.45 |
| 1 | 54.00 | 3054.00 | 12 | 65.71 | 3716.16 |
| 2 | 54.97 | 3108.97 | 13 | 66.89 | 3783.05 |
| 3 | 55.96 | 3164.93 | 14 | 68.10 | 3851.15 |
| 4 | 56.97 | 3221.90 | 15 | 69.32 | 3920.47 |
| 5 | 58.00 | 3279.90 | 16 | 70.57 | 3991.04 |
| 6 | 59.03 | 3338.93 | 17 | 71.84 | 4062.88 |
| 7 | 60.11 | 3399.04 | 18 | 73.12 | 4136.01 |
| 8 | 61.18 | 3460.22 | 19 | 74.45 | 4210.46 |
| 9 | 62.28 | 3522.50 | 20 | 75.78 | 4286.24 |
| 10 | 63.41 | 3585.91 | | | |

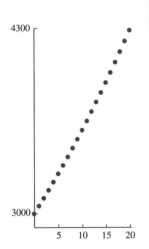

**FIGURE 5–1** Growth of $3000 at 7.2% compounded quarterly.

Note that the interest earned each quarter grows steadily, from $54.00 the first quarter to $75.78 in the 20th quarter.

The growth of a compound interest investment illustrated in Table 5–1 suggests why there is some truth to the statement: "Compound interest is a way to get rich slowly."

The value of an investment increases faster the longer it is invested at compound interest. While the increase in the 20th quarter (5 years) was $75.78, the increase in the 40th quarter (10 years) is $108.28, and the increase in the 60th quarter (15 years) is $154.71. After 40 years the quarterly increase is about $920 and the value of the investment is about $52,000. It may be that for you to "get rich" with compound interest is too slow and too long, but it indicates that it is wise to begin your investment for retirement as soon as possible so it has more time to grow.

A credit card account or store charge account that charges interest when you miss payments will likely charge you compound interest. For an account that bills monthly and charges "1.25% of the unpaid balance" is charging compound interest. For example, if your balance is $275 and you miss a payment, they will add 1.25% of $275 for the new balance. If you miss several payments, each month's new balance will be charged an additional 1.25%, making it 1.25% compounded monthly.

**Example 6**  A department store charges 1% per month on the unpaid balance of a charge account. This makes it compound interest. A customer owes $135.00, and the bill is unpaid for 4 months. What is the amount of the bill at the end of 4 months?

**Solution**
The interest rate is 1% per month and is compounded monthly. The value of $i$ is then 0.01, and the number of periods is 4. The amount of the bill is

$$A = 135(1.01)^4$$
$$= 135(1.04060)$$
$$= 140.48$$

Note that $i$ and $n$ were given in this example, so that they can be substituted directly into the formula.

—————●

Money invested at compound interest grows faster for shorter compound periods. Table 5–2 shows how $1000 grows over 10 years at 7.2% interest with different compounding periods. Note that the future value increases with more frequent compounding.

Although the future value increases with more frequent compounding, it increases more slowly. In fact, even with more frequent compounding such as each hour, minute, or second, the future value will never exceed $2054.44.

| TABLE **5–2** | **$1000 invested at 7.2% for 10 years** | |
|---|---|---|
| **Compound Frequency** | **Number of Periods** | **Future Value** |
| Simple interest (no compounding) | | $1000[1 + 0.072(10)] = 1720$ |
| Annually | 10 | $1000(1.072)^{10} = 2004.23$ |
| Quarterly | 40 | $1000\left(1 + \dfrac{0.072}{4}\right)^{40} = 2041.32$ |
| Monthly | 120 | $1000\left(1 + \dfrac{0.072}{12}\right)^{120} = 2050.02$ |
| Weekly | 520 | $1000\left(1 + \dfrac{0.072}{52}\right)^{520} = 2053.41$ |
| Daily | 3650 | $1000\left(1 + \dfrac{0.072}{365}\right)^{3650} = 2054.29$ |

## Present Value

We use the term **present value** to designate the principal that must be invested now to accumulate an amount at a specified time in the future.

**Example 7** How much should Josh invest at 8%, compounded quarterly, so that he will have $5000 at the end of 7 years?

**Solution**

In this case, we are given $A = 5000$, $r = 0.08$, $m = 4$, and $t = 7$. Then, $i = \dfrac{0.08}{4} = 0.02$ and $n = 4(7) = 28$. Using the compound interest formula, we have

$$5000 = P(1.02)^{28}$$
$$= 1.74102P$$

Solve the equation for $P$:

$$P = \frac{5000}{1.74102} = 2871.87$$

Josh should invest $2871.87 to have $5000 in 7 years. In other words, the present value of $5000 due in 7 years is $2871.87.

## Doubling an Investment

An investor wishes to know how fast an investment will grow. One measure is the time it takes for an investment to double in value.

**Example 8**   An investor has visions of doubling her money in 6 years. What interest rate is required for her to do so if the investment draws interest compounded quarterly?

**Solution**
$P$ dollars are invested in order to have $2P$ dollars in 6 years. The future value formula for interest compounded quarterly becomes

$$2P = P(1 + i)^{24}$$

where we wish to find $i$. We do not need to know the value of $P$ because we can divide both sides by $P$ and have $2 = (1 + i)^{24}$.
   You can solve this by taking the 24th root of both sides of the equation

$$2 = (1 + i)^{24}$$

to get

$$2^{1/24} = 1 + i$$

to four decimal places, $2^{1/24} = 1.0293$, or $i = 2.93\%$ per quarter or an 11.72% annual rate.

## Effective Rate

For a given annual rate, a more frequent compounding of interest gives a larger value of an investment at the end of the year. A 10% annual rate, compounded monthly, gives a larger amount at the end of the year than does 10%, compounded semiannually. However, a lower rate, compounded more frequently, may or may not give a larger return. For example, which yields the better return, 9% compounded semiannually or 8.8% compounded quarterly? To answer this, let's assume that $1000 is invested in each case and compute the amount at the end of 1 year.

1. $r = 9\%$ is compounded semiannually. Let $P = 1000$, $t = 1$ year, $m = 2$, and
   $i = \dfrac{0.09}{2} = 0.045$. Then,
   $$A = 1000(1.045)^2 = 1000(1.09203)$$
   $$= 1092.03$$

2. $r = 8.8\%$ is compounded quarterly. Let $P = 1000$, $t = 1$ year, $m = 4$, and
   $i = \dfrac{0.088}{4} = 0.022$. Then,
   $$A = 1000(1.022)^4$$
   $$= 1000(1.090947) = 1090.95$$

Thus, 9% compounded semiannually is a slightly better investment.

However, 8.85% compounded monthly is slightly better than 9% compounded semiannually. In that case

$$i = \frac{0.0885}{12} = 0.007375 \quad \text{and} \quad A = 1000(1.007375^{12}) = 1092.18$$

Different rates and frequency of compounding can be put on a comparable basis by finding **effective rate**. The effective rate is the percentage increase of an investment in 1 year—that is, the simple interest rate that gives the same annual increase as the compound rate. If an investment increases by 5.9%, for example, the effective rate is 5.9%.

**DEFINITION**
**Effective Rate**

The **effective rate** of an annual interest rate $r$ compounded $m$ times per year is the simple interest rate that produces the same total value of investment per year as the compound interest.

Here's how we find the effective rate.

**Example 9**     Find the effective rate of 8% compounded quarterly.

**Solution**
If we invest $P$ dollars, the amount of the investment at the end of the year is

$$A = P(1.02)^4$$

If we invest the same amount, $P$, at a simple interest rate, $r$, the amount of the investment at the end of the year is $A = P + Pr(1)$.

Now $r$ is the unknown simple interest rate that gives the same amount $A$ as does compound interest, so

$$P + Pr = P(1.02)^4$$

We can divide throughout by $P$ to get

$$1 + r = (1.02)^4$$

and so

$$r = (1.02)^4 - 1$$
$$= 1.08243 - 1 \quad \text{(rounded)}$$
$$= 0.08243$$

In percentage form, $r = 8.243\%$ is the effective rate of 8% compounded quarterly.

This method works generally, so we can make the following statement:

**Effective Rate**

If money is invested at an annual rate $r$ and compounded $m$ times per year, the effective rate, $x$, in decimal form is

$$x = (1 + i)^m - 1 \quad \text{where} \quad i = \frac{r}{m}$$

Now let's look at an example with a given effective rate and find the annual interest rate, compounded periodically.

**Example 10**    The Mattson Brothers Investment Firm advertises certificates of deposit paying a 7.2% effective rate. Find the annual interest rate, compounded quarterly, that gives the effective rate.

**Solution**
We let $i$ = quarterly rate. Then,

$$0.072 = (1 + i)^4 - 1$$
$$1.072 = (1 + i)^4$$
$$\sqrt[4]{1.072} = 1 + i$$
$$1.017533 = 1 + i$$
$$i = 0.017533$$

The annual rate = $4(0.017533) = 0.070133 = 7.013\%$ (rounded). The annual rate just found is also called the **nominal rate**.

## Zero-Coupon Bonds

An investor can purchase a **zero-coupon bond** for an amount less than the maturity value, paying (say) $9,000 for a bond that has a maturity value of $10,000. At maturity, the investor receives $10,000. The bond pays no interest, but the difference, $10,000 - \$9,000 = \$1,000$, can be considered interest that accumulates and is paid at maturity.

**Example 11**    Erwin plans to buy a 10-year zero-coupon bond with a maturity value of $10,000. How much should he pay to get a return of 5.6% compounded quarterly?

**Solution**
We use the compound interest formula $A = P(1 + i)^n$, where $A = 10,000$, $i = 0.056/4 = 0.014$, and $n = 40$.

$$10,000 = P(1.014)^{40}$$

$$P = \frac{10,000}{(1.014)^{40}} = 5,734.32$$

Erwin should pay $5,734.32.

## 5.2    EXERCISES

Access end-of-section exercises online at **www.webassign.net**    ENHANCED **WebAssign**

# Using Your TI Graphing Calculator

## Making a Table

In Section 1.1, we illustrated how to make a table of values of a function for several values of $x$. We can make tables of period-by-period amounts of an investment in a similar way. A built-in table has a column for a list of values of the variable $x$, and columns for $y_1, y_2, \ldots$ that contain the values of $y$ calculated from the corresponding formulas in the $\boxed{Y =}$ menu.

We illustrate by making a year-by-year table for two investments, both investing $1000 at 5% interest for 10 years. For the first investment, the interest is compounded annually, and the second draws simple interest.

Briefly, here's how.

- Press $\boxed{\text{TblSet}}$ to display the **TABLE SETUP** screen.
- Enter 0 for **TblStart** and press $\boxed{\text{ENTER}}$.
- Enter 1 for **ΔTbl** and press $\boxed{\text{ENTER}}$.
- Select the **Auto** option for **Indpnt** and press $\boxed{\text{ENTER}}$.

  (This sequence will enter 0, 1, 2, . . . in $x$. To enter 0, 2, 4, . . . use 2 for **ΔTbl**.)
- Select **Auto** for **Depend** and press $\boxed{\text{ENTER}}$.

## NOTE

This will give a list of $x$ values from zero to the end of the table, so you pick the ones you want.

To calculate the values of compound interest in $y_1$ and simple interest in $y_2$:

- Select $\boxed{Y =}$ and enter $1000(1.05)^\wedge x$ as the $y_1$ function,
- To calculate the simple interest amounts, enter $1000(1 + .05x)$ for the $y_2$ function.
- Press $\boxed{\text{Table}}$ to view the tables.
- You may also graph the two functions by setting the Window to 0–20 for the $x$-range and 0–2000 for the $y$-range and press $\boxed{\text{GRAPH}}$.

## Using Intersect

We can use the **intersect** command (see Section 1.3) to determine, for example, when an investment of $700 doubles in value if it is invested at 4.4% compounded quarterly. Do so by plotting the two functions $y_1 = 700(1.011)^x$ and $y_2 = 1400$, and then selecting the **intersect** command.

## Exercises

1. Make a table of the year-by-year amounts of a $200 investment at 5.1% compounded annually for 10 years.
2. Make a table, for eight years, of the year-by-year amounts of two $100 investments, one at 4.5% compounded annually and the second at 5.5% compounded annually.
3. Find how long it takes for $5,000 to increase to $12,000 when invested at 5.5% compounded quarterly.

## Using Excel

We will show how to graph functions using Excel. Let's illustrate by graphing the amount of compound interest and the amount of simple interest. We use an initial investment of $1000, interest of 7%, for a period of 20 years. The compound interest investment is compounded annually. Thus, we graph the equations

Compound interest: $A = 1000(1.07)^x$ where $x$ ranges from 0 to 20.
Simple interest: $A = 1000(1 + 0.07x)$ where $x$ ranges from 0 to 20.

Excel makes a graph by plotting points on the graph and connecting them with a smooth curve. This means we compute a list of points to be used by Excel. Here's how:

- In column A, list values of $x$ to be used. We use 0, 2, 4, . . . , 20, listed in A2:A12.

  *Note:* We could use 0, 1, 2 . . . , 20 or some other spacing for $x$.

- In column B, calculate the amount of compound interest corresponding to the number of periods, $x$. We use $=1000*(1.07)^\wedge A2$ in B2 and drag the formula down to B12.

- In column C, we calculate the amounts of simple interest using $=1000*(1+.07*A2)$ in C2 and dragging the formula down to C12.

| | A | B | C |
|---|---|---|---|
| 1 | x | A=1000(1.07)^x | A=1000(1+0.07x) |
| 2 | 0 | 1000.00 | 1000 |
| 3 | 2 | 1144.90 | 1140 |
| 4 | 4 | 1310.80 | 1280 |
| 5 | 6 | 1500.73 | 1420 |
| 6 | 8 | 1718.19 | 1560 |
| 7 | 10 | 1967.15 | 1700 |
| 8 | 12 | 2252.19 | 1840 |
| 9 | 14 | 2578.53 | 1980 |
| 10 | 16 | 2952.16 | 2120 |
| 11 | 18 | 3379.93 | 2260 |
| 12 | 20 | 3869.68 | 2400 |

We now use these points to graph both functions on the same graph.

- Select the cells A2:C12.
- Click on the **Insert** tab and move to the **Charts** group.
- Select **Scatter**, then **Scatter with Smooth Line and Marker** subtype.
- A graph will be inserted in the worksheet.

In case you want to adjust the $x$-scale or $y$-scale, double-click on the axis to be adjusted, make the changes in the dialog box that appears, and click on **OK**. The Series 1 and Series 2 legends that appear with the graph refer to the function defined in column B (for Series 1), and to the function defined in Column C (for Series 2).

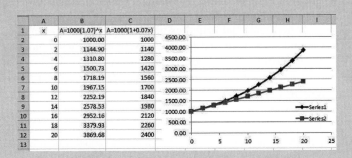

*(continued)*

We can use **Goal Seek** to determine when an investment reaches a certain value, doubles in value, increases 60%, and so forth. Let's illustrate with the following. $1500 is invested at 4.8% compounded quarterly. How long will it take for it to reach $2000 in value? We let $x$ = the required number of quarters, $i = 0.012$, and $A = 1500(1.012)^x$. We enter zero for $x$ in A2, and $=1500*(1.012)^\wedge A2$, in B2.

- Select the **Data** tab; then choose **Goal Seek** under the **What-If Analysis** option.
- In the **Goal Seek** dialog box, enter B2 for **Set Cell,** 2000 for **To Value,** and A2 for **By changing cell**.
- Click **OK;** then A2 shows 24.11 and B2 shows 2000. Thus, the investment reaches $2000 in the 24th quarter.

## Exercises

1. Make a table of year-by-year amounts of $500 invested at 6% compounded annually for six years.

2. Make a table of quarter-by-quarter amounts of $800 invested at 5.6% compounded quarterly for four years.

3. Make a table of year-by-year amounts of two investments of $2000 each. The first is invested at 6% compounded annually and the second is invested at 5.5% compounded annually. Both are invested for five years.

4. Graph the two functions in Exercise 3 on one graph.

5. Find how long it will take for $2500 invested at 5.2% compounded quarterly to reach $4000.

6. Find the interest rate, compounded quarterly, that will enable an $800 investment to double in eight years.

---

## 5.3    ANNUITIES AND SINKING FUNDS

- Ordinary Annuity
- Sinking Funds

### Ordinary Annuity

Most of us have encountered or will encounter debt several times. You may now have a student loan or owe some credit card charges. You may be looking forward to life after graduation when you can begin a good job or upgrade your present job. Early in your career, you may experience a struggle to pay off a student loan, to make payments on a car, and to keep up the monthly payments on your dream house. Then there might be a time when you decide to invest money periodically into a fund to provide a college education for your children. Later in your career, you will focus more on your retirement and the income you might expect during that time.

These financial activities have a common characteristic, that of periodic payments. Car and house loans have periodic payments, usually monthly, of a specified amount that must be paid until the loan is paid off. A college fund receives periodic payments to accumulate an amount for use at a future date. A retiree receives monthly payments from a retirement fund.

A series of equal periodic payments like these provide an example of what we call an annuity. Whether the payments are used to pay off a loan, build a college fund, or provide retirement income, equal payments paid at equal time intervals form an **annuity**.

We use some standard terminology when discussing annuities. We call the time between successive payments the **payment period** and the amount of each payment the

**periodic payment**. The interest on an annuity is compound interest. The payments may be made annually, semiannually, quarterly, or at any specified time interval. Monthly payments are common. The annuity may take on different forms, but we will study just one form, the **ordinary annuity**. Furthermore, we assume the payment period and the interest compounding period are the same.

**Ordinary Annuity**

An **ordinary annuity** is an annuity with periodic payments made at the end of each payment period.

We first look at annuities in which payments are invested to build up a fund for future use.

**Example 1**

A family enters a savings plan whereby they will invest $1000 at the end of each year for 5 years. The annuity will pay 7% interest, compounded annually. Find the value of the annuity at the end of the 5 years.

**Solution**

Because $1000 is deposited each year, the first deposit will draw interest longer than subsequent deposits. The value of each deposit at the end of 5 years is the original $1000 plus the compound interest for the time it draws interest. Use the formula for the amount at compound interest. Here is a summary of how payments of $1000 deposited at the end of each year grow over a 5-year period. The interest rate, $r = 0.07$, is compounded annually.

| Year deposited | Length of time deposit draws interest | Value of deposit at end of 5 years |
|:---:|:---:|:---|
| 1 | 4 years | $1000(1.07)^4 = 1310.80$ |
| 2 | 3 years | $1000(1.07)^3 = 1225.04$ |
| 3 | 2 years | $1000(1.07)^2 = 1144.90$ |
| 4 | 1 year | $1000(1.07) = 1070.00$ |
| 5 | 0 year | $1000(1) = 1000.00$ |
|   |   | Final value:    5750.74 |

You obtain the final value by adding the five payments and the interest accumulated on each one. We call the final value the **amount**, or **future value**, of the annuity.

Let's use this simple example to observe some general relationships. In the example, five payments are made, and the accumulated value is obtained by adding five amounts, the first of which is $1000(1.07)^4$ and the last is 1000, the annual payment. If $1000 were deposited for 10 years, a similar pattern holds with ten amounts, the first being $1000(1.07)^9$ and the last 1000. In general, the first amount has an exponent on 1.07 that is one less than the number of payments. The total accumulated value of five payments is

$$1000 + 1000(1.07) + 1000(1.07)^2 + 1000(1.07)^3 + 1000(1.07)^4$$

**NOTE**

For other interest rates and amounts of periodic payments, use that interest rate instead of 0.07 and the amount invested instead of 1000. You will obtain a similar expression for the total, $A$.

In a like manner, the total of ten payments is

$$1000 + 1000(1.07) + 1000(1.07)^2 + \cdots + 1000(1.07)^9$$

where we write the terms in reverse order. In general, the sum of the values of $n$ payments is

$$A = 1000 + 1000(1.07) + 1000(0.07)^2 + \cdots + 1000(1.07)^{n-1}$$

Clearly, computing the total value (future value) of an annuity can be tedious for a number of payments. Let's examine a way to avoid a period-by-period computation by using the 5-year future value in Example 1. The sum of the five terms is

$$A = 1000 + 1000(1.07) + 1000(1.07)^2 + 1000(1.07)^3 + 1000(1.07)^4$$

Now multiply both sides of this equation by 1.07 to obtain

$$A(1.07) = 1000(1.07) + 1000(1.07)^2 + 1000(1.07)^3 + 1000(1.07)^4 + 1000(1.07)^5$$

We subtract the first equation from this to obtain

$$A(1.07) - A = 1000(1.07) + 1000(1.07)^2 + 1000(1.07)^3 + 1000(1.07)^4$$
$$+ 1000(1.07)^5 - 1000 - 1000(1.07) - 1000(1.07)^2$$
$$- 1000(1.07)^3 - 1000(1.07)^4$$

and obtain the form

$$A(1.07) - A = -1000 + 1000(1.07)^5$$

From this we get

$$A(1.07 - 1) = 1000(1.07^5 - 1)$$
$$A = \frac{1000(1.07^5 - 1)}{1.07 - 1} = \frac{1000(1.07^5 - 1)}{0.07}$$

Check this computation to verify that it gives the same total as in Example 1, $5750.74. Note in the final result that 1000 represents the periodic payment, 1.07 is 1 plus the periodic interest rate, the exponent 5 is the number of time periods, and the denominator 0.07 is the periodic interest rate.

We would like a formula that computes the future value for any amount for the periodic payment—say, $R$—for any periodic interest rate — say, $i$ — and any number of periods — say, $n$. If we follow the steps above using $R$, $i$, and $n$ instead of 1000, 0.7, and 5, we will end up with a similar formula giving the value of the amount accumulated, or future value. The general form of the formula is the following.

**Future Value (Amount) of an Ordinary Annuity**    (Payments are made at the end of each period)

$$A = R\left[\frac{(1 + i)^n - 1}{i}\right]$$

where

$i$ = interest rate per period
$n$ = number of periods
$R$ = amount of each periodic payment
$A$ = future value or amount

The next example gives you an idea how an annuity grows over time.

**Example 2**   Tiffany invests $1000 at the end of each year into an ordinary annuity that pays 6% compounded annually. Figure 5–2 and Table 5–3 show how the annuity grows year by year for 15 years.

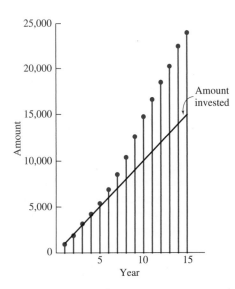

**FIGURE 5–2**   Growth of an annuity with $1000 invested annually at 6%.

| TABLE **5–3** | | | |
| --- | --- | --- | --- |
| **Year** | **Payment** | **Interest** | **Total** |
| 1 | 1000 | 0 | 1,000.00 |
| 2 | 1000 | 60.00 | 2,060.00 |
| 3 | 1000 | 123.60 | 3,183.60 |
| 4 | 1000 | 191.02 | 4,374.62 |
| 5 | 1000 | 262.47 | 5,637.09 |
| 6 | 1000 | 338.23 | 6,975.32 |
| 7 | 1000 | 418.52 | 8,393.84 |
| 8 | 1000 | 503.63 | 9,897.47 |
| 9 | 1000 | 593.85 | 11,491.32 |
| 10 | 1000 | 689.47 | 13,180.79 |
| 11 | 1000 | 790.85 | 14,971.64 |
| 12 | 1000 | 898.30 | 16,869.94 |
| 13 | 1000 | 1012.20 | 18,882.14 |
| 14 | 1000 | 1132.93 | 21,015.07 |
| 15 | 1000 | 1260.90 | 23,275.97 |

Using Table 5–3, notice the growth over 5-year periods. In the first 5 years, a total of $5000 was invested and grew to a value of $5637, a gain of $637 over the amount invested. During the sixth through tenth year, a total of $5000 was invested, and the value of the investment grew to $13,181—a gain of $7544, of which $2544 was interest earned. The last 5 years a total of $5000 was invested, and the value of the investment increased to $23,276—a gain of $10,095, of which $5095 was interest earned. This acceleration in the increase in value occurs because more money becomes available to earn interest. The compounding effect of interest contributes greatly to the growth. This illustrates that regular investment over a long period of time is a smart way to invest.

Figure 5–2 shows how the total value of the investment increases with time relative to the amount invested.

We need not compute a period-by-period table like Table 5–3 to find the value at some future date. We use the future value formula as illustrated in the next example.

**Example 3**   How much money will you have when you retire if you save $50 each month from graduation until retirement? We need some definite numbers to answer this question, so let's assume that you start saving at age 23 until age 70, 47 years later, and the interest rate averages 4.2% annual rate compounded monthly.

**Solution**

Because the payments are made monthly, the periodic rate, $= 0.042/12 = 0.0035$, the number of periods is $n = 12(47) = 564$, and the periodic payments are $R = 50$.

We substitute these values into the future value formula

$$A = R\left[\frac{(1 + i)^n - 1}{i}\right]$$

to obtain

$$A = 50\left[\frac{1.0035^{564} - 1}{0.0035}\right] = 88,209.24$$

You will accumulate $88,209.24 in 47 years.

## Sinking Funds

There are times when a company or an individual expects to need a specified amount at a time in the future. For example, a company may expect to replace a machine costing $150,000 in 6 years, a family expects to need a new car in 4 years, or new parents wish to accumulate $450,000 for their child's college expenses in 18 years. In cases like this when an amount of money will be needed at some future date, the company or person can systematically accumulate a fund that will build to the desired amount at the time needed. We call such a fund a **sinking fund**.

**Example 4**    Susie wants to deposit her savings at the end of every three months so that she will have $7500 available in four years. The account will pay 8% interest per annum, compounded quarterly. How much should she deposit every quarter?

**Solution**

Susie is accumulating a sinking fund with a future value of $7500, periodic rate $i = 0.08/4 = 0.02$, and $n = 16$ periods. Use the formula for the future value of an annuity,

$$A = R\left[\frac{(1 + i)^n - 1}{i}\right]$$

to find the periodic payments, $R$:

$$7500 = R\left[\frac{(1.02)^{16} - 1}{0.02}\right]$$
$$7500 = R(18.63929)$$
$$R = \frac{7500}{18.63929} = 402.38$$

Susie should deposit $402.38 every quarter to accumulate the desired $7500.

We obtain the formula for periodic payments into a sinking fund from the future value formula for an annuity by solving for $R$.

| Formula for Periodic Payments of a Sinking Fund | where | $$R = \dfrac{Ai}{(1 + i)^n - 1}$$ |
|---|---|---|

$A$ = value of the annuity after $n$ payments
$n$ = number of payments
$i$ = periodic interest rate
$R$ = amount of each periodic payment

**Example 5**  Darden Publishing Company plans to replace a piece of equipment at an expected cost of $65,000 in 10 years. The company establishes a sinking fund with annual payments. The fund draws 7% interest, compounded annually. Find the periodic payments.

**Solution**
$A$ = 65,000, $i$ = 0.07, and $n$ = 10, so

$$R = \frac{65000(0.07)}{(1.07)^{10} - 1} = \frac{4550}{0.9671513} = 4704.54$$

The annual payment is $4704.54.

A sinking fund problem is a variation of an annuity problem. The growth of an annuity can generally be described as "I can save $25 a month. How much will I have after 5 years?" The sinking fund problem is like "I will need $15,000 in five years to buy a new car. How much should I save each month to have the amount I need?" In the growth of an annuity, we know the amount of the periodic payments, and we want to know the future value. In the sinking fund problem, we know the future value, and we want to know the periodic payments that will accrue to the known future value.

## 5.3  EXERCISES

Access end-of-section exercises online at **www.webassign.net**     WebAssign

 **Using Your TI Graphing Calculator**

### Growth of Annuities

We have formulas to compute the amount of an annuity after a length of time. To see how it grows month by month when a fixed amount is invested on a regular monthly basis requires tedious computing. A TI graphing calculator program, **ANNGRO,** will do the tedious calculating. The program requires that you enter the amount

*(continued)*

invested periodically ($R$), the periodic interest rate ($I$), and the number of periods ($N$). The program computes the amount of the annuity period by period. Here are the program instructions.

```
: Lbl 1
: Prompt R,I,N
: 0 → J
: R → A
: Disp "PERIOD AMOUNT"
: Lbl 2
: J + 1 → J
: round(A,2) → B
: Disp J,B
: Pause
: A(1 + I) + R → A
: If J < N
: Goto 2
: Goto 1
: End
```

After you have entered the program into your calculator, use it as follows. We will illustrate with monthly payments of $100, 9% interest (.0075 monthly) for 12 months.

Press PRGM <ANNGRO> ENTER, and R = ? will appear on the screen, askng you to enter the periodic payment. Then press ENTER and continue entering $I$ and $N$.

```
prgmANNGRO
R=?100
I=?.0075
N=?12
```

The calculator will pause after showing each period number and amount. Press ENTER to continue to the next period.

```
PERIOD AMOUNT    :
              1
           100
              2
        200.75
              3
        302.26
```

When you reach the end, the program will start over, requesting new values of $R$, $I$, and $N$.

```
        927.48
            10
       1034.43
            11
       1142.19
            12
       1250.76
R=?
```

You can exit from the program by pressing ON, and this screen will appear:

```
ERR:BREAK
1∎Quit
2:Goto
```

Press ENTER to quit.

## Exercises

1. Find the month-by-month amounts of an annuity with monthly payments of $200 and annual interest rate of 9%, for 15 months.

2. Find the quarterly amounts of an annuity with quarterly payments of $500 and 8% annual interest rate, for 5 years.

3. Find the annual amounts of an annuity with annual payments of $1000 and annual interest rate of 7.2%, for 10 years.

## Using Excel

We illustrate how to show the period-by-period amounts of an annuity. We use a monthly investment of $100, 6.6% interest (0.0055 per month) for 12 months. Column A will list the months, and column B will list the corresponding amounts of the annuity.

| | A | B |
|---|---|---|
| 1 | Month | Amount |
| 2 | 0 | 0.00 |
| 3 | 1 | 100.00 |
| 4 | 2 | 200.55 |
| 5 | 3 | 301.65 |
| 6 | 4 | 403.31 |
| 7 | 5 | 505.53 |
| 8 | 6 | 608.31 |
| 9 | 7 | 711.66 |
| 10 | 8 | 815.57 |
| 11 | 9 | 920.06 |
| 12 | 10 | 1025.12 |
| 13 | 11 | 1130.75 |
| 14 | 12 | 1236.97 |

- Enter zero in A2
- Enter $= A2+1$ in A3 and drag through A14.
- Enter $= 100*(1.0055^{A2}-1)/.0055$ in B3 and drag through B14 and you get

## Exercises

1. Find the annual amounts of an annuity with annual payments of $1500 and an annual interest rate of 5.7% for 10 years.

2. Find the quarterly amounts of an annuity with quarterly payments of $250 and a 5.6% annual interest, for 2 years.

3. Find the monthly amounts of an annuity with monthly payments of $75 and an annual interest rate of 5.4% for 12 months.

---

## 5.4  PRESENT VALUE OF AN ANNUITY AND AMORTIZATION

- Present Value
- Amortization

### Present Value

Let's look at some variations of the ordinary annuity. We have studied the result of accumulating equal payments made at regular intervals with the object to provide a fund at a future date. Now we want to look at two variations of this problem.

1. How much should be put into a savings account in *one lump sum* at compound interest so that the amount accumulated at the end of 5 years equals the amount accumulated by an annuity, say investing $25 each month for 5 years?

    The lump sum payment that yields the same total amount as that obtained through equal periodic payments made over the same period of time is called the **present value of the annuity**.

2. An example of a second variation finds the monthly payments that will pay off a 4-year car loan.

    A similar problem seeks to find the amount that grandparents should place in a college fund to provide, say $5000 each year, to help with a grandchild's college expenses. We note that $20,000 is not required to provide $5000 each year for four years, because the interest earned on the unused part of the fund provides part of the $5000 annual payments.

Solve the first problem, finding the present value of an annuity, as follows. Recall that the amount of an annuity is

$$A = R\left[\frac{(1 + i)^n - 1}{i}\right]$$

and the amount of compound interest is

$$A = P(1 + i)^n$$

where $i$ is the periodic interest rate, $n$ is the number of periods, $R$ is the periodic payment for the annuity, and $P$ is the lump sum invested at compound interest. We want to find $P$ so that the amount of compound interest equals the amount of the annuity; that is

$$P(1 + i)^n = R\left[\frac{(1 + i)^n - 1}{i}\right]$$

Solving for $P$ gives

$$P = R\left[\frac{(1 + i)^n - 1}{i(1 + i)^n}\right]$$

When we divide the numerator and denominator by $(1 + i)^n$, we obtain the equivalent form

$$P = R\left[\frac{1 - (1 + i)^{-n}}{i}\right]$$

We have obtained two equivalent formulas for present value of an annuity.

**Present Value of an Annuity**

$$P = R\left[\frac{(1 + i)^n - 1}{i(1 + i)^n}\right] \text{ or } P = R\left[\frac{1 - (1 + i)^{-n}}{i}\right]$$

where

$i$ = periodic rate
$n$ = number of periods
$R$ = periodic payments
$P$ = present value of the annuity

You might ask which form should be used to find present value

$$P = R\left[\frac{(1 + i)^n - 1}{i(1 + i)^n}\right] \quad \text{or}$$

$$P = R\left[\frac{1 - (1 + i)^{-n}}{i}\right]$$

Either form is acceptable. The second form has the advantage that it uses fewer steps to evaluate it, but be sure you use a negative $n$ for the exponent. You will see both forms used in examples.

**Example 1** Find the present value of an annuity with periodic payments of $2000, semiannually, for a period of 10 years at an interest rate of 6% compounded semiannually.

**Solution**

Here $R = 2000$, $i = \frac{0.06}{2} = 0.03$, and $n = 20$. We use these values in the formula for present value

$$P = R\left[\frac{(1 + i)^n - 1}{i(1 + i)^n}\right]$$

and obtain

$$P = 2000\left[\frac{(1.03)^{20} - 1}{0.03(1.03)^{20}}\right]$$

or the form

$$P = 2000\left[\frac{1 - (1.03)^{-20}}{0.03}\right]$$

can be used.

$$P = 2000(14.8774749)$$
$$= 29754.95$$

The present value of the annuity is $29,754.95. This lump sum will accumulate the same amount in 10 years as investing $2000 semiannually for 10 years.

The person who invests $2000 semiannually will have the same total investment at the end of 10 years as the person who invests a lump sum of $29,755 at the beginning of the 10 years. However, the person who invests semiannually will invest a total of $40,000 over 10 years.

Another form of periodic equal payments occurs when an established fund pays out money periodically. For example, a college student's grandparents established a fund to pay $5000 each semester to help pay the student's education expenses. This "annuity in reverse" does not require a fund that is 8 times $5000 for eight semesters of payments, because the money still in the fund draws interest. Let's see how much money we need to make equal periodic payments for a specified number of times with no money left when the last payment is made.

To analyze this kind of problem, we use a simpler example from which we draw a general formula: What amount of money is required to make payments of $100 each month for 4 months? The fund draws 12% annual interest and is compounded monthly.

Let's think of the money needed as divided into four parts that we label $A_1$ for the investment that will grow to the amount needed to make the first payment, $A_2$ for the investment that will grow to the amount needed for the second payment, and $A_3$ and $A_4$ likewise for the third and fourth payments. These investments are not equal because $A_4$, for example, draws interest longer than the other investments, so it can be smaller. The total $A_1 + A_2 + A_3 + A_4$ is the amount needed initially.

At the beginning of the plan, the amount $A_1$ is needed to enable a payment of $100 at the end of one month. Because the monthly interest rate is 1%, $A_1$ will draw interest for one month and will be worth $1.01A_1$ when the payment is made. Thus,

$$1.01A_1 = 100 \quad \text{and} \quad A_1 = \frac{100}{1.01}$$

The amount $A_2$ in the original investment will draw interest for two months before it makes the second payment, so

$$(1.01)^2A_2 = 100 \quad \text{and} \quad A_2 = \frac{100}{(1.01)^2}$$

Likewise,

$$(1.01)^3 A_3 = 100 \quad \text{and} \quad A_3 = \frac{100}{(1.01)^3}$$

and

$$(1.01)^4 A_4 = 100 \quad \text{and} \quad A_4 = \frac{100}{(1.01)^4}$$

The investment needed in the fund at the beginning of the fund is

$$S = A_1 + A_2 + A_3 + A_4 = \frac{100}{1.01} + \frac{100}{(1.01)^2} + \frac{100}{(1.01)^3} + \frac{100}{(1.01)^4}$$

We find $S$ by first computing

$$1.01S - S = 100 + \frac{100}{1.01} + \frac{100}{(1.01)^2} + \frac{100}{(1.01)^3}$$
$$- \frac{100}{1.01} - \frac{100}{(1.01)^2} - \frac{100}{(1.01)^3} - \frac{100}{(1.01)^4}$$

which reduces to

$$(1.01 - 1)S = 100 - \frac{100}{(1.01)^4}$$

$$0.01S = 100\left[1 - \frac{1}{(1.01)^4}\right]$$

$$= 100\left[\frac{(1.01)^4 - 1}{(1.01)^4}\right]$$

$$S = 100\left[\frac{(1.01)^4 - 1}{0.01(1.01)^4}\right]$$

By dividing numerator and denominator by $(1.01)^4$, this can be reduced to an alternate form

$$S = 100\left[\frac{1 - (1.01)^{-4}}{0.01}\right]$$

We can now do the arithmetic computations to obtain

$$100\left[\frac{1 - 1.01^{-4}}{0.01}\right] = 100(3.9019656)$$
$$= 390.20 \qquad \text{(Rounded)}$$

A total of \$390.20 will provide \$100 a month for 4 months.

If we let $P$ = amount needed at the beginning, we have

$$P = 100\left[\frac{1 - 1.01^{-4}}{0.01}\right]$$

Now observe the following:

100 is the amount of a monthly payment (in general, periodic payment).

4 is the number of payments.

0.01 is the monthly interest rate (in general, the periodic rate).

This is an example of the general form.

**Amount Needed to Provide Equal Periodic Payments**

$$P = R\left[\frac{(1 + i)^n - 1}{i(1 + i)^n}\right] \text{ or, equivalently, } P = R\left[\frac{1 - (1 + i)^{-n}}{i}\right]$$

where

$P$ = amount needed in the fund
$R$ = amount of periodic payments
$i$ = periodic interest rate
$n$ = number of payments

Note that this gives us the present value formula for an annuity.

Thus, this type of problem does not require a new formula. We can call the amount needed in the fund at the beginning the *present value of an annuity*.

**Example 2**    Find the present value of an annuity (lump sum investment) that will pay $1000 per quarter for 4 years. The annual interest rate is 10%, compounded quarterly.

**Solution**
$R = 1000$, $i = \frac{0.10}{4} = 0.025$, and $n = 16$ quarters.

$$P = 1000\left[\frac{1 - (1.025)^{-16}}{0.025}\right]$$

$$P = 1000(13.0550027) = 13{,}055$$

A lump sum investment of $13,055 will provide $1000 per quarter for 4 years.

## Amortization

We now analyze the problem of paying a debt with equal periodic payments. A standard method of paying off a car or house loan requires that the borrower pay equal monthly payments until the debt is paid. This is called **amortization**. Each monthly payment pays all of the interest charged for that month and repays a part of the loan. By reducing the loan each month, the interest decreases a little each month, and the amount repaid increases by the same amount.

To illustrate the process, we use a simplified example of a $10,000 car loan with equal payments at the end of each year for 4 years. The interest rate is 8%. We let $R$ represent the annual payments.

To find the annual payments, we will find the amount still owed, the **balance**, at the end of each year. The following observations are helpful.

**Observation 1.**   The original amount of the loan is $10,000.

**Observation 2.**   As the end of the first year approaches, the amount owed increases from $10,000 to $10,800, because 8% interest, 0.08($10,000) = $800, is now to be paid.

**Observation 3.**   The balance owed, $10,800, is reduced by the payment $R$. Clearly, $R$ must be greater than $800 so that all of the interest and some of the loan are paid.

**Observation 4.**

*First year:* We can summarize the first year balance as

$$\text{Balance} = 10{,}000 + 800 - R$$

which can also be written as

$$\text{Balance} = 10{,}000 + 0.08(10{,}000) - R$$

and as

$$\text{Balance} = 10{,}000(1.08) - R$$

**Observation 5.**

*Second year:* The balance at the end of the second year is

$$\text{First year's balance} + \text{interest on first year's balance} - R$$

This can be written as

$$\begin{aligned}
\text{Balance} &= 10{,}000(1.08) - R + 0.08\,[\,10{,}000(1.08) - R\,] - R\\
&= [10{,}000(1.08) - R]\,(1.08) - R\\
&= 10{,}000(1.08)^2 - 1.08R - R
\end{aligned}$$

**Observation 6.** The balances for the third and fourth year are

*Third year:*

$$\begin{aligned}
\text{Balance} &= \text{second year's balance} + \text{interest} - R\\
&= 10{,}000(1.08)^2 - 1.08R - R + [10{,}000(1.08)^2 - 1.08R - R]\,(0.08) - R\\
&= [10{,}000(1.08)^2 - 1.08R - R](1.08) - R\\
&= 10{,}000(1.08)^3 - (1.08)^2 R - 1.08R - R
\end{aligned}$$

*Fourth year:*

$$\begin{aligned}
\text{Balance} &= [10{,}000(1.08)^3 - (1.08)^2 R - 1.08R - R](1.08) - R\\
&= 10{,}000(1.08)^4 - (1.08)^3 R - (1.08)^2 R - 1.08R - R\\
&= 10{,}000(1.08)^4 - R[(1.08)^3 + (1.08)^2 + 1.08 + 1]
\end{aligned}$$

We now find the sum in the second term above:

$$S = (1.08)^3 + (1.08)^2 + (1.08) + 1$$

It will help in more general cases if we find this sum in the following way. First, write $1.08S$ as follows:

$$1.08S = (1.08)^4 + (1.08)^3 + (1.08)^2 + 1.08$$

Then,

$$\begin{aligned}
1.08S - S &= (1.08)^4 + (1.08)^3 + (1.08)^2 + 1.08\\
&\quad - (1.08)^3 - (1.08)^2 - (1.08) - 1\\
1.08S - S &= (1.08)^4 - 1\\
(1.08 - 1)S &= (1.08)^4 - 1\\
S &= \frac{(1.08)^4 - 1}{1.08 - 1} = \frac{(1.08)^4 - 1}{0.08}
\end{aligned}$$

We can now write the fourth year's balance as

$$\text{Balance} = 10{,}000(1.08)^4 - R\left[\frac{(1.08)^4 - 1}{0.08}\right]$$

This is an example of the general formula for the balance of a loan. The general form is

$$\text{Balance} = P(1 + i)^n - R\left[\frac{(1 + i)^n - 1}{i}\right]$$

where $P$ = the amount of the loan, $i$ = periodic interest rate, $n$ = number of periods, and $R$ = periodic payments.

Now we are ready to find $R$, the annual payment. Because the loan is to be repaid in 4 years, the fourth year's balance must be zero. We then have

$$10{,}000(1.08)^4 - R\left[\frac{(1.08)^4 - 1}{0.08}\right] = 0$$

and

$$10{,}000(1.08)^4 = R\left[\frac{(1.08)^4 - 1}{0.08}\right]$$

We now calculate $\dfrac{(1.08)^4 - 1}{0.08} = 4.506112$

and

$$10{,}000(1.08)^4 = R\left[\frac{(1.08)^4 - 1}{0.08}\right] = R[4.506112]$$

so

$$R = \frac{10{,}000(1.08)^4}{4.506112} = 3019.21$$

The annual payments are $3019.21.

Let's observe that the result

$$10{,}000(1.08)^4 = R\left[\frac{(1.08)^4 - 1}{0.08}\right]$$

when written as

$$10{,}000 = R\left[\frac{(1.08)^4 - 1}{0.08(1.08)^4}\right]$$

is exactly the form of present value of an annuity. In this case,

$P$ = 10,000, the amount of the loan
$i$ = 0.08, the periodic interest rate
$R$ = periodic payments
$n$ = 4, the number of time periods

This relationship between the amount of a loan and the periodic payments holds when payments are made monthly, quarterly, or any other time period provided that the interest rate and number of periods are also monthly rates and number of months, and so on.

We have gone to some effort to show the following:

The amortization of a debt (repayment of a debt) requires no new formula because *the amount borrowed is just the present value of an annuity.*

This method usually applies to car payments and house payments. We can use the present value formula of an annuity to find the periodic payments.

**Debt Payments**   **Amortization of a Loan**

The amount borrowed, $P$, is related to the periodic payments, $R$, by the formula

$$P = R\left[\frac{(1 + i)^n - 1}{i(1 + i)^n}\right] \quad \text{or} \quad P = R\left[\frac{1 - (1 + i)^{-n}}{i}\right]$$

where

$i$ = periodic interest rate   and   $n$ = number of payments

(*Note:* This is the present value formula for an ordinary annuity.)

**Example 3**   An employee borrows $8000 from the company credit union to purchase a car. The interest rate is 12%, compounded monthly, with payments every month. The employee wants to pay off the loan in 3 years. (The loan is amortized over 3 years.) How much are the monthly payments?

**Solution**
Here we have $P = \$8000$, $i = \frac{0.12}{12} = 0.01$, $n = 36$ months. Substituting into the present value formula, we have

$$8000 = R\left[\frac{1 - (1.01)^{-36}}{0.01}\right]$$

$$8000 = R(30.107505)$$

Solving for $R$, we have

$$R = \frac{8000}{30.107505} = 265.71$$

The monthly payments are $265.71 each.

In general, when we solve for $R$ in the present value formula, we have the amortization payment formula.

You should be aware that the amount of the periodic payment may involve a fraction of a cent. In that case the bank rounds up to the next cent. Consequently, the final payment may be a little less than the other payments.

**Example 4**   A student obtained a 24-month loan on a car with monthly payments of $366.64 based on a 4.5% interest rate. Find the amount borrowed.

**Solution**

The amount borrowed is just the present value of the annuity. We then have

$$R = 366.64$$

$$i = \frac{0.045}{12} = 0.00375 \qquad \text{(monthly rate)}$$

$$n = 24 \qquad\qquad\qquad \text{(number of months)}$$

So,

$$P = 366.64 \left[ \frac{1 - 1.00375^{-24}}{0.00375} \right] = 8399.96$$

It is reasonable to round this to $8400, the amount borrowed.

---

**Example 5**  Habitat for Humanity helps low-income families build affordable homes. In one Southwest area, they can build a house for $40,000. The family makes monthly payments of $160, with no interest, until the loan is paid.

**(a)** How long will it take a family to pay off the loan?

**(b)** What monthly payments would it take to pay off the loan if they were charged 6% interest and the length of the loan was the same as in part (a)?

**NOTE**

In order to qualify for a Habitat House, a family must put in 300 hours of "sweat equity" in building their house and the houses of other families.

**Solution**

**(a)** It will take $\dfrac{40{,}000}{160} = 250$ months to complete the payments.

**(b)** If $P = 40{,}000$, $i = \dfrac{0.06}{12} = 0.005$, and $n = 250$, then

$$40{,}000 = R \left[ \frac{1 - 1.005^{-250}}{0.005} \right]$$

$$R = \frac{40{,}000}{142.5202916} = 280.66$$

To pay the loan in 250 months at a 6% interest rate requires monthly payments of $280.66.

---

We can gain other information related to amortization as illustrated in the following examples.

---

**Example 6**  A family borrowed $60,000 to buy a house. The loan was for 30 years at a 12% interest rate. The monthly payments were $617.17.

**(a)** How much of the first month's payment was interest and how much was principal?

**(b)** What total amount did the family pay over the 30 years?

**Solution**

**(a)** The monthly interest rate was $1\% = 0.01$, so the first month's interest was $60{,}000(0.01) = 600.00$. The family paid $600 interest the first month. The rest of the payment, $17.17, went to repay part of the principal.

**(b)** The family paid $617.17 each month for 360 months, so the total amount paid was 617.17(360) = $222,181.20. You may be surprised at this figure, but it is true. Note that the total amount paid for interest was

$$\$222,181.20 - 60,000 = \$162,181.20$$

When a family makes monthly payments on a house mortgage, some of each month's payment goes to reduce the loan. In Example 6, the first payment reduced the loan by $17.17. To find the balance of the loan after a period of time—say, 5 years—you can find the amount repaid each month for 5 years and deduct these from the loan. For example, here is an amortization schedule for the first 12 months of the loan:

**$60,000 Loan for 30 Years at 12%**

| Month | Monthly payment | Interest paid | Principal paid | Balance |
|---|---|---|---|---|
| 0 | | | | $60,000.00 |
| 1 | $617.17 | $600.00 | $17.17 | 59,982.83 |
| 2 | 617.17 | 599.83 | 17.34 | 59,965.49 |
| 3 | 617.17 | 599.65 | 17.52 | 59,947.97 |
| 4 | 617.17 | 599.48 | 17.69 | 59,930.28 |
| 5 | 617.17 | 599.30 | 17.87 | 59,912.41 |
| 6 | 617.17 | 599.12 | 18.05 | 59,894.36 |
| 7 | 617.17 | 598.94 | 18.23 | 59,876.13 |
| 8 | 617.17 | 598.76 | 18.41 | 59,857.72 |
| 9 | 617.17 | 598.58 | 18.59 | 59,839.13 |
| 10 | 617.17 | 598.39 | 18.78 | 59,820.35 |
| 11 | 617.17 | 598.20 | 18.97 | 59,801.38 |
| 12 | 617.17 | 598.01 | 19.16 | 59,782.22 |

We see that the balance declines very slowly during the first year of the loan. However, it gradually declines faster, and by the end of the 29th year the balance is $6938.72. The amortization schedule for the last 12 months is the following:

| Month from end | Monthly payment | Interest paid | Principal paid | Balance |
|---|---|---|---|---|
| 12 | $617.17 | $69.39 | $547.78 | $6390.94 |
| 11 | 617.17 | 63.91 | 553.26 | 5837.68 |
| 10 | 617.17 | 58.38 | 558.79 | 5278.89 |
| 9 | 617.17 | 52.79 | 564.38 | 4714.51 |
| 8 | 617.17 | 47.15 | 570.02 | 4144.49 |
| 7 | 617.17 | 41.44 | 575.73 | 3568.76 |
| 6 | 617.17 | 35.69 | 581.48 | 2987.28 |
| 5 | 617.17 | 29.87 | 587.30 | 2399.98 |
| 4 | 617.17 | 24.00 | 593.17 | 1806.81 |
| 3 | 617.17 | 18.07 | 599.10 | 1207.71 |
| 2 | 617.17 | 12.08 | 605.09 | 602.62 |
| 1 | 608.65 | 6.03 | 602.62 | — |

Note that the last payment totals less than $617.17 because payments were rounded to the nearest penny.

A chart like this helps to see how fast a debt is reduced (not very fast in the early stages). This approach becomes too tedious in finding the balance after a number of payments. Actually, we found a formula for the balance early in the discussion of amortization. It is

$$\text{Balance} = P(1 + i)^n - R\left(\frac{(1 + i)^n - 1}{i}\right)$$

Note that the balance after $n$ periods is

$$(\text{amount of compound interest}) - (\text{amount of an annuity})$$

---

**The Balance of an Amortization**

$$\text{Balance} = P(1 + i)^n - R\left[\frac{(1 + i)^n - 1}{i}\right]$$

where

$P$ = the amount borrowed
$i$ = periodic interest rate
$n$ = number of time periods elapsed
$R$ = monthly payments

---

**Example 7**   What is the balance of the loan in Example 6 after 2 years?

**Solution**
In Example 6,

$$P = 60{,}000$$
$$i = 1\% = 0.01 \text{ per month}$$
$$R = 617.17$$
$$n = 24 \text{ months} \quad (\text{The balance after 24 months is desired.})$$

To find the balance after 2 years, substitute these values in the Balance of an Amortization formula above.

$$\begin{aligned}
\text{Balance} &= 60{,}000(1.01)^{24} - 617.17\left[\frac{(1.01)^{24} - 1}{0.01}\right] \\
&= 60{,}000(1.01)^{24} - 617.17(26.9734649) \\
&= 60{,}000(1.26973) - 617.17(26.9734649) \\
&= 76{,}184.08 - 16{,}647.21 \\
&= 59{,}536.87
\end{aligned}$$

So the balance owed after two years is $59,536.87. We call the part of the loan repaid the **equity**:

$$\text{Equity} = \text{loan} - \text{balance}$$

In this case, the equity after 2 years is

$$\text{Equity} = 60{,}000 - 59{,}536.87 = \$463.13$$

The results of these computations might seem wrong. Hardly any principal is repaid each month. In two years, the 24 payments total $14,812.08, but only $463.41 of the $60,000 loan has been repaid. However, the amount repaid increases a little each month (about 18¢; see the amortization table for this loan on page 298). Although an 18¢ per month increase hardly seems worthwhile, it eventually becomes a significant increase, which then pays off the loan faster. Notice from Example 6 that the total interest on the loan of the example exceeds $160,000.

Now let's look at a situation that combines the future value and present value of an annuity.

**Example 8**   The parents of a baby want to provide for the child's college education. How much should be deposited on each of the child's first 17 birthdays to be able to withdraw $10,000 on each of the next four birthdays? Assume an interest rate of 8%.

**Solution**
First, compute the amount that must be in the account on the child's 17th birthday to withdraw $10,000 per year for 4 years. This is the present value of an annuity, where

$$P \text{ is to be found}$$
$$i = 0.08$$
$$n = 4$$
$$R = 10,000$$

So, substituting in the present value formula

$$P = R\left[\frac{1 - (1 + i)^{-n}}{i}\right]$$

we have

$$P = 10,000\left[\frac{1 - (1.08)^{-4}}{0.08}\right]$$
$$= 10,000(3.3121268) = \$33,121.27$$

the total necessary on the 17th birthday.

Next, find the annual payments that will yield a future value of $33,121.27 in 17 years at 8% using the formula for the future value of an annuity

$$A = R\left[\frac{(1 + i)^n - 1}{i}\right]$$

Given this formula, we have:

$$A = 33,121.27$$
$$i = 0.08$$
$$n = 17$$
$$33,121.27 = R\left[\frac{(1.08)^{17} - 1}{0.08}\right]$$
$$33,121.27 = 33.7502257R$$
$$R = \frac{33,121.27}{33.7502257}$$
$$= 981.36$$

So, $981.36 should be deposited every birthday for 17 years to provide $10,000 per year for 4 years.

## 5.4   EXERCISES

Access end-of-section exercises online at **www.webassign.net**

# Using Your TI Graphing Calculator

### Monthly Loan Payments

To see how much of a monthly loan payment goes to interest and how much goes to reduce the loan, we have a TI graphing calculator program that shows this and the balance of the loan as well. It requires that you enter the amount of the loan, the annual interest rate, and the number of years. It assumes monthly payments. The program is named AMLN.

**An Amortization Schedule of a Loan**

: Lbl 1
: Disp "AMOUNT BORROWED"
: Input P
: Disp "ANNUAL RATE"
: Input K
: Disp "NO. YEARS"
: Input M
: $K/12 \rightarrow I$
: $12*M \rightarrow N$
: $(P*I)/(1-(1+I) \wedge (-N)) \rightarrow R$
: Disp "MONTHLY PAY"
: $Round(R,2) \rightarrow R$
: Pause R
: $1 \rightarrow J$
: Lbl 2
: Disp "MO I PR BAL"

: $I*P \rightarrow A$
: $Round(A,2) \rightarrow A$
: $R-A \rightarrow B$
: $P-B \rightarrow P$
: Disp J,A,B,P
: Pause
: $J + 1 \rightarrow J$
: If $J \leq N$
: Goto 2
: Pause
: Goto 1
: End

Enter the program into your calculator. We illustrate its use with a loan of $1500 at 7.5% annual interest for 1 year.

Press PRGM <AMLN> ENTER , and AMT BORROWED will appear on the screen, asking you to enter the amount of the loan. Then press ENTER and continue entering *I* and number of years.

Each time the calculator pauses showing a calculation, press ENTER to continue to the next step. The screen shows (a) the amount of each monthly payment, $130.14. It shows for month 1 the amount paid for interest,

*(continued)*

$9.38; the amount of principal repaid, $120.76; and the balance of the loan, $1379.24. Screen (b) shows the interest, principal repaid, and balance for month 6.

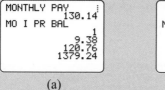

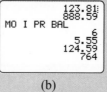

(a)                              (b)

## Exercises

1. Find the monthly interest and principal payments and the monthly balance for a loan of $2000 at 9% annual interest for 1 year.

2. Find the monthly interest and principal payments and the monthly balance for a loan of $4500 at 8.5% annual interest for 2 years.

3. Find the monthly interest and principal payments and the monthly balance for a loan of $85,000 at 7.8% annual interest for 15 years. This loan schedule is 180 months long, so don't step through the entire schedule. Run the steps for only the first year.

# Using Excel

We illustrate how much of a monthly payment goes for interest and how much goes to repay the loan. We also show how to find the balance for each month.

We illustrate with a loan of $2500 at 6.6% annual interest for one year. We use cell B2 for the amount borrowed, C2 for the annual interest rate, D2 for the monthly interest rate, E2 for the number of years, F2 for the number of months, and G2 for the monthly payments.

## Cell Entries

Numbers entered are: 2500 in B2, 0.066 in C2, 1 in E2, 2500 in D4, and 0 in A4. As you change these numbers for other problems, the spreadsheet will automatically compute the appropriate new values.

Formulas entered: In D2, $= C2/12$; in F2, $= E2*12$; in G2, the formula $\dfrac{Pi(1 + i)^n}{(1 + i)^n - 1}$, i.e., $= B2*D2*(1 + D2)^\wedge F2/((1 + D2)^\wedge F2-1)$.

Formulas entered and dragged through row 16: In A5, $= A4 + 1$; in B5, $= \$D\$2*D4$; in C5, $= \$G\$2-B5$; in D5, $= D4-C5$.

We show the completed table below.

| | A | B | C | D | E | F | G |
|---|---|---|---|---|---|---|---|
| 1 | | Borrowed | Annual % | Monthly % | Years | Months | Month. Pay. |
| 2 | | 2500 | 0.066 | 0.0055 | 1 | 12 | 215.86 |
| 3 | Month | Interest | Repaid | Balance | | | |
| 4 | 0 | | | 2500 | | | |
| 5 | 1 | 13.75 | 202.11 | 2297.89 | | | |
| 6 | 2 | 12.64 | 203.22 | 2094.68 | | | |
| 7 | 3 | 11.52 | 204.34 | 1890.34 | | | |
| 8 | 4 | 10.40 | 205.46 | 1684.88 | | | |
| 9 | 5 | 9.27 | 206.59 | 1478.29 | | | |
| 10 | 6 | 8.13 | 207.73 | 1270.57 | | | |
| 11 | 7 | 6.99 | 208.87 | 1061.70 | | | |
| 12 | 8 | 5.84 | 210.02 | 851.68 | | | |
| 13 | 9 | 4.68 | 211.17 | 640.51 | | | |
| 14 | 10 | 3.52 | 212.33 | 428.18 | | | |
| 15 | 11 | 2.35 | 213.50 | 214.68 | | | |
| 16 | 12 | 1.18 | 214.68 | 0.00 | | | |

**5**

# IMPORTANT TERMS

**5.1**
Simple Interest
Principal
Present Value
Interest Rate
Future Value Amount
Simple Discount
Proceeds
Discount Rate
Maturity Value

**5.2**
Compound Interest
Periodic Interest Rate Amount

Future Value of Compound Interest
Nominal Rate
Present Value
Effective Rate
Zero-Coupon Bond

**5.3**
Annuity
Payment Period
Periodic Payment
Ordinary Annuity Amount
Future Value
Sinking Fund
Annuity Due

**5.4**
Present Value of an Annuity
Amortization
Balance
Equity
APR

**5**

# SUMMARY OF FORMULAS

**Simple interest**
    **Interest**

$$I = Prt$$

    **Future Value**

$$A = P + I = P(I + rt)$$

**Simple Discount**
    **Discount**

$$D = Mdt$$

    **Proceeds**

$$PR = M - D = M(1 - dt)$$

**Compound Interest**
    **Future Value**

$$A = P(1 + i)^n$$

    **Effective Rate**

$$x = (1 + i)^m - 1$$

**Annuity**

    **Future Value**

$$A = R\left[\frac{(1 + i)^n - 1}{i}\right]$$

    **Present Value**

$$P = R\left[\frac{(1 + i)^n - 1}{i(1 + i)^n}\right] = R\left[\frac{1 - (1 + i)^{-n}}{i}\right]$$

**Amortization**

    **Periodic Payments**

$$R = \frac{P}{\left[\dfrac{(1 + i)^n - 1}{i(1 + i)^n}\right]} = \frac{P}{\left[\dfrac{1 - (1 + i)^{-n}}{i}\right]}$$

$$= \frac{P(i)(1 + i)^n}{(1 + i)^n - 1} = \frac{Pi}{1 - (1 + i)^{-n}}$$

    **Balance of a Loan**

$$\text{Bal} = P(1 + i)^n - R\left[\frac{(1 + i)^n - 1}{i}\right]$$

## IMPORTANT CONCEPTS

| | |
|---|---|
| **Simple Interest** | The interest $I$ paid to borrow the amount $P$ at a rate per time period $r$ for $t$ time periods. $I = Prt$. |
| **Compound Interest** | When $P$ dollars are invested at a periodic rate $r$, the amount at the end of $t$ periods is $A = P(1 + r)^t$. |
| **Future Value** | The amount of compound interest at the end of the compound periods. |
| **Effective Rate** | For an annual rate, $r$, compounded $m$ times a year, the effective rate, $x$, in decimal form is $x = (1 + i)^m - 1$ where $i = r/m$. |
| **Annuity** | An investment with equal payments paid at equal times. |
| **Future Value of an Annuity** | The value of an annuity at the end of the payments. |
| **Present Value of an Annuity** | A lump sum invested over the same time as an annuity that yields the future value of the annuity. |
| **Amortization** | The payment of a debt using equal periodic payments. |

## REVIEW EXERCISES

1.  A loan is made for $500 at 9% simple interest for 2 years. How much interest is paid?

2.  How much simple interest was paid on an $1100 loan at 6% for 10 months?

3.  A loan, principal and interest, was paid with $1190.40. The loan was made at 8% simple interest for 3 years. How much was borrowed?

4.  The interest on a loan was $94.50, the simple interest rate was 7.5%, and the loan was for 1.5 years. What was the amount borrowed?

5.  A sum of $3000 is borrowed for a period of 5 years at simple interest of 9% per annum. Compute the total interest paid over this period.

6.  Find the future value of a $6500 loan at 7.5% simple interest for 18 months.

7.  A simple discount note at 9% discount rate for 2 years has a maturity value of $8500. What are the discount and the proceeds?

8.  How much should be borrowed on a discount note with 6.1% discount rate for 1.5 years so that the borrower obtains $900?

9.  Compute the interest earned on $5000 in 3 years if the interest is 7% compounded annually.

10. Compute the interest and the amount of $500 after 2 years if the interest is 10% compounded semi-annually.

11. Missy invested $1000 at 6.8% compounded quarterly. Can she expect it to double in value after ten years?

12. A sum of $1000 is deposited in a savings account that pays interest of 5%, compounded annually. Determine the amount in the account after 4 years.

13. Find the effective rate of 4.6% compounded semiannually.

14. Find the effective rate of 5% compounded quarterly.

15. Find the effective rate of 7.2% compounded monthly.

16. Which is the better investment, one that pays 8% compounded quarterly or one that pays 8.3% compounded annually?

17. Which account gives the better interest, one that gives 5.6% per annum compounded quarterly or one that gives 5.7% compounded annually?

18. The price of an automobile is now $15,000. What would be the anticipated price of that automobile in 2 years' time if prices are expected to increase at an annual rate of 8%?

**Find (a) future values and (b) the total interest earned on the amounts in Exercises 19 through 21.**

19. $8000 at 6% compounded annually for 4 years

20. $3000 at 5% compounded annually for 2 years

21. $5000 at 5.8% compounded quarterly for 6 years

22. A company sets aside $120,000 cash in a special building fund to be used at the end of 5 years to construct a new building. The fund will earn 6% compounded semiannually. How much money will be in the fund at the end of the period?

23. On January 1, a Chicago firm purchased a new machine to be used in the plant. The list price of the machine was $15,000, payable at the end of 2 years with interest of 5.5% compounded annually. How much was due on the machine on January 1 two years later?

24. Anticipating college tuition for their child in 10 years, the Heggens want to deposit a lump sum of money into an account that will provide $96,000 at the end of that 10-year period. The account selected pays 6.2%, compounded semiannually. How much should they deposit?

25. An executive thinks that it is a good time to sell a certain stock. She wants to take a leave of absence in 5 years. How much stock should she sell and invest at 8% compounded quarterly so that $50,000 will be available in 5 years?

26. Will a $1000 investment increase to $3000 in 15 years if the interest rate is 7.5% compounded quarterly?

27. Jonel is advised that her investment of $2000 will grow to $3500 in 6 years if it is invested at 6.6% compounded quarterly. Determine if this advice is correct.

**Find the future value of each of the ordinary annuities in Exercises 28 through 30.**

28. $R = \$500, i = 0.05, n = 8$

29. $R = \$1000, i = 0.06, n = 5$

30. $R = \$300, i = 0.025, n = 20$

31. Find the future value of $600 paid into an annuity at the end of every 6 months for 5 years. The annual interest rate is 6.4%.

32. Find the future value of $1000 paid into an annuity at the end of every year for 6 years. The interest rate is 5.9%.

33. $250 is invested in an annuity at the end of every 3 months at 6.4%. Find the value of the annuity at the end of 6 years.

34. An annuity consists of payments of $1000 at the end of each year for a period of 5 years. Interest is paid at 6% per year compounded annually. Determine the amount of the annuity at the end of 5 years.

35. Midway School District sold $12 million of bonds to construct an elementary school. The bonds must be paid in 10 years. The district establishes a sinking fund that pays 4.8% interest. Find the annual payments into the sinking fund that will provide the necessary funds at the end of 10 years.

36. A couple wants to start a fund with annual payments that will give them $20,000 cash on retirement in 10 years. The proposed fund will give interest of 7% compounded annually. What will be the annual payments?

**In Exercises 37 through 40, determine the present value of the given amounts using the given interest rate and length of time.**

37. $5000 over 5 years at 6% compounded annually

38. $4750 over 4 years at 5% compounded semi-annually

39. $6000 over 5 years at 6.2% compounded quarterly

40. $1000 over 8 years at 5.8% compounded quarterly

41. A company wants to deposit a certain sum at the present time into an account that pays compound interest at 8% compounded quarterly to meet an expected expense of $50,000 in 5 years' time. How much should it deposit?

42. A corporation is planning plant expansion as soon as adequate funds can be accumulated. The corporation has estimated that the additions will cost approximately $105,000. At the present time it has $70,935 cash on hand that will not be needed in the near future. A local savings institution will pay 8% compounded semiannually. Can $70,935 accumulate to approximately $105,000 in 5 years?

43. A medical supplier is making plans to issue $200,000 in bonds to finance plant modernization. The interest rate on the bonds will be 8% compounded quarterly. The company estimates that it can pay up to a maximum of $97,000 in total interest on the bonds. The company wants to stretch

out the period of payment as long as possible. Is it possible for the supplier to take longer than 5 years to pay off the bonds?

**44.** A corporation owed a $40,000 debt. Its creditor agreed to let the corporation pay the debt in five equal annual payments at 6.5% interest. Compute the annual payments.

**45.** A student borrowed $3000 from a credit union toward purchasing a car. The interest rate on such loans is 6.4% compounded quarterly with payments due every quarter. The student wants to pay off the loan in 3 years. Find the quarterly payments.

**46.** The interest on a house mortgage of $53,000 for 30 years is 6.9%, compounded monthly, with payments made monthly. Compute the monthly payments.

**47.** Andrew borrows $4800 at 7.4% interest. It is to be paid in five annual payments. Find the annual payments.

**48.** Holt's Clothing Store borrowed $125,000 to be repaid in semiannual payments. Find the semiannual payments if the loan is for 4 years and the interest rate is 7.4% compounded semiannually.

**49.** The city of McGregor borrowed $7.8 million to build a zoo. The interest rate is 4% compounded annually, and the loan is to be paid in 20 years. Find the annual payments.

**50.** The Thaxton family purchased a franchise with a loan of $25,000 at 8% compounded quarterly for 6 years. Find the quarterly payments.

**51.** The Bar X Ranch purchased additional land with a loan of $98,000 at 9% compounded annually for 8 years. Find the annual payments.

**52.** Fashion Floors borrowed $65,000 at 8.7% compounded semiannually for 3 years. Find the semiannual payments.

**53.** An investor invests $1000 at 6.2% compounded quarterly for 5 years. At the end of the 5 years, the total amount in the account is reinvested at 6.4% compounded quarterly for another 5 years. How much is in the account at the end of 10 years?

**54.** A family borrowed $85,000 at 8.7% for 30 years to buy a house. Their payments are $665.66 per month. How much of the first month's payment is interest? How much of the first month's payment is principal? What is the total amount they will pay over the 30 years?

**55.** If $1700 is invested at 5.8% compounded quarterly, find the amount at the end of 10 years.

**56.** Find the amount of $20,000 invested at 5.2% compounded quarterly for 50 years.

**57.** How much should be invested at 6.4% compounded quarterly in order to have $500,000 at the end of 40 years?

**58.** If $100 per month is deposited in an annuity earning 5.7% interest, find the amount at the end of 10 years.

**59.** How much should be deposited monthly into an annuity paying 6% in order to have $100,000 at the end of 6 years?

**60.** The unemployment rate for Japan was 2.1 in 1990 and 4.8 in 2009. Determine approximately the average annual rate of increase in unemployment in these 19 years.

**61.** A *Forbes* magazine article estimated that the net worth of the Sam Walton family heirs to be $37.4 billion in 1998 and $92 billion in 2010. Find the average annual rate of increase in their net worth for these 12 years.

# SETS AND COUNTING

An understanding of set theory provides a basis for understanding logic, enumeration problems, and probability. Through the centuries, set theory evolved slowly as an area of logic and mathematics. Although a number of mathematicians contributed to the development of set theory, an Englishman, George Boole, extended and formalized the area into an algebra that developed a well-defined structure of set theory and logic. His book *Investigation of the Laws of Thought*, published in 1854, was significant in the evolution of mathematics.

We call the process of counting **enumeration**, such as in counting the number of people attending a concert, the possible number of phone numbers per area code, the number of ways to select three paintings from a collection of 12 paintings, the number of student parking spaces on campus, or the number of ways a student can select three articles from a reading list of 12 articles.

The theory of enumeration, permutation, and combinatorics evolved over wide geographical areas and for a long period of time. Traces of the theory in China date back to the twelfth century. About the same time, the Hindus found rules to determine the number of "changes upon apertures of a building," "the scheme of musical permutations," and "the combination of different savors" in medicine (Bhäskara in *Lilavati*). The early Hebrews and Arabs studied permutations because of their belief in the mysticism of arrangements and the combinations of planets.

The first major publication on the theory of enumeration was the work of a Swiss mathematician, Jacques Bernoulli (1654–1705). His *Ars Conjectandi* was published in 1713 after his death.

Today, permutations and combinations have moved out of the realm of mysticism and find applications in the sciences, engineering, social science, statistics, and computer science.

The process of counting the number of objects in a set seems elementary, and it is in many cases. However, in some instances, a correct

count is obtained with much difficulty, if at all. For example, a law required a census be taken in the year 2010 to count all U.S. residents. Problems occurred because some people refused to return the census form, census workers did not find some people at home, and people moved to new locations. Another example: Astronomers know that other solar systems similar to ours exist, but the question of "how many?" can be answered at best with a rough estimate.

The ease of counting may range from easy to impossible. We are interested in counting the harder, but possible, cases. In this chapter, we study some ways to determine the number of objects or activities without counting them one by one.

## 6.1    SETS

- Set-Builder Notation
- Empty Set
- Venn Diagrams
- Equal Sets
- Subset

- The Union of Two Sets
- The Intersection of Two Sets
- Complement
- Disjoint Sets

One of the basic concepts in mathematics is that of a **set**. Even in everyday life, we talk about sets. We talk about a set of dishes, a set of spark plugs, a reading list (a set of books), the people invited to a party (a set of people), the vegetables served at the cafeteria (a set of vegetables), and the teams in the NFL play-offs (a set of football teams).

The concept of a set is so basic that proposed definitions of it tend to go in circles, using words that mean essentially the same thing. We may speak of a set as a collection of objects. If we attempt to define a collection, we tend to think of the same words as we use in defining a set. Rather than go in circles trying to define a set, we will not give a definition. However, you should have an intuitive feeling of the meaning of a set.

Although we do not define the concept of a set, any particular set should be well defined.

---

**Set**    A set is well defined when there is no ambiguity as to the objects that make up the set.

---

For example, the collection of all letters of the English alphabet forms a set of 26 elements. No ambiguity arises as to which elements belong to the set and which do not, so the set is well defined. However, the collection of the ten greatest American authors is not well defined; it depends on who makes the selection and what they mean by "great."

Let's look at some of the traditional terminology used in speaking about sets. The objects that form a set may be varied. We can have a set of people, a set of numbers, a set of ideas, or a set of raindrops. In mathematics, the general term for an object in a set is **element**. A set may have people, numbers, ideas, or raindrops as elements. So that the contents of a set will be clearly understood, we need to be rather precise in describing a set of elements. One way to describe a set is to explicitly list all its elements. We usually enclose the list with braces. When we write

$$A = \{\text{Tom, Dick, Harry}\}$$

we are stating that we have named the set "$A$," with elements Tom, Dick, and Harry. Customarily, capital letters designate sets, and lowercase letters represent elements in a set. We read the notation

$$x \in A$$

as "$x$ is an element of $A$." We may also say that "$x$ is a member of the set $A$." The statement "$x$ is not an element of the set $A$" is written

$$x \notin A$$

For the set $B = \{2, 4, 6, 8\}$, we may write $4 \in B$ to specify that 4 is one of the elements of the set $B$, and $5 \notin B$ to specify that 5 is not one of the elements of $B$. Here are some examples of other sets.

**Example 1**   The set of positive integers less than 7 is $\{1, 2, 3, 4, 5, 6\}$.

**Example 2**   The vowels of the English alphabet form the set $\{a, e, i, o, u\}$.

**Example 3**   The countries of North America form the set $\{$Canada, USA, Mexico$\}$.

**Example 4**   The set of all natural numbers is $\{1, 2, 3, 4, \ldots\}$. Note that we use three dots to indicate that the natural numbers continue, without ending, beyond 4. This indicates an **infinite set**. Use the three dots to indicate that the preceding pattern continues. This notation is the mathematical equivalent of *et cetera*. The context in which we use the three dots determines which elements are missing from the list. The notation $\{20, 22, 24, \ldots, 32\}$ indicates that 26, 28, and 30 are missing from the list. This set is finite because it contains a finite number of elements. When "$\ldots$" is used in representing a finite set, the last element of the set is usually given. An infinite set gives no indication of a last element, as in the set of positive even numbers $\{2, 4, 6, \ldots\}$.

### Set-Builder Notation

Another way of describing a set uses the **implicit** or **set-builder notation**. The set of the vowels of the English alphabet may be described by

$$\{x \mid x \text{ is a vowel of the English alphabet}\}$$

Read this as "The set of all $x$ such that $x$ is a vowel of the English alphabet." Note that the vertical line | is read "such that." The symbol preceding the vertical line, $x$ in this instance, designates a typical element of the set. The statement to the right of the vertical line *describes* a typical element of the set; it tells how to find a specific instance of $x$.

**Example 5**    **(a)** The elements of the set $A = \{x \mid x$ is an odd integer between 10 and 20$\}$ are 11, 13, 15, 17, and 19.

**(b)** The set $\{x \mid x$ is a male student at Community College and $x$ is over 6 ft tall$\}$ is the set of all male students at Community College who are over 6 ft tall.

**(c)** The set $\{x \mid x$ is a member of the United States Senate$\}$ is the set of all senators of the United States.

**(d)** The set $\{x \mid x = 3n - 2$ where $n$ is a positive integer$\}$ contains 1, 4, 7, 10, . . .

### Empty Set

Sometimes a set can be described but it has no elements. For example, the set of golfers who made nine consecutive holes-in-one at the Richland Golf Course has no elements. Such a set, consisting of no elements, plays an important role in set theory. We call it the **empty set**, denoted by $\varnothing$. We call a set *nonempty* if it contains one or more elements. The set of people who are 10 feet tall is empty. The set of blue-eyed people is not empty. The set of numbers that belong to both $\{1, 3, 5\}$ and $\{2, 4, 6\}$ is empty.

### Venn Diagrams

It usually helps to use a diagram to represent an idea. For sets, we use a rectangular area to represent the **universe** to which a set belongs. We have not used the term *universe* before. The universe is not some all-inclusive set that contains everything. When I talk about a set of students, I am generally thinking of college students, so the elements of my set of students are restricted to college students. In that context, the universe becomes the set of college students. When a student at Midway High speaks of a set of students, she restricts the elements of her set to students at her school, so the set of all students at Midway High School forms her universe. In a given context, the sets under discussion have elements that come from some restricted set. We call the set from which the elements come the **universe** or the **universal set**. We usually represent the universe by a rectangle. A circular area within the rectangle represents a set of elements in the universe. The diagram in Figure 6–1 represents a set $A$ in a universe $U$. We call it a **Venn diagram**.

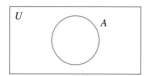

**FIGURE 6–1**    A Venn diagram.

### Equal Sets

Two sets $A$ and $B$ are said to be **equal** if they consist of exactly the same elements; this is denoted by $A = B$. The sets $A = \{2, 4, 6, 8\}$ and $B = \{6, 6, 2, 8, 4\}$ are equal because they consist of the same elements. *Listing the elements in a different order or with repetition does not create a different set.*

We use the symbol $\neq$ to indicate that two sets are not equal:

$$\{1, 2, 3, 4\} \neq \{2, 5, 6\}$$

### Subset

The set $B$ is said to be a **subset** of $A$ if every element of $B$ is also an element of $A$. This is written $B \subseteq A$.

**Example 6**     $\{2, 5, 9\} \subseteq \{1, 2, 4, 5, 7, 9\}$ because every element of $\{2, 5, 9\}$ is in the set $\{1, 2, 4, 5, 7, 9\}$. The set $\{3, 5, 9\}$ is not a subset of $\{1, 2, 4, 5, 7, 9\}$ because 3 is in the first set but not in the second. Figure 6–2 shows the Venn diagram.

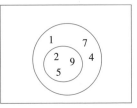

When $B$ is not a subset of $A$, we write $B \nsubseteq A$:

$$\{3, 5, 9\} \nsubseteq \{1, 2, 4, 5, 7, 9\}$$

**FIGURE 6–2**
$\{2, 5, 9\} \subseteq \{1, 2, 4, 5, 7, 9\}$.

**Example 7**     The subsets of $\{1, 2, 3\}$ are

$$\varnothing, \{1\}, \{2\}, \{3\}, \{1, 2\}, \{1, 3\}, \{2, 3\}, \{1, 2, 3\}$$

> **NOTE**
>
> The collection of all subsets of $\{1, 2, 3\}$ is called the **power set** of $\{1, 2, 3\}$.

Note that a set is a subset of itself; $\{1, 2, 3\}$ is a subset of $\{1, 2, 3\}$. Because $\varnothing$ contains no elements, it is true that every element of $\varnothing$ is in $\{1, 2, 3\}$, so $\varnothing \subseteq \{1, 2, 3\}$. In the same sense, $\varnothing$ is a subset of every set. The set $\{1, 2\}$ is a **proper subset** of $\{1, 2, 3\}$ because it is a subset of and not equal to $\{1, 2, 3\}$. We point out that if $A \subseteq B$ and $B \subseteq A$, then $A = B$.

### The Union of Two Sets

Sometimes, we wish to construct a set using elements from two given sets. One way to obtain a new set combines the two given sets into one.

> **DEFINITION**
> **Union of Sets**
>
> The **union** of two sets, $A$ and $B$, is the set whose elements are from $A$ or from $B$, or from both. Denote this set by $A \cup B$.
> In set-builder notation,
>
> $$A \cup B = \{x \mid x \in A \text{ or } x \in B \text{ or } x \text{ is in both}\}$$

In ordinary use, the word *or* is ambiguous, and its meaning may be confusing. The statement "To be admitted to an R-rated movie, a person must be at least 17 years old or accompanied by a parent" allows a 20-year-old to attend an R-rated movie alone. It also allows a 20-year-old to attend with a parent. This is an example of the **inclusive or**; it allows one or both options to occur. We interpret the statement "I will take history or language at 10:00 MWF" to allow only one course at 10:00 MWF. This is called the **exclusive or**; it does not allow both options. We interpret the word *or* in mathematics as the inclusive or.

In standard mathematical terminology, the phrase "$x \in A$ or $x \in B$" includes the possibility of $x$ existing in both. Therefore, the phrase "or $x$ is in both" is mathematically redundant.

Figure 6–3 shows a Venn diagram that represents the union of two sets. The shaded area represents $A \cup B$.

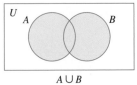

**FIGURE 6–3**     The union of sets $A$ and $B$, $A \cup B$.

**Example 8**    Given: $A = \{2, 4, 6, 8, 10\}$ and $B = \{6, 7, 8, 9, 10\}$. To determine $A \cup B$, list all elements of $A$ and add those from $B$ that are not already listed to obtain

$$A \cup B = \{2, 4, 6, 8, 10, 7, 9\}$$

**Example 9**    **(a)** Given the sets

$$A = \{\text{Tom, Danny, Harry}\}$$
$$B = \{\text{Sue, Ann, Jo, Carmen}\}$$

then,

$$A \cup B = \{\text{Tom, Danny, Harry, Sue, Ann, Jo, Carmen}\}$$

**(b)** Given the sets

$$A = \{x \mid x \text{ is a letter of the word } radio\}$$
$$B = \{x \mid x \text{ is one of the first six letters of the English alphabet}\}$$

then,

$$A \cup B = \{a, b, c, d, e, f, r, i, o\}$$

> **NOTE**
>
> The order in which the elements in a set are listed is not important. It might seem more natural to list the elements in Example 8 as $\{2, 4, 6, 7, 8, 9, 10\}$. Note that an element that appears in both $A$ and $B$ is listed only once in $A \cup B$.

## The Intersection of Two Sets

A set can be constructed from two sets in another way by performing the operation called the **intersection** of two sets.

**DEFINITION**    The **intersection** of two sets, $A$ and $B$, is the set of all elements contained in both sets
**Intersection of Sets**    $A$ and $B$—that is, those elements that $A$ and $B$ have in common. The intersection of $A$ and $B$ is denoted by

$$A \cap B = \{x \mid x \in A \text{ and } x \in B\}$$

**Example 10**    Figure 6–4 shows the Venn diagram of the intersection of two sets. The shaded area represents $A \cap B$.

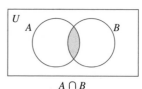

$A \cap B$

**FIGURE 6–4**    The intersection of sets $A$ and $B$, $A \cap B$.

**Example 11**    **(a)** If $A = \{2, 4, 5, 8, 10\}$ and $B = \{3, 4, 5, 6, 7\}$, then

$$A \cap B = \{4, 5\}$$

**(b)** If $A = \{a, b, c, d, e, f, g, h\}$ and $B = \{b, d, e\}$, then

$$A \cap B = \{b, d, e\}$$

**Example 12**    Given the sets

$$A = \text{set of all positive even integers}$$
$$B = \text{set of all positive multiples of 3}$$

describe and list the elements of $A \cap B$ and $A \cup B$.

**Solution**
Because an element of $A \cap B$ must be even (a multiple of 2) and a multiple of 3, each such element must be a multiple of 6. Thus $A \cap B$ is the set of all positive multiples of 6, that is, $\{6, 12, 18, 24, \ldots\}$. As this is an infinite set, only the pattern of numbers can be listed. $A \cup B$ is the set of positive integers that are multiples of 2 or multiples of 3.

Note that the sets $A$ and $B$ relate to $A \cup B$ and $A \cap B$ in the following way.

$$A \subseteq A \cup B$$
$$B \subseteq A \cup B$$
$$A \cap B \subseteq A$$
$$A \cap B \subseteq B$$

In a finite set, you can count the elements and finish at a definite number. The set of positive integers gives us an example where you cannot find the last element. A largest positive integer cannot exist because you can add 1 to the "largest" and have a larger number. The positive integers are endless. We call a set with an endless supply of elements an **infinite** set. The set of whole numbers between 10 and 20 is finite; there are 9 of them. The set of fractions between 10 and 20 is infinite; there is no end to them.

**Example 13**    Let

$$U = \text{set of letters of the English alphabet}$$
$$A = \text{set of vowels}$$
$$B = \text{set of the first nine letters of the alphabet}$$

Place each of the letters $a, c, d, e, i, u, x, y$ in the appropriate region in a Venn diagram.

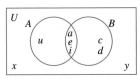

**FIGURE 6–5**

**Solution**
The letters $a$, $e$, $i$, and $u$ are elements of $A$; $a$, $c$, $d$, $e$, and $i$ are elements of $B$; and $x$ and $y$ are in neither set. Note that $a$, $e$, and $i$ are in both $A$ and $B$, so they are in their intersection. We may indicate this as shown in Figure 6–5.

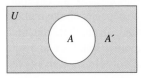

**FIGURE 6–6**   The shaded area, $A'$, is the complement of $A$.

## Complement

The elements in the universe $U$ that lie outside $A$ form a set called the **complement** of $A$, denoted by $A'$. (See Figure 6–6.)

**Example 14**    Let $U = \{1, 2, 3, 4, 5, 6, 7, 8\}$ and $A = \{1, 2, 8\}$. Then $A' = \{3, 4, 5, 6, 7\}$.

From the definition of complement, it follows that for any set $A$ in a universe $U$,

$$A \cap A' = \varnothing \qquad \text{and} \qquad A \cup A' = U$$

A set and its complement have no elements in common. The union of a set and its complement is the universal set.

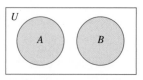

**FIGURE 6–7**    Because $A$ and $B$ are disjoint, $A \cap B$ has no elements; it is the empty set: $A \cap B = \varnothing$. The shaded area is $A \cup B$.

### Disjoint Sets

If $A$ and $B$ have no elements in common, we call them **disjoint sets**, and $A \cap B = \varnothing$. The sets $A = \{2, 4, 6, 8\}$ and $B = \{1, 3, 5, 7\}$ are disjoint.

Figure 6–7 shows two disjoint sets. Although their intersection is empty, their union is not, and it is shown by the shaded area.

---

## 6.1    EXERCISES

Access end-of-section exercises online at **www.webassign.net**

---

## 6.2    COUNTING ELEMENTS IN A SUBSET USING A VENN DIAGRAM

- Number of Elements in a Subset
- Venn Diagrams Using Three Sets

### Number of Elements in a Subset

We now turn our attention to counting the elements in a set. Sometimes we want to be able to use information about a set and determine the number of elements in it but avoid actually counting the elements one by one.

We use the notation $n(A)$, read "**$n$ of $A$**," to indicate the number of elements in set $A$. If $A$ contains 23 elements, we write $n(A) = 23$.

Suppose you count 10 people in a group that like brand $X$ cola and 15 that like brand $Y$. We denote this by $n(\text{brand } X) = 10$ and $n(\text{brand } Y) = 15$. We may ask how many people we have when they are combined into a single group. The answer depends on the number who like both brands. Let's assume that four people like both brands. Note that "like both brands" is $X \cap Y$. The 10 who like Brand $X$ consist of 6 who like only Brand $X$ and 4 who like both brands. The 15 who like Brand $Y$ consist of 11 who like only Brand $Y$ and 4 who like both. Figure 6–8 shows this in a Venn diagram. From the diagram we see that $n(X \cup Y)$ = 6 + 4 + 11 = 21. We write this is in an equivalent, but slightly more complicated form:

$$6 + 4 + 11 = (6 + 4) + (4 + 11) - 4$$

In this case this equation is

$$n(X \cup Y) = n(X) + n(Y) - n(X \cap Y)$$

We obtained the above equation using specific numbers of people. However, the equation applies to any two sets. When we count the number of elements in two sets, $A$ and $B$, the totality of all elements in the two sets forms the union of the two sets, $A \cup B$, and those elements in both sets forms the intersection of the two sets, $A \cap B$. If we attempt to determine the total count by adding the number in $A$ to the number in $B$, we count those in both, the intersection, twice. We need to subtract the number in both sets, $A \cap B$, from the sum of those in $A$ and those in $B$ to obtain the total number of elements involved.

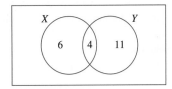

**FIGURE 6–8**

$$n(X) = 10, n(Y) = 15, n(X \cap Y) = 4, n(X \cup Y) = 21$$
$$n(X \cup Y) = n(X) + n(Y) - n(X \cap Y)$$

In general, the number of elements in the union of two sets is given by the following theorem called the **Inclusion–Exclusion Principle**.

| **THEOREM** Inclusion–Exclusion Principle $n(A \cup B)$ | $n(A \cup B) = n(A) + n(B) - n(A \cap B)$ <br> where $n(A)$ represents the number of elements in set $A$, $n(B)$ represents the number of elements in set $B$, and $n(A \cap B)$ represents the number of elements of $A \cap B$. |
|---|---|

**Example 1**   $A = \{a, b, c, d, e, f\}, B = \{a, e, i, o, u, w, y\}$. Compute $n(A)$, $n(B)$, $n(A \cap B)$, and $n(A \cup B)$.

**Solution**
$n(A) = 6$ and $n(B) = 7$. In this case, $A \cap B = \{a, e\}$, so $n(A \cap B) = 2$. $A \cup B = \{a, b, c, d, e, f, i, o, u, w, y\}$, so $n(A \cup B) = 11$. This checks with the formula, $n(A \cup B) = 6 + 7 - 2 = 11$.

**Example 2**   Set $A$ is the 9 o'clock English class of 15 students, so $A$ contains 15 elements. Set $B$ is the 11 o'clock history class of 20 students, so $B$ contains 20 elements. $A \cap B$ is the set of

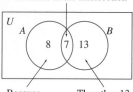

7 elements in the intersection
U
A
B
8
7
13
Because $n(A) = 15$, $15 - 7 = 8$ go here
The other 13 elements of B go here

**FIGURE 6–9**

students in both classes (there are 7), so $A \cap B$ contains 7 elements. The number of elements in $A \cup B$ (a joint meeting of the classes) is

$$n(A \cup B) = n(A) + n(B) - n(A \cap B)$$

so

$$n(A \cup B) = 15 + 20 - 7 = 28$$

The sets $A$ and $B$ divide their union into three regions. The number in each region is shown in the Venn diagram in Figure 6–9.

We find the number of elements that are in $A$ only by taking the number in $A$, $n(A) = 15$, and then subtracting the number that are in both $A$ and $B$, $n(A \cap B) = 7$; thus $15 - 7 = 8$, the number in $A$ only.

**Example 3** The union of two sets, $A \cup B$, has 48 elements. Set $A$ contains 27 elements, and set $B$ contains 30 elements. How many elements are in $A \cap B$?

**Solution**

Using the Inclusion–Exclusion Principle, we have

$$\begin{aligned} 48 &= 27 + 30 - n(A \cap B) \\ 48 - 27 - 30 &= -n(A \cap B) \\ -9 &= -n(A \cap B) \end{aligned}$$

So, $n(A \cap B) = 9$.

**Example 4** One hundred students were asked whether they were taking psychology ($P$) or biology ($B$). The responses showed that

61 were taking psychology; that is, $n(P) = 61$.

18 were taking both; that is, $n(P \cap B) = 18$.

12 were taking neither.

**(a)** How many were taking biology? [Find $n(B)$.]

**(b)** How many were taking psychology but not biology? [Find $n(P \cap B')$.]

**(c)** How many were not taking biology? [Find $n(B')$.]

**(d)** Find $n[(B \cap P)']$.

**Solution**

**(a)** Because 12 were taking neither, the rest, 88, were taking at least one of the courses, so $n(P \cup B) = 88$. We can find $n(B)$ from

$$\begin{aligned} n(P \cup B) &= n(P) + n(B) - n(P \cap B) \\ 88 &= 61 + n(B) - 18 \\ n(B) &= 45 \end{aligned}$$

**(b)** Because 18 students were taking both psychology and biology, the remainder of the 61 psychology students were taking only psychology. So, $61 - 18 = 43$ students were taking psychology but not biology.

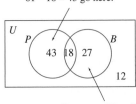

Because $n(P) = 61$,
$61 - 18 = 43$ go here.

Because $n(B) = 45$,
27 go here.

**FIGURE 6–10**

**(c)** The students not taking biology were those 12 taking neither and the 43 taking only psychology, a total of 55 not taking biology. (See Figure 6–10.)

**(d)** Figure 6–10 shows that the number outside $B \cap P$; that is, $n(B \cap P)'$, is $43 + 27 + 12 = 82$.

## Venn Diagrams Using Three Sets

A general Venn diagram of three sets divides a universe into as many as eight nonoverlapping regions. (See Figure 6–11.) We can use information about the number of elements in some of the regions (subsets) to obtain the number of elements in other subsets.

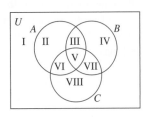

**FIGURE 6–11**   Three sets may divide the universe into eight regions.

**Example 5**   The sets $A$, $B$, and $C$ intersect as shown in Figure 6–12. The numbers in each region indicate the number of elements in that subset.

The number of elements in other subsets may be obtained from this diagram. For example,

$$n(A) = 9 + 2 + 3 + 7 = 21$$
$$n(B) = 2 + 3 + 1 + 4 = 10$$
$$n(A \cap B) = 2 + 3 = 5$$
$$n(A \cap B \cap C) = 3$$
$$n(A \cup B) = 9 + 2 + 3 + 7 + 4 + 1 = 26$$

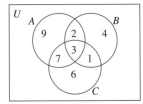

**FIGURE 6–12**

**Example 6**   A survey yields the following information about the musical preferences of students:

30 like classical.

24 like country.

31 like jazz.

9 like country and classical.

12 like country and jazz.

10 like classical and jazz.

4 like all three.

6 like none of the three.

Draw a diagram that shows this breakdown of musical tastes. Determine the total number of students interviewed.

### Solution

Begin by drawing a Venn diagram as shown in Figure 6–13a.

The universe is the set of college students interviewed. We want to determine the number of students in each region of the diagram. Because some students may like more than one kind of music, these sets may overlap. We begin where the three sets intersect, and because we know that 4 students like all three types of music, we place a 4 in the region where all three sets intersect. Of the 9 students who like both country and classical, we have already recorded 4 (those who like all three).

The other 5 lie in the intersection of classical and country that lies outside jazz (Figure 6–13b). In a similar fashion, the number who like both jazz and country breaks down into the 4 who like all three and the 8 who like jazz and country but lie outside the region of all three. Because these three regions account for 17 of those who like country, the other 7 who like country lie in the region where country does not intersect the jazz and classical. Fill in the rest of the regions; the results are shown in the diagram in Figure 6–13c.

Obtain the total number of students interviewed by adding the number in each region of the Venn diagram. The total is 64.

*Note:* We first determined the number in the innermost region, where all three sets intersect, and then determined the number in the regions as we moved out. This procedure works well for problems like this.

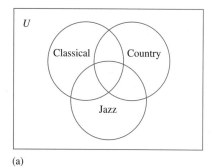

(a)

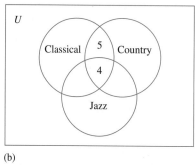

(b)

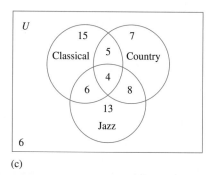

(c)

**FIGURE 6–13**    (a) A Venn diagram representing students grouped by musical preference. (b) Four students like all three types of music. A total of nine students like both classical and country. (c) The number in each category of musical preference.

## 6.2    EXERCISES

Access end-of-section exercises online at **www.webassign.net**

## 6.3 BASIC COUNTING PRINCIPLES

- Tree Diagrams
- Multiplication Rule
- Addition Rule for Counting Elements in Disjoint Sets

Elizabeth Barrett Browning answered her own question "How do I love thee? Let me count the ways" in a poem that seems to be immortal. Few "how many?" questions are answered poetically, but such questions often need answers.

For example:

- The telephone company, for planning purposes, may ask "How many telephone exchanges will Memphis need in ten years?"
- The state highway department may ask "How many vehicles can I-70 carry safely?"
- A person interested in his or her chances of winning might ask "How many combinations of numbers are possible in a lottery drawing?"
- On most college campuses, someone asks "How many parking places are needed on campus?"
- A professor making out quizzes might ask "How many ways can I select five problems from a review sheet?"

In many "How many?" questions, it may not be possible, or practical, to list all possible ways and count them one by one. If possible, we would like to have a way to arrive at a total count without a tedious one-by-one listing.

Methods do exist; for example, we learned how to count the number of elements in the union of two sets (Section 6.2). We learn about some other methods in this section.

We illustrate one method by first making a list in a systematic manner and drawing conclusions from the list.

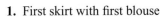

**Example 1** A teenager asked how many different outfits she could form if she had two skirts and three blouses. Let's list the different ways.

1. First skirt with first blouse
2. First skirt with second blouse
3. First skirt with third blouse
4. Second skirt with first blouse
5. Second skirt with second blouse
6. Second skirt with third blouse

Notice the pattern of the list. For each skirt, she can obtain three outfits by selecting each of the different blouses, so the total number of outfits is simply three times the number of skirts.

If you are to select one book from a list of five books and a second book from a list of seven books, you can determine the number of possible selections of the two books by writing the first book from list one with each of the seven books from list two. You will list the seven books five times, one with each of the books from list one, so you will have a list that is $5 \times 7 = 35$ long.

We can list all possible selections in another way, using what are called **tree diagrams**.

## Tree Diagrams

We illustrate the construction and use of tree diagrams in the next two examples.

**Example 2**     Let's look at another problem. Suppose there are two highways from Speegleville to Crawford and three highways from Crawford to McGregor. How many different routes can we choose to go from Speegleville to McGregor through Crawford? A tree diagram provides a visual means to list all possible routes (Figure 6–18).

   Reading from left to right, starting at 0, draw two branches representing the two highways from Speegleville to Crawford (use 1 and 2 to designate the highways). At the end of each of these two branches, draw three branches representing the three highways from Crawford to McGregor. (Use $A$, $B$, and $C$ to designate the highways.) A choice of a first-level branch and a second-level branch determines a route from Speegleville to McGregor. Note that the total number of possible routes (branches that end at McGregor) is six because each first-level branch is followed by three second-level branches.

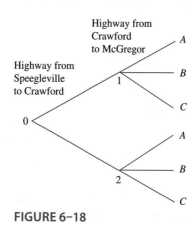

**FIGURE 6–18**

**Example 3**     Cox's Department Store has two positions to fill, those of a department manager and an assistant manager. Three people are eligible for the manager position, and four people are eligible for the assistant manager position. Use a tree diagram to show the different ways in which the two positions can be filled.

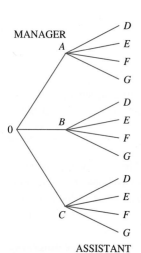

**FIGURE 6–19**

**Solution**
Label the candidates for department manager as $A$, $B$, and $C$. Label the candidates for assistant manager as $D$, $E$, $F$, and $G$.

   The tree diagram in Figure 6–19 illustrates the 12 possible ways in which the positions can be filled. Reading from left to right, starting at 0, you find three possible "branches" (managers). A branch for each possible assistant manager is attached to the end of each branch representing a manager. In all, 12 paths begin at 0 and go to the end of a branch. For example, the path $0BD$ represents the selection of $B$ as the manager and $D$ as the assistant manager.

A tree diagram shows all possible ways to make a sequence of selections, and it shows the number of different ways the selections can be made.

We can find the total number of selections from the end points of branches, which is the product of the number of first-level branches and the number of second-level branches; that is, the number of ways in which the first selection can be made times the number of ways in which the second selection can be made.

These three examples are fundamentally the same kind of problem. Let's make a general statement that includes each one.

## Multiplication Rule

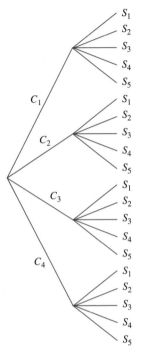

Call two activities $A_1$ and $A_2$. Each activity can be carried out in several ways. We want to determine in how many different ways the first activity followed by the second activity can be performed.

In the three preceding examples, $A_1$ and $A_2$ are the following:

*Selections of an outfit:* $A_1$ is the activity of selecting a skirt, and $A_2$ is the activity of selecting a blouse.

*Routes from Speegleville to McGregor:* $A_1$ is the selection of a highway from Speegleville to Crawford, and $A_2$ is the selection of a highway from Crawford to McGregor.

*Filling two positions:* $A_1$ is the selection of a manager, and $A_2$ is the selection of an assistant manager.

In many cases, we do not need a list of all possible selections, but we need their number. In such a case, the problem reduces to a question of the number of ways in which we can carry out the activity $A_1$ followed by the activity $A_2$. The solution is simple: Multiply the number of ways in which activity $A_1$ can be performed by the number of ways in which activity $A_2$ can be performed. We often call this the **Multiplication Rule**.

**FIGURE 6–20**

| **THEOREM** | Two activities $A_1$ and $A_2$ can be performed in $n_1$ and $n_2$ different ways, respectively. The |
|---|---|
| **Multiplication Rule** | total number of ways in which $A_1$ followed by $A_2$ can be performed is |

$$n_1 \times n_2$$

Now let's apply the Multiplication Rule to some examples.

**Example 4**   A taxpayers' association is to elect a chairman and a secretary. There are four candidates for chairman and five candidates for secretary. In how many different ways can a slate of officers be elected?

**Solution**
This problem can be analyzed using a tree diagram. (See Figure 6–20.) The first activity, selecting a chairman, is represented by the four first-stage branches, $C_1$, $C_2$, $C_3$, and $C_4$. To each of these branches we attach the five possible selections of the second activity, selecting a secretary. This results in $4 \times 5 = 20$ branches representing the different slates possible.

### NOTE

Two slates of officers differ if one or more officers is replaced by another person or if two or more officers exchange positions.

The analysis of a problem using a tree diagram can quickly become unwieldy. You quickly see that you *don't* want to use a tree diagram to analyze the next example, even though it is theoretically possible to do so.

**Example 5**   Moody Library wishes to display two rare books, one from the history collection and one from the literature collection. If the library has 50 history and 125 literature books to select from, how many different ways can the books be selected for display?

**Solution**
A tree diagram of this problem would have 50 branches at the first stage representing the 50 history books. Attached to the end of each of the history branches are 125 branches representing the literature books. The number of branches finally is the number of choices of the first-stage branch times the number of choices of the second-stage branch. In this case, there are $50 \times 125 = 6250$ possible displays.

Don't attempt to draw a tree with 6250 branches; use the Multiplication Rule to analyze the problem. The first activity is a selection of a history book, and the second activity is the selection of a literature book. This sequence can be performed in $50 \times 125 = 6250$ ways.

**Example 6**   Jane selects one card from a deck of 52 different cards. The first card is *not* replaced before Joe selects the second one. In how many different ways can they select the two cards?

**Solution**
Jane selects from a set of 52 cards, so $n_1 = 52$. Joe selects from the remaining cards, so $n_2 = 51$. Two cards can be drawn in $52 \times 51 = 2652$ ways.

Some problems ask for the number of ways that a sequence of more than two activities can be performed. The Multiplication Rule can be applied to sequences of more than two activities.

**Corollary**   Activities $A_1, A_2, \ldots, A_k$ can be performed in $n_1, n_2, \ldots, n_k$ different ways, respectively. The number of ways in which one can perform $A_1$ followed by $A_2 \ldots$ followed by $A_k$ is

$$n_1 \times n_2 \times \cdots \times n_k$$

**Example 7**   A quiz consists of four multiple-choice questions with five possible responses to each question. In how many different ways can the quiz be answered?

**Solution**
In this case, there are four activities—that is, answering each of four questions. Each activity (answering a question) can be performed (choosing a response) in five different ways. The answers can be given in

$$5 \times 5 \times 5 \times 5 = 625$$

different ways.

**Example 8**  Three couples attend a movie and are seated in a row of six seats. How many different seating arrangements are possible if couples are seated together?

**Solution**
Think of this as a sequence of six activities, that of assigning a person to sit in each of seats 1 through 6. The Multiplication Rule states that the number of ways in which this can be done is the product of the numbers of ways in which each selection can be made.

*First seat:*  Any one of the six people may be chosen, so there are six choices.

*Second seat:*  Only the partner of the person in the first seat may be chosen, so there is one choice.

*Third seat:*  One couple is seated, so any one of the remaining four people may be chosen. Thus there are four choices.

*Fourth seat:*  There is one choice only because it must be the partner of the person in the third seat.

*Fifth seat:*  There are two choices because either of the two remaining people may be seated.

*Sixth seat:*  The one remaining person is the only choice.

By the multiplication rule, the number of possible arrangements is the product of the number of choices for each seat:

$$6 \times 1 \times 4 \times 1 \times 2 \times 1 = 48$$

**Example 9**  The Cameron Art Gallery has several paintings by each of five artists. A wall has space to hang four paintings in a row. How many different arrangements by artists are possible if

**(a)** The paintings are by different artists?

**(b)** More than one painting by an artist may be displayed, but they may not be hung next to each other?

**Solution**
These problems may be viewed as a sequence of four activities, choosing an artist for each of the four spaces. The total number of ways is the product of the number of choices for each space.

**(a)** Because all artists must be different, there are five choices for the artist in the first space, four for the second, three for the third, and two for the fourth. The number of arrangements is $5 \times 4 \times 3 \times 2 = 120$.

**(b)** The first painting to be hung can be selected from any one of the five artists. A painting by that artist may not be used in the second space, but a painting by any of the remaining four artists may be used. The artist chosen for the second space may not be chosen for the third space, but the artist chosen for the first space may be used. (The first and third paintings are not hung next to each other.) Thus, four choices exist for the third space. Likewise, four choices exist for the fourth space. The total number of arrangements is

$$5 \times 4 \times 4 \times 4 = 320$$

## Addition Rule for Counting Elements in Disjoint Sets

In Section 6.2, we have the Inclusion–Exclusion Principle,

$$n(A \cup B) = n(A) + n(B) - n(A \cap B)$$

which gives the number of elements in the union of two sets. When the sets are disjoint, $n(A \cap B) = 0$, so the rule reduces to

$$n(A \cup B) = n(A) + n(B)$$

We can apply this counting principle to a situation in which you wish to determine the number of ways to perform one activity or another, but not both. For example, there are six movies and four talk shows on cable TV at the same time. You plan to watch either a movie or a talk show. How many choices do you have? Think of listing your choices. List the six movies and then the talk shows (or vice versa) and you have a list of $6 + 4 = 10$ choices. This illustrates a basic principle, the **Addition Rule**.

| | |
|---|---|
| **Addition Rule** | If an activity $A_1$ can be performed in $n_1$ ways and activity $A_2$ can be performed in $n_2$ ways, then either $A_1$ or $A_2$, but not both simultaneously, can be performed in $n_1 + n_2$ ways. |

**Example 10**

**NOTE**

We emphasize that the Addition Rule applies only to cases where there is no overlap of the sets $A$ and $B$. When there is overlap, the Inclusion–Exclusion Priniciple applies.

Rhonda plans to buy a hybrid car. The Temple dealer has 15 hybrids in stock, and the Austin dealer has 23 in stock. How many choices does Rhonda have if she buys a hybrid from one of these dealers?

**Solution**
Because Rhonda will buy from one dealer or the other (and not both), the Addition Rule applies. She has $15 + 23 = 38$ choices.

The next example illustrates a problem where both the Addition Rule and the Multiplication Rule apply.

**Example 11**

In how many different ways can 3 science and 2 history books be arranged on a shelf if books on each subject are kept together?

**Solution**
First, observe that the subjects can be arranged in two ways, science on the left and history on the right or vice versa.

Next, count the number of ways in which the books can be arranged within the subject arrangements:

*Science–history:* There are five positions to be filled, science first and history second. The number of possible arrangements is $3 \times 2 \times 1 \times 2 \times 1 = 12$.

*History–science:* Again there are five positions to be filled, but history books are placed first and science books second. The number of possible arrangements is $2 \times 1 \times 3 \times 2 \times 1 = 12$.

Because the arrangements can be either science–history or history–science, the Addition Rule states there are $12 + 12 = 24$ arrangements.

We point out that this can also be solved using the Multiplication Rule exclusively with six arrangements of the science books within the science group, two arrangements of the history books within the history group, and two arrangements of the subject groups, giving $6 \times 2 \times 2 = 24$ arrangements.

Some of the exercises of this section, and the sections that follow, refer to a bridge deck or a deck of 52 cards. The following figure shows the composition of such a deck of cards.

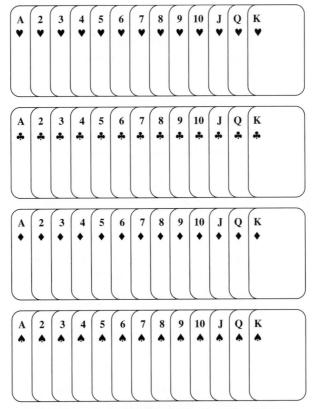

A standard deck or bridge deck has 52 cards with 13 cards in each suit, hearts (♥), clubs (♣), diamonds (♦), and spades (♠). Each suit contains an ace (A), 2 through 10, jack (J), queen (Q), and king (K). The hearts and diamonds are red cards. The clubs and spades are black cards. The jack, queen, and king are called face cards.

## 6.3   EXERCISES

Access end-of-section exercises online at **www.webassign.net**

## 6.4   PERMUTATIONS

- Permutations
- Notations for Number of Permutations
- Permutation of Objects with Some Alike (Optional)

### Permutations

The Multiplication Rule counts the number of ways a sequence of activities can be performed. A common application counts the ways a selection can be made from two or more sets such as the number of ways a person can select a salad, an entrée, and a dessert from a menu. The Multiplication Rule can also apply to counting the number of ways a sequence of objects can be selected from *one* set. We next focus on such an application that occurs frequently enough to be given a name, a *permutation*.

The next example provides an illustration from which we will form a general conclusion.

**Example 1**   The Art Department received a gift of seven paintings from an alumnus artist. An art major selects three of the paintings and arranges them in three given locations in the foyer of the art building. How many different arrangements are possible?

**Solution**

This problem can be viewed as a Multiplication Rule problem with three activities.

- *Activity 1:* Select a painting from the set of seven for the first location. This can be done in seven ways because any of the seven paintings can be selected.
- *Activity 2:* Select one of the remaining paintings for the second location. This can be done in six ways because any of the six remaining paintings can be selected.
- *Activity 3:* This can be done in five ways because any of the five remaining paintings can be selected.

Therefore, there are $7 \times 6 \times 5 = 210$ possible arrangements.

Note that this is not the same problem as selecting three paintings to send to a gallery for exhibition when just the collection, not the order of arrangement, form the activities. This example has properties that we identify with a permutation.

A permutation problem can be viewed in two ways:

1. Count the number of sequences in which elements can be selected from a set, removing an element at each step and making the next selection from the elements that remain. This fits the Multiplication Rule directly, with the number of choices decreasing by one at each step.

2. Select all subsets of elements with the designated number of elements from a set and count the number of ways the elements can be arranged.

Many problems are presented from the second viewpoint.

In either case, the number of sequences or arrangements of, say, four selections from a set of $n$ elements is

$$n(n - 1)(n - 2)(n - 3) \text{ ways}$$

**DEFINITION**     A subset of distinct elements selected from a given set and arranged in a specific order is called a **permutation**.

Because a problem may not be identified as a permutation problem, you need to be able to recognize it as such. Here are three keys that can help in recognizing a permutation.

**Keys to Recognizing a Permutation**

1. A permutation is an arrangement or sequence of selections of elements from a single set.

2. Repetitions are not allowed, which means that once an element is selected it is not available for a subsequent selection. The same element may not appear more than once in a particular arrangement.

3. The order in which the elements are selected or arranged is significant.

Students sometimes ask how to tell when order is significant. A different order occurs when two items exchange positions or the order of two selections is exchanged. If such an exchange makes a difference in the selection, then order is significant. If the exchange makes no difference, then order is not significant.

**Example 2**     Ten students each submit one essay for competition. In how many ways can first, second, and third prizes be awarded?

**Solution**
This is a permutation problem because

1. Each essay selected is from the same set.

2. No essay can be submitted more than once; that is, an essay cannot be awarded two prizes (no repetition).

3. The order (prize given) of the essays is important.

Any of the ten essays may be chosen for first prize. Then any of the remaining nine may be chosen for the second prize, and any of the other eight may be chosen for third prize. According to the Multiplication Rule, the three prizes may be awarded in

$$10 \times 9 \times 8 = 720 \text{ different ways}$$

**Example 3**   In how many different ways can a penny, a nickel, a dime, and a quarter be given to four children if one coin is given to each child?

**Solution**
Each child may be considered a "position" that receives a coin. The number of ways a coin may be given to each child follows:

>   *First child:*  four possibilities of a coin
>   *Second child:*  three possibilities of a coin
>   *Third child:*  two possibilities of a coin
>   *Fourth child:*  one possibility of a coin

Therefore, the coins may be distributed in

$$4 \times 3 \times 2 \times 1 = 24 \text{ different ways}$$

**Example 4**   At the Cumberland River Festival, four young women (called "belles" in the South) are stationed at historic Fort House; one stands at the entrance, one in the living room, one in the dining room, and one on the back veranda. If there are 10 belles, in how many different ways can 4 be selected for the stations?

**Solution**
This is a permutation, because a woman can be selected for, at most, one station, and the order (place stationed) is significant. The permutation can be made in

$$10 \times 9 \times 8 \times 7 = 5040 \text{ different ways}$$

If the problem in Example 4 had been to select 4 belles to be present in the living room with no particular station for each one, then Example 4 would not have been a permutation problem, because the belles would not be arranged in any particular order. (The number of selections in this case uses a technique that we discuss later.)

## Notations for Number of Permutations

The notation commonly used to represent the number of permutations for a set is written $P(8, 3)$, which is read "permutation of eight things taken three at a time." This notation represents the number of permutations of three elements from a set of eight elements. $P(10, 4)$ represents the number of permutations of four elements selected from a set of ten elements. [$P(7, 3)$ is the answer to Example 1, and $P(10, 4)$ is the answer to Example 4.]

We want you to understand the pattern for calculating numbers like $P(10, 4)$ so that you can do it routinely. Let's look at some examples.

>   $P(10, 4) = 10 \times 9 \times 8 \times 7$ permutations of four elements
>       taken from a set of ten elements.
>   $P(5, 3) = 5 \times 4 \times 3$
>   $P(7, 2) = 7 \times 6$ permutations of two elements
>       selected from a set of seven elements.
>   $P(21, 3) = 21 \times 20 \times 19$

In each case, the calculation begins with the first number in the parentheses, 21 in $P(21, 3)$ and 7 in $P(7, 2)$. The second number in $P(21, 3)$, $P(7, 2)$, and so on determines the number of terms in the product. Because the terms decrease by 1 to the next term, you need to know only the first term and how many are needed to calculate the answer.

This reasoning lets us know that $P(30, 5)$ is a product of five terms beginning with 30, each term thereafter decreasing by 1, so

$$P(30, 5) = 30 \times 29 \times 28 \times 27 \times 26$$
$$P(105, 4) = 105 \times 104 \times 103 \times 102$$

and

$$P(4, 4) = 4 \times 3 \times 2 \times 1$$

Can you give the last term in $P(52, 14)$ without writing out all the terms? The preceding examples give a pattern that helps. To give a general description of the last term in the calculation of $P(n, r)$, we write the last terms of the examples in a way that seems unnecessarily complicated. However, it will help us obtain the general form.

| Example | Last term | May be written as |
|---|---|---|
| $P(5, 3) = 5 \times 4 \times 3$ | $3 = 5 - 2$ | $5 - 3 + 1$ |
| $P(10, 4) = 10 \times 9 \times 8 \times 7$ | $7 = 10 - 3$ | $10 - 4 + 1$ |
| $P(105, 4) = 105 \times 104 \times 103 \times 102$ | $102 = 105 - 3$ | $105 - 4 + 1$ |

so we expect the last term of

$$P(52, 14) \quad \text{to be} \quad 52 - 13 = 39 \quad \text{(also written } 52 - 14 + 1\text{)}$$

In general, $P(n, k)$ gives the number of arrangements that can be formed by selecting $k$ elements from a set of $n$ elements. Following the observed pattern, it may be written

$$P(n, k) = n(n - 1)(n - 2) \cdots (n - k + 1)$$

---

**Number of Permutations, $P(n, k)$**

$$P(n, k) = n(n - 1)(n - 2) \cdots (n - k + 1)$$

---

We use a special notation for the case when a permutation uses all elements of a set. Note that $P(4, 4)$ is just the product of the integers 4 through 1, that is, $4 \times 3 \times 2 \times 1$. In general, $P(n, n)$ is the product of the integers $n$ through 1. The following notation is used.

**DEFINITION**
**$n!$**

The product of the integers $n$ through 1 is denoted by $n!$ (called **$n$ factorial**).

$$1! = 1$$
$$n! = n(n - 1)(n - 2) \times \cdots \times 2 \times 1 = n(n - 1)! \quad (\text{for } n > 1)$$
$$0! = 1$$

Note that $n!$ is not defined for negative values of $n$.

**Example 5**

$$7! = 7 \times 6 \times 5 \times 4 \times 3 \times 2 \times 1 = 5040$$
$$2! = 2 \times 1 = 2$$
$$6! = 6 \times 5 \times 4 \times 3 \times 2 \times 1 = 720$$

Notice that $6! = 6 \times 5!$ and $4! = 4 \times 3!$, and so on.

$$1! = 1$$

$0!$ is defined to be 1

The definition of $0! = 1$ allows us to use $n! = n(n - 1)!$ when $n = 1$, and thus use $n! = n(n - 1)!$ for $n \geq 1$.

Arithmetic involving factorials can be carried out easily if you are careful to use the factorial as defined.

$$\frac{5!}{3!} = \frac{5 \times 4 \times \cancel{3} \times \cancel{2} \times \cancel{1}}{\cancel{3} \times \cancel{2} \times \cancel{1}}$$
$$= 5 \times 4$$
$$= 20$$
$$3! \, 4! = 3 \times 2 \times 1 \times 4 \times 3 \times 2 \times 1 = 144$$

**Example 6**   How many different ways can six people be seated in a row of six chairs?

**Solution**

This is a permutation because

1. The people are all selected from the same set.
2. Repetitions are not allowed, a person may not occupy two different seats at the same time.
3. Order is significant, because a different seating arrangement occurs when two people exchange seats.

   Because six positions are to be filled from a set of six people, the number of arrangements is

$$P(6, 6) = 6! = 6 \times 5 \times 4 \times 3 \times 2 \times 1 = 720$$

Factorials allow us to write the expression for the number of permutations in another form that is sometimes useful. For example,

$$P(8, 3) = 8 \times 7 \times 6$$
$$= \frac{8 \times 7 \times 6 \times 5!}{5!}$$

Because $8 \times 7 \times 6 \times 5! = 8 \times 7 \times 6 \times 5 \times 4 \times 3 \times 2 \times 1 = 8!$, we can write

$$P(8, 3) = \frac{8!}{5!}$$

Be sure you understand that

$$P(6, 4) = \frac{6!}{2!} \quad \text{(2! came from (6 − 4)!)}$$

In general, we can write

$$P(n, k) = \frac{n!}{(n - k)!}$$

**Example 7**  Many auto license plates have three letters followed by three digits. How many different license plates are possible if

**(a)** Letters and digits are not repeated on a license plate?

**(b)** Repetitions of letters and digits are allowed?

**Solution**

**(a)** First of all, this may be viewed as a Multiplication Rule problem with two activities. The first activity is the selection and arrangement of letters; the second activity is the selection and arrangement of digits. We find the number of license plates by multiplying the number of selections of letters and the number of selections of digits. The selection of letters is a permutation, $P(26, 3)$ in number, and the selection of digits is a permutation, $P(10, 3)$ in number. The number of license plates is then $P(26, 3) \times P(10, 3) = 15{,}600 \times 720 = 11{,}232{,}000$.

We can also view this as a problem with six activities with the selection of a letter for each of the first three activities, and the selection of a digit for the last three activities giving the number of license plates as $26 \times 25 \times 24 \times 10 \times 9 \times 8 = 11{,}232{,}000$.

**(b)** This is an ordered arrangement that is not a permutation, because a letter or digit may appear more than once on a license plate. This is a Multiplication Rule problem with six activities, the selection of three letters followed by the selection of three digits. This can be done in $26 \times 26 \times 26 \times 10 \times 10 \times 10 = 17{,}576{,}000$ ways.

Example 7(a) could also be worked as a Multiplication Rule problem with six activities, giving $26 \times 25 \times 24 \times 10 \times 9 \times 8 = 11{,}232{,}000$ ways.

## Permutations of Objects with Some Alike *(Optional)*

So far the permutation problems have involved objects that are all different. Sometimes we arrange objects when some are alike. For example, we may ask for all arrangements of the letters of the word AGREE. Generally, we have said that we can arrange five objects in $P(5, 5) = 5! = 120$ ways. However, when we interchange the two E's in a word, we obtain the same word. Each time we "spell" a word, the E's are placed in certain positions. We can arrange the E's in those positions in 2! ways and still have the same "word." Therefore, the number of different "words" (arrangements) is 120/2!.

For example, in the word EAGER we can think of the two E's as $E_1$ and $E_2$ to distinguish them momentarily. One spelling is $E_1$ AGE$_2$ R, and another is $E_2$ AGE$_1$ R. The number of "different" spellings that give the same words depends on the number of arrangements of the identical letters. In this case the two E's can be arranged in 2! ways. In general, $k$ identical objects can be arranged in $k!$ ways that leave the overall arrangement unchanged.

We find the number of distinguishable arrangements by the total number of arrangements (distinguishable and undistinguishable) divided by $k!$.

**Example 8**     How many different words can be formed using all the letters of DEEPEN?

**Solution**
Because three of the six letters (E's) are identical, the number of permutations is
$$\frac{6!}{3!} = 120.$$

---

**THEOREM**

**Permutation of Identical Objects**

**(a)** The number of permutations of $n$ objects with $r$ of the objects identical is $\dfrac{n!}{r!}$.

**(b)** If a set of $n$ objects contains $k$ subsets of objects in which the objects in each subset are identical and objects in different subsets are not identical, the number of different permutations of all $n$ objects is
$$\frac{n!}{r_1! r_2! \ldots r_k!}$$
where $r_1$ is the number of identical objects in the first subset, $r_2$ is the number of identical objects in the second subset, and so on.

---

Part (b) of the theorem tells how to compute the number of permutations for two or more categories of identical objects.

**Example 9**     In how many ways can the letters of REARRANGE be permuted?

**Solution**
There are nine letters, with three R's, two A's, and two E's. The number of permutations is

$$\frac{9!}{3!\,2!\,2!} = \frac{9 \times 8 \times 7 \times 6 \times 5 \times 4 \times 3 \times 2 \times 1}{3 \times 2 \times 1 \times 2 \times 1 \times 2 \times 1} = 15,120$$

**Example 10**     Basketball teams X and Y are in a playoff. The team that wins three out of a possible five games is the winner. Denote the sequence of winners by a sequence of letters such as XXYYY. This indicates that X won the first two games and Y won the last three. How many different sequences are possible if X wins the playoff?

**Solution**
In cases where a team wins the playoff in fewer than five games, such as XXYX, only one Y appears, so we can let Y "win" the unplayed game and have 3 X's and 2 Y's, XXYXY. This way we can represent all possible sequences of playoff games with 3 X's and 2 Y's. The number of games is

$$\frac{5!}{3!\,2!} = 10 \text{ different ways}$$

Similarly, there are ten different sequences possible for Y to win, giving $10 + 10 = 20$ possible sequences for the playoff to occur.

## 6.4 EXERCISES

Access end-of-section exercises online at **www.webassign.net**     WebAssign

## Using Your TI Graphing Calculator

### Permutations

Using the **nPr** command, the number of permutations—say, $P(7, 4)$—can be calculated on the TI graphing calculator as follows.

First, enter 7, then select **nPr**, using [MATH] <PRB> <2:nPr>, which gives the screen:

Press [ENTER], then enter 4:

Press [ENTER], which gives 840:

```
MATH NUM CPX PRB
1:rand
2:nPr
3:nCr
4:!
5:randInt(
6:randNorm(
7:randBin(
```

```
7 nPr 4█
```

```
7 nPr 4
             840
█
```

### Exercises

Calculate the following:

**1.** $P(12, 5)$     **2.** $P(22, 13)$     **3.** $P(8, 4)P(6, 3)$     **4.** $\dfrac{P(15, 7)}{P(7, 5)}$

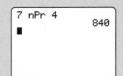

## Using Excel

### Permutations

The number of permutations, say $P(7, 4)$, can be calculated in Excel using the PERMUT function, written as PERMUT(7,4). The formula = PERMUT(A2,A3) will calculate the number of permutations using the numbers in A2 and A3.

### Exercises

**1.** Find $P(6, 3)$.     **2.** Find $P(12, 7)$.     **3.** Find $P(50, 18)$.

## 6.5 COMBINATIONS

- Combinations
- Special Cases
- Problems Involving More Than One Counting Technique
- Binomial Theorem

### Combinations

When you pay your bill at the Pizza Place, the cashier is interested in the collection of coins and bills you give her, not the order in which you present them. When you are asked to answer six out of eight test questions, the collection of questions is important, not the arrangement. Therefore, if the professor wishes to compute the number of different ways in which students can choose six questions from eight, she does not deal with permutations. She wants the number of ways in which a subset of six elements can be obtained.

**DEFINITION**
**Combination**

A subset of elements chosen from a given set without regard to their arrangement is called a **combination**.

The notation $C(n, k)$, read "combinations of $n$ things taken $k$ at a time," represents the number of subsets consisting of $k$ elements taken from a set of $n$ elements.

$C(8, 3)$ denotes the number of ways in which three elements can be selected from a set of eight distinct elements with no repetition of elements and the order of elements does not matter. $C(52, 6)$ denotes the number of ways in which 6 elements can be selected from a set of 52 distinct elements.

The keys to recognizing a combination are the following.

**Combinations**
1. A combination selects elements from a single set.
2. Repetitions are not allowed.
3. The order in which the elements are arranged is *not* significant.

Note that a combination differs from a permutation only because order is not significant in a combination, whereas it is important in a permutation.

**Example 1**

Given the set $A = \{a, b, c, d, e, f\}$, the subset $\{b, d, f\}$ is a combination of three elements taken from a set of six elements.

Because the elements of the subset $\{b, d, f\}$ can be arranged in several ways, we expect there to be several permutations for each subset. This indicates that you should expect more permutations than combinations in a given set.

**Example 2**

List all combinations of two elements taken from the set $\{a, b, c\}$.

**Solution**

Because of the small number of elements involved, it is rather easy to list all subsets consisting of two elements. They are $\{a, b\}$, $\{a, c\}$, and $\{b, c\}$. Therefore, $C(3, 2) = 3$.

If we want lists like all 5-letter subsets of our 26-letter alphabet, the listing rapidly increases in difficulty. If we want only the number of such subsets, not the list, then the problem becomes easier. Let's look at two examples that illustrate how to determine the number of combinations.

We will see a definite relationship between the number of permutations and the number of combinations.

**Example 3**    Select all 2-letter combinations of letters from the set of 4 letters {A, B, C, D}.

**Solution**

We use a tree diagram to show all the ways that we can select a pair of letters. (See Figure 6–21.)

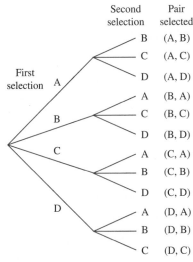

**FIGURE 6–21**    Pairs selected from the set {A, B, C, D}.

Because we are interested only in the pair of objects, not their order of selection, notice that (A, B) and (B, A) both appear and are equal sets, so (A, B) occurs twice. The list of all possible pairs then is

$$(A, B), (A, C), (A, D), (B, C), (B, D), (C, D)$$

each appearing twice in the tree diagram. It is no accident that the number of pairs, 6, is

$$6 = \frac{12}{2} = \frac{P(4, 2)}{2!}$$

where $12 = P(4, 2)$ is the number of permutations, and 2! gives the number of arrangements of each pair. Thus,

$$C(4, 2) = \frac{P(4, 2)}{2!}$$

This shows 6 ways a subset of 2 letters can be chosen; that is, there are 6 combinations of 2 letters selected from a set of 4.

**Example 4**    Let's look at the relationship between permutations and combinations again by making a list of combinations and corresponding permutations using subsets with more than two elements.

We use a set of 4 letters, $\{a, b, c, d\}$, and form all 3-element subsets. There are 4, each of which may be obtained by removing 1 letter from the set of 4 letters. We list the combinations and all permutations that can be formed from the letters in the combination.

| Subset | Arrangements | Subset | Arrangements |
|--------|-------------|--------|-------------|
| Combinations | Permutations | Combinations | Permutations |
| $\{a, b, c\}$ | abc | $\{a, c, d\}$ | acd |
| | acb | | adc |
| | bac | | cad |
| | bca | | cda |
| | cab | | dac |
| | cba | | dca |
| $\{a, b, d\}$ | abd | $\{b, c, d\}$ | bcd |
| | adb | | bdc |
| | bad | | cbd |
| | bda | | cdb |
| | dab | | dbc |
| | dba | | dcb |

Note the 6 permutations for each combination. We expect this because 3 letters can be arranged in $3! = 6$ ways. We can get the total number of permutations with

(Number of combinations) $\times$ (Number of permutations per combination)

In the above, that is,

$$C(4, 3) \times 6 = P(4, 3)$$

which we write as

$$C(4, 3) \times 3! = P(4, 3)$$

dividing by 3! we have

$$C(4, 3) = \frac{P(4, 3)}{3!}$$

This gives us what we need to compute $C(4, 3)$:

$$C(4, 3) = \frac{P(4, 3)}{3!} = \frac{4 \times 3 \times 2}{3 \times 2 \times 1} = \frac{24}{6} = 4$$

In general, the same type of result applies to other sizes of the set and the other numbers in the subset, so

$$C(11, 5) = \frac{P(11, 5)}{5!}, \quad C(9, 2) = \frac{P(9, 2)}{2!}, \quad C(101, 14) = \frac{P(101, 14)}{14!}$$

and generally we have the following theorem:

**THEOREM**

$$C(n, r) = \frac{P(n, r)}{r!}$$

or

$$P(n, r) = r! \, C(n, r)$$

Because $P(n, r)$ can be written as

$$\frac{n!}{(n - r)!}$$

$C(n, r)$ can also be written as

$$\frac{n!}{r!(n - r)!}$$

We now have a convenient way of calculating the number of combinations.

**Example 5**

$$C(5, 2) = \frac{P(5, 2)}{2!} = \frac{5 \times 4}{2 \times 1} = 10$$

$$C(5, 3) = \frac{P(5, 3)}{3!} = \frac{5 \times 4 \times 3}{3 \times 2 \times 1} = 10$$

$$C(10, 4) = \frac{P(10, 4)}{4!} = \frac{10 \times 9 \times 8 \times 7}{4 \times 3 \times 2 \times 1} = 210$$

$$C(8, 6) = \frac{8!}{6! \, 2!} = 28$$

$$C(15, 3) = \frac{15!}{3! \, 12!} = 455$$

**Example 6**  A student has seven books on his desk. In how many different ways can he select a set of three?

Solution
Because the order is not important, we have a combination problem:

$$C(7, 3) = \frac{P(7, 3)}{3!} = \frac{7 \times 6 \times 5}{3 \times 2 \times 1} = 35$$

**Example 7**  **(a)**  In how many ways can a committee of 4 be selected from a group of 10 people?

**(b)**  In how many ways can a slate of officers consisting of a president, vice-president, and secretary be selected from a group of 10 people?

**Solution**

**(a)** The order of selection is not important in the selection of a committee, so we have a combination problem of taking 4 elements from a set of 10:

$$C(10, 4) = \frac{P(10, 4)}{4!} = 210$$

**(b)** In selecting a slate of officers, President Jones, Vice-President Smith, and Secretary Allen is a different slate than President Allen, Vice-President Smith, and Secretary Jones. We can view each office as a position to be filled, so order is significant. The number of slates is $P(10, 3) = 720$.

Note the pattern used in computing combinations. To compute $C(10, 4)$, begin with 10 and write 4 integers decreasing by 1. Then divide by 4!. This holds true in general. To compute $C(15, 5)$, form the numerator using the 5 integers beginning with 15 and decreasing by 1. The denominator is 5!. In general, we can write $C(n, r)$ by forming the numerator from the product of $r$ integers that begin with $n$ and decrease by 1. The denominator is $r!$.

## Special Cases

The form for the number of combinations can be written as

$$C(n, r) = \frac{n!}{r!(n - r)!}$$

We find this form to be useful. Let's use it to look at some special cases.

**1.** In how many ways can one element be selected from a set? $C(6, 1)$ is the number of ways that one element can be selected from a set of six. It is

$$C(6, 1) = \frac{6!}{1!\,5!} = \frac{6 \times 5!}{1!\,5!} = 6$$

In general,

$$
\begin{aligned}
C(n, 1) &= \frac{n!}{1!(n - 1)!} \\
&= \frac{n(n - 1)!}{1!(n - 1)!} = n
\end{aligned}
$$

So one item can be selected from a set of $n$ items in $n$ ways.

**2.** In how many ways can zero items be selected from a set? We write $C(6, 0)$ to represent the number of ways in which no elements can be selected from a set of six. The formula gives

$$C(6, 0) = \frac{6!}{0!\,6!}$$

Because $0! = 1$, this reduces to $C(6, 0) = 1$. In general,

$$C(n, 0) = \frac{n!}{0!\, n!} = 1$$

Does your intuition tell you that there is just one way to select zero elements from a set? The one way is to take none.

3. In how many ways can all the elements be selected from a set? Our intuition tells us there is just one way, namely, take all of them. The formula agrees.

$$C(6, 6) = \frac{6!}{6!\, 0!} = 1$$

and

$$C(n, n) = \frac{n!}{n!\, 0!} = 1$$

4. For positive integers, $n$, $P(n, 1) = n$, so when selecting one element from a set, the number of permutations equals the number of combinations, $n$.

## Problems Involving More Than One Counting Technique

The solution to a problem may involve more than one counting technique. Often the first level is the Multiplication Rule with two or more activities involved. To count the number of ways in which each of these activities can occur may require permutations, combinations, or the Multiplication Rule again. The examples that follow involve more than one counting technique.

**Example 8**    A cafeteria offers a selection of four meats, six vegetables, and five desserts. In how many ways can you select a meal consisting of two different meats, three different vegetables, and two different desserts?

**Solution**

Basically, this is a problem whose solution first uses the Multiplication Rule. We obtain the possible number of meals by multiplying the number of ways in which you can select two meats, the number of ways in which you can select three vegetables, and the number of ways in which you can select two desserts.

Each of the numbers of ways in which you can select meats, vegetables, and desserts forms a combination problem. Therefore, we obtain the number of meals as

(Number of meat selections) $\times$ (number of vegetable selections)

$$\times \text{ (number of dessert selections)} = C(4, 2) \times C(6, 3) \times C(5, 2)$$

$$= \frac{4 \times 3}{2 \times 1} \times \frac{6 \times 5 \times 4}{3 \times 2 \times 1} \times \frac{5 \times 4}{2 \times 1}$$

$$= 6 \times 20 \times 10 = 1200$$

**Example 9**    The Beta Club has 14 male and 16 female members.Three men and 3 women form a committee. In how many ways can this be done?

**Solution**
Because we want to determine the number of ways a first event (selecting 3 males) *and* a second event (selecting 3 females) can occur, we need to compute the number of ways that each can occur and then multiply these values.
    The male members can be chosen in

$$C(14, 3) = \frac{14 \times 13 \times 12}{3 \times 2 \times 1} = 364 \text{ different ways}$$

The female members can be chosen in

$$C(16, 3) = \frac{16 \times 15 \times 14}{3 \times 2 \times 1} = 560 \text{ different ways}$$

By the Multiplication Rule, the committee can be chosen in $364 \times 560 = 203{,}840$ ways.

Instead of counting the number of outcomes for a sequence of activities, some counting problems seek the number of possible outcomes when the outcome selected is from one activity *or* another. We can also state this as a selection from one of two disjoint sets. This type of problem makes a selection from either set A or from set B, but not from both, where A and B are disjoint sets. It uses a variation of the *Addition Rule* (Section 6.3).

**Addition Rule II**    The number of ways that a selection can be made from just one of two disjoint sets, $A$ and $B$, is the number of ways that the selection can be made from $A$ plus the number of ways the selection can be made from $B$.

**Example 10**    How many different committees can be selected from 8 men and 10 women for a committee composed of 3 men *or* 3 women?

**Solution**
For a moment, think of listing all possible selections of a committee. The list has two parts, a list of committees composed of 3 women and a list of committees composed of 3 men. The total number of possible committees can be obtained by adding the number of all-female to the number of all-male committees. We get each of these by the following:

$$\begin{aligned}
\text{Number of all-female committees} &= C(10, 3) = 120 \\
\text{Number of all-male committees} &= C(8, 3) = \underline{\phantom{1}56} \\
\text{Total number of committees} &= 176
\end{aligned}$$

    Do not confuse this problem with the number of ways in which a committee of three men *and* a committee of three women can be chosen. That calls for the selection of a *pair* of committees, one from *each* of two disjoint sets.
    This example calls for the selection of *one* committee from *one* of two disjoint sets. So, this example uses the Addition Rule.

**Example 11**    One freshman, 3 sophomores, 4 juniors, and 6 seniors apply for 5 positions on an Honor Council. If the council must have at least 2 seniors, in how many different ways can the council be selected?

Solution

The council has at least 2 seniors when it has 2, 3, 4, or 5 seniors. Because this situation asks for the number of ways one event or another event can occur, we need to compute the number of ways each event can occur and then *add* them. We must compute the number of councils possible with two seniors, with three seniors, and so on, and add:

| | |
|---|---|
| 2 seniors and 3 others: | $C(6, 2) \times C(8, 3) = 15 \times 56 = 840$ |
| 3 seniors and 2 others: | $C(6, 3) \times C(8, 2) = 20 \times 28 = 560$ |
| 4 seniors and 1 other: | $C(6, 4) \times C(8, 1) = 15 \times 8 = 120$ |
| 5 seniors: | $C(6, 5) = 6$ |

The total is $840 + 560 + 120 + 6 = 1526$.

**Example 12**    The Huck Manufacturing firm forms a six-person advisory committee. The committee is composed of a chair, vice-chair, and secretary from the administrative staff and three members from the plant workers. Seven members from the administrative staff and eight plant workers are eligible for the committee positions. In how many different ways can the committee be formed?

Solution

At the first level, we view this as a Multiplication Rule problem because two activities are involved: selecting officers from the administrative staff and selecting committee members from the plant workers. We compute the number of ways in which each can occur and then we multiply. The selection of officers is a permutation because repetitions are not allowed (a person may not hold two offices) and the different offices impose an order. The number of slates of officers is $P(7, 3)$. The selection of committee members from the plant workers is a combination because no distinction is made between those positions, and repetitions are not allowed. The number of selections is $C(8, 3)$.

The total number of ways in which the administrative committee can be selected is $P(7, 3) \times C(8, 3) = 11{,}760$.

## Binomial Theorem

Perhaps you remember from your high school algebra that

$$(x + y)^2 = x^2 + 2xy + y^2$$

You are less likely to remember that

$$(x + y)^3 = x^3 + 3x^2y + 3xy^2 + y^3$$

and few remember that

$$(x + y)^4 = x^4 + 4x^3y + 6x^2y^2 + 4xy^3 + y^4$$

At this point, you may be asking why we discuss the expansion of $(x + y)^n$ in a section on counting combinations. We do so because we can use combinations to find the expansion of, say, $(x + y)^5$ without memorizing the coefficients and without multiplying

$$(x + y)(x + y)(x + y)(x + y)(x + y)$$

To see how to do this, let's make some helpful observations. In any term of $(x + y)^4$, the exponents on $x$ and $y$ add to 4. In any term of $(x + y)^3$, the exponents on $x$ and $y$ add to 3. In general, in any term of $(x + y)^n$, the exponents on $x$ and $y$ add to $n$.

In the expansion $(x + y)^4 = x^4 + 4x^3y + 6x^2y^2 + 4xy^3 + y^4$, the coefficients 1, 4, 6, 4, and 1 are equal, respectively, to

$$1 = C(4, 0)$$
$$4 = C(4, 1)$$
$$6 = C(4, 2)$$
$$4 = C(4, 3)$$
$$1 = C(4, 4)$$

In another example, $(x + y)^3 = x^3 + 3x^2y + 3xy^2 + y^3$, the coefficient of $x^2y$ is $C(3, 1)$, the coefficient of $x^8y^5$ in $(x + y)^{13}$ is $C(13, 5)$, and in general, the coefficient of $x^{n-4}y^4$ in $(x + y)^n$ is $C(n, 4)$. Even more general, the coefficient of $x^{n-k}y^k$ in $(x + y)^n$ is $C(n, k)$. This enables us to write $(x + y)^n$ in the following way using what we call the **Binomial Theorem**.

---

**Binomial Theorem**    Expansion of $(x + y)^n$:

$$(x + y)^n = C(n, 0)x^n + C(n, 1)x^{n-1}y + C(n, 2)x^{n-2}y^2 + \cdots$$
$$+ C(n, k)x^{n-k}y^k + \cdots + C(n, n-1)xy^{n-1} + C(n, n)y^n$$

---

**Example 13**

$$(x + y)^6 = C(6, 0)x^6 + C(6, 1)x^5y + C(6, 2)x^4y^2 + C(6, 3)x^3y^3$$
$$+ C(6, 4)x^2y^4 + C(6, 5)xy^5 + C(6, 6)y^6$$
$$= x^6 + 6x^5y + 15x^4y^2 + 20x^3y^3 + 15x^2y^4 + 6xy^5 + y^6$$

**NOTE**

We refer to this as the Binomial Theorem because an expression of two terms, $x + y$, is called a binomial.

**Application to Subsets**    We can apply the Binomial Theorem to the counting of all subsets of a set. First, we observe that we can write $2^n = (1 + 1)^n$, so we then can write, for example,

$$2^4 = (1 + 1)^4 = C(4, 0)(1)^4 + C(4, 1)(1)^3(1)$$
$$+ C(4, 2)(1)^2(1)^2 + C(4, 3)(1)(1)^3 + C(4, 4)(1)^4$$

Because $1 = 1^2 = 1^3 = \cdots$,

$$2^4 = (1 + 1)^4 = C(4, 0) + C(4, 1) + C(4, 2) + C(4, 3) + C(4, 4)$$

We recognize the terms in the sum above as

A set of 4 elements has $C(4, 0)$ subsets with 0 elements.

A set of 4 elements has $C(4, 1)$ subsets with 1 element.

A set of 4 elements has $C(4, 2)$ subsets with 2 elements.

A set of 4 elements has $C(4, 3)$ subsets with 3 elements.

A set of 4 elements has $C(4, 4)$ subsets with 4 elements.

Thus, the sum of $(1 + 1)^4$ gives the number of all subsets of a 4-element set, namely, $2^4 = 16$.

The example of the number of subsets of a 4-element set illustrates a more general principle.

| Number of Subsets of a Set | The number of all subsets of a set with $n$ elements is $2^n$. |
|---|---|

**Example 14**    In how many ways can Suzie invite two or more of her five friends to her birthday party?

**Solution**

This problem essentially asks for the number of subsets with two or more elements that can be formed from a five-element set. The number of subsets with two or more elements is the *number of all subsets minus the number of subsets with zero or one element*, which is:

$$\text{Number with two or more elements} = 2^5 - C(5, 0) - C(5, 1)$$
$$= 2^5 - 1 - 5 = 32 - 6 = 26$$

Two or more friends can be selected in 26 ways.

We conclude this section with an explanation of why, for example, $C(6, 4)$ gives the correct coefficient of $x^2 y^4$ in the expansion of $(x + y)^6$. First, we write $(x + y)^6$ as

$$(x + y)(x + y)(x + y)(x + y)(x + y)(x + y)$$

Now, for the moment, think of each factor, $x + y$, as a pair of two elements, $(x, y)$. We have six such pairs from the product.

When you obtain $(x + y)^6$ by multiplying the factors of

$$(x + y)(x + y)(x + y)(x + y)(x + y)(x + y)$$

different people may vary some in the way they find the product, but all will essentially take either an $x$ or a $y$ from each of the six factors and multiply. This is done in all possible ways and the terms added. If a $y$ is taken from four of the factors and an $x$ from the other two, then $x^2 y^4$ is obtained. In the process, this term is obtained several times. We want to determine exactly what we mean by "several times." We have six pairs $(x, y)$, and we select four of them from which to take a $y$; then an $x$ is automatically taken from the other two. We can select the four pairs from the six pairs in $C(6, 4)$ ways. Thus, when we add all the $x^2 y^4$ terms, we will have $C(6, 4)x^2 y^4$. If we use an integer $n$ instead of 6 and an integer $k$ instead of 4, a similar argument implies that the coefficient of $x^{n-k} y^k$ in $(x + y)^n$ is $C(n, k)$.

## 6.5   EXERCISES

Access end-of-section exercises online at **www.webassign.net**    Web**Assign**

### Using Your TI Graphing Calculator

#### Combinations

Using the **nCr** command, the number of combinations—say, C(7, 4)—can be calculated on the TI Graphing Calculator as follows. First, enter 7; then select **nCr,** using

MATH <PRB> <3:nCr>
which gives the screen

Press ENTER,
then enter 4:

Press ENTER,
which gives 35:

```
MATH NUM CPX PRB
1:rand
2:nPr
3:nCr
4:!
5:randInt(
6:randNorm(
7:randBin(
```

```
7 nCr 4
```

```
7 nCr 4
              35
```

#### Exercises

Calculate the following:

**1.** $C(14, 6)$     **2.** $C(33, 20)$     **3.** $C(16, 9)C(10, 5)$     **4.** $C(54, 5)$     **5.** $\dfrac{C(23, 18)}{C(11, 5)}$

### Using Excel

#### Combinations

The number of combinations, say $C(7, 4)$, can be calculated in Excel using the COMBIN function, written as COMBIN(7,4).

The formula = COMBIN(A2,A3) will calculate the number of combinations using the numbers in A2 and A3.

#### Exercises

**1.** Find $C(6, 3)$.     **2.** Find $C(12, 7)$.     **3.** Find $C(50, 18)$.

# 6.6    A MIXTURE OF COUNTING PROBLEMS

We have studied four kinds of counting techniques: the Multiplication Rule, the Addition Rule, permutations (a special case of the Multiplication Rule), and combinations. One of the more difficult steps of counting problems is the determination of the appropriate counting technique to be used.

Usually, you expect to solve the exercises in the permutation section using the permutation counting technique. Because we want you to be able to analyze a problem and determine the counting technique, or techniques, to be used, this section has a mixture of problems for which you determine the technique to be used.

We offer a sequence of questions that should help you determine the appropriate counting technique. The basic questions to ask are

"How many sets are selections made from?"

"Is repetition of elements allowed?"

"Is the order of selection or arrangement significant?"

"Are elements selected from each set or is the selection made from one of several sets?"

Figure 6–22 shows how the answers to these questions lead to the appropriate counting technique. Let's apply it to some examples.

In some problems you will need to apply the procedure more than once, as in Examples 3 through 6.

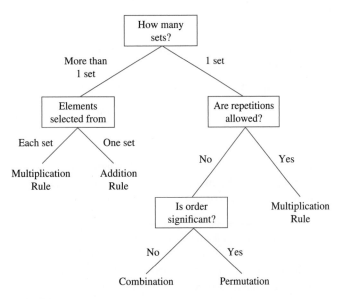

**FIGURE 6–22**    Procedure to determine the appropriate counting technique.

**Example 1**    Billy's mother allows him to select three Dr. Seuss books from eight such books in the Children's Library. How many ways can Billy make the selection?

**Solution**
To determine the counting technique, start at the top of the diagram in Figure 6–22 and take the path indicated by your answer. At each junction, continue according to your answer. The end of a path indicates the appropriate counting technique.
Here are the questions to ask, and the answers that lead us through the diagram:

1. How many sets are involved? One set of eight books.
2. Are repetitions allowed? No, we expect Billy to select three different books.
3. Is order significant? No, the order of selection is irrelevant.

The tree diagram indicates this is a combination problem with a selection of three objects from a set of eight. The number of possible selections is

$$C(8, 3) = \frac{8 \times 7 \times 6}{3 \times 2 \times 1} = 56$$

**Example 2**    Billy is allowed to select one book from eight Dr. Seuss books and one book from six "What If . . . ?" books. How many ways can he make the selection?

**Solution**
We ask the questions:

"How many sets are involved?" Two, the Dr. Seuss books and the "What If . . . ?" books.
"Is a selection made from each set?" Yes, so this is a Multiplication Rule problem.

Billy can make the selection in $8 \times 6 = 48$ ways.

Some problems are multilevel problems, where a determination of the appropriate counting technique must be made at each level, so you may need to use the tree diagram in Figure 6–22 more than once.

**Example 3**    The Gamma Club has a membership of 10 freshmen, 15 sophomores, 22 juniors, and 12 seniors. The club has an executive group consisting of a chair, vice-chair, and secretary, all of whom must be seniors, and a three-member committee composed of juniors. How many different ways can the executive group be formed?

**Solution**
The answer to the question "How many sets are involved?" is two because, for the purposes of this problem, one kind of selection is made from the group of juniors and another kind from the seniors so the set of juniors and the set of seniors are the only relevant sets.

Because selections are made from each of the two sets, this is a Multiplication Rule problem with

Number of ways the executive group can be selected
$$= \text{(number of ways the officers can be selected)}$$
$$\times \text{(number of ways the committee of juniors can be selected)}$$

This is a multilevel problem, because we still must determine the number of ways the committee and the officers can be selected.

The officers come from one set, the seniors. We have no repetitions, because one person does not hold two offices, and order is significant because when two people exchange offices, a different slate is formed. The number of slates equals $P(12, 3) = 1320$.

For the committee, the selections come from one set, the juniors. There are no repetitions because one person cannot hold two committee positions, and order is not significant because no distinction is made in committee positions. This is a combination problem with the number of possible committee formations equaling $C(22, 3) = 1540$.

Finally, the executive group can be formed in $1320 \times 1540 = 2,032,800$ different ways.

**Example 4**  The Mathematics Department hosts a high school mathematics contest. Professors Tidmore, Johns, and Cannon have volunteered students in their classes to help with the contest. The coordinator of the contest asks Prof. Tidmore to enlist four students, two female and two male, to serve as a welcoming committee. Prof. Johns is asked to enlist five students to host the lounge where visiting teachers and parents will wait, and Prof. Cannon is asked to provide three students who will help grade: one to grade algebra, one to grade geometry, and one to grade calculator problems. The composition of the classes is

*Tidmore's class:*  13 males and 15 females

*Johns's class:*  7 males and 5 females

*Cannon's class:*  9 females and 9 males

In how many ways can the requested students be selected?

**Solution**

This is a multilevel selection process, so you may use the decision tree in Figure 6–22 at each level.

Students are selected from each of three sets—Tidmore's class, Johns's class, and Cannon's class—so at the first level, this is a Multiplication Rule problem.

We now need to determine the number of ways each selection can be made.

*Tidmore's class:*  Two males are selected from the 13 males, and 2 females are selected from the 15 females, so the Multiplication Rule applies. The selection of 2 males is a combination, $C(13, 2) = 78$. The selection of 2 females is a combination, $C(15, 2) = 105$. The 4 students can be selected in $78 \times 105 = 8190$ ways.

*Johns's class:*  The 5 students are selected from the 12 students in the class, and there is no assignment of position or order, so this is a combination, $C(12, 5) = 792$ ways.

*Cannon's class:*  The 3 students are selected from the 18 students in the class, and they are assigned a position, or activity, so this is a permutation, $P(18, 3) = 4896$ ways.

Finally, the 12 students can be selected in

$$8190 \times 792 \times 4896 = 31{,}757{,}806{,}080 \text{ ways}$$

Some problems can be correctly worked in more than one way. Sometimes we want just some of all possible outcomes. The outcomes can be classified as those we want and those we don't want. Sometimes it is easier to determine the number we don't want. We use that number to determine the number we do want, as illustrated in the following example.

**Example 5**    The PIN number used at an ATM is a sequence of six digits, using the digits 0 through 9 with repetitions allowed.

(a) How many PIN numbers are possible?

(b) How many PIN numbers are possible with no repetition of digits?

(c) How many PIN numbers are possible with at least one digit repeated?

**Solution**

(a) Each digit can be selected in 10 ways, so the number of PIN numbers is $10 \times 10 \times 10 \times 10 \times 10 \times 10 = 1{,}000{,}000$.

(b) Since digits are not repeated and the sequence of digits is significant, this is a permutation with $P(10, 6) = 151{,}200$.

(c) The PIN numbers with at least one digit repeated are all possible PIN numbers *except* those with *no* repetition. Thus, we subtract the number with no repetition from the number of all possible to obtain the number with at least one repeated digit:

$$\begin{aligned} \text{Number with at least one repeat} &= 1{,}000{,}000 - 151{,}200 \\ &= 848{,}800 \end{aligned}$$

Recall the Addition Rule that states when making selections from one of two or more disjoint sets, the number of possible selections is the sum of the number of selections from each set.

**Example 6**    The PTA program committee decides to ask for 4 students to participate in the next PTA program. They want all 4 of the students from either Mr. Dudley's class of 14 students, from Miss DeWeese's class of 17 students, or from Mrs. Bowden's class of 15 students. How many selections are possible?

**Solution**
One group of 4 students is selected from 1 of the 3 classes, so the Addition Rule applies. The number of selections is $C(14, 4) + C(17, 4) + C(15, 4) = 1001 + 2380 + 1365 = 4746$.

## 6.6    EXERCISES

Access end-of-section exercises online at **www.webassign.net**

## 6.7    PARTITIONS (OPTIONAL)

- Ordered Partitions
- Number of Ordered Partitions
- Special Case: Partition into Two Subsets

- Unordered Partitions
- Number of Unordered Partitions

In this section we discuss an idea called the **partitioning** of a set. We want to determine the number of ways a set can be partitioned. We will look at two kinds of partitions, **ordered** and **unordered** partitions. Let's begin with an example to lead us into the ideas.

**Example 1**    A group of 15 students is to be divided into 3 groups to be transported to a game. The three vehicles will carry 4, 5, and 6 students, respectively. In how many different ways can the three groups be formed?

**Solution**
Select the 4 students that ride in the first vehicle. This can be done in

$$C(15, 4) = \frac{15!}{4!\ 11!}$$

different ways. (Note the form we use for $C(15, 4)$. It is more useful in this case.) After this selection, 5 students may be selected for the second vehicle in

$$C(11, 5) = \frac{11!}{5!\ 6!}$$

different ways. (There are 11 students left after the first vehicle is filled.) There are 6 students left for the last vehicle, and they can be chosen in

$$C(6, 6) = \frac{6!}{6!\ 0!}$$

different ways.

By the Multiplication Rule, the total number of different ways is

$$C(15, 4) \times C(11, 5) \times C(6, 6) = \frac{15!}{4!\ 11!} \times \frac{11!}{5!\ 6!} \times \frac{6!}{6!\ 0!}$$

$$= \frac{15!}{4!\ 5!\ 6!} = 630{,}630$$

This partition problem has the following properties that make it a partition.

1. The set is divided into disjoint subsets (no two subsets intersect).
2. Each member of the set is in one of the subsets.

The following is a more formal definition of a partition.

---

**DEFINITION**
**Partition**

A set $S$ is **partitioned** into $k$ nonempty subsets $A_1, A_2, \ldots, A_k$ if:

1. Every pair of subsets is disjoint: that is, $A_i \cap A_j = \varnothing$ when $i \neq j$.
2. $A_1 \cup A_2 \cup \cdots \cup A_k = S$.

---

## Ordered Partitions

We first discuss ordered partitions.

---

**THEOREM**
**Ordered Partition**

A partition is **ordered** if different subsets of the partition have characteristics that distinguishes one from the other.

---

The characteristics that distinguish subsets may vary widely. For example, one subset may be males, another females; one subset may be A students, another C students; one subset is awarded a million dollar contract, another a $1000 contract; one subset is the first team, another the second team, another the third team; one subset contains ten elements, another eight elements.

A partition is ordered if an exchange of two subsets gives a different partition. For example, let 15 basketball players be divided into three teams of five players each, and let the teams be designated as first, second, and third teams. If the division of players is left unchanged, but the first team is now designated as the second team and the second team becomes the first team, then a different partition is obtained. Thus, we have an ordered partition.

## Number of Ordered Partitions

We now determine the number of ways in which a set can be partitioned.

From Example 1, we see that the number of ways in which a set of 15 elements can be partitioned into subsets of four, five, and six elements may be expressed as

$$\frac{15!}{4!\,5!\,6!}$$

A commonly used notation for this quantity is

$$\binom{15}{4,\,5,\,6}$$

This is generalized in the following theorem.

---

**THEOREM**
**Ordered Partitions**

A set with $n$ elements can be partitioned into $k$ ordered subsets of $r_1, r_2, \ldots, r_k$ elements $(r_1 + r_2 + \cdots + r_k = n)$ in the following number of ways:

$$\binom{n}{r_1, r_2, \ldots, r_k} = \frac{n!}{r_1!\,r_2! \ldots r_k!}$$

---

**Example 2**   A set of 12 people ($n = 12$) can be divided into three groups of three, four, and five ($r_1$, $r_2$, and $r_3$) in

$$\binom{12}{3, 4, 5} = \frac{12!}{3!\ 4!\ 5!} = 27{,}720$$

different ways.

**Example 3**   The United Way Allocations Committee has 14 members. In how many ways can they be divided into the following subcommittees so that no member serves on 2 subcommittees?

      Scouting subcommittee: 2 members

      Salvation Army subcommittee: 4 members

      Health Services subcommittee: 5 members

      Summer Recreational Program subcommittee: 3 members.

**Solution**

The subcommittees form a partition, because no one is on 2 subcommittees and all 14 members are used. The partitions are ordered for 2 reasons: The subcommittees are of different sizes, and they have different functions. The number of partitions is

$$\binom{14}{2, 4, 5, 3} = \frac{14!}{2!\ 4!\ 5!\ 3!} = 2{,}522{,}520$$

**Example 4**   A college basketball squad has 15 players. In how many ways can the coach form a first, second, and third team of 5 players each?

**Solution**

We have an ordered partition because of a distinction between teams. The number of partitions is

$$\binom{15}{5, 5, 5} = \frac{15!}{5!\ 5!\ 5!} = 756{,}756$$

## Special Case: Partition into Two Subsets

Let's look at a special case of partitions. Suppose a set of 8 objects is partitioned into two subsets of 3 and 5 objects. The formula for partitions gives

$$\binom{8}{3, 5} = \frac{8!}{3!\ 5!}$$

Note that the formulas for $C(8, 3)$ and $C(8, 5)$ both give

$$C(8, 3) = \frac{8!}{5!\,3!} = C(8, 5)$$

so the number of partitions into two subsets is just the number of ways in which a subset of one size can be selected. This result occurs because when one subset of 3 objects is selected, the remaining 5 objects automatically form the other subset in the partition.

In general the following is true:

> The number of partitions of a set into two ordered subsets is the number of ways in which one of the subsets can be formed.

## Unordered Partitions

We now look at partitions that are not ordered.

**DEFINITION**
**Unordered Partition**

A partition is **unordered** when no distinction is made between subsets.

For a partition to be unordered, all subsets must be the same size, otherwise, the different sizes would distinguish between subsets. When a teacher partitions a class into four equal groups, all groups working on the same problem, an **unordered partition** has been formed. If the four equal groups work on different problems, the partition is **ordered**. If eight members of a traveling squad are paired to room together on the trip, an unordered partition is formed. If the pairs are assigned to rooms 516, 517, 518, and 519, an ordered partition is formed.

## Number of Unordered Partitions

A basketball squad of 15 members can be divided into first, second, and third teams of five players each in $\dfrac{15!}{5!\,5!\,5!}$ ways. Because a distinction is made between teams, this is an ordered partition. We ask in how many ways an unordered partition can be made; that is, no distinction is made between teams. We can find the number by relating the number of ordered and unordered partitions.

First, divide the 15 players into 3 teams of 5 each, with no distinction made between teams. Call these teams A, B, and C. These teams can be ordered into first, second, and third teams in six ways: ABC, ACB, BAC, BCA, CAB, and CBA. You recognize this as the 3! permutations of the 3 groups. In general, the ordered partitions can be obtained by forming 3 groups (an unordered partition) and then arranging them in 3! ways. If we let $N$ be the number of unordered partitions, then

$$3!N = \text{number of ordered partitions} = \binom{15}{5, 5, 5}$$

This gives

$$N = \frac{1}{3!}\binom{15}{5, 5, 5} = \frac{15!}{3!\,5!\,5!\,5!}$$

This generalizes to the following theorem:

**THEOREM** A set of $n$ elements can be partitioned into $k$ **unordered subsets** of $r$ elements each ($kr = n$) in the following number of ways:

$$\frac{1}{k!}\binom{n}{r, r, \ldots, r} = \frac{n!}{k!\, r!\, r! \ldots r!} = \frac{n!}{k!(r!)^k}$$

**Example 5** A set of 12 elements can be partitioned into 3 unordered subsets of 4 each in

$$\frac{12!}{3!\,4!\,4!\,4!} = 5775 \text{ ways}$$

We now show an example of partitioning a set with no distinction made between some subsets and a distinction made between others.

**Example 6** Find the number of partitions of a set of 12 elements into subsets of three, three, four, and two elements. No distinction is to be made between subsets except for their size.

Solution
Because the two subsets of three elements are the same size, no distinction is made between them. Because they are of different sizes, a distinction is made between subsets of size 2 and 4 (or 2 and 3). The number of ordered partitions is $\dfrac{12!}{3!\,3!\,4!\,2!}$. The number of unordered partitions is found by dividing by 2! because two sets (of size 3) are indistinct. Thus there are $\dfrac{12!}{2!\,3!\,3!\,4!\,2!}$ unordered partitions.

**Example 7** Find the number of unordered partitions of a set of 23 elements that is partitioned into 2 subsets of 4 elements and 3 subsets of 5 elements.

Solution
Because there are 2 indistinct subsets of 4 elements and 3 indistinct subsets of 5 elements, we divide the number of *ordered* subsets by 2! and 3! to obtain $\dfrac{23!}{2!\,3!\,4!\,4!\,5!\,5!\,5!}$.

In general, the number of unordered partitions is given by the following theorem:

**THEOREM** A set of $n$ elements is partitioned into unordered subsets with $k$ subsets of $r$ elements each and $j$ subsets of $t$ each ($kr + jt = n$). The number of such partitions is

$$\frac{\text{number of ordered partitions}}{k!\,j!} = \frac{n!}{k!\,j!(r!)^k(t!)^j}$$

## 6.7 EXERCISES

Access end-of-section exercises online at **www.webassign.net**

## IMPORTANT TERMS

**6.1**
Set
Element of a Set
Set-Builder Notation
Equal Sets
Empty Set
Universe
    (Universal Set)
Venn Diagram
Subset
Proper Subset
Union
Intersection

Complement
Disjoint Sets

**6.2**
$n(A)$
Inclusion–Exclusion
    Principle

**6.3**
Tree Diagram
Multiplication Rule
Addition Rule

**6.4**
Permutation
Factorial
Permutation with Identical Objects

**6.5**
Combination

**6.7**
Partition
Ordered Partition
Unordered Partition

## IMPORTANT CONCEPTS

| | |
|---|---|
| **Venn Diagram** | A pictorial representation of a universe (a rectangle) and sets (circles) in the universe. |
| **Union of Two Sets** | The set, $A \cup B$, whose elements are from $A$, or $B$, or both. |
| **Intersection of Two Sets** | The set, $A \cap B$, whose elements are those in both $A$ and $B$. |
| **Inclusion-Exclusion Principle** | $n(A \cup B) = n(A) + n(B) - n(A \cap B)$ |
| **Multiplication Rule** | Two activities $A_1$ and $A_2$ can be performed in $n_1$ and $n_2$ ways, respectively. The number of ways in which $A_1$ followed by $A_2$ can be performed is $n_1 \times n_2$. |
| **Addition Rule** | Two activities $A_1$ and $A_2$ can be performed in $n_1$ and $n_2$ ways, respectively. The number of ways in which either $A_1$ or $A_2$, but not both simultaneously, can be performed is $n_1 + n_2$. |

| Permutation | A subset of distinct elements selected from a given set and arranged in a specific order. |
|---|---|
| Combination | A subset of distinct elements chosen from a given set without regard to their arrangement. |
| Partition of a Set $S$ | $S$ is divided into $k$ subsets $A_1, A_2, A_3, ... A_k$, where every pair of subsets is disjoint and $A_1 \cup A_2 \cup A_3 \cup ... \cup A_k = S$. |
| Ordered Partition | Different subsets in the partition have characterstics that distinguish one from the other. |
| Unordered Partition | No distinction is made between subsets of the partition. |

## REVIEW EXERCISES

**1.** Let $A = \{6, 10, 15, 21, 30\}$, $B = \{6, 12, 24, 48\}$, $C = \{x \mid x$ is an integer divisible by 3$\}$. Identify the following as true or false.

(a) $21 \in A$      (b) $21 \in B$

(c) $25 \in C$      (d) $30 \notin A$

(e) $16 \notin B$      (f) $24 \notin C$

(g) $6 \in A \cap B \cap C$      (h) $12 \in A \cap B$

(i) $10 \in A \cup B$      (j) $A \subseteq B$

(k) $B \subseteq C$      (l) $C \subseteq A$

(m) $\varnothing \subseteq B$      (n) $A \subseteq C$

(o) $A$ and $B$ are disjoint

**2.** Let the universe set $U = \{-2, -1, 0, 1, 2, 3, 4\}$, $A = \{-2, 0, 2, 4\}$, $B = \{-2, -1, 1, 2\}$. Find the following.

(a) $A'$      (b) $B'$

(c) $(A \cap B)'$      (d) $A' \cap B'$

(e) $A' \cup B'$      (f) $A \cup A'$

**3.** Which of the following pairs of sets are equal?

(a) $A = \{x \mid x$ is a digit in the number 25102351$\}$
$B = \{x \mid x$ is a digit in the number 5111023$\}$

(b) $A = \{x \mid x$ is a letter in the word PATTERN$\}$
$B = \{x \mid x$ is a letter in the word REPEAT$\}$

(c) $A = \{2, 4, 9, 8\} \cap \{6, 7, 20, 22, 23\}$
$B = \{x \mid x$ is a letter in both words STRESS and HAPPY$\}$

**4.** $n(A) = 27$, $n(B) = 30$, and $n(A \cap B) = 8$. Find $n(A \cup B)$.

**5.** $n(A \cup B) = 58$, $n(A) = 32$, and $n(B) = 40$. Find $n(A \cap B)$.

**6.** $A$ and $B$ are sets in a universe $U$ with $n(U) = 42$, $n(A) = 15$, $n(B) = 24$, and $n(A \cup B)' = 8$. Find $n(A \cup B)$ and $n(A \cap B)$.

**7.** Draw a tree diagram showing the ways in which you can select a meat and then a vegetable from roast, fish, chicken, peas, beans, and squash.

**8.** The freshman class traditionally guards the school mascot the night before homecoming. There are five key locations where a freshman is posted. Nine freshmen volunteer for the 2:00 A.M. assignment. In how many different ways can they be assigned?

**9.** How many different license plates can be made using four digits followed by two letters

(a) If repetitions of digits and letters are allowed?

(b) If repetitions are not allowed?

**10.** Strecker Museum has a display case with four display compartments. Eight antique vases are available for display. How many ways can the display be arranged with one vase in each compartment?

**11.** A medical research team selects 5 patients at random from a group of 15 patients for special treatment. In how many different ways can the patients be selected?

**12.** In how many ways can Andrew invite one or more of his four friends to come to his house to play?

**13.** One student representative is selected from each of 4 clubs. In how many different ways can the 4 students be selected, given the following number of members in each club: Rodeo Club, 40 members; Kite Club, 27 members; Frisbee Club, 85 members; and Canoeing Club, 34 members.

**14.** In the finale of the university sing, there are 10 people in the first row. Club A has 3 members on the left end, club B has 4 members in the center, and Club C has 3 members on the right end. In how many different ways can the line be arranged?

15. A program consists of four musical numbers and three speeches. In how many ways can the program be arranged so that it begins and ends with a musical number?

16. Students take four exams in Sociology 101. On each exam the possible grades are A, B, C, D, and F. How many sequences of grades can a student receive?

17. An advertising agency designs 11 full-page ads for Uncle Dan's Barbecue. In how many ways can 1 ad be selected for each of 3 different magazines

    (a) If the 3 ads are different?

    (b) If the ads need not be different?

18. A computer password is composed of six alphabetic characters. How many different passwords are possible?

19. In how many different ways can a chairman, a secretary, and 4 other committee members be formed from a group of 10 people?

20. The KOT club has 12 pledges. On a club workday, 4 pledges are assigned to the Red Cross, 6 are assigned to the Salvation Army, and 2 are not assigned. In how many ways can the groups be selected?

21. A survey of 60 people gave the following information:

    25 jog regularly.

    26 ride a bicycle regularly.

    26 swim regularly.

    10 both jog and swim.

    6 both swim and ride a bicycle.

    7 both jog and ride a bicycle.

    1 does all three.

    3 do none of the three.

    Show that there is an error in this information.

22. The Spirit Shop had a sale on records, books, and T-shirts. A cashier observed the purchases of 38 people and found that

    16 bought records.

    15 bought books.

    19 bought T-shirts.

    5 bought books and records.

7 bought books and T-shirts.

6 bought records and T-shirts.

3 bought all three.

(a) How many bought records and T-shirts but no books?

(b) How many bought records but no books?

(c) How many bought T-shirts but no books and no records?

(d) How many bought none of the three?

23. A poll was conducted among a group of teenagers to see how many have televisions, radios, and computers. The results were as follows: T denotes television, R denotes radio, and C denotes computer.

| Item | Number of teenagers having this item |
|------|--------------------------------------|
| T | 39 |
| R | 73 |
| C | 10 |
| T and R | 22 |
| C and R | 3 |
| T and C | 4 |
| T and R and C | 2 |

Determine the following.

(a) How many had a radio and TV but no computer?

(b) How many had a computer and had no TV?

(c) How many had exactly two of the three items?

24. During the summer, 110 students toured Europe. Their language skills were as follows: 46 spoke German, 56 spoke French, 8 spoke Italian, 16 spoke French and German, 3 spoke French and Italian, 2 spoke German and Italian, and 1 spoke all three.

(a) How many spoke only French?

(b) How many spoke French or German?

(c) How many spoke French or Italian but not both?

(d) How many spoke none of the languages?

25. Mrs. Bass has 5 bracelets, 8 necklaces, and 7 sets of earrings. In how many ways can she select 1 of each to wear?

**26.** The Labor Day Raft Race has 110 entries. In how many ways is it possible to award prizes for the fastest raft, the slowest raft, and the most original raft?

**27.** From a group of five people, two are to be selected to be delegates to a conference. How many selections are possible?

**28.** In how many different ways can a group of 15 people select a president, vice-president, and secretary?

**29.** Twenty people attend a meeting at which three different door prizes are awarded by drawing names.

   **(a)** If a name is drawn and replaced for the next drawing, in how many ways can the door prizes be awarded?

   **(b)** If a name is drawn and not replaced, in how many ways can the door prizes be awarded?

**30.** A bag contains 6 white balls, 4 red balls, and 3 green balls. In how many ways can a person draw out 2 white balls, 3 red balls, and 2 green balls?

**31.** An Honor Council consists of 4 seniors, 4 juniors, 3 sophomores, and 1 freshman. Fifteen seniors, 20 juniors, 25 sophomores, and 11 freshmen apply. In how many ways can the Honor Council be selected? Leave your answer in symbolic form.

**32.** An art gallery has 8 oil paintings and 4 watercolors. A display of 5 oil paintings and 2 watercolors arranged in a row is planned. How many different displays are possible with a watercolor at each end and the oils in the center?

**33.** A club agrees to provide five students to work at the school carnival. One sells balloons, one sells popcorn, one sells cotton candy, one sells candied apples, and one sells soft drinks. Nine students agree to help. In how many ways can the assignments be made?

**34.** Prof. Goode gives a reading list of six books. A student is to read three. In how many ways can the selection be made?

**35.** Five students are to be chosen from a high school government class of 22 students to meet the governor when he visits the school. In how many ways can this be done?

**36.** How many different 5-card hands can be obtained from a deck of 52 cards?

**37.** Compute:

   **(a)** $P(8, 4)$   **(b)** $C(9, 5)$   **(c)** $P(7, 7)$   **(d)** $C(5, 5)$

   **(e)** $4!$   **(f)** $\dfrac{7!}{3!\,4!}$   **(g)** $\dfrac{8!}{4!}$

   **(h)** $\begin{pmatrix} 15 \\ 4, 5, 6 \end{pmatrix}$   **(i)** $\begin{pmatrix} 9 \\ 3, 3, 3 \end{pmatrix}$ (unordered)

**38.** One day a machine produced 50 good circuit boards and 8 defective ones.

   **(a)** In how many ways can 2 defective circuit boards be selected?

   **(b)** In how many ways can 3 good circuit boards be selected?

   **(c)** In how many ways can 2 defective and 3 good circuit boards be selected?

**39.** A club has 80 members of whom 20 are seniors, 15 are juniors, 25 are sophomores, and 20 are freshmen. A chair, vice-chair, secretary, and treasurer are to be selected. The chair and vice-chair must be seniors, the treasurer must be a junior, and the secretary must be a sophomore. How many different slates of officers can be formed?

**40.** The Campus Deli offers caffeinated and decaffeinated regular coffee, and caffeinated and decaffeinated French roast coffee. Coffee creamer is available in plain, Irish Creme, and hazelnut flavors. Sugar and two brands of sweetener are available.

   **(a)** How many ways can a student select a coffee and a creamer?

   **(b)** How many ways can a student select a decaffeinated coffee and a creamer, including the choice of no creamer?

   **(c)** How many ways can a student select a coffee, one creamer, and sugar or a sweetener?

**41.** A social organization and a service club held a joint meeting. Of the 83 people present, 46 belonged to the social organization, and 51 belonged to the service club. How many belonged to both?

**42.** The digits $\{2, 3, 4, 5, 6, 7\}$ are used to form three-digit numbers.

   **(a)** How many can be formed if repetitions are allowed?

**(b)** How many can be formed if repetitions are not allowed?

**(c)** How many larger than 500 can be formed with repetitions allowed?

**43.** The Sports Mart store has 10 sportswear outfits for display purposes. In how many ways can a group of 4 outfits be selected for display?

**44.** Nye Printing has 16 female employees and 14 male employees. How many different advisory committees consisting of 2 males and 2 females are possible?

**45.** List all the subsets of {red, white, blue}.

**46.** A panel of four is selected from eight businessmen and seated in a row behind a table. In how many different orders can they be seated?

**47.** The physics club has 10 freshmen and 8 sophomore members. At a club picnic, the cook and entertainment leader are freshmen, and the cleanup crew consists of 3 sophomores. In how many different ways can these 5 be selected?

**48.** Draw a tree diagram showing the ways that a girl and then a boy can be selected from the children Carlos, Betty, Darla, Gary, and Natasha.

**49.** A student is allowed to check out 4 books from the reserve room. All the books must come from one collection of 6 books or from another collection of 8 books. In how many different ways can the selection be made?

**50.** Three married couples are seated in a row. How many different seating arrangements are possible:

**(a)** if there is no restriction of seating order?

**(b)** if the men sit together and the women sit together?

**(c)** if a husband and wife sit together?

**51.** How many different words are possible using all the letters of

**(a)** RELAX? **(b)** PUPPY? **(c)** OFFICIAL?

**52.** Individuals and businesses who can profit from decisions made by politicians will contribute to political campaigns, sometimes to all politicians running for the same office. Financial reports for Ray Meador and Billy Bob Garner, candidates for county commissioner, showed 113 contributions to those candidates. Of those contributions, 82 were to Ray Meador and 52 were to Billy Bob. How many contributions were made to both candidates?

# PROBABILITY

We may hear a question like "What are the chances of a quiz?", "What are the chances of rain tomorrow?", or "What are the chances of winning the game tomorrow?" The answer to the third question may be something like "I don't know," "We should win," or "We will crush them." These responses don't give much information on the likely outcome of the game. On the other hand, a weather forecaster gives more helpful answers to the second question, such as "No rain tomorrow," "The chances of rain are 20%," or "We have a 70% chance of rain tomorrow." While these responses do not guarantee that it will or will not rain, you have information that helps you decide to take or leave your umbrella.

Probability theory seeks to answer questions like the second to measure the likelihood of an outcome of phenomena and events. In addition to weather forecasters, people in a variety of disciplines, such as business, industry, science, politics, and insurance, use probability in making decisions.

The weather forecaster assigns a number to the likelihood, or *probability*, of rain. In general, probability assigns a number to indicate the level of likelihood of a desired outcome. Two methods are used to determine probability: the theoretical method and the empirical method. The theoretical method begins with certain assumptions about the activity or phenomenon and uses logical reasoning to arrive at the way to determine the probability. The empirical method sets up an experiment and observes the number of times the desired outcome occurs. The number of outcomes serves to estimate a probability.

This chapter will introduce you to these methods and suggest many of their uses.

## 7.1 INTRODUCTION TO PROBABILITY

- Terminology
- Empirical Probability
- Properties of Probability
- Probability Assignments
- A Visual Model of Probability

An area of mathematics known as **probability theory** provides a measure of the likelihood of the outcome of phenomena and events. The government uses it to determine fiscal and economic policies; theoretical physicists use it to understand the nature of atomic-sized systems in quantum mechanics; and public-opinion polls, such as the Harris Poll, have their theoretical acceptability based on probability theory.

The theory of probability is said to have originated from the following gambling question: Two gamblers play for a stake that goes to the player who first wins a specified number of points. The game is interrupted before either player has won enough points to win the stake. (We don't know whether the game was raided.) How does one determine a fair division of the stakes based on the number of points won by each player at the time of the interruption?

Probability turned out to have applications far beyond interrupted gambling games. Today, insurance companies depend on probability to determine competitive and profitable rates for their policies. Quality control in manufacturing and product development decisions are based on probability, and politicians rely heavily on opinion polls. We will learn some elementary applications in this chapter.

### Terminology

We use the terms *probability, experiment, outcome,* and *trial* in our discussions. When we ask the likelihood that a tossed coin will turn up heads, we call the activity of tossing the coin an **experiment**, the result of tossing the coin an **outcome**. For this experiment, we have two possible outcomes, "heads" and "tails." With each coin toss we have a **trial**. We measure the likelihood of "heads" with a number, which we call the **probability** of heads, and we use the notation $P(\text{heads})$ to denote that number.

| | |
|---|---|
| **DEFINITION**<br>Experiment, Outcome, Trial | An activity or phenomenon under consideration is called an **experiment**. The experiment can produce a variety of observable results, called **outcomes**. We study activities that can be repeated or phenomena that can be observed a number of times. We call each observation or repetition of the experiment a **trial**. |

We can apply the terms *experiment, outcome,* and *trial* in a wide variety of ways, as illustrated in the following:

**(a)** Drawing a number out of a hat is an experiment with the number drawn as an outcome. Each draw of a number is a trial.

**(b)** A test to determine the germination of flower seeds is an experiment with "germinated" and "not germinated" as possible outcomes. Each test conducted is a trial.

**(c)** A drawing of lottery numbers is an experiment with the numbers drawn as an outcome. Each draw is a trial.

In general, experiments involve chance or **random** results. This means that the outcomes do not occur in a set pattern but vary depending on impartial chance, and the outcome cannot be determined in advance. The order in which leaves fall off a tree, the number of cars that pass a checkpoint on the freeway, and the selection of a card from a well-shuffled deck are examples of experiments that have random outcomes.

Probability deals with random outcomes that have some long-term pattern for which we cannot predict what happens next, but a prediction can be made of what will happen in the long term.

Life insurance companies know quite accurately how many people will die by a certain age. They cannot tell when a certain individual will die, but they can predict general, long-term numbers of deaths.

The determination of probabilities can be a difficult and expensive process, but we can't escape the desire to measure the likelihood, or probability, that a certain outcome occurs. What is the probability of getting a ticket if I exceed the speed limit? What are the chances of my book being stolen if I leave it on a shelf in the cafeteria? What is the probability of a walk in history class today?

The losing basketball team may intentionally foul a member of the winning team in the closing minutes of a game in hopes that the player will miss the free throw. This gives the losing team a chance of getting the ball. The winning team wants Mike to have the ball, because he has made 80% of his free throws for the season. The losing team wants to foul Art, because he has made 55% of his free throws. Based on this past history, each team assumes that Mike is more likely to make the free throw than Art.

In order to compare the likelihood of two different outcomes, we assign a number to each outcome. We consider the outcome assigned the larger number to be more likely. In the basketball game, it seems natural to use the numbers 80% and 55% as a measure of the probability of making free throws.

Let's look at one way to measure probability, called **empirical probability**.

## Empirical Probability

The empirical method estimates the probability of an outcome by observing a number of trials and counting the number of times the outcome occurs. The **relative frequency** of the outcome then denotes the probability of the outcome. We find the relative frequency of an outcome by dividing the number of occurrences of the outcome by the number of trials.

---

**Relative Frequency**    The relative frequency of an outcome of an experiment:

$$\text{Relative frequency} = \frac{\text{Number of occurrences of the outcome}}{\text{Total number of trials}}$$

---

Let's illustrate with a coin tossing experiment. Our intuition suggests that a fair coin tossed a lot of times will turn up heads half the time; that is, the relative frequency of heads = 0.5. Experience tells us that we do not expect one head and one tail every time we toss a coin twice. Could we expect heads half the time if the coin is tossed a lot of times? Let's look at four historical coin-tossing experiments with relatively large numbers of trials. These experiments were conducted by the Comte de Buffon (1701–1788),

Karl Pearson (1857–1936), and John Kerrich, a prisoner of war during World War II. Each tossed a coin many times. The results of their efforts follow:

| | **Number of tosses (trials)** | **Frequency of heads** | **Relative frequency** |
|---|---|---|---|
| Buffon | 4,040 | 2,048 | $\frac{2048}{4040} = 0.5069$ |
| Pearson | 12,000 | 6,019 | $\frac{6019}{12000} = 0.5016$ |
| Pearson | 24,000 | 12,012 | $\frac{12012}{24000} = 0.5005$ |
| Kerrich | 10,000 | 5,067 | $\frac{5067}{10000} = 0.5067$ |

In each of the experiments, we use the relative frequency to estimate the probability of heads occurring when a coin is tossed. We use the notation $P(\text{heads})$ for the probability of heads. From above we have the estimates:

For Buffon, $P(\text{heads}) = 0.5069$

For Pearson, $P(\text{heads}) = 0.5016$ and $P(\text{heads}) = 0.5005$

For Kerrich, $P(\text{heads}) = 0.5067$

Although these results fail to exactly give our intuitive value of 0.5000 for the probability of heads, they do suggest that 0.5000 is a reasonable probability. Let's put the four results in order by the number of trials.

| **Trials** | **P(Heads)** |
|---|---|
| 4,040 | 0.5069 |
| 10,000 | 0.5067 |
| 12,000 | 0.5016 |
| 24,000 | 0.5005 |

Note that as the number of trials increases, the probability is closer to 0.5000. This suggests that we can get better estimates of the probability by repeating the trials more times. As the number of trials increases, we expect to obtain better estimates of the probability. We call this concept the **law of large numbers**.

**THEOREM**
**Law of Large Numbers** As more and more trials of an experiment are repeated, the relative frequency obtained approaches the actual probability.

We give examples of observing a number of trials and using relative frequency to estimate probability.

**Example 1**  The operator of a concession stand at a park keeps a record of the kinds of drinks children buy. Her records show the following:

| Drink | Frequency |
|-------|-----------|
| Cola | 150 |
| Lemonade | 275 |
| Fruit juice | 75 |
| Total | 500 |

To estimate the probability that a child will buy a certain kind of drink, we compute the **relative frequency** of each drink. We do this by dividing the frequency of each drink by the total number of drinks.

| Drink | Frequency | Relative frequency |
|-------|-----------|--------------------|
| Cola | 150 | $\dfrac{150}{500} = 0.30$ |
| Lemonade | 275 | $\dfrac{275}{500} = 0.55$ |
| Fruit juice | $\dfrac{75}{500}$ | $\dfrac{75}{500} = 0.15$ |

This experiment has three outcomes, $S = \{$cola, lemonade, fruit juice$\}$. We use relative frequency to estimate probability. The probability a child will buy lemonade is 0.55, which we write symbolically as $P(\text{lemonade}) = 0.55$; also, $P(\text{cola}) = 0.30$ and $P(\text{fruit juice}) = 0.15$.

The determination of empirical probability requires gathering data to obtain relative frequencies. In practice, this may be a highly sophisticated operation such as a national political poll.

**Example 2**  A college has an enrollment of 1210 students. The number in each class is as shown in the following table:

| Class | Number of students |
|-------|--------------------|
| Freshman | 420 |
| Sophomore | 315 |
| Junior | 260 |
| Senior | 215 |
| Total | 1210 |

A student is selected at random. Determine the empirical probability that the student is

**(a)** a freshman.  **(b)** a sophomore.
**(c)** a junior.  **(d)** a senior.
**(e)** a freshman or sophomore.

(A **random selection** means that each student has a chance of being selected and any two students have equal chances of being selected. To select people in a random manner usually requires careful planning and methodology, often a difficult process.)

**Solution**

Estimate the probability of each as the relative frequency.

| Class | Number of students | Relative frequency |
|-------|--------------------|--------------------|
| Freshman | 420 | $\dfrac{420}{1210} = 0.35$ |
| Sophomore | 315 | $\dfrac{315}{1210} = 0.26$ |
| Junior | 260 | $\dfrac{260}{1210} = 0.21$ |
| Senior | 215 | $\dfrac{215}{1210} = 0.18$ |

This gives $P(\text{freshman}) = 0.35$, $P(\text{sophomore}) = 0.26$, $P(\text{junior}) = 0.21$, $P(\text{senior}) = 0.18$, and $P(\text{freshman or sophomore}) =$

$$\frac{420 + 315}{1210} = \frac{420}{1210} + \frac{315}{1210} = 0.35 + 0.26 = 0.61$$

## Properties of Probability

Let's observe the properties of relative frequency that correspond to properties needed for probability assignments in general. First, let's discuss the terminology we use.

An experiment need not classify outcomes in a unique way. It depends on how the results are interpreted. When a multiple-choice test of 100 questions is given, the instructor wants to know the number of correct answers given by each student. For this purpose, an outcome can be any of the numbers 0 through 100. When the tests are returned to the students, they tend to ask, "What is an A?" They are interested in the outcomes A, B, C, D, and F. Then there might be the student who only asks, "What is passing?" To that student there are just two outcomes of interest, pass and fail.

When asked "What is today?" a person may respond in several ways, such as "It is April 1," "It is Friday," "It is payday," or any one of numerous responses. Depending on the focus of the individual, the set of possible responses may be all the days in a year, all the days of the week, or the two outcomes payday and not payday.

Because the outcomes of an experiment can be classified in a variety of ways, it is important that the appropriate set of outcomes be selected and that everyone understand which set of outcomes is used. We call the set of outcomes used a **sample space**.

**DEFINITION**
**Sample Space**    A **sample space** is the set of all possible outcomes of an experiment. Each element of the sample space is called a **sample point** or **simple outcome**.

The next example illustrates some sample spaces. Note that the sample space can be defined in more than one way for the same experiment.

**Example 3**  **(a)** If the experiment is tossing a coin, the sample space is {heads, tails}.

**(b)** If the experiment is drawing a card from a bridge deck, one sample space is the set of 52 cards.

**(c)** If the experiment is drawing a number from the numbers 1 through 10, the sample space can be {1, 2, 3, 4, 5, 6, 7, 8, 9, 10}. Sometimes people are assigned a number and those assigned an odd number are placed in one group and those assigned an even number are placed in another group. In this case, the sample space of interest is {even, odd}.

**(d)** If the experiment is tossing a coin twice, a sample space is {HH, HT, TH, TT}.

We do not insist on just one correct sample space for an experiment because the situation dictates how to interpret the results. However, we do insist that a sample space conform to two properties.

---

**Properties of a Sample Space**

Let $S$ be the sample space of an experiment.

**1.** Each element in the set $S$ is an outcome of the experiment.

**2.** Each outcome of the experiment corresponds to exactly one element in $S$.

---

If a student is selected from a group of university students and the class standing of the student is the outcome of interest, then {freshman, sophomore, junior, senior} is a valid sample space. If the gender of the student is the outcome of interest, then {male, female} is a valid sample space. You can form other sample spaces using age, GPA, and so on as the outcomes of interest.

In defining the outcomes of an experiment, care must be taken that the properties of a sample space hold. In an experiment involving the GPA of students, the second property of a sample space is violated if we define the outcomes of interest as

| Outcome | GPA |
|---|---|
| Unacceptable | 0.0–0.9 |
| Marginal | 1.0–1.9 |
| Acceptable | 2.0–2.9 |
| Superior | 3.0–4.0 |

This definition provides no outcome for a GPA such as 1.95 or 2.97, so property 2 does not hold.

Property 2 is violated if we define the outcomes so that a GPA belongs to two outcomes:

| Outcome | GPA |
|---|---|
| Unacceptable | 0.0–1.0 |
| Marginal | 1.0–2.0 |
| Acceptable | 2.0–3.0 |
| Superior | 3.0–4.0 |

In this case, the GPA scores of 1.0, 2.0, and 3.0 are indicated to be in two different outcomes, so property 2 does not hold.

Property 1 would be violated if we defined the unacceptable outcome as −1.0 through −0.9, because there are no negative values of GPA.

In some instances our interest lies in a *collection* of outcomes in the sample space, not just one outcome. If I toss a coin twice, I may be interested in the likelihood that the coin will land with the same face up both times. I am interested in the subset of outcomes {HH, TT}, not just one of the possible outcomes. We call such a collection of simple outcomes an **event**. In Example 2, we might be interested in the event that a freshman or sophomore is selected.

| | |
|---|---|
| **DEFINITION**<br>Event | An **event** is a subset of a sample space.<br>An event can be a subset consisting of a single outcome. Such an event is called a **simple event**. An event also can be as much as the entire sample space. The empty set can be an event. |

We focus primarily on the probability of events that includes the probability of a simple event. An event often can be formed in more than one way from a set of simple outcomes. The next example shows some variations in events.

**Example 4**    **(a)** In the experiment of drawing a number from the numbers 1 through 10, the sample space is

$$S = \{1, 2, 3, 4, 5, 6, 7, 8, 9, 10\}$$

The event of drawing an odd number is the subset {1, 3, 5, 7, 9}. The event of drawing an even number is the subset {2, 4, 6, 8, 10}. The event of drawing a prime number is the subset {2, 3, 5, 7}.

**(b)** A teacher selects one student from a group of six students. The sample space is {Scott, Jane, Mary, Kaye, Ray, Randy}. The event of selecting a student with first initial R is {Ray, Randy}. The event of selecting a student with first initial J is {Jane}. The event of selecting a student with first initial A is the empty set.

**An Event Occurs**    An **event occurs** if the trial yields an outcome that belongs to the event set.

## Probability Assignments

To help understand the properties of a probability assignment, we look again at Examples 1 and 2 to observe some properties that hold for probability assignments generally.

We want to accomplish two goals when making a probability assignment:

1. Assign a probability, a number, to each simple outcome in the sample space that will indicate the relative frequency with which the simple outcome occurs.

2. Assign a probability to each set of outcomes, that is, to each event, that also indicates the relative frequency with which the event occurs.

In Examples 1 and 2, note the following:

- Each relative frequency can be as small as 0 or as large as 1. In general, a probability is in the interval 0 through 1.
- For a sample space, the relative frequencies add to 1. In general, the probabilities of all simple outcomes add to 1.
- The relative frequency of an event is the sum of the relative frequencies of the simple outcomes making up the event.

Although a general probability assignment need not be formed using relative frequencies, the assignment cannot be arbitrary. It must satisfy some standard conditions similar to those seen in relative frequency.

**Properties of Probability**    Let $S = \{e_1, e_2, \ldots, e_n\}$ be a sample space with simple outcomes $e_1, e_2, \ldots, e_n$.

1. Each simple outcome in a sample space, $e_i$, is assigned a probability denoted by $P(e_i)$.
2. The probability of an event $E$ is determined by the simple outcomes making up $E$. $P(E)$ is the sum of the probabilities of all simple outcomes making up $E$. For example, if $E = \{e_1, e_2, e_3\}$, then $P(E) = P(e_1) + P(e_2) + P(e_3)$.
3. Each probability is a number that is not negative and is no larger than 1; for simple events $e_i$, $0 \leq P(e_i) \leq 1$ and for each event $E$, $0 \leq P(E) \leq 1$.
4. $P(S) = 1$, that is, $P(S) =$ the sum of probabilities of all simple events in a sample space

$$P(S) = P(e_1) + P(e_2) + \cdots + P(e_n) = 1$$

The next three examples help to understand the properties of probability.

**Example 5**    As part of a class assignment, Annette, Ben, and Casie go to different classroom buildings to poll students on the number of siblings they have. They classify them into five categories: 0, 1, 2, 3, and 4 or more. They report their findings as empirical probability. Here are their reports.

Annette reported:

$$P(0) = 0.20, \quad P(1) = 0.10, \quad P(2) = 0.15$$
$$P(3) = 0.30, \quad P(4 \text{ or more}) = 0.25$$

Is this a valid probability assignment? It is, because

1. Each outcome is assigned a probability.
2. Each probability is nonnegative and not larger than 1.
3. The sum of the probabilities of all simple events is 1.

Possibly, Annette made errors in counting so her report may not be an *accurate* empirical probability, but it is *valid* because it satisfies the properties of probability.

Ben reported:

$$P(0) = 0.2, \quad P(1) = 0.2, \quad P(2) = 0.3$$
$$P(3) = 0.25, \quad P(4 \text{ or more}) = 0.4$$

Is this a valid probability assignment? It is not valid, because the sum of all probabilities is 1.35.

Casie reported:

$$P(0) = 0.3, \quad P(1) = -0.1, \quad P(2) = 0.2$$
$$P(3) = 0.3, \quad P(4 \text{ or more}) = 0.3$$

Is this a valid probability assignment? The sum of all probabilities is 1.00, but the assignment is not valid because one of the probabilities, $P(1)$, is negative.

**Example 6**    The sample space of an experiment is $\{A, B, C, D\}$, where $P(A) = 0.35$, $P(B) = 0.15$, $P(C) = 0.22$, $P(D) = 0.28$. The properties of probability enable us to find the probabilities of the following events:

$$P(\{A, B\}) = 0.35 + 0.15 = 0.50$$
$$P(\{A, C\}) = 0.35 + 0.22 = 0.57$$
$$P(\{B, C, D\}) = 0.15 + 0.22 + 0.28 = 0.65$$
$$P(\{A, B, C, D\}) = 0.35 + 0.15 + 0.22 + 0.28 = 1.00$$

**Example 7**    An ice chest at a junior high picnic contains three brands of soft drinks—Pepsi, Coke, and Dr Pepper—and there are some regular and some diet drinks of each brand. The mathematics teacher had filled the chest and counted the number of each kind of drink, so she was able to tell the students, "If you get a drink from the chest without looking, then for each kind of drink, here is the probability of its being selected."

|         | Pepsi | Coke | Dr Pepper |
|---------|-------|------|-----------|
| Regular | 0.05  | 0.15 | 0.23      |
| Diet    | 0.10  | 0.17 | 0.30      |

"For example, note the probability of selecting a regular Dr Pepper is 0.23. Using our probability notation of $P(E)$, we write it as $P$(regular Dr Pepper). Now before you get your drinks, verify that the following are correct."

**(a)** What is the probability of drawing a Coke? In this example, the event is the subset {regular Coke, diet Coke}. We ask for the probability that the drink selected is in that subset. According to the second condition in the properties of probability, the probability of selecting a Coke is $0.15 + 0.17 = 0.32$, the sum of the probabilities of the simple outcomes making up the event.

In a similar manner, we obtain the probability of the following event:

**(b)** The probability of drawing a diet drink is $0.10 + 0.17 + 0.30 = 0.57$ [$P$(diet drink) = 0.57].

**(c)** The probability of selecting a regular Pepsi or a diet Dr Pepper is $0.05 + 0.30 = 0.35$.

(d) The probability of selecting a drink that is not a Dr Pepper is $0.05 + 0.15 + 0.10 + 0.17 = 0.47$ [$P(\text{not Dr Pepper}) = 0.47$].

**Example 8** A new student asks about the chances of finding a parking place on campus. His roommate, who is taking finite mathematics, responds with the following statement: "The probability of finding a parking place at remote parking is twice the probability of finding a place in the parking garage, and the probability of finding a place near your classroom building is one half the probability of finding a place in the parking garage." Assuming that the roommate was correct, find the probability of each outcome: nearby, parking garage, or remote parking.

**Solution**
Let $P(G) = $ the probability of finding a place in the parking garage. Then, for remote parking, $P(R) = 2P(G)$, and for nearby, $P(N) = 0.5P(G)$. Because these probabilities must add to 1, we have

$$2P(G) + P(G) + 0.5P(G) = 1$$
$$3.5P(G) = 1$$
$$P(G) = \frac{1}{3.5} = 0.286 \quad \text{(rounded)}$$

From this we have $P(R) = 0.572$, $P(G) = 0.286$, and $P(N) = 0.143$. (Because we rounded to three decimal places, these add to 1.001.)

Recall that a probability is a number in the interval from 0 through 1, with 0 representing the probability that the event *cannot* occur and 1 representing the probability that the event *must* occur. If a selection is made from a sample space, it cannot come from the empty set because the empty set has no elements. Thus, the probability of the empty set is zero. The probability of the sample space, $P(S)$, is the probability that the outcome will come from the sample space. Because the selection is made from the sample space, the outcome must be in the sample space, so that makes $P(S) = 1$. We highlight these special cases.

**Reminder: Two Important Special Cases**

1. If an event is the empty set, the probability that an outcome is in the event is zero; that is, $P(\varnothing) = 0$.
2. If an event is the entire sample space, then $E = S$ and $P(E) = P(S) = 1$.

**NOTE**

When we represent an event with a circle, we do not attempt to draw it to scale because the figure is intended only to show relationships and is not intended to be used to estimate probabilities.

## A Visual Model of Probability

We can use area in a Venn diagram to visualize the probability of an event. We let a rectangle, the universe, represent the sample space and a circle represent an event (Figure 7–1). We let the probability of an event be represented by the area of the figure representing it. Thus, we let the rectangle representing the sample space have area 1. If the probability of an event is 0.20, then we think of the area of circle $E$ as 20% of the sample space.

Intuitively, a Venn diagram suggests that when an event $A$ is a subset of an event $B$ (Figure 7–2), then the probability of $A$ cannot be larger than the probability of $B$.

Likewise, the probability that an outcome is in the intersection of events cannot be larger than the probability of either of the events (Figure 7–3) and the probability of an event $A$ and the probability of an event $B$ must not exceed the probability of their union (Figure 7–4).

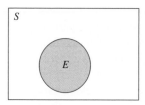

**FIGURE 7–1**    An event in sample space $S$.

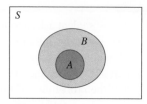

**FIGURE 7–2**
$P(A) \le P(B)$

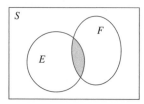

**FIGURE 7–3**
$P(E \text{ and } F) = P(E \cap F) \le P(E)$
$P(E \text{ and } F) = P(E \cap F) \le P(F)$

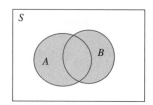

**FIGURE 7–4**
$P(A) \le P(A \cup B) = P(A \text{ or } B)$
$P(B) \le P(A \cup B) = P(A \text{ or } B)$

---

## **7.1**    EXERCISES

Access end-of-section exercises online at **www.webassign.net**

---

## **7.2**    EQUALLY LIKELY EVENTS

In general, no unique method exists for assigning probabilities to the outcomes of an experiment. The conditions, processes, and interactions involved vary depending on the experiment. The skill of the players largely determines the outcome of a baseball game, but other factors such as the weather or a decision by a coach can alter the outcome. The type of soil, the amount and the timing of rainfall, temperature, and insects affect the probability that a new hybrid of corn will be productive. One important class of experiments that have well-defined probability assignments are called **equally likely** models.

Such experiments have the characteristic that the individual outcomes are equally likely; that is, each outcome has the *same* chance, or probability, of occurring as any other outcome. Generally, we say that a tossed coin is just as likely to turn up a head as a tail, so heads and tails are equally likely to happen. If we alter a coin so that it comes up heads two-thirds of the time, then heads and tails are not equally likely. Unless stated otherwise, we assume that a tossed coin is fair; that is, heads and tails are equally likely.

For one card randomly selected from a well-shuffled deck, each card has the same chance of being the one selected. We say that the outcomes are equally likely.

When a person draws a coin from a purse with coins of different denominations, it doesn't seem reasonable to say that the outcomes are equally likely because of the variations of size. The small size of a dime makes it less likely to be drawn than a half dollar, for example.

For experiments with a finite number of equally likely outcomes, the probability of each simple outcome is $\frac{1}{n}$, where $n$ is the number of outcomes in the sample space.

If we toss a coin, the sample space is $\{H, T\}$. We intuitively agree that heads and tails are equally likely, so

$$P(H) = \frac{1}{2} \quad \text{and} \quad P(T) = \frac{1}{2}$$

because the sample space has two elements. If we select a name at random from 25 different names, then each name has $\frac{1}{25}$ probability of being drawn.

Because the probability of an event is the sum of probabilities of all simple outcomes in the event, we can compute probabilities of events like the following:

Select a number at random from the set

$$S = \{1, 2, 3, 4, 5, 6, 7, 8, 9\}$$

Determine the probability that it is even, that is, the number is in the event

$$E = \{2, 4, 6, 8\}$$

We assume equally likely outcomes, so the probability of each number is $\frac{1}{9}$. Then,

$$\begin{aligned}
P(E) &= P(2) + P(4) + P(6) + P(8) \\
&= \frac{1}{9} + \frac{1}{9} + \frac{1}{9} + \frac{1}{9} \\
&= \frac{4}{9}
\end{aligned}$$

It is no accident that

$$\frac{4}{9} = \frac{\text{number of elements in } E}{\text{number of elements in } S}$$

| NOTE |
| --- |

The event $E$ occurs when any of the simple outcomes in $E$ occurs.

For an experiment with equally likely outcomes, compute the probability of an event by counting the number of elements in the event and then dividing by the number of elements in the sample space. We sometimes call the outcomes in the event **successes** and outcomes not in the event **failures**.

| **THEOREM** | For an event $E$ where |
| --- | --- |

 **(a)** $E$ contains $s$ simple outcomes (successes),

 **(b)** the sample space $S$ contains $n$ simple outcomes, and

 **(c)** the simple outcomes are equally likely (have the same probabilities), the probability of $E$ is

*(continued)*

$$P(E) = \frac{s}{n} = \frac{\text{number of outcomes of interest (success)}}{\text{total number of outcomes possible}}$$

$$= \frac{n(E)}{n(S)}$$

If the event can fail in $f$ ways, the probability of failure is

$$P(\text{failure}) = \frac{f}{n}$$

Because we admit only success or failure, $n = s + f$. We can conclude from this that

$$P(\text{success}) + P(\text{failure}) = \frac{s}{n} + \frac{f}{n} = \frac{s + f}{n} = \frac{n}{n} = 1$$

so the probability of success and the probability of failure always add to 1.

For the rest of this section, we assume equally likely simple events, and we look at examples that illustrate the probability of an event in such cases.

**Example 1**  Draw a number at random from the integers 1 through 10. Find the probability that a prime is drawn.

**Solution**
In this case, $n = 10$ and $E = \{2, 3, 5, 7\}$, so $s = 4$. This gives $P(\text{prime}) = \frac{4}{10}$.

**Example 2**  A clearance sale table at the Southside Bookstore contains a pile of 17 mystery books and 23 romance novels. A customer selects a book at random. Find the probability that it is a mystery book.

**Solution**
As there is a total of 40 books, $n = 40$. An outcome is successful in 17 ways, the selection of any one of the 17 mystery books, so $s = 17$. Thus,

$$P(\text{mystery}) = \frac{17}{40} = 0.425$$

Examples often occur in mathematics textbooks that ask for probabilities involving dice. The next example introduces some events involving dice.

**Example 3**  A pair of dice, one blue and one white, is rolled. What is the probability of rolling an 8 (the two numbers that turn up add to 8)? Of rolling a 3 (the two numbers add to 3)?

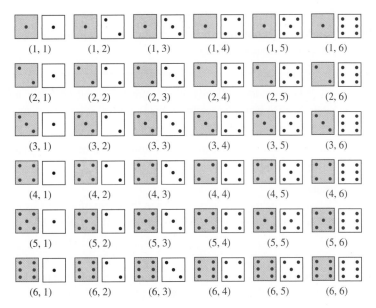

**FIGURE 7–5**  Sample space for two dice.

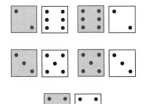

**FIGURE 7–6**  The event of rolling an 8.

**Solution**

The blue die can turn up in six ways and the white die can turn up in six ways, so the Multiplication Rule gives $6 \times 6 = 36$ ways the two dice can turn up, so $n = 36$.

Each outcome in the sample space may be thought of as a pair of dice with a pair of numbers showing. We will identify each pair of dice with a pair of numbers like (3, 2) where the first number, 3, shows on the blue die and the second number, 2, shows on the white die. The pairs of dice and numbers shown in Figure 7–5 compose the sample space.

We say an 8 is rolled when the two numbers showing add to 8. An 8 may be obtained in five different ways:

blue die 2, white die 6; that is, (2, 6)

blue die 6, white die 2; that is, (6, 2)

blue die 3, white die 5; that is, (3, 5)

blue die 5, white die 3; that is, (5, 3)

blue die 4, white die 4; that is, (4, 4)

The event is also shown in Figure 7–6.

The probability of rolling an 8 is then

$$P(8) = \frac{5}{36}$$

Likewise, there are just two ways to roll a 3, so

$$P(3) = \frac{2}{36} = \frac{1}{18}$$

**Random Selection**    **Random selection** or **random outcomes** imply that individual outcomes in the sample space are equally likely.

When we have equally likely outcomes, the counting methods we encountered in Chapter 6 can be used to determine the probability of an event.

**Example 4**    Two students are selected at random from a class of eight boys and nine girls. Find the probability that both students selected are girls.

**Solution**
Determine the number of outcomes in the event of interest by the number of different ways in which two girls can be selected from a group of nine. This is

$$n(2 \text{ girls}) = C(9, 2) = \frac{9 \times 8}{2 \times 1} = 36$$

The number of outcomes, $n$, in the sample space is the number of different ways two students can be selected from the whole group of 17. This is

$$n(2 \text{ students}) = C(17, 2) = \frac{17 \times 16}{2 \times 1} = 136$$

Thus, $P(\text{two girls}) = \dfrac{n(2 \text{ girls})}{n(2 \text{ students})} = \dfrac{36}{136} = \dfrac{9}{34}.$

**Example 5**    Find the probability of at least two heads appearing in a sequence of three tosses of a coin.

**Solution**
We use a tree diagram to obtain the necessary information. (See Figure 7–7.)

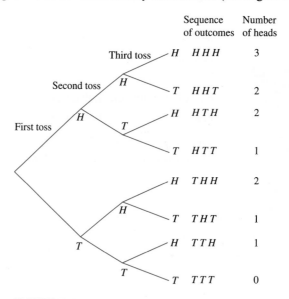

**FIGURE 7–7**    Sequence of heads and tails on three tosses of a coin.

We observe that there are a total of eight possible equally likely outcomes with four of them showing two or more heads. Thus,

$$P(\text{at least 2 heads}) = \frac{4}{8} = \frac{1}{2}$$

**Example 6**    A basket of tomatoes contains 9 that were grown using pesticides and 11 that were grown organically. If a customer randomly selects three tomatoes, find the probability of two of these having been grown using pesticides and one having been grown organically.

**Solution**
For 20 tomatoes the total number of outcomes is $C(20, 3)$. A success occurs with the selection of two tomatoes grown with pesticides and one grown organically. These can be selected in $C(9, 2) \times C(11, 1)$ ways. The probability is

$$\frac{C(9, 2) \times C(11, 1)}{C(20, 3)} = \frac{36 \times 11}{1140} = \frac{396}{1140} = \frac{99}{285} = 0.347$$

Next is an example using permutations.

**Example 7**    A corporation has eight men and six women on its board of directors. At a stockholders' meeting, seven directors are seated in a row on the platform. The directors are chosen and assigned a seat in a random manner.

**(a)** What is the probability that the directors are arranged in a sequence with men in seats 1 through 3 and women in seats 4 through 7?

**(b)** What is the probability that the directors are arranged in a sequence so that men and women alternate seats with a man in seat 1, a woman in seat 2, and so on?

**(c)** Find the probability that three men are seated together and four women are seated together in the sequence.

**Solution**
Because the directors are arranged in a sequence, the order of arrangement is significant.

**(a)** The number of possible ways to arrange any seven directors in a row is $P(14, 7)$. (This is the total number of outcomes possible.) The number of successes is the number of arrangements with three men first and four women next in the row. This can be done $P(8, 3) \times P(6, 4)$ ways. Thus, the probability of men in seats 1 through 3 and women in seats 4 through 7 is

$$\frac{P(8, 3) \times P(6, 4)}{P(14, 7)} = \frac{8 \times 7 \times 6 \times 6 \times 5 \times 4 \times 3}{14 \times 13 \times 12 \times 11 \times 10 \times 9 \times 8} = \frac{1}{143} = 0.007$$

**(b)** Again the total number of outcomes is $P(14, 7)$. To find the number of successes, we need to find the number of ways in which seven people can be arranged with a man first, a woman second, a man third, and so on. To do so, we multiply the number of choices for seat 1, for seat 2, ..., for seat 7. That product is

$$8 \times 6 \times 7 \times 5 \times 6 \times 4 \times 5$$

so the probability is

$$\frac{8 \times 6 \times 7 \times 5 \times 6 \times 4 \times 5}{14 \times 13 \times 12 \times 11 \times 10 \times 9 \times 8} = \frac{5}{429} = 0.012$$

**(c)** This arrangement can succeed in two ways: men in seats 1 through 3 and women in seats 4 through 7, or women in seats 1 through 4 and men in seats 5 through 7. We compute the number of each and add to obtain the total number of successes, that is,

$$P(8, 3)P(6, 4) + P(6, 4)P(8, 3) = 2P(8, 3)P(6, 4)$$

The probability is

$$\frac{2P(8, 3)\,P(6, 4)}{P(14, 7)} = \frac{2}{143} = 0.014$$

## 7.2   EXERCISES

Access end-of-section exercises online at **www.webassign.net**

## 7.3   COMPOUND EVENTS: UNION, INTERSECTION, AND COMPLEMENT

- Keys to Recognizing Compound Events
- Probability of Compound Events
- Probability of $E'$
- Probability of $E \cup F$
- Mutually Exclusive Events

Because events are subsets of a sample space, we can use set operations to form other events. In particular, we may form the union or intersection of two events $E$ and $F$ to form another event. We may also take the complement of an event to form another event. We will see how the probabilities of events $E$ and $F$ relate to the probabilities of their union, intersection, and complement. An understanding of these relationships often helps to analyze a complex problem by breaking it down into simpler problems. We call an event that can be described in terms of the union, intersection, or complement of some events a **compound event**.

**Compound Events**    Let $E$ and $F$ be events in a sample space $S$.

1. The event $E \cup F$ is the event consisting of those outcomes that are in $E$ or $F$ or both (**union**).
2. The event $E \cap F$ is the event consisting of those outcomes that are in both $E$ and $F$ (**intersection**).
3. The event $E'$ (**complement** of $E$) is the event consisting of those elements in the sample space that are not in $E$.

**Example 1**    Let the sample space $S = \{1, 2, 3, 4, 5, 6, 7, 8, 9, 10\}$ (see Figure 7–8). Let the event $E$ be "the number is even." Then $E = \{2, 4, 6, 8, 10\}$. Let the event $F$ be "the number is prime." Then $F = \{2, 3, 5, 7\}$. From these,

$$E \cup F = \{2, 3, 4, 5, 6, 7, 8, 10\}$$
$$E \cap F = \{2\}$$
$$F' = \{1, 4, 6, 8, 9, 10\}$$
$$E' = \{1, 3, 5, 7, 9\}$$

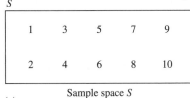

**(a)** Sample space $S$

**(b)** $E$ = Set of even numbers in $S$

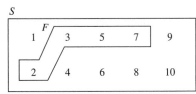

**(c)** $F$ = Set of prime numbers in $S$

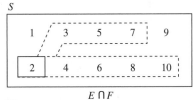

**(d)** $E \cap F$

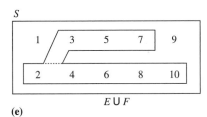

**(e)** $E \cup F$

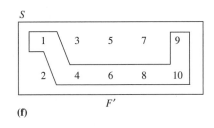

**(f)** $F'$

**FIGURE 7–8**

**Example 2**    For the experiment of selecting a student at random, let $E$ be the event "the student is taking art," and let $F$ be the event "the student is taking history." Then,

$E \cup F$ is the event "the student is taking art or history or both."

$E \cap F$ is the event "the student is taking both art and history."

$E'$ is the event "the student is not taking art."

$F'$ is the event "the student is not taking history."

## Keys to Recognizing Compound Events

Let's emphasize the key words and phrases that describe compound events. They will help in recognizing the approach to take in solving those problems.

$E \cup F$.    An outcome belongs to $E \cup F$ if it belongs to $E$ or $F$ or both. The key word for describing and recognizing $E \cup F$ is *or*. When the word *or* is used in mathematics, it means $E$ or $F$ or both unless stated otherwise. Another way to state that an event is in $E$ or $F$ is to state that the event belongs to *at least one* of them.

The statements

"Find the probability of $E \cup F$"

"Find the probability that an outcome belongs to $E$ or $F$"

"Find the probability that an outcome belongs to at least one of $E$ and $F$"

are equivalent statements.

$E \cap F$.    An outcome is in $E \cap F$ if it belongs to both $E$ and $F$. A key word for recognizing and describing $E \cap F$ is *and*.

The statements

"Find $P(E \cap F)$"

"Find the probability that an outcome belongs to $E$ and $F$"

"Find the probability of $E$ and $F$"

are equivalent statements.

$E'$.    An outcome belongs to $E'$ if it is not in $E$. A key word for recognizing and describing $E'$ is *not*.

The statements

"Find $P(E')$"

"Find the probability that an outcome is not in $E$"

"Find the probability that $E$ fails"

are equivalent statements.

## Probability of Compound Events

We can sometimes determine the probability of a compound event by using the probabilities of the individual events making up the compound event.

## Probability of $E'$

You recognize that if 10% of a class receives an A grade, then $100\% - 10\% = 90\%$ do not receive an A grade. If $\frac{2}{3}$ of a store's customers prefer brand A, then $1 - \frac{2}{3} = \frac{1}{3}$ of the customers do not prefer brand A. These statements are similar to the probability statement in the following theorem:

| **THEOREM** Complement Theorem | For an event $E$, |
|---|---|

$$P(E') = 1 - P(E)$$
$$P(E) = 1 - P(E')$$
$$P(E) + P(E') = 1$$

where $E'$ is the complement of $E$ in the sample space $S$.

**Example 3** If the probability that Smith wins the door prize at a club meeting is 0.1, then the probability that Smith does not win is $1 - 0.1 = 0.9$.

If the probability that a part is defective is 0.08, then the probability that it is not defective is $1 - 0.08 = 0.92$.

If the probability that Jones fails to get a promotion is 0.35, then the probability that Jones does get a promotion is 0.65.

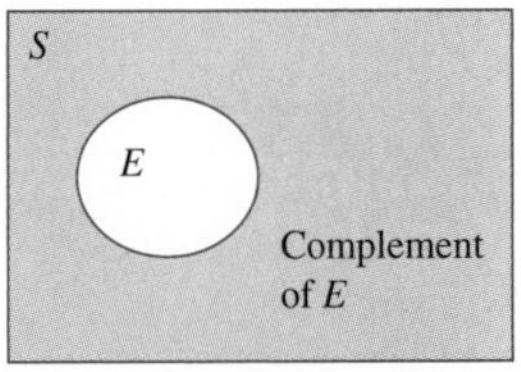

**FIGURE 7–9** Subtract the area of $E$ from the area of $S$ to obtain the area of the complement of $E$.

The area model of probability illustrates the Complement Theorem. In Figure 7–9, the sample space has area $= 1$, and the event $E$ has area $= P(E)$. If we take $E$ out of the sample space, we have the complement of $E$ left. We subtract the area of $E$, $P(E)$, from the area of the sample space, 1, to obtain the area that remains, $P(E')$. This illustrates that $P(E') = 1 - P(E)$.

You will encounter problems in which it may be difficult to find the probability of an event, but rather easy to find the probability of the complement of the event. If so, the Complement Theorem enables us to more easily find the probability of the event. The following example illustrates this idea.

**Example 4** A branch office of a corporation employs six women and five men. If four employees are selected at random to help open a new branch office, find the probability that at least one is a woman.

**Solution**

Selection of at least one woman occurs when one woman or two women or three women or four women are selected.

We count the number of successes by counting the number of ways in which we can select

one woman and three men.

two women and two men.

three women and one man.

four women.

Because we want the probability that one of these four events occurs, the Addition Rule applies. We add the number of ways each event can occur and find that the number of successes is

$$C(6, 1)C(5, 3) + C(6, 2)C(5, 2) + C(6, 3)C(5, 1) + C(6, 4)$$

We divide this quantity by $C(11, 4)$ to obtain the probability of at least one woman being selected. You can carry out the computation on the previous page if you like, but let's look at an easier way.

The only way the company can *fail* to select at least one woman is to select *all* men. That probability is

$$\frac{C(5, 4)}{C(11, 4)} = \frac{5}{330} = \frac{1}{66}$$

By the Complement Theorem, the probability of success is 1 minus the probability of failure. So the probability of at least one woman being selected is $1 - \frac{1}{66} = \frac{65}{66}$.

## Probability of $E \cup F$

We use the following example to illustrate a useful theorem to determine $P(E \cup F)$.

**Example 5** In a group of 200 students, 40 take English, 50 take math, and 10 take both. If a student is selected at random, the following probabilities hold:

$$P(\text{English}) = \frac{\text{number taking English}}{\text{number in the group}} = \frac{40}{200}$$

$$P(\text{math}) = \frac{\text{number taking math}}{\text{number in the group}} = \frac{50}{200}$$

$$P(\text{English and math}) = \frac{\text{number taking English and math}}{\text{number in the group}} = \frac{10}{200}$$

$$P(\text{English or math}) = \frac{\text{number taking English or math}}{\text{number in the group}} = \frac{80}{200}$$

The number 80 in the last equation was obtained by using the Inclusion–Exclusion Principle from Section 6.2 on sets:

$$n(E \cup F) = n(E) + n(F) - n(E \cap F)$$

If we let $S$ be the group of 200 students, let $E$ be those taking English, and let $F$ be those taking math, then the preceding equations may be written as

$$P(\text{English}) = \frac{n(E)}{n(S)} = \frac{40}{200}$$

$$P(\text{math}) = \frac{n(F)}{n(S)} = \frac{50}{200}$$

$$P(\text{English and math}) = \frac{n(E \cap F)}{n(S)} = \frac{10}{200}$$

$$P(\text{English or math}) = \frac{n(E \cup F)}{n(S)}$$

$$= \frac{n(E) + n(F) - n(E \cap F)}{n(S)}$$

$$= \frac{40 + 50 - 10}{200} = \frac{80}{200}$$

(See Figure 7–10.)

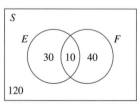

$n(S) = 200, n(E) = 40, n(F) = 50,$
$n(E \cap F) = 10, n(E \cup F) = 80$

**FIGURE 7–10**

You might wonder why we seem to have complicated the last equation. We want to illustrate a basic property of probability, so we need to carry the last equation a little further. We may write

$$P(E \cup F) = \frac{40 + 50 - 10}{200}$$

as

$$P(E \cup F) = \frac{40}{200} + \frac{50}{200} - \frac{10}{200}$$

The right-hand side of the last equation is

$$P(E) + P(F) - P(E \cap F)$$

This holds even when the outcomes are not equally likely.

**THEOREM**
$P(E \cup F)$

**(a)** $P(E \text{ or } F) = P(E \cup F) = P(E) + P(F) - P(E \cap F)$

**(b)** If the outcomes are equally likely, then

$$P(E \text{ or } F) = P(E \cup F) = \frac{n(E) + n(F) - n(E \cap F)}{n(S)}$$

We point out that the preceding example, Example 5, assumes equally likely outcomes. However, part (a) of the theorem holds in other cases as well.

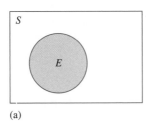 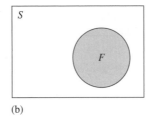 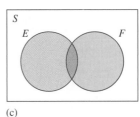

(a)          (b)          (c)

**FIGURE 7–11**

Let's look at the area model of probability to illustrate the $P(E \cup F)$ theorem.

In Figure 7–11a, $P(E)$ is represented by the area of $E$; in Figure 7–11b, $P(F)$ is represented by the area of $F$. When we form $E \cup F$ (Figure 7–11c), the region $E \cap F$ occurs in both $E$ and $F$. So if we add the area of $E$ to the area of $F$, then the area of $E \cap F$ is used twice. To find the area of $E \cup F$, that is, $P(E \cup F)$, we need to subtract the area in $E \cap F$ from the sum of areas of $E$ and $F$. Thus,

Area of $E \cup F$ = area of $E$ + area of $F$ − area of $E \cap F$

In terms of the probabilities represented by the areas, we have the $P(E \cup F)$ Theorem:

$$P(E \cup F) = P(E) + P(F) - P(E \cap F)$$

**Example 6**   In a remote jungle village, the probability of a child contracting malaria is 0.45, the probability of contracting measles is 0.65, and the probability of contracting both is 0.20. What is the probability of a child contracting malaria or measles?

**Solution**
Note the information given: $P(\text{malaria}) = 0.45$, $P(\text{measles}) = 0.65$, and $P(\text{malaria and measles}) = P(\text{malaria} \cap \text{measles}) = 0.20$. You are asked to find $P(\text{malaria or measles}) = P(\text{malaria} \cup \text{measles})$.

By the $P(E \cup F)$ theorem,

$$P(\text{malaria or measles}) = P(\text{malaria}) + P(\text{measles}) - P(\text{malaria and measles})$$

$$= 0.45 + 0.65 - 0.20 = 0.90$$

**Example 7**   A survey of couples in a certain country found the following:

The probability that the husband has a college degree is 0.65.

The probability that the wife has a college degree is 0.70.

The probability that both have a college degree is 0.50.

A couple is selected at random.

**(a)** Find the probability that at least one has a college degree.

**(b)** Find the probability that neither has a college degree.

**(c)** Find the probability the wife has a college degree and the husband doesn't.

**Solution**
Let $M$ represent the event that the husband has a college degree and $F$ the event that the wife has a college degree.

**(a)** Then, $M \cup F$ represents the event that at least one of them has a college degree, and $P(M \cup F) = 0.65 + 0.70 - 0.50 = 0.85$.

**(b)** The event that "neither has a college degree" is the complement of "at least one has a college degree," so

$$P(\text{neither has a degree}) = 1 - P(\text{at least one has a degree})$$
$$= 1 - 0.85 = 0.15$$

**(c)** From the information given, we can draw a Venn diagram (Figure 7–12) with probabilities entered in the appropriate region. The region that represents the wife with a degree and the husband without a degree is the region with 0.20 probability.

**FIGURE 7–12**

## Mutually Exclusive Events

The $P(E \cup F)$ theorem might not be helpful in some problems because we might not know how to compute $P(E \cap F)$. That will come in the next section. However, the special situation in which $E$ and $F$ have no outcomes in common can be solved.

**DEFINITION**
**Mutually Exclusive**
**(Disjoint)**

We call two events $E$ and $F$ **mutually exclusive** if they have no outcomes in common. When two events are mutually exclusive, an outcome in one is excluded from the other. As sets, $E$ and $F$ are **disjoint**. (See Figure 7–13.)

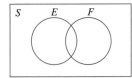

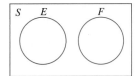

E and F are not
mutually exclusive.

(a)

E and F are
mutually exclusive.

(b)

**FIGURE 7–13**

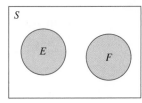

Area of $E \cup F =$
Area of $E$ + Area of $F$

**FIGURE 7–14**

In case the events $E$ and $F$ are mutually exclusive, the area model of probability illustrates that $P(E \cup F) = P(E) + P(F)$ (Figure 7–14).

**Example 8** When a coin is tossed, heads and tails are mutually exclusive because each one excludes the other.

Rolling a 7 with a pair of dice is mutually exclusive with rolling a 9 because they cannot occur at the same time.

Taking English and taking art are not mutually exclusive because both courses can be taken if the classes meet at different times.

For two mutually exclusive events, the computation of the probability of $E$ or $F$ simplifies because $E \cap F = \varnothing$. Recall that $P(\varnothing) = 0$, so when $E$ and $F$ are mutually exclusive,

$$P(E \cup F) = P(E) + P(F) - P(E \cap F)$$

becomes $P(E \cup F) = P(E) + P(F)$.

We now restate the $P(E \cup F)$ theorem, to emphasize this special case of mutually exclusive events.

**THEOREM**
$P(E \cup F)$

For any events $E$ and $F$,

$$P(E \text{ or } F) = P(E \cup F) = P(E) + P(F) - P(E \cap F)$$

If $E$ and $F$ are mutually exclusive events, the above reduces to

$$P(E \text{ or } F) = P(E \cup F) = P(E) + P(F)$$

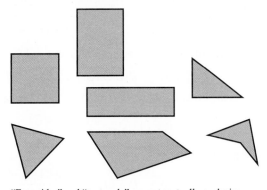

"Four sides" and "rectangle" are not mutually exclusive.
"Triangle" and "square" are mutually exclusive.

**Example 9**    Two people are selected at random from a group of seven men and five women.

(a) Find the probability that both are men or both are women.

(b) Find the probability that at least one is a man.

**Solution**

(a) The events "both men" and "both women" are mutually exclusive because the existence of one excludes the other. Thus,

$$P(\text{both men or both women}) = P(\text{both men}) + P(\text{both women})$$

$$= \frac{C(7, 2)}{C(12, 2)} + \frac{C(5, 2)}{C(12, 2)}$$

$$= \frac{21}{66} + \frac{10}{66} = \frac{31}{66}$$

(b) The complement of "at least one man" is "two women." Thus,

$$P(\text{at least one man}) = 1 - P(\text{two women})$$

$$= 1 - \frac{C(5, 2)}{C(12, 2)}$$

$$= 1 - \frac{10}{66}$$

$$= \frac{56}{66} = \frac{28}{33}$$

---

**Summary of Theorems**    **Complement Theorem**

For an event $E$,

$$P(E') = 1 - P(E)$$
$$P(E) = 1 - P(E')$$
$$P(E) + P(E') = 1$$

where $E'$ is the complement of $E$ in the sample space $S$.

**$P(E \cup F)$**

For any events $E$ and $F$,

$$P(E \text{ or } F) = P(E \cup F) = P(E) + P(F) - P(E \cap F)$$

If $E$ and $F$ are mutually exclusive events, the above reduces to

$$P(E \text{ or } F) = P(E \cup F) = P(E) + P(F)$$

If the outcomes are equally likely, $P(E \text{ or } F)$ can also be stated as

$$P(E \text{ or } F) = P(E \cup F) = \frac{n(E) + n(F) - n(E \cap F)}{n(S)}$$

and when $E$ and $F$ are mutually exclusive,

$$P(E \text{ or } F) = P(E \cup F) = \frac{n(E) + n(F)}{n(S)}$$

## 7.3 EXERCISES

Access end-of-section exercises online at **www.webassign.net**

---

## 7.4 CONDITIONAL PROBABILITY

- Conditional Probability
- Multiplication Rule

We have computed the probability of events involving equally likely outcomes by determining the number of outcomes in the event and in the sample space. We have learned to compute the probability when events are used to form a compound event. We now turn our attention to some instances when these computations are affected by a related event or by additional conditions imposed. These may modify the sample space and thereby change the probability.

### Conditional Probability

The following simple example illustrates how to adjust the probability based on additional information. Suppose you are taking a test with multiple-choice questions. A question has four possible answers listed, and you have no idea of the correct answer. If you make a wild guess, the probability of selecting the correct answer is $\frac{1}{4}$. However, if you know one of the answers cannot be correct, then your chance of guessing the correct answer improves because the sample space has been reduced to three elements. You now choose from three answers, increasing the probability of guessing correctly to $\frac{1}{3}$.

We denote this situation by **conditional probability**. In general terms we describe conditional probability as follows. We seek the probability of an event $E$. A related event $F$ occurs, giving reason to change the sample space and thereby potentially changing the probability of $E$.

We say that we want to determine the probability of $E$ given that $F$ occurred and use the notation $P(E \mid F)$, which we read "the probability of $E$ given $F$."

We can state the multiple-choice question example as "The probability of guessing the correct answer given that one answer is known to be incorrect is $\frac{1}{3}$." Or, we can write $P(\text{correct} \mid \text{one answer known incorrect}) = \frac{1}{3}$.

If a student guesses wildly at the correct answer from the four given ones, the sample space consists of the four possible answers. When one answer is ruled out, the sample space reduces to three possible answers. Sometimes it helps to look at a conditional probability problem as one in which the sample space changes when certain conditions exist or related information is given. Let's look at the following from that viewpoint.

---

**Example 1**   A student has a job testing microcomputer chips. The chips are produced by two machines, I and II. It is known that 5% of the chips produced by machine I are defective and 15% of the chips produced by machine II are defective. The student has a batch of chips that she assumes is a mixture from both machines. If she selects one at random, what is the probability that it is defective? You cannot give a precise answer to this question unless you know the proportion of chips from each machine. It does seem reasonable to say that the probability lies in the interval from 0.05 through 0.15.

386 Chapter 7    Probability

Now suppose that the student obtains more information: The chips all come from machine II. This certainly changes her estimate of the probability of a defective chip; she knows that the probability is 0.15. The sample space changes from a set of chips from both machines to a set of chips from machine II. This illustrates the point that when you gain information about the state of the experiment, you may need to change the probabilities assigned to the outcomes.

Here's how we write some of the information from the preceding examples.

**Example 2**  **(a)** Machines I and II produce microchips, with 5% of those from machine I being defective and 15% of those from machine II being defective. This can be stated as

$$P(\text{defective chip} \mid \text{machine I}) = 0.05$$
$$P(\text{defective chip} \mid \text{machine II}) = 0.15$$

**(b)** There are four possible answers to a multiple-choice question, one of which is correct. The probability of guessing the right answer is $\frac{1}{4}$. However, if one incorrect answer can be eliminated, the sample space is reduced from four to three answers, and the probability of guessing correctly becomes $\frac{1}{3}$. This is stated as

$$P(\text{guessing correct answer}) = \frac{1}{4}$$

$$P(\text{guessing correct answer} \mid \text{one incorrect answer eliminated}) = \frac{1}{3}$$

Let's look at an example of how to compute $P(E \mid F)$.

**Example 3**  Professor Baird teaches two sections of philosophy. The regular section has 35 students, and the honors section has 25 students. The professor gives both sections the same test, and 14 students make an A, 5 in the regular section and 9 in the honors section.

**(a)** If a test paper is selected at random from all papers, what is the probability that it is an A paper?

**(b)** A test paper is selected at random. If it is known that the paper is from the honors section, what is the probability that it is an A paper?

**Solution**
We will refer to Venn diagrams as we work this problem to visualize the solution and to illustrate some general principles of conditional probability.

In this example, the sample space is the collection of all 60 papers, the event $E$ is the set of all A papers, and $F$ is the set of papers from the honors class (Figure 7–15a).

**(a)** $P(E) = $ the probability a paper selected from the 60 papers is an A paper (Figure 7–15b).

$$P(E) = \frac{14}{60}$$

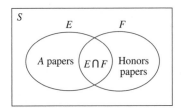

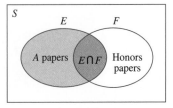

$S$ = All of the exams, 60 total
$E$ = All A exam papers, 14 total
$F$ = Papers from honors section, 25 total
$E \cap F$ = A papers from honors section, 9 total

(a)

$P(E)$ can be represented by the area of $E$

(b)

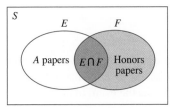

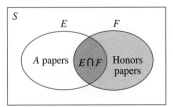

The paper is an honors paper, so the paper must come from $F$. Thus, $F$ becomes the reduced sample space.

(c)

The event $E|F$ focuses on the outcomes in $E \cap F$
$n(E \cap F) = 9$, $n(F) = 25$

(d)

**FIGURE 7–15**

**(b)** The knowledge that the paper is from the honors section restricts the outcomes to $F$ (Figure 7–15c), so $F$ becomes the **reduced sample space**. If the paper selected is an A paper, it then must come from the A papers in the honors section, from $E \cap F$. Because the honors section contains 9 A papers and 25 papers total,

$$P(E|F) = \frac{9}{25}$$

We can express this as

$$P(E|F) = \frac{9}{25} = \frac{n(E \cap F)}{n(F)}$$

Let's take this a step farther and divide the numerator and denominator of the last fraction by $n(S)$, giving

$$P(E|F) = \frac{\dfrac{n(E \cap F)}{n(S)}}{\dfrac{n(F)}{n(S)}} = \frac{P(E \cap F)}{P(F)}$$

In terms of using area as a visual probability model, the probability $P(E|F)$ is the ratio

$$\frac{\text{area of } E \cap F}{\text{area of } F}$$

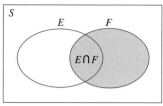

For $P(E|F)$
$F$ = reduced sample space
$E \cap F$ = the event
$$P(E|F) = \frac{\text{Area of } E \cap F}{\text{Area of } F} = \frac{P(E \cap F)}{P(F)}$$

**FIGURE 7–16**

Be sure you understand that, in essence, $F$ *becomes* the sample space when we compute $P(E \mid F)$, and we focus our attention on the contents of $F$. We sometimes call $F$ the **reduced sample space**. Then, $E \cap F$ becomes the event of successful outcomes. (See Figure 7–16.)

**DEFINITION**
**Conditional Probability $P(E|F)$**

$E$ and $F$ are events in a sample space $S$, with $P(F) \neq 0$. The conditional probability of $E$ given $F$, denoted by $P(E \mid F)$, is

**(a)** $P(E|F) = \dfrac{P(E \cap F)}{P(F)}$

This holds whether or not the outcomes of $S$ are equally likely.

**(b)** If the outcomes of $S$ are equally likely, $P(E \mid F)$ may be written as

$$P(E|F) = \frac{n(E \cap F)}{n(F)}$$

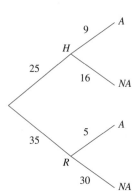

**FIGURE 7–17**

Let's use a tree diagram to illustrate conditional probability using the honors and regular classes again.

We have 60 students divided into two categories, 25 in an honors class ($n(H) = 25$) and 35 in a regular class ($n(R) = 35$). Each class is divided into two categories, A and not A papers. The tree diagram in Figure 7–17 shows the number in each category. We summarize the numbers shown:

$$n(\text{Honors}) = n(H) = 25$$
$$n(\text{Honors and A}) = n(H \cap A) = 9$$
$$n(\text{Honors and not A}) = n(H \cap NA) = 16$$
$$n(\text{Regular}) = n(R) = 35$$
$$n(\text{Regular and A}) = n(R \cap A) = 5$$
$$n(\text{Regular and not A}) = n(R \cap NA) = 30$$

From these numbers we can obtain the conditional probabilities:

$$P(A|H) = \frac{n(A \cap H)}{n(H)} = \frac{9}{25}$$

$$P(NA|H) = \frac{n(NA \cap H)}{n(H)} = \frac{16}{25}$$

$$P(A|R) = \frac{n(A \cap R)}{n(R)} = \frac{5}{35}$$

$$P(NA|R) = \frac{n(NA \cap R)}{n(R)} = \frac{30}{35}$$

**Example 4**   In a group of 200 students, 40 are taking English, 50 are taking mathematics, and 12 are taking both.

**(a)** If a student is selected at random, what is the probability that the student is taking English?

**(b)** A student is selected at random from those taking mathematics. What is the probability that the student is taking English?

**(c)** A student is selected at random from those taking English. What is the probability that the student is taking mathematics?

**(d)** A student is selected at random from those taking English. What is the probability that the student is not taking mathematics?

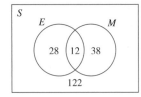

**FIGURE 7–18**

**Solution**
Figure 7–18 represents the given information with the Venn diagrams.

**(a)** $P(\text{English}) = \dfrac{40}{200} = \dfrac{1}{5}$.

**(b)** This problem is that of finding $P(\text{English} \mid \text{math})$, so

$$P(\text{English} \mid \text{math}) = \frac{n(\text{English and math})}{n(\text{math})} = \frac{12}{50} = \frac{6}{25} = 0.24$$

This may also be expressed in terms of probability:

$$P(\text{English} \mid \text{math}) = \frac{P(\text{English and math})}{P(\text{math})} = \frac{\frac{12}{200}}{\frac{50}{200}} = \frac{0.06}{0.25} = 0.24$$

**(c)** This asks for

$$P(\text{math} \mid \text{English}) = \frac{n(\text{math and English})}{n(\text{English})} = \frac{12}{40} = \frac{3}{10}$$

Parts (b) and (c) illustrate that $P(E \mid F)$ and $P(F \mid E)$ might not be equal.

**(d)** This asks for

$$P(\text{not math} \mid \text{English}) = \frac{n(\text{not math and English})}{n(\text{English})}$$

$$= \frac{28}{40} = \frac{7}{10}$$

Note that this is also

$$P(\text{not math} \mid \text{English}) = 1 - P(\text{math} \mid \text{English})$$

$$= 1 - \frac{3}{10}$$

because {not math | English} is the complement of {math | English}.

In Section 7.3 we studied some useful properties of probability. These properties hold for conditional probability if the conditions are applied consistently. The following are properties that follow from the three theorems of Section 7.3.

**Properties of $P(E|F)$**

$$P(E \mid F) + P(E' \mid F) = 1$$
$$P([A \cup B] \mid F) = P(A \mid F) + P(B \mid F) - P((A \cap B) \mid F)$$

If $A$ and $B$ are mutually exclusive events, then

$$P((A \cup B) \mid F) = P(A \mid F) + P(B \mid F)$$

We now can find how to compute the probability of $E$ and $F$, using the **Multiplication Rule**.

## Multiplication Rule

We obtain a useful formula for the probability of $E$ and $F$ by multiplying the equation (a) in the definition of conditional probability throughout by $P(F)$:

$$P(E \mid F) = \frac{P(E \cap F)}{P(F)}$$

Multiply by $P(F)$ to obtain

$$P(F)P(E \mid F) = P(E \cap F)$$

Because

$$P(F \mid E) = \frac{P(F \cap E)}{P(E)} = \frac{P(E \cap F)}{P(E)}$$

we also have

$$P(E)P(F \mid E) = P(E \cap F)$$

Thus, we have two forms by which to compute $P(E \cap F)$.

**THEOREM**
**Multiplication Rule for Conditional Probability**

$E$ and $F$ are events in a sample space $S$.

$$P(E \text{ and } F) = P(E \cap F) = P(F)P(E|F)$$

or

$$P(E \text{ and } F) = P(E \cap F) = P(E)P(F|E)$$

This theorem states that we can find the probability of $E$ and $F$ by multiplying the probability of $E$ by the conditional probability of $F$ given $E$. Let's use the area model of probability again to visualize the Multiplication Rule.

First, look at $P(E \cap F)$ as the fraction of $S$ occupied by $E \cap F$. We would like to find that fraction from information given about $E$ and $F$.

If we know $P(F)$, we know the fraction of $S$ occupied by $F$. (If $P(F) = 0.40$, then $F$ occupies 40% of $S$ in Figure 7–19a). If we know the fraction of $F$ occupied by $E \cap F$, then we know $P(E|F)$. (If $E \cap F$ occupies 35% of $F$, $P(E|F) = 0.35$ in Figure 7–19b). Then, $F$ occupies 40% of $S$ and $E \cap F$ occupies 35% of $F$ (35% of the 40% occupied by $F$). So, $E \cap F$ occupies $(0.35)(0.40) = 0.14$ of $S$, or $E \cap F$ occupies 14% of $S$.

Looking back over these computations, we have computed

$$P(E \cap F) = (0.35)(0.40) = P(E|F)P(F)$$

which is our Multiplication Rule.

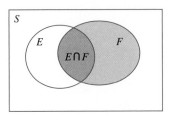

Say, $F$ occupies 40% of $S$

(a)

Say, $E \cap F$ occupies 35% of $F$

(b)

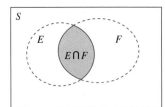

Then, $E \cap F$ occupies $(0.35)(0.40)$ of $S$

(c)

**FIGURE 7–19**

**Example 5** Two cards are drawn from a bridge deck, without replacement. What is the probability that the first is an ace and the second is a king?

**Solution**
According to the Multiplication Rule for conditional probability, we want to find

$$P(\text{ace first and king second}) = P(\text{ace first}) \times P(\text{king second} | \text{ace first})$$

Because the first card is drawn from the full deck of 52 cards,

$$P(\text{ace first}) = \frac{4}{52}$$

This first card is not replaced, so the sample space for the second card reduces to 51 cards. As we are assuming that the first card was an ace, four kings remain in the deck. Then,

$$P(\text{king second} \mid \text{ace first}) = \frac{4}{51}$$

It then follows that

$$P(\text{ace first and king second}) = \left(\frac{4}{52}\right) \times \left(\frac{4}{51}\right) = \frac{4}{663}$$

**Example 6**    A box contains 12 light bulbs, 3 of which are defective. If 3 bulbs are selected at random without replacement, what is the probability that all 3 are defective?

**Solution**

$$P(\text{first defective}) = \frac{3}{12}$$

Given that the first bulb is defective, there are 2 defective bulbs left, so

$$P(\text{second defective}) = \frac{2}{11}$$

For the third selection, there is 1 defective bulb left, so

$$P(\text{third defective}) = \frac{1}{10}$$

According to the Multiplication Rule, applied twice,

$$P(\text{first defective and second defective and third defective})$$

$$= \left(\frac{3}{12}\right)\left(\frac{2}{11}\right)\left(\frac{1}{10}\right) = \frac{1}{220}$$

Here are two examples in which selections are made *with replacement*.

**Example 7**    Two cards are drawn from a bridge deck. The first card is drawn, the outcome is observed, and the card is replaced and the deck shuffled before the second card is drawn. Find the probability that the first is an ace and the second is a king.

**Solution**
The fact that the first card is replaced before the second card is drawn makes this example different from Example 5. We still use the Multiplication Rule, but notice the difference when we compute

$$P(\text{ace first and king second})$$
$$= P(\text{ace first}) \times P(\text{king second} \mid \text{ace first})$$

When we compute $P(\text{ace first})$, we get $\frac{4}{52}$, just as we did in Example 5.

When the second card is drawn, the deck still contains 52 cards, because the first card drawn was replaced. This gives

$$P(\text{king second} \mid \text{ace first}) = \frac{4}{52}$$

We now have

$$P(\text{ace first and king second})$$
$$= P(\text{ace first}) \times P(\text{king second} \mid \text{ace first})$$
$$= \frac{4}{52} \times \frac{4}{52} = \frac{1}{169}$$

Be sure you understand how this example differs from Example 5. They both use the basic property that

$$P(E \cup F) = P(E)P(F \mid E)$$

but replacing the first card in this example makes the sample space for the second draw different from the sample space in Example 5. In Example 5, the sample space for the second draw contains 51 cards, an ace having been removed. In this example, the sample space for the second draw contains 52 cards because the ace drawn first was replaced before the second draw occurred.

These two examples illustrate that you need to be sure you understand what effect the first action has on the second when you compute conditional probability.

The next two examples use the principles introduced in this section.

**Example 8**   A box contains five red balls, six green balls, and two white balls. Three balls are drawn, but each one is replaced before the next one is drawn. Find the probability the first is red, the second is green, and the third is white.

**Solution**

$$P(\text{red first and green second and white third})$$
$$= P(\text{red first}) \times P(\text{green second} \mid \text{red first})$$
$$\times P(\text{white third} \mid \text{red first and green second})$$
$$= \frac{5}{13} \times \frac{6}{13} \times \frac{2}{13} = \frac{60}{2197}$$

Note that the sample space has 13 elements for each draw because the balls are replaced. If the balls are not replaced after each draw, the probability is

$$\frac{5}{13} \times \frac{6}{12} \times \frac{2}{11} = \frac{60}{1716}$$

**Example 9**   A large corporation has 1500 male and 1200 female employees. In a fitness survey, it was found that 40% of the men and 30% of the women are overweight. An employee is selected at random. Find the probability that

**(a)** the person is male.

**(b)** the person is overweight, given the person is a male.

**(c)** the person is overweight, if the person is selected from the females.

**(d)** the person is overweight.

**(e)** the person is male if the person is selected from the overweight persons.

**Solution**

It can be helpful to summarize the given information like this:

|        | Overweight            | Not overweight         | Total |
|--------|-----------------------|------------------------|-------|
| Male   | 600 (0.4 × 1500)      | 900 (0.6 × 1500)       | 1500  |
| Female | 360 (0.3 × 1200)      | 840 (0.7 × 1200)       | 1200  |
| Total  | 960                   | 1740                   | 2700  |

Let $O$ represent the event of an overweight person, let $F$ represent the event of a female, and let $M$ represent the event of a male.

**(a)** We want $P(M)$, which is $\dfrac{1500}{2700} = \dfrac{5}{9}$.

**(b)** $P(O \mid M)$ is the statement that 40% of the males are overweight, so $P(O \mid M) = 0.40$.

**(c)** This asks to find $P(\text{overweight} \mid \text{female})$ and it is given that 30% of the females are overweight, so $P(O \mid F) = 0.30$.

**(d)** The information given tells us that 600 males are overweight (40% of 1500) and 360 females are overweight (30% of 1200), giving a total of 960 overweight persons. Then,

$$P(O) = \frac{960}{2700} = \frac{32}{90}$$

**(e)** We want to find $P(M \mid O)$:

$$P(M \cap O) = \frac{600}{2700} = \frac{2}{9} \quad \text{and} \quad P(O) = \frac{32}{90}$$

So,

$$P(M|0) = \frac{P(M \cap O)}{P(O)} = \frac{\frac{2}{9}}{\frac{32}{90}} = \frac{10}{16} = 0.625$$

We now summarize the main ideas in this section.

**Conditional Probability**

If $P(F) \neq 0$,

$$P(E|F) = \frac{P(E \cap F)}{P(F)}$$

If the outcomes are equally likely, this may be written as

$$P(E|F) = \frac{n(E \cap F)}{n(F)}$$

**Properties of P**

$$+ P(E'|F) = 1$$
$$P(A \cup B|F) = P(A|F) + P(B|F) - P(A \cap B|F)$$

If $A$ and $B$ are mutually exclusive events, then

$$P(A \cup B|F) = P(A|F) + P(B|F)$$

**Multiplication Rule**

$$P(E \cap F) = P(F)P(E|F) = P(E)P(F|E)$$

## 7.4 EXERCISES

Access end-of-section exercises online at **www.webassign.net**  Web**Assign**

## 7.5 INDEPENDENT EVENTS

- Independent Events
- Multiplication Rule for Independent Events
- Mutually Exclusive and Independent Events

### Independent Events

One evening, Andy tosses a coin to determine which subject he will study. At the same time across campus, Bill tosses a coin to determine which movie he will see. Does the outcome of Andy's toss have any influence on how Bill's coin turns up? We have no difficulty in stating that whether Andy's coin comes up heads or tails has no influence on how Bill's coin turns up. We say that Andy's toss and Bill's toss are **independent**, because the outcome of one toss has no effect on the outcome of the other. In a similar situation when Andy tosses a coin twice, the fact that it turns up heads on the first toss has no effect on whether the second toss turns up heads. The outcome of the first toss is independent of the outcome of the second toss.

Generally speaking, sometimes the occurrence of an event $E$ affects whether or not an event $F$ will occur. In other cases, the occurrence of an event $E$ has no effect whatever on the occurrence of event $F$. If one person selects a card from one deck and another person selects one from another deck, then an ace (or any other card) drawn from the first deck has no effect on whether or not a king is drawn from the second deck. If two people select one card each from the *same* deck, then the outcome of the first selection affects the outcome of the second person's selection because the first card drawn is excluded as a possible outcome of the second selection. In this case, the outcome of the second person's draw is not independent of the first draw, so we call the two draws **dependent**.

People often confuse *independent* with *mutually exclusive*. They are not the same. Mutually exclusive indicates that two events have no outcomes in common. Events are independent when their probabilities relate in a definite way. In fact, mutually exclusive events with nonzero probabilities are always dependent, whereas events having outcomes in common may or may not be independent.

For two independent events, the knowledge of one event provides no information on the other event. More precisely, we may determine the independence of two events by how their probabilities relate.

Situations arise in which our intuition may not be able to decide if two events are independent or not. We have a means to help in those situations. Again, we appeal to our intuition to make a general definition.

If two events $E$ and $F$ have no effect on the occurrence of each other, we intuitively expect that the occurrence of $F$ has no effect on the probability that $E$ occurs, and vice versa. In terms of probability, this means that $P(E|F) = P(E)$. Similarly, $P(F|E) = P(F)$. This leads to the following definition of independence.

**DEFINITION**
**Independent Events**

The events $E$ and $F$ are **independent** if

$$P(E|F) = P(E) \quad \text{and} \quad P(F|E) = P(F)$$

Otherwise, $E$ and $F$ are dependent. (*Note:* If $P(E|F) = P(E)$, then $P(F|E) = P(F)$ also holds, and vice versa.)

In the next example, we check $P(E|F)$ and $P(E)$ for two situations, one in which intuition suggests that the events are independent and one in which intuition suggests that the events are dependent.

**Example 1**    Two children play with three toys: a red, a green, and a blue one. We denote the toys by the set $\{R, G, B\}$. We consider two situations.

**(a)** Each child selects a toy with the first toy replaced before the second one is selected.

**(b)** Each child selects a toy with the first toy *not* replaced before the second one is selected.

In each situation, determine if the events "first toy is red" and "second toy is green" are independent or not. Do so by comparing $P$(second is green) with $P$(second is green | first is red).

**Solution**
We will use tree diagrams to help analyze this problem. The first stage of the tree diagram represents the possible selections by the first child and the second stage represents the possible selections by the second child.

**(a)** In this case, three outcomes are possible for the first selection, and for each of these selections, three possible outcomes exist for the second selection (Figure 7–20a). This gives a total of nine possible sequences of selection. To check the independence of selecting $R$ first and $G$ second, compare $P(G$ second$)$ and $P(G$ second | $R$ first$)$.

From the tree diagram, we see that three of the nine branches have $G$ as the second selection. Thus,

$$P(G \text{ second}) = \frac{3}{9} = \frac{1}{3}$$

To find $P(G \text{ second} | R \text{ first})$, we focus only on the part of the tree where the first branch is red (Figure 7–20b). We see that one of the three branches has green as the second selection, so

$$P(G \text{ second} | R \text{ first}) = \frac{1}{3}$$

Intuitively, we expected that "$R$ first" and "$G$ second" would be independent because the same sets of toys are available for each selection. We have further verified our intuition by finding $P(G \text{ second}) = P(G \text{ second} | R \text{ first})$. We point out that we can also verify independence by comparing $P(R \text{ first})$ and $P(R \text{ first} | G \text{ second})$.

**(b)** Intuitively, we expect that "$R$ first" and "$G$ second" will be dependent because the toy drawn first determines the toys left for the second draw. Again, we compare $P(G \text{ second})$ with $P(G \text{ second} | R \text{ first})$.

Figure 7–21a shows the possible sequence of selections, six in all. Observe that two of the six branches have $G$ as the second selection, so

$$P(G \text{ second}) = \frac{2}{6} = \frac{1}{3}$$

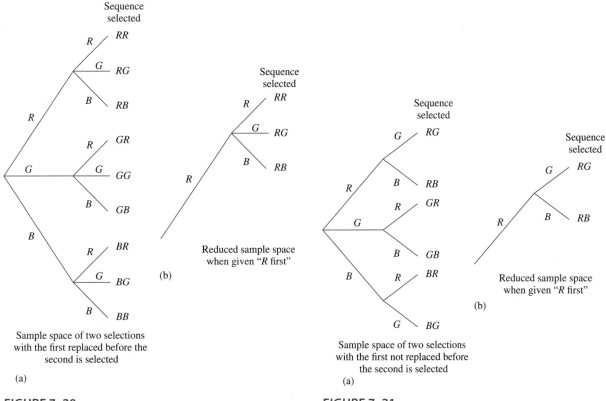

**FIGURE 7–20**

Sample space of two selections with the first replaced before the second is selected

(a)

Reduced sample space when given "R first"

(b)

**FIGURE 7–21**

Sample space of two selections with the first not replaced before the second is selected

(a)

Reduced sample space when given "R first"

(b)

To determine $P(G \text{ second} | R \text{ first})$, we focus on the part of the tree where the first branch is $R$ (Figure 7–21b). One of the two possible sequences has $G$ as the second selection, so

$$P(G \text{ second} | R \text{ first}) = \frac{1}{2}$$

$P(G \text{ second}) = \frac{1}{3}$ is not equal to $P(G \text{ second} | R \text{ first}) = \frac{1}{2}$, so we conclude that "$G$ second" and "$R$ first" are dependent.

In Example 1, our intuition and the probabilities behaved in a way that was consistent with the definition of independent events. Now we go to an extreme and look at an example in which our intuition gives no indication of whether or not two events are independent.

**Example 2**    A survey of 100 psychology majors revealed that 25 had taken an economics course ($E$), 35 had taken a French course ($F$), and 5 had taken both economics and French ($E \cap F$). Are the events "had taken economics" and "had taken French" independent or dependent?

**Solution**

It helps to draw a Venn diagram of this information (see Figure 7–22). The sample space contains 100 elements, $E$ contains 25, $F$ contains 35, and $E \cap F$ contains 5. From the Venn diagram, we get the following probabilities:

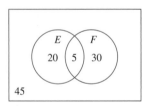

**FIGURE 7–22**

$$P(E) = \frac{25}{100} = 0.25$$

$$P(F) = \frac{35}{100} = 0.35$$

$$P(E|F) = \frac{5}{35} = 0.1429$$

$$P(F|E) = \frac{5}{25} = 0.20$$

As $P(E) \neq P(E|F)$, $E$ and $F$ are dependent. We could reach the same conclusion by the observation that $P(F) \neq P(F|E)$.

Next we look at an example in which two events are independent, but our intuition provides no help in making that determination. We must find $P(E)$ and $P(E|F)$ and see whether they are equal or not.

**Example 3**    A survey of 165 students revealed that 60 had read John Grisham's latest book, 22 had read Tom Clancy's latest book, and 8 had read both. Determine if the events "read Grisham's book" ($G$) and "read Clancy's book" ($C$) are independent or not.

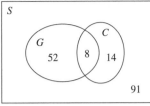

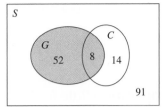

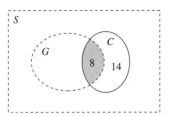

(a)

(b)
$P(G)$ = fraction of $S$ occupied by $G$
$$P(G) = \frac{60}{165}$$

(c)
For $P(G|C)$, $C$ is the sample space
$$P(G|C) = \frac{8}{22}$$

**FIGURE 7–23**

### Solution

We visualize this information with a Venn diagram (Figure 7–23) and compare $P(G)$ with $P(G|C)$. We could use $P(C)$ and $P(C|G)$ just as well. As an area, $P(G)$ is the fraction of $S$ that is occupied by $G$. Because $n(G) = 60$ and $n(S) = 165$,

$$P(G) \;=\; \frac{60}{165} \;=\; \frac{4}{11} \quad \text{(Figure 7–23b)}$$

For $P(G|C)$, $C$ becomes the sample space and $G \cap C$ the event, so

$$P(G|C) \;=\; \frac{8}{22} \;=\; \frac{4}{11} \quad \text{(Figure 7–23c)}$$

Because $P(G) = P(G|C) = \dfrac{4}{11}$, the events "read Grisham" and "read Clancy" are independent.

These examples illustrate that sometimes our intuition correctly tells us whether or not two events are independent, and sometimes it does not tell us. We can tell by computing $P(E)$ and $P(E|F)$ to see whether they are equal or not.

Now, let's use the area model of probability to make a general statement of what we did in Examples 2 and 3.

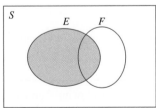

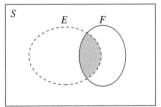

$P(E)$ = fraction of $S$ occupied by $E$

$P(E|F)$ = fraction of $F$ occupied by $E \cap F$

(a)

(b)

**FIGURE 7–24**

To determine the independence of events $E$ and $F$, we compare $P(E)$ and $P(E|F)$. We use the area of $E$ in Figure 7–24a to determine $P(E)$. $P(E)$ is the fraction of $S$ occupied by $E$. For $P(E|F)$, we determine the area of $F$ occupied by $E$. As only those elements of $E$ that are also in $F$ can occupy space in $F$, you recognize this as the area of $F$ occupied by $E \cap F$ (Figure 7–24b).

The events $E$ and $F$ are independent when the fraction of $S$ occupied by $E$ equals the fraction of $F$ occupied by $E \cap F$.

## Multiplication Rule for Independent Events

For independent events $E$ and $F$, the Multiplication Rule $P(E \cap F) = P(E)P(F|E)$ simplifies to the following theorem.

**THEOREM**
**Multiplication Rule for Independent Events**

If $E$ and $F$ are **independent** events, then

$$P(E \text{ and } F) = P(E \cap F) = P(E)P(F)$$

We can also use this theorem to determine whether two events are independent. For the events in Example 2,

$$P(E \cap F) = \frac{5}{100} = 0.05$$
$$P(E)P(F) = (0.25)(0.35) = 0.0875$$

As $P(E \cap F) \neq P(E)P(F)$, $E$ and $F$ are not independent—the same conclusion we reached by using the definition of independent events.

Although it may be intuitively clear that two tosses of a coin are independent, in other situations independence may be determined only by using the definition of independent events or the Multiplication Rule for independent events (as in Example 2). Actually, *any* one of the three conditions provides a test for independence.

**Test for Independence of Events**

If any one of the following holds, then events $E$ and $F$ are independent.

**NOTE**

If any one of the three holds, then they all hold.

**1.** $P(E|F) = P(E)$
**2.** $P(F|E) = P(F)$
**3.** $P(E \cap F) = P(E)P(F)$

Now let's look at the relation between independent events and mutually exclusive events.

**Example 4**    A card is selected at random from the following four cards: 10 of diamonds, 10 of spades, 8 of hearts, and 6 of clubs. Let $E$ be the event "select a red card" and let $F$ be the event "select a 10."

**(a)** Are $E$ and $F$ mutually exclusive?

**(b)** Are $E$ and $F$ independent?

*Note:* Diamonds and hearts are red cards; spades and clubs are black cards.

### Solution

**(a)** $E$ and $F$ are not mutually exclusive because selecting a red card and selecting a 10 can occur at the same time—for example, selecting the 10 of diamonds.

**(b)** From the information given,

$$P(E) = P(\text{red card}) = \frac{1}{2}$$

$$P(F) = P(10) = \frac{1}{2}$$

$$P(E \cap F) = P(\text{red 10}) = \frac{1}{4}$$

As $P(E \cap F) = P(E)P(F) = \frac{1}{4}$, $E$ and $F$ are independent.

This example illustrates two points:

1. **Two independent events need not be mutually exclusive**. In fact, mutually exclusive events with nonzero probabilities are always dependent.

2. Sometimes the only way to determine whether events are independent is to perform a computation in the tests for independence.

We can now compute the compound probability, $P(E \cap F)$, by using the properties of $P(E \mid F)$ for independent events and the Multiplication Rule for independent events.

**Example 5**  Ms. Bowden brought two bags of cookies to her second-grade class. The first bag contained 10 chocolate chip cookies and 14 oatmeal cookies. The second bag contained 8 peanut butter cookies and 12 sugar cookies. A student took one cookie from the first bag, and another student took one cookie from the second bag. Find the probability that the first was a chocolate chip cookie and the second was a sugar cookie.

### Solution
Because the cookies were taken from different bags, the events are independent, so

$$P(\text{first chocolate chip and second sugar}) = \left(\frac{10}{24}\right) \times \left(\frac{12}{20}\right) = \frac{1}{4}$$

**Example 6**  Jack and Jill work on a problem independently. The probability that Jack solves it is $\frac{2}{3}$, and the probability that Jill solves it is $\frac{4}{5}$.

**(a)** What is the probability that both solve it?

**(b)** What is the probability that neither solves it?

**(c)** What is the probability that exactly one of them solves it?

**(d)** For each possible value of $X$, construct a table summarizing the probability that $X$ people solve the problem.

**Solution**

**(a)** Because they work independently, the probability that both solve the problem is

$$P(\text{Jack solves it and Jill solves it}) = \left(\frac{2}{3}\right) \times \left(\frac{4}{5}\right) = \frac{8}{15}$$

**(b)** The probability that Jack does not solve the problem is $1 - \frac{2}{3} = \frac{1}{3}$, and the probability that Jill does not is $1 - \frac{4}{5} = \frac{1}{5}$. Then,

$$P(\text{Jack doesn't and Jill doesn't}) = \left(\frac{1}{3}\right) \times \left(\frac{1}{5}\right) = \frac{1}{15}$$

**(c)** There are two ways in which one of them will solve the problem, namely, Jack does and Jill doesn't or Jack doesn't and Jill does. These two ways are mutually exclusive events, so we need to compute the probability of each of these outcomes and then add.

$$P(\text{Jack does and Jill doesn't}) = \left(\frac{2}{3}\right) \times \left(\frac{1}{5}\right) = \frac{2}{15}$$

$$P(\text{Jack doesn't and Jill does}) = \left(\frac{1}{3}\right) \times \left(\frac{4}{5}\right) = \frac{4}{15}$$

Then, $P(\text{one of them solves the problem}) = \dfrac{2}{15} + \dfrac{4}{15} = \dfrac{6}{15} = \dfrac{2}{5}$

**(d)** Because two people are involved, two, one, or no people solve the problem ($X = 2, 1,$ or $0$).

In part (a), we found $P(X = 2) = \dfrac{8}{15}$.

In part (b), we found $P(X = 0) = \dfrac{1}{15}$.

In part (c), we found $P(X = 1) = \dfrac{6}{15}$.

| X | P(X) |
|---|------|
| 0 | 1/15 |
| 1 | 6/15 |
| 2 | 8/15 |

Note that the probabilities in the table add to 1, because the table contains all possible values of $X$.

*Note:* A table (such as this) that gives the probability of each element in the sample space forms a **probability distribution**.

————————●

Recall that $P(E \cup F) = P(E) + P(F) - P(E \cap F)$. When $E$ and $F$ are independent, $P(E \cap F) = P(E)P(F)$, so we can substitute for $P(E \cap F)$ to obtain the following theorem:

| **THEOREM**<br>**Union of Independent**<br>**Events** | When $E$ and $F$ are independent,<br><br>$$P(E \cup F) = P(E) + P(F) - P(E)P(F)$$ |

In some problems, you have the choice of more than one approach to solving the problem. We illustrate with the following example.

**Example 7**  Jack and Jill work on a problem independently. The probability that Jack solves it is $\frac{2}{3}$, and the probability that Jill solves it is $\frac{4}{5}$. What is the probability that at least one of them solves it?

**Solution**
We will show you three ways to work this problem:

> *Case I:*  Using the Union of Independent Events
>
> *Case II:*  Using the Complement Theorem (Section 7.3)
>
> *Case III:*  Using mutually exclusive events

*Case I:*  If we let $A$ = the event that Jack solves the problem and $B$ = the event that Jill solves the problem, then $A \cup B$ is one or both (at least one) solves the problem. From the Union of Independent Events Theorem in this section,

$$\begin{aligned} P(A \cup B) &= P(A) + P(B) - P(A)P(B) \\ &= \frac{2}{3} + \frac{4}{5} - \frac{2}{3} \times \frac{4}{5} \\ &= \frac{10 + 12 - 8}{15} = \frac{14}{15} \end{aligned}$$

*Case II:*  The complement of "at least one solves the problem" is "neither solves the problem," so

$$P(\text{at least one solves}) = 1 - P(\text{neither solves})$$

by the Complement Theorem (Section 7.3). This is

$$1 - \frac{1}{3} \times \frac{1}{5} = 1 - \frac{1}{15} = \frac{14}{15}$$

*Case III:*  At least one of them solving the problem is equivalent to exactly one of them solving the problem ($E$) or both of them solving the problem ($F$). These events, $E$ and $F$, are mutually exclusive, so

$$P(E \cup F) = P(E) + P(F)$$

$$P(E) = \frac{6}{15} \qquad \text{(From Example 6c)}$$

$$P(F) = \frac{8}{15} \qquad \text{(From Example 6a)}$$

so $P(\text{at least one}) = P(E \cup F) = \dfrac{6}{15} + \dfrac{8}{15} = \dfrac{14}{15}.$

Let's illustrate some of these concepts using a tree diagram.

**Example 8**    The city council has money for one public service project: a recreation–sports complex, a performing arts center, or a branch library. The council polled 200 citizens for their preference, including 120 men and 80 women. The men responded as follows: 45% preferred the recreation–sports complex, 20% the performing arts center, and 35% the branch library. The women responded as follows: 15% favored the recreation–sports complex, 40% the performing arts center, and 45% the branch library.

**(a)** Represent this information on a tree diagram.

**(b)** What is the probability that a person selected at random prefers the performing arts center?

**(c)** What is the probability that a person selected at random is a woman who prefers the performing arts center or the branch library?

**(d)** Are the events "male" and "prefers the recreation–sports complex" independent?

**Solution**
Use the following abbreviations: $M$ for male and $F$ for female; $RS$, $AR$, and $BL$ for recreation–sports complex, performing arts center, and branch library, respectively.
The information provided gives the following probabilities:

$$P(M) = 0.6, \qquad P(F) = 0.4$$
$$P(RS|M) = 0.45, \qquad P(AR|M) = 0.20, \qquad P(BL|M) = 0.35$$
$$P(RS|F) = 0.15, \qquad P(AR|F) = 0.40, \qquad P(BL|F) = 0.45$$

**(a)** Figure 7–25 shows the tree diagram with this information.

**(b)** The two branches that terminate at $AR$ are $M \cap AR$ and $F \cap AR$, so

$$P(AR) = 0.6 \times 0.20 + 0.4 \times 0.40 = 0.12 + 0.16 = 0.28$$

**(c)** The branches $F \cap AR$ and $F \cap BL$ form the two outcomes in this event, so

$$P(F \text{ who prefers } AR \text{ or } BL) = 0.16 + 0.18 = 0.34$$

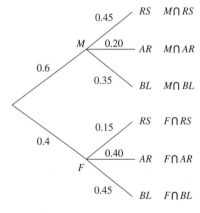

**FIGURE 7–25**

**(d)** If the events "male" and "prefers the recreation–sports complex" are independent, then $P(RS\,|\,M)$ must equal $P(RS)$. As $P(RS\,|\,M) = 0.45$ and $P(RS) = 0.27 + 0.06 = 0.33$, the events are not independent.

## Mutually Exclusive and Independent Events

Sometimes students find it difficult to understand the concepts of independent, dependent, and mutually exclusive events. If you determine two events to be dependent, you still might not know whether they are mutually exclusive or not because two dependent events may or may not be mutually exclusive. Let's look at some examples to get a feeling of the concepts.

**Example 9**  Nikki and Alice are two students at the university who do not know each other and have no contact with each other. In such a case, it seems reasonable that Nikki's decision to take art history or not has no influence on whether or not Alice takes art history. We say the events "Nikki takes art history" and "Alice takes art history" are independent because the probability that one of them takes the course has no influence on the probability that the other takes it. On the other hand, both could take the course, so the events are *not* mutually exclusive. To be mutually exclusive, the occurrence of one event *must exclude* the possibility of the other occurring.

Under different circumstances, as when Nikki and Alice are roommates, it seems reasonable to think that whether Nikki takes art history or not might influence whether or not Alice takes it; that is, the events "Nikki takes art history" and "Alice takes art history" might be dependent. Even if they are roommates, their decisions might still be independent, so we need information on the probability of each person's decision before we can say.

Even if their decisions are dependent, the events are not mutually exclusive because both could take art history.

Keep in mind that we test for mutually exclusive events by determining whether or not the two events can occur at the same time. The test for the independence or dependence of two events depends on how their probabilities are related.

**Example 10**  Some people's eyes become irritated and water when the pollen count rises above a certain threshold value. A similar reaction occurs when the ozone pollution rises above a threshold value. The city's Air Quality Board monitors the pollution levels and knows that 30% of the days the ozone will be above threshold value, 20% of the days pollen will be above threshold, and 6% of the days both will be above threshold.

**(a)** Are the events "ozone above threshold" and "pollen above threshold" mutually exclusive?

**(b)** Are the events "ozone above threshold" and "pollen above threshold" independent?

**Solution**

**(a)** Both events can occur at the same time. In fact, they occur at the same time 6% of the time, so the occurrence of one does not exclude the possibility of the other. They are not mutually exclusive.

**(b)** To determine independence or dependence, we need to check the probabilities of the events. One test given earlier in this section is to determine whether

$$P(\text{ozone pollution}) \times P(\text{pollen pollution})$$

is equal to $P(\text{both ozone and pollen pollution})$.

From the given information, we have

$$P(\text{ozone pollution}) \times P(\text{pollen pollution}) = 0.30 \times 0.20 = 0.06$$

and

$$P(\text{both ozone and pollen pollution}) = 0.06$$

As these computations are equal, the events are independent.

**Example 11**    Look at the three situations in Figure 7–26 and determine whether the events are mutually exclusive and whether they are independent or dependent.

**(a)** Because $E$ and $F$ overlap, they have outcomes in common, which tells us that they are not mutually exclusive.

To check for independence, we must compute the probabilities

$$P(E) \times P(F) = \frac{48}{132} \times \frac{22}{132} = \frac{8}{132}$$

and

$$P(E \cap F) = \frac{8}{132}$$

$E$ and $F$ are independent because $P(E) \times P(F) = P(E \cap F)$.

**(b)** The events $E$ and $F$ are not mutually exclusive, because they have events in common.

Now compute

$$P(E) \times P(F) = \frac{45}{320} \times \frac{65}{320} = \frac{117}{4096}$$

and

$$P(E \cap F) = \frac{5}{320} = \frac{1}{64}$$

$E$ and $F$ are dependent because $P(E) \times P(F) \neq P(E \cap F)$.

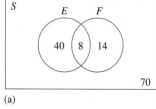
(a)

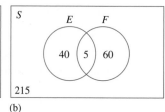
(b)

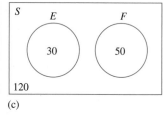
(c)

**FIGURE 7–26**

**(c)** The events $E$ and $F$ are mutually exclusive because they have no outcomes in common. $E$ and $F$ are dependent because $P(E) \times P(F) \neq P(E \cap F)$ as the following computations show:

$$P(E) \times P(F) = \frac{30}{200} \times \frac{50}{200} = \frac{3}{80}$$

and

$$P(E \cap F) = \frac{0}{200} = 0$$

---

## 7.5 EXERCISES

Access end-of-section exercises online at **www.webassign.net**

---

## 7.6 BAYES' RULE

- An Application Using a Tree Diagram

Conditional probability typically deals with the probability of an event when you have information about something that happened earlier. Let's look at a situation that reverses the information. Imagine the following:

A club separates a stack of bills for a drawing and places some in box A and the rest in box B. Each box contains some $50 bills and other bills. At the drawing, a person selects one box and draws a bill from that box. Conditional probability answers questions such as "If box A is selected, what is the probability that a $50 bill is drawn?" Symbolically, this is "What is $P$($50 bill, given that box A is selected)?" This question assumes that the first event (selecting a box) is known and asks for the probability of the second event.

Bayes' Rule deals with a reverse situation. It answers a question such as "If the person ends up with a $50 bill, what is the probability that it came from box A?" This assumes that the second event is known and asks for the probability of the first event. Bayes' Rule determines the probability of an earlier event based on information about an event that happened later.

Let's look at an example with some probabilities given to see when and how to use Bayes' Rule. We will then make a formal statement of the rule.

The student body at a college is 60% male and 40% female. The registrar's records show that 30% of the men attended private high schools and 70% attended public high schools. Furthermore, 75% of the women attended private high schools, and 25% attended public schools.

Before we go further, let's summarize this information in probability notation. We use $M$ and $F$ for male and female, and we abbreviate private and public with $PRI$ and $PUB$. Then, for a student selected at random,

$$P(M) = 0.6, \qquad\qquad P(F) = 0.4$$
$$P(PRI\,|\,M) = 0.3, \qquad P(PUB\,|\,M) = 0.7$$
$$P(PRI\,|\,F) = 0.75, \qquad P(PUB\,|\,F) = 0.25$$

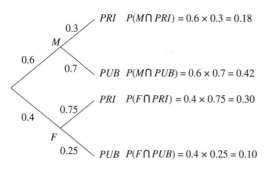

**FIGURE 7-28**

Figure 7–28 shows a tree diagram of this information, with its four branches terminating at $M \cap PRI$, $M \cap PUB$, $F \cap PRI$, and $F \cap PUB$. Their respective probabilities are

$$P(M \cap PRI) = P(M)P(PRI \,|\, M) = 0.6 \times 0.3 = 0.18$$

$$P(M \cap PUB) = P(M)P(PUB \,|\, M) = 0.6 \times 0.7 = 0.42$$

$$P(F \cap PRI) = P(F)P(PRI \,|\, F) = 0.4 \times 0.75 = 0.30$$

$$P(F \cap PUB) = P(F)P(PUB \,|\, F) = 0.4 \times 0.25 = 0.10$$

We can find the probability that a randomly selected student attended a private school by locating all branches that terminate in *PRI* and adding the probabilities. Thus,

$$P(PRI) = 0.18 + 0.30 = 0.48$$
$$P(PUB) = 0.42 + 0.10 = 0.52$$

Notice that in symbolic notation,

$$P(PRI) = P(M \cap PRI) + P(F \cap PRI)$$

and

$$P(PUB) = P(M \cap PUB) + P(F \cap PUB)$$

All the above information was developed for reference throughout the example. Now let's look at a problem in which Bayes' Rule is helpful.

Suppose a student selected at random is known to have attended a private school. What is the probability that the student selected is female; that is, what is $P(F \,|\, PRI)$? Note that this information is missing from the above. The definition of conditional probability gives

$$P(F|PRI) = \frac{P(F \cap PRI)}{P(PRI)}$$

You will find both $P(F \cap PRI)$ and $P(PRI)$ listed above. They were computed, not given originally. We obtained $P(PRI)$ from the two branches ending in *PRI*, so

$$P(PRI) = P(M \cap PRI) + P(F \cap PRI)$$

Thus, we can write

$$P(F|PRI) = \frac{P(F \cap PRI)}{P(M \cap PRI) + P(F \cap PRI)}$$

This is one form of Bayes' Rule. We get a more complicated-looking form, but one that uses the given information more directly, when we substitute

$$P(M \cap PRI) = P(M)P(PRI \mid M)$$

and

$$P(F \cap PRI) = P(F)P(PRI \mid F)$$

to get

$$P(F|PRI) = \frac{P(F)P(PRI|F)}{P(M)P(PRI|M) + P(F)P(PRI|F)}$$

$$= \frac{0.4 \times 0.75}{0.6 \times 0.3 + 0.4 \times 0.75} = \frac{0.30}{0.18 + 0.30} = \frac{0.30}{0.48} = 0.625$$

This last form has the advantage of using the information that was given directly in the problem.

We need to observe one more fact before making a general statement. We used $P(PRI) = P(M \cap PRI) + P(F \cap PRI)$. The events $M$ and $F$ make up *all* the branches in the first stage of the tree diagram, so $M \cup F$ gives all of the sample space ($M \cup F = S$). Furthermore, $M$ and $F$ are mutually exclusive. These two conditions on $M$ and $F$ are needed for Bayes' Rule.

Now we are ready for the general statement.

> **NOTE**
>
> This means that $M$ and $F$ form a partition of $S$.

---

**Bayes' Rule**    Let $E_1$ and $E_2$ be mutually exclusive events whose union is the sample space ($E_1 \cup E_2 = S$). Let $F$ be an event in $S$, $P(F) \neq 0$. Then,

**(a)** $P(E_1|F) = \dfrac{P(E_1 \cap F)}{P(F)}$

**(b)** $P(E_1|F) = \dfrac{P(E_1 \cap F)}{P(E_1 \cap F) + P(E_2 \cap F)}$

**(c)** $P(E_1|F) = \dfrac{P(E_1)P(F|E_1)}{P(E_1)P(F|E_1) + P(E_2)P(F|E_2)}$

---

Let's look at some Venn diagrams to help clarify these rules. The events $E_1$ and $E_2$ divide $S$ into two disjoint parts, and except when $F$ lies completely within $E_1$ or $E_2$, they divide $F$ into two parts (Figure 7–29).

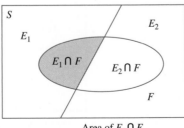

$$P(E_1|F) = \frac{\text{Area of } E_1 \cap F}{\text{Area of } F}$$

Area of $F$ = Area of $E_1 \cap F$ + Area of $E_2 \cap F$

**FIGURE 7–29**

The conditional probability $P(E_1|F)$ can be viewed as the fraction of $F$ that is covered by $E_1 \cap F$. Rule (a) states this as

$$P(E_1|F) = \frac{P(E_1 \cap F)}{P(F)}$$

We can also view $F$ as the part of $F$ that is in $E_1$ plus the part in $E_2$, symbolically stated as

$$F = (E_1 \cap F) \cup (E_2 \cap F)$$

Because these two parts are disjoint,

$$P(F) = P(E_1 \cap F) + P(E_2 \cap F)$$

Making this substitution in the denominator of rule (a) gives rule (b). Rule (c) then follows by replacing each term with its equivalent using the Multiplication Rule.

**Example 1**     A microchip company has two machines that produce the chips. Machine I produces 65% of the chips, but 5% of its chips are defective. Machine II produces 35% of the chips, and 15% of its chips are defective. A chip is selected at random and found to be defective. What is the probability that it came from machine I?

**Solution**

Let I be the set of chips produced by machine I and II be those produced by machine II. We want to find $P(\text{I} | \text{defective})$. We are given $P(\text{I}) = 0.65$, $P(\text{II}) = 0.35$, $P(\text{defective} | \text{I}) = 0.05$, and $P(\text{defective} | \text{II}) = 0.15$. We may use the second form of Bayes' Rule directly:

$$P(\text{I}|\text{defective}) = \frac{P(\text{I})P(\text{defective}|\text{I})}{P(\text{I})P(\text{defective}|\text{I}) + P(\text{II})P(\text{defective}|\text{II})}$$

$$= \frac{0.65 \times 0.05}{0.65 \times 0.05 + 0.35 \times 0.15} = \frac{0.0325}{0.085} = 0.38$$

Locate all of these computations on the tree diagram in Figure 7–30.

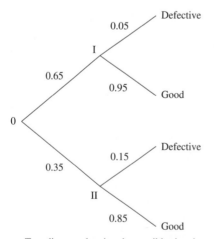

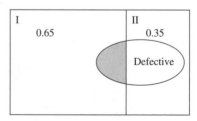

Tree diagram showing the possible situations of defective and good chips from machines I and II

(a)

The shaded area represents defective chips from machine I.

(b)

**FIGURE 7–30**

Bayes' Rule is not restricted to the situation in which just two mutually exclusive events form all of $S$. There can be any finite number of mutually exclusive events, as long as their union is the sample space. A more general form of Bayes' Rule is the following:

---

**Bayes' Rule (General)** Let $E_1, E_2, \ldots, E_n$ be mutually exclusive events whose union is the sample space $S$ (we say that they partition $S$), and let $F$ be any event where $P(F) \neq 0$. Then,

**(a)** $P(E_i|F) = \dfrac{P(E_i \cap F)}{P(F)}$

**(b)** $P(E_i|F) = \dfrac{P(E_i \cap F)}{P(E_1 \cap F) + P(E_2 \cap F) + \ldots + P(E_n \cap F)}$

**(c)** $P(E_i|F) = \dfrac{P(E_i)\,P(F|E_i)}{P(E_1)P(F|E_1) + P(E_2)P(F|E_2) + \ldots + P(E_n)P(F|E_n)}$

---

As with the Bayes' Rule given earlier, we can represent the general Bayes' Rule with a diagram. We do so with $E_1$, $E_2$, and $E_3$.

If we seek $P(E_2|F)$ for example, we want the ratio

$$\frac{\text{area of } E_2 \text{ in } F}{\text{area of } F}$$

This ratio can be obtained by using the area of all of $F$, or by finding the area of pieces of $F$ and putting them together (Figure 7–31) as the sum of the parts of $F$ that are in $E_1$, $E_2$, and $E_3$, symbolically stated as

$$F = \left(E_1 \cap F\right) \cup \left(E_2 \cap F\right) \cup \left(E_3 \cap F\right)$$

As these three parts are disjoint,

$$P(F) = P\left(E_1 \cap F\right) + P\left(E_2 \cap F\right) + P\left(E_3 \cap F\right)$$

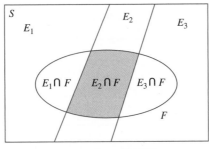

$$P(E_2|F) = \frac{\text{Area of } E_2 \cap F}{\text{Area of } F}$$

Area of $F$ = Area of $E_1 \cap F$ + Area of $E_2 \cap F$ + Area of $E_3 \cap F$

**FIGURE 7–31**

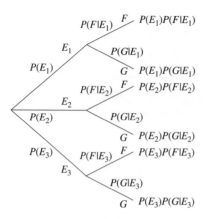

**FIGURE 7–32**

Making this substitution in the denominator of rule (a) gives rule (b). Rule (c) then follows by replacing each term with its equivalent using the Multiplication Rule.

We find it helpful to also represent Bayes' Rule with a tree diagram. Again we look at the case of a sequence of two events. The first event has three outcomes $E_1$, $E_2$, and $E_3$ with respective probabilities $P(E_1)$, $P(E_2)$, and $P(E_3)$. The second event has two outcomes, $F$ and $G$. The probability of $F$ occurring depends on which first event it follows. We designate this by $P(F \mid E_1)$ if $F$ follows $E_1$, by $P(F \mid E_2)$ if $F$ follows $E_2$, and by $P(F \mid E_3)$ if $F$ follows $E_3$. We show all of this with the tree diagram in Figure 7–32. The probability of terminating at an endpoint ($F$ or $G$) is shown for each endpoint. Now recall that the conditional probability is

$$P(E_1 \mid F) = \frac{P(E_1 \cap F)}{P(F)}.$$

Now look at Figure 7–32 to see how these probabilities are related to the tree diagram.

Observe that $E_1 \cap F$ follows the branches $E_1$ and then $F$; $E_2 \cap F$ follows the branches $E_2$ and then $F$; and $E_3 \cap F$ follows the branches $E_3$ and then $F$. We find $P(F)$ by adding the probabilities of terminating at each of these end points.

We summarize the probabilities as follows:

| Path of branches | Probability of path |
|---|---|
| $E_1$ followed by $F$, $E_1 \cap F$ | $P(E_1 \cap F) = P(E_1)\, P(F \mid E_1)$ |
| $E_2$ followed by $F$, $E_2 \cap F$ | $P(E_2 \cap F) = P(E_2)\, P(F \mid E_2)$ |
| $E_3$ followed by $F$, $E_3 \cap F$ | $P(E_3 \cap F) = P(E_3)\, P(F \mid E_3)$ |

Since $F = (E_1 \cap F) \cup (E_2 \cap F) \cup (E_3 \cap F)$, it follows that

$$P(F) = P(E_1 \cap F) + P(E_2 \cap F) + P(E_3 \cap F)$$
$$= P(E_1)\, P(F \mid E_1) + P(E_2)\, P(F \mid E_2) + P(E_3)\, P(F \mid E_3)$$

and so we have Bayes' Rule:

$$P(E_1 \mid F) = \frac{P(E_1 \cap F)}{P(F)} = \frac{P(E_1)P(F \mid E_1)}{P(E_1)P(F \mid E_1) + P(E_2)P(F \mid E_2) + P(E_3)P(F \mid E_3)}$$

In terms of branches of a tree diagram, we can summarize Bayes' Rule as:

$$P(E_1|F) = \frac{\text{Multiply the probabilities of the branches through } E_1 \text{ and ending at } F}{\text{Add the probabilities of all branches ending at } F}$$

The next two exercises illustrate some applications of Bayes' Rule.

**Example 2**   A manufacturer buys an item from three subcontractors, A, B, and C. A has the better quality control; only 2% of its items are defective. A furnishes the manufacturer with 50% of the items. B furnishes 30% of the items, and 5% of its items are defective. C furnishes 20% of the items, and 6% of its items are defective. The manufacturer finds an item defective (D) and would like to know which subcontractor supplied it.

**(a)** What is the probability that it came from A? (Find $P(A|D)$.)

**(b)** What is the probability that it came from B? (Find $P(B|D)$.)

**(c)** What is the probability that it came from C? (Find $P(C|D)$.)

**(d)** Which subcontractor was the most likely source of the defective item?

**Solution**
Let

$A$ represent the set of items produced by A.

$B$ represent the set of items produced by B.

$C$ represent the set of items produced by C.

$D$ represent the set of defective items.

The following probabilities are given:

$$P(A) = 0.50, \qquad P(B) = 0.30, \qquad P(C) = 0.20$$
$$P(D|A) = 0.02, \qquad P(D|B) = 0.05, \qquad P(D|C) = 0.06$$

Let's use the second form of Bayes' Rule to solve the problem. We need the following probabilities, which are computed by using the Multiplication Rule:

$$P(D \cap A) = P(A)P(D|A) = 0.50 \times 0.02 = 0.010$$
$$P(D \cap B) = P(B)P(D|B) = 0.30 \times 0.05 = 0.015$$
$$P(D \cap C) = P(C)P(D|C) = 0.20 \times 0.06 = 0.012$$

These areas are shown in Figure 7–33a. Then,

**(a)** $P(A|D) = \dfrac{0.010}{0.010 + 0.015 + 0.012} = \dfrac{0.010}{0.037} = 0.27$

**(b)** $P(B|D) = \dfrac{0.015}{0.010 + 0.015 + 0.012} = \dfrac{0.015}{0.037} = 0.41$

**(c)** $P(C|D) = \dfrac{0.012}{0.010 + 0.015 + 0.012} = \dfrac{0.012}{0.037} = 0.32$

**(d)** The largest probability, $P(B|D) = 0.41$, suggests that subcontractor B is the most likely source.

Trace the computations on the tree diagram in Figure 7–33b.

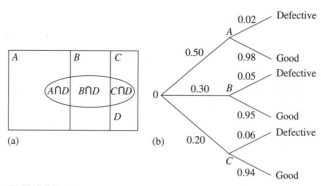

(a)                    (b)

**FIGURE 7–33**

**Example 3**   Studies show that a pregnant woman who contracts German measles is more likely to bear a child with certain birth defects. In a certain country, the probability that a pregnant woman contracts German measles is 0.2. If a pregnant woman contracts the disease, the probability that her child will have the defect is 0.1. If a pregnant woman does not contract German measles, the probability that her child will have the defect is 0.01. A child is born with this defect. What is the probability that the child's mother contracted German measles while pregnant? Restrict the analysis to pregnant women.

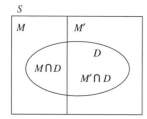

**FIGURE 7–34**

**Solution**

Let

$M$ = the set of pregnant women who contracted measles

$M'$ = the set of pregnant women who did not contract measles

$S$ = the sample space of pregnant women ($M \cup M'$)

$D$ = the set of mothers who bear a child with this defect

The Venn diagram of these events is shown in Figure 7–34 and the tree diagram in Figure 7–35.

We are given $P(M) = 0.2$, $P(M') = 0.8$, $P(D|M) = 0.1$, and $P(D|M') = 0.01$; and we are to find $P(M|D)$:

$$P(M|D) = \frac{P(M \cap D)}{P(M \cap D) + P(M' \cap D)}$$

$$= \frac{P(M)P(D|M)}{P(M)P(D|M) + P(M')P(D|M')}$$

$$= \frac{0.2 \times 0.1}{0.2 \times 0.1 + 0.8 \times 0.01} = \frac{0.02}{0.028} = 0.71$$

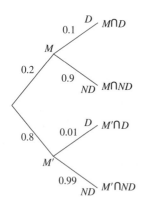

**FIGURE 7–35**

## An Application Using a Tree Diagram

For practicing physicians, much of their work involves diagnosis and treatment of diseases. For some diseases a diagnosis may be quite elusive. For other diseases, tests have been developed that greatly assist in diagnosis. However, such tests usually are not perfect. They may miss when a person actually has a disease and may wrongly indicate the disease when it is not present. Let's look at some probabilities of a hypothetical disease, the bad crud disease.

**Example 4**

It is known that the bad crud disease affects 1 in 100,000 people. That is, for a randomly selected person, the probability the person has bad crud is $P(BC) = \frac{1}{100,000} = 0.00001$. Likewise, the probability a person does not have bad crud is $P(NBC) = 1 - 0.00001 = 0.99999$.

Experience has shown that the test's accuracy can be characterized by the following probabilities. If a person has bad crud, the test so indicates 99% of the time (the test is *positive* 99% of the time). We can state this as the probability of a positive test for a person who has bad crud is $P(P \mid BC) = 0.99$. The other 1% of the time, the test is (falsely) *negative*, $P(N \mid BC) = 0.01$. Further, if the person does not have bad crud then the test so indicates (the test is negative) 99.5% of the time; that is, $P(N \mid NBC) = 0.995$. Then the other 0.5% of the time the test is (falsely) positive, so $P(P \mid NBC) = 0.005$.

We represent these probabilities with a tree diagram in Figure 7–36.

Now let's calculate some probabilities for a randomly selected person tested.

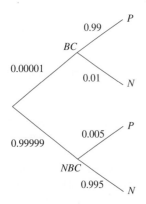

**FIGURE 7–36**

**(a)** Find the probability it is a person with bad crud and tests positive, $P(BC \cap P)$.

**(b)** Find the probability it is a person without bad crud and tests negative, $P(NBC \cap N)$.

**(c)** A person tests positive. Find the probability it is a person with bad crud, $P(BC \mid P)$.

**(d)** A person tests negative. Find the probability it is a person without bad crud, $P(NBC \mid N)$.

**Solution**

**(a)** In the tree diagram, this is the path $BC$ followed by $P$.

$$P(BC \cap P) = P(BC)P(P \mid BC) = 0.00001(0.99) = 0.0000099$$

A small probability indeed.

**(b)** In the tree diagram this is the path $NBC$ followed by $N$.

$$P(NBC \cap N) = P(NBC)P(N \mid NBC) = 0.99999(0.995) = 0.99499005$$

**(c)** In this case we want to find $P(BC \mid P)$, so we use Bayes' Rule:

$$P(BC \mid P) = \frac{P(BC \cap P)}{P(P)}$$

We found $P(BC \cap P)$ in part (a), $P(BC \cap P) = 0.0000099$. We find $P(P)$ by adding the probabilities for those paths that end at $P$ in the tree diagram. That is, the path through $BC$ and $P$ and the path through $NBC$ and $P$. From the tree diagram we get

$$0.00001(0.99) + 0.99999(0.005) = 0.00500985$$

so $P(BC|P) = \frac{0.0000099}{0.00500985} = 0.0019761$. This indicates that the chances the person selected has $BC$ is less than two tenths of a percent.

**(d)** We want to find $P(NBC \mid N)$, so we use Bayes' Rule:

$$P(NBC \mid N) = \frac{P(NBC \cap N)}{P(N)}$$

We use $P(NBC \cap N)$ from part (b) and find $P(N)$ from the tree diagram by adding the probabilities of those paths that end at $N$; that is,

$$\frac{0.99999 \times 0.995}{0.00001 \times 0.01 + 0.99999 \times 0.995} = \frac{0.99499005}{0.99499015} = 0.9999998995$$

which is close to 1. This indicates that the person selected almost certainly does not have bad crud.

———————●

Looking at the probabilities obtained in parts (b) and (c), doesn't it seem reasonable to conclude that the test for bad crud is meaningless? It doesn't matter if the tests are positive or negative, the probability of having bad crud is pretty small.

We remind you that the probabilities we obtained were for a *randomly selected person.* For a randomly selected person, the probability of having bad crud is 1 in 100,000 regardless of the test's outcome. We would expect a low probability that we found a person with the disease and a high probability we found a person without it. Thus, there seems little merit in giving this test on a random or routine basis. Generally, a physician will test for bad crud (or other diseases) when symptoms suggest the patient might actually have bad crud, or possibly another disease or two.

We note that sometimes a test is given on a random or routine basis. Some companies and organizations will conduct tests for illegal drug use on a random or routine basis. When the use of illegal drugs was rare, no one performed random drug tests; only when illegal drug use became prevalent did random testing take place, since the probability of illegal drug use had become higher.

Let's summarize this by using the bad crud example. When the probability of bad crud is low, there is little merit in random or routine testing. If the probability of bad crud is relatively high, the merit of random or routine testing increases.

A test is appropriate when a patient demonstrates symptoms that indicate the presence of bad crud—say, 60% of the time. This information indicates the person is no longer a random selection from the general population. The person becomes a random selection from the group of people with those symptoms. This changes $P(BC)$ and $P(NBC)$ to $P(BC)$ = 0.60 and $P(NBC)$ = 0.40, so the tree diagram would show 0.60 for the $BC$ branch and 0.40 for the $NBC$ branch. In this case, the probability that the patient actually has bad crud when the test is positive is

$$P(BC \mid P) = \frac{P(BC \cap P)}{P(P)} = \frac{0.60 \times 0.99}{0.60 \times 0.99 + 0.40 \times 0.005} = \frac{0.594}{0.596} = 0.997$$

This probability indicates a high possibility of $BC$.

Now, the patient may be in denial and argues the test isn't perfect so no bad crud exists. Let's find the probability of no bad crud when the test is positive, that is find $P(NBC \mid P)$. We have

$$P(NBC \mid P) = \frac{P(NBC \cap P)}{P(P)} = \frac{0.40 \times 0.005}{0.40 \times 0.005 + 0.60 \times 0.99} = 0.0034$$

which is a low probability that the patient does not have the bad crud. Both probabilities support the presence of the bad crud.

This example illustrates that the process of random selection is a fundamental requirement of probability. It also illustrates that, by representing the given information in a tree diagram, Bayes' Rule can be calculated using the key ideas

1. $P(E \mid F) = \dfrac{P(E \cap F)}{P(F)}$

2. $P(E \cap F)$ is the probability of the path from $E$ through $F$.

3. $P(F)$ is the sum of the probabilities of the paths ending at $F$.

---

## 7.6 EXERCISES

Access end-of-section exercises online at **www.webassign.net**

---

## 7.7 MARKOV CHAINS

- Transition Matrix
- Markov Chain
- Steady State
- Finding the Steady-State Matrix

### Transition Matrix

We now study a type of problem that uses probability, matrix operations, and systems of linear equations (you may wish to review those topics). We introduce you to this type of problem with the following simplified example.

Sedco Inc. has developed a quit smoking program. They find that 75% who complete the program quit smoking and 25% continue to smoke. They realize that the success of the program depends on the number who remain long-term nonsmokers, so they annually follow the participants of the program. They find that each year 90% of smokers continue to smoke while 10% become nonsmokers. Of the nonsmokers, 80% remain nonsmokers and 20% become smokers again. Sedco wants to know the expected number of nonsmokers after 1 year, after 5 years, and after 10 years.

To solve a problem like this, we need to define some terminology and some basic concepts. First, the participants can be placed in one of exactly two categories: they either smoke or do not smoke. We call these categories **states**.

> **DEFINITION** A **state** is a category, situation, outcome, or position that a process can occupy at any given time. The states are disjoint and cover all possibilities.

For example, the participants are either in the state of being a smoker or in the state of being a nonsmoker. A patient is ill or well. A person resides in an urban, suburban, or rural location. An employee works part time or full time. In a Markov process, the system may move from one state to another at any given time. When the process moves from one state to the next, we say that a transition is made from the **present state** to the **next state**. A participant can make a transition from the smoker state to the nonsmoker state or vice versa.

A transition matrix can represent the information on the proportion of participants who smoke and who do not smoke. Let $S$ represent those who smoke and $N$ represent those who do not:

$$\begin{array}{c}\\ \\ \text{Present} \\ \text{State}\end{array}\begin{array}{c}\text{Next State} \\ \begin{array}{cc} S & N \end{array} \\ \begin{array}{c} S \\ N \end{array}\begin{bmatrix} 0.90 & 0.10 \\ 0.20 & 0.80 \end{bmatrix}\end{array}$$

The entries in the transition matrix give the probabilities that a person will move from one state (present state) to another state (next state) the following year. For example, the probability that a person who smokes will smoke next year is 0.90. We get this from the statement "90% of smokers continue to smoke."

The headings to the left of the matrix identify the present state, and the headings above the matrix identify the state at the next stage. Interpret each entry in the matrix as follows:

0.90 is the probability that person passes from present state $S$ to state $S$ next; that is, a smoker remains a smoker.

0.10 is the probability that a person passes from present state $S$ to state $N$ next; that is, a smoker becomes a nonsmoker.

0.20 is the probability that a person passes from present state $N$ to state $S$ next; that is, a nonsmoker becomes a smoker.

0.80 is the probability that a person passes from present state $N$ to state $N$ next; that is, a nonsmoker remains a nonsmoker.

Because each row lists the probabilities of going from that state to each of all possible states, the entries in a row always add to 1.

**DEFINITION**
**Transition Matrix**
A **transition matrix** is a square matrix with each entry a number from the interval 0 through 1. The entries in each row add to 1.

In the Quit Smoking example, a row matrix may be used to represent the proportion of people in each state. For the example, the matrix [0.25   0.75] indicates that 25% are in state $S$ and 75% are in state $N$. The same row matrix may also be interpreted as indicating that the probability of a person being in state $S$ is 0.25 and the probability of being in state $N$ is 0.75.

We call these row matrices **state matrices** or **probability-state matrices**. For each state, they show the probability of a person being in that state.

Why do we put the information in this form? We do so because we can use matrix operations to provide useful information. Here's how.

Multiply the state matrix and the transition matrix:

$$\begin{array}{cc} S & N \end{array}$$
$$[0.25 \quad 0.75]\begin{bmatrix} 0.90 & 0.10 \\ 0.20 & 0.80 \end{bmatrix}$$

giving the result

$$[0.25(0.90) + 0.75(0.20) \quad 0.25(0.10) + 0.75(0.80)]$$

$$\begin{array}{cc} S & N \end{array}$$
$$= [0.375 \quad 0.625] \quad \text{(the state matrix at the next stage)}$$

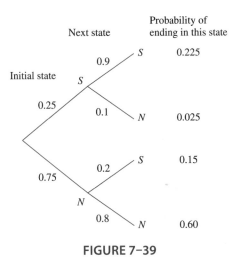

**FIGURE 7–39**

You should interpret this as follows: In one year, the participants moved from 25% in $S$ to 37.5% in $S$ and from 75% in $N$ to 62.5% in $N$.

Let's look at the tree diagram in Figure 7–39 to help justify these claims.

The first stage shows the two possible states, $S$ and $N$, with the probability of each. The second stage shows the states and the probability of entering those states. At the end of each branch is the probability of terminating there.

Note that two branches terminate in $S$. We simply add the two probabilities there to obtain the probability of ending in state $S$. That probability is

$$0.25(0.90) + 0.75(0.20) = 0.225 + 0.150 = 0.375$$

Note that this is exactly the computation used in the matrix product that gave the first entry in the state matrix for the next stage. The second entry in the state matrix for the next stage is

$$0.25(0.10) + 0.75(0.80) = 0.625$$

which gives the probability for entering state $N$ as obtained from the tree diagram.

This illustrates how the present state matrix and transition matrix can be used to obtain the next state matrix.

**Example 1**   Using $[0.25 \quad 0.75]$ as the first year's state matrix (sometimes called the **initial state matrix**) and

$$\begin{bmatrix} 0.90 & 0.10 \\ 0.20 & 0.80 \end{bmatrix}$$

as the transition matrix for the quit smoking problem, find the percentage of participants in each category for the second, third, and fourth years.

**Solution**

As illustrated on the previous page, the second-year breakdown is

$$
\begin{array}{cc} S & N \end{array} \\
[0.25 \quad 0.75] \begin{bmatrix} 0.90 & 0.10 \\ 0.20 & 0.80 \end{bmatrix} = \begin{array}{cc} S & N \end{array} \\
[0.375 \quad 0.625]
$$

1st year                                   2nd year

We obtain the third-year matrix by multiplying the second-year state matrix (it is now the present matrix) by the transition matrix:

$$
[0.375 \quad 0.625] \begin{bmatrix} 0.90 & 0.10 \\ 0.20 & 0.80 \end{bmatrix} = \left[ 0.375(0.90) + 0.625(0.20) \quad 0.375(0.10) + 0.625(0.80) \right]
$$

$$
= [0.4625 \quad 0.5375]
$$

The fourth-year state matrix is

$$
[0.4625 \quad 0.5375] \begin{bmatrix} 0.90 & 0.10 \\ 0.20 & 0.80 \end{bmatrix} = \left[ 0.4625(0.90) + 0.5375(0.20) \quad 0.4625(0.10) + 0.5375(0.80) \right]
$$

$$
= [0.52375 \quad 0.47625]
$$

This process may be continued for years 5, 6, and so on.

We apply the same process when there are more than two states.

**Example 2**   At the end of each fiscal year, the student loan program gathers information on the payment status of the loans. The loans are divided into three categories: payments up to date, with payments made within 15 days of the due date considered current (labeled 0–15); payments that are 16 to 90 days late (labeled 16–90); and payments over 90 days late (labeled 90+). Each year, some of the students change categories because they get behind in payments or catch up. A study of past years gives the following transition matrix showing the fraction of students that change from one category to another or stay in the same category:

|  |  | Move to category | | |
|---|---|---|---|---|
|  |  | 0–15 | 16–90 | 90+ |
| Move | 0–15 | 0.86 | 0.08 | 0.06 |
| from | 16–90 | 0.62 | 0.29 | 0.09 |
| Category | 90+ | 0.17 | 0.37 | 0.46 |

One year, the percentage in each category was 0–15, 80%; 16–90, 11%; and 90+, 9%.

**(a)** Find the percentage in each category the next year.

**(b)** Find the percentage in each category three years later.

**Solution**

At the end of each fiscal year, each student loan is in one of three states: paid up (0–15), in arrears 90 days or less (16–90), or over 90 days in arrears (90+). The present state matrix is [0.80   0.11   0.09].

**(a)** The next-state matrix is

$$[0.80 \ 0.11 \ 0.09] \begin{bmatrix} 0.86 & 0.08 & 0.06 \\ 0.62 & 0.29 & 0.09 \\ 0.17 & 0.37 & 0.46 \end{bmatrix}$$
$$= [0.772 \ 0.129 \ 0.099]$$

with the entries rounded to three decimals. A year later, there are 77.2% in the 0–15 category, 12.9% in the 16–90 category, and 9.9% in the 90+ category.

**(b)** Two years later, the state matrix is

$$[0.772 \ 0.129 \ 0.099] \begin{bmatrix} 0.86 & 0.08 & 0.06 \\ 0.62 & 0.29 & 0.09 \\ 0.17 & 0.37 & 0.46 \end{bmatrix}$$
$$= [0.760 \ 0.136 \ 0.104]$$

Two years later, there are 76.0% in the 0–15 category, 13.6% in the 16–90 category, and 10.4% in the 90+ category.

Three years later, the state matrix is

$$[0.760 \ 0.136 \ 0.104] \begin{bmatrix} 0.86 & 0.08 & 0.06 \\ 0.62 & 0.29 & 0.09 \\ 0.17 & 0.37 & 0.46 \end{bmatrix}$$
$$= [0.755 \ 0.139 \ 0.106]$$

Three years later, there are 75.5% in the 0–15 category, 13.9% in the 16–90 category, and 10.6% in the 90+ category.

## Markov Chain

Let's summarize the ideas of a Markov chain.

A **Markov chain**, or **Markov process**, is a sequence of experiments with the following properties:

1. An experiment has a finite number of discrete outcomes, called **states**. The process, or experiment, is always in one of these states.
2. With each additional trial, the experiment can move from its present state to any other state or remain in the same state.
3. The probability of going from one state to another on the next trial depends only on the present state and not on past states.

*(continued)*

**4.** The probability of moving from any one state to another in one step is represented in a transition matrix.

   **(a)** The transition matrix is square, because all possible states are used for rows and columns.

   **(b)** Each entry is between 0 and 1, inclusive.

   **(c)** The entries in each row add to 1.

**5.** The state matrix times the transition matrix gives the state matrix for the next stage.

## Steady State

A study of Markov processes enables us to determine the probability-state matrix for a sequence of trials. Sometimes it helps to know the long-term trends of a population, of the market of a product, or of political processes. A Markov chain may provide some useful long-term information because some Markov processes will tend toward a **steady state**, or **equilibrium**. Here is a simple example.

**Example 3**    The transition matrix of a Markov process is

$$T = \begin{bmatrix} 0.6 & 0.4 \\ 0.1 & 0.9 \end{bmatrix}$$

and an initial-state matrix is [0.50    0.50].

   If we compute a sequence of state matrices for subsequent stages, we obtain the following information:

| Step | State matrix | | |
|---|---|---|---|
| Initial | [0.50    0.50] | | |
| 1 | [0.35    0.65] | = [0.50    0.50]$T$ | |
| 2 | [0.275    0.725] | = [0.35    0.65]$T$ | = [0.50    0.50]$T^2$ |
| 3 | [0.238    0.762] | = [0.275    0.725]$T$ | = [0.50    0.50]$T^3$ |
| 4 | [0.219    0.781] | = [0.238    0.762]$T$ | = [0.50    0.50]$T^4$ |
| 5 | [0.209    0.791] | = [0.219    0.781]$T$ | = [0.50    0.50]$T^5$ |
| 6 | [0.204    0.796] | = [0.209    0.791]$T$ | = [0.50    0.50]$T^6$ |
| 7 | [0.202    0.798] | = [0.204    0.796]$T$ | = [0.50    0.50]$T^7$ |
| 8 | [0.201    0.799] | = [0.202    0.798]$T$ | = [0.50    0.50]$T^8$ |

It appears that the state matrix is approaching [0.20    0.80] as the sequence of trials progresses. In fact, that is the case. Furthermore, the state matrix [0.20    0.80] has an interesting property, which we can observe when we find the state matrix for the next stage:

$$[0.20 \ 0.80] \begin{bmatrix} 0.6 & 0.4 \\ 0.1 & 0.9 \end{bmatrix} = [(0.20)(0.6) + (0.80)(0.1) \ \ (0.20)(0.4) + (0.80)(0.9)]$$

$$= [0.12 + 0.08 \ \ 0.08 + 0.72]$$

$$= [0.20 \ 0.80]$$

There is no change in the next state matrix. The process has reached a **steady**, or **equilibrium**, state.

---

**DEFINITION**
**Steady-State Matrix**

A state matrix $X = [p_1 \quad p_2 \ldots p_n]$ is a **steady-state**, or **equilibrium**, **matrix** for a transition matrix $T$ if $XT = X$.

---

**Example 4**

The steady-state matrix for the Quit Smoking problem is $\begin{bmatrix} \frac{2}{3} & \frac{1}{3} \end{bmatrix}$ because an initial state matrix will eventually approach $\begin{bmatrix} \frac{2}{3} & \frac{1}{3} \end{bmatrix}$ and

$$\begin{bmatrix} \frac{2}{3} & \frac{1}{3} \end{bmatrix}\begin{bmatrix} 0.90 & 0.10 \\ 0.20 & 0.80 \end{bmatrix} = \begin{bmatrix} \frac{1.8}{3} + \frac{0.2}{3} & \frac{0.2}{3} + \frac{0.8}{3} \end{bmatrix} = \begin{bmatrix} \frac{2}{3} & \frac{1}{3} \end{bmatrix}$$

This indicates that, as long as the transition matrix represents the smoking practices of the participants, those practices will stabilize at two thirds smokers and one third nonsmokers.

---

## Finding the Steady-State Matrix

To find the steady-state matrix of the quit smoking problem, let $X = [x \quad y]$ be the desired, but unknown, steady-state matrix. We want to find $x$ and $y$ such that

$$[x \quad y]\begin{bmatrix} 0.90 & 0.10 \\ 0.20 & 0.80 \end{bmatrix} = [x \quad y]$$

The matrix product on the left gives

$$[0.90x + 0.20y \quad 0.10x + 0.80y]$$

so

$$0.90x + 0.20y = x$$
$$0.10x + 0.80y = y$$

which is equivalent to

$$-0.10x + 0.20y = 0$$
$$0.10x - 0.20y = 0$$

Since these two equations are equivalent, we can drop one of them.

Because $[x \quad y]$ is a probability matrix, we must have $x + y = 1$. This, together with the previous equation, gives the system

$$x + \quad y = 1$$
$$-0.10x + 0.20y = 0$$

If we use an augmented matrix to solve this system, we have

$$\begin{bmatrix} 1 & 1 & | & 1 \\ -0.1 & 0.2 & | & 0 \end{bmatrix}$$

which reduces to

$$\begin{bmatrix} 1 & 0 & | & \frac{2}{3} \\ 0 & 1 & | & \frac{1}{3} \end{bmatrix}$$

so $x = \frac{2}{3}$, $y = \frac{1}{3}$ gives $\begin{bmatrix} \frac{2}{3} & \frac{1}{3} \end{bmatrix}$ as the steady-state matrix. This approach will work in general.

**Example 5**   Find the steady-state matrix of the transition matrix

$$T = \begin{bmatrix} 0.3 & 0.2 & 0.5 \\ 0.1 & 0.4 & 0.5 \\ 0.4 & 0 & 0.6 \end{bmatrix}$$

**Solution**
Solve the equation

$$\begin{bmatrix} x & y & z \end{bmatrix} \begin{bmatrix} 0.3 & 0.2 & 0.5 \\ 0.1 & 0.4 & 0.5 \\ 0.4 & 0 & 0.6 \end{bmatrix} = \begin{bmatrix} x & y & z \end{bmatrix}$$

for a probability matrix $\begin{bmatrix} x & y & z \end{bmatrix}$, which is the system

$$\begin{aligned} 0.3x + 0.1y + 0.4z &= x \\ 0.2x + 0.4y &= y \\ 0.5x + 0.5y + 0.6z &= z \end{aligned}$$

plus the equation $x + y + z = 1$. Write this as

$$\begin{aligned} x + y + z &= 1 \\ -0.7x + 0.1y + 0.4z &= 0 \\ 0.2x - 0.6y &= 0 \\ 0.5x + 0.5y - 0.4z &= 0 \end{aligned}$$

Solve this system using the Gauss-Jordan Method. It gives the following sequence of augmented matrices:

$$\begin{bmatrix} 1 & 1 & 1 & | & 1 \\ -0.7 & 0.1 & 0.4 & | & 0 \\ 0.2 & -0.6 & 0 & | & 0 \\ 0.5 & 0.5 & -0.4 & | & 0 \end{bmatrix}$$

Multiply the last three rows by 10 to obtain integer entries and perform row operations indicated

$$\begin{bmatrix} 1 & 1 & 1 & | & 1 \\ -7 & 1 & 4 & | & 0 \\ 2 & -6 & 0 & | & 0 \\ 5 & 5 & -4 & | & 0 \end{bmatrix} \qquad \begin{aligned} 7R1 &+ R2 \rightarrow R2 \\ -2R1 &+ R3 \rightarrow R3 \\ -5R1 &+ R4 \rightarrow R4 \end{aligned}$$

$$\begin{bmatrix} 1 & 1 & 1 & | & 1 \\ 0 & 8 & 11 & | & 7 \\ 0 & -8 & -2 & | & -2 \\ 0 & 0 & -9 & | & -5 \end{bmatrix} \quad \text{R2 + R3} \rightarrow \text{R3}$$

$$\begin{bmatrix} 1 & 1 & 1 & | & 1 \\ 0 & 8 & 11 & | & 7 \\ 0 & 0 & 9 & | & 5 \\ 0 & 0 & -9 & | & -5 \end{bmatrix} \quad \text{R3 + R4} \rightarrow \text{R4}$$

$$\begin{bmatrix} 1 & 1 & 1 & | & 1 \\ 0 & 8 & 11 & | & 7 \\ 0 & 0 & 9 & | & 5 \\ 0 & 0 & 0 & | & 0 \end{bmatrix} \quad \begin{array}{l} \frac{1}{8}\text{R2} \rightarrow \text{R2} \\ \frac{1}{9}\text{R3} \rightarrow \text{R3} \end{array}$$

$$\begin{bmatrix} 1 & 1 & 1 & | & 1 \\ 0 & 1 & \frac{11}{8} & | & \frac{7}{8} \\ 0 & 0 & 1 & | & \frac{5}{9} \\ 0 & 0 & 0 & | & 0 \end{bmatrix} \quad \begin{array}{l} -\text{R3 + R1} \rightarrow \text{R1} \\ \\ -\frac{11}{8}\text{R3 + R2} \rightarrow \text{R2} \end{array}$$

$$\begin{bmatrix} 1 & 1 & 0 & | & \frac{4}{9} \\ 0 & 1 & 0 & | & \frac{1}{9} \\ 0 & 0 & 1 & | & \frac{5}{9} \\ 0 & 0 & 0 & | & 0 \end{bmatrix} \quad -\text{R2 + R1} \rightarrow \text{R1}$$

$$\begin{bmatrix} 1 & 0 & 0 & | & \frac{3}{9} \\ 0 & 1 & 0 & | & \frac{1}{9} \\ 0 & 0 & 1 & | & \frac{5}{9} \\ 0 & 0 & 0 & | & 0 \end{bmatrix}$$

This gives $x = \frac{3}{9}, y = \frac{1}{9}, z = \frac{5}{9}$ and the steady-state matrix $\begin{bmatrix} \frac{3}{9} & \frac{1}{9} & \frac{5}{9} \end{bmatrix}$.

**Example 6**   A sociologist made a regional study of the shift of population between rural and urban areas. The transition matrix of the annual shift from one area to another was found to be

|  | To | |
|---|---|---|
|  | R | U |
| From   R | 0.76 | 0.24 |
| U | 0.08 | 0.92 |

indicating that 76% of rural residents remain in rural areas, 24% move from rural to urban areas, 8% of urban residents move from urban to rural areas, and 92% remain in the urban areas. Find the percentage of the population in rural and urban areas when the population stabilizes.

**Solution**

Let $[x \quad y]$ be the state matrix of the population, with $x$ the proportion in rural areas and $y$ the proportion in urban areas. We want to find the steady-state matrix, that is, the solution to

$$[x \quad y]\begin{bmatrix} 0.76 & 0.24 \\ 0.08 & 0.92 \end{bmatrix} = [x \quad y]$$

This condition, with $x + y = 1$, gives the system

$$\begin{aligned} x + \quad y &= 1 \\ -0.24x + 0.08y &= 0 \\ 0.24x - 0.08y &= 0 \end{aligned}$$

The solution to the system is $x = 0.25$ and $y = 0.75$ (solve it), so the steady-state matrix is $[0.25 \quad 0.75]$, indicating that the population will stabilize at 25% in rural areas and 75% in urban areas.

Let's look at the steady-state situation with two states, and thus a $2 \times 2$ transition matrix, to see how the steady-state solution can be reduced to a linear equation. We use the transition matrix

$$\begin{bmatrix} 0.76 & 0.24 \\ 0.08 & 0.92 \end{bmatrix}$$

from Example 6.

The steady-state solution is found by solving the system

$$\begin{aligned} x + \quad y &= 1 \\ 0.76x + 0.08y &= x \\ 0.24x + 0.92y &= y \end{aligned}$$

In a steady-state problem with two states, the first equation always appears, and the other equations (like the last two equations here) are always equivalent. Thus, $x$ and $y$ are related with $y = 1 - x$. The steady-state solution can be found by substituting $x = t$ and $y = 1 - t$ into one of the last two equations. Using the middle equation, we have $0.76t + 0.08(1 - t) = t$, which reduces to $0.32t - 0.08 = 0$, which gives $t = 0.25$ and $1 - t = 0.75$. The steady-state solution then is $[0.25 \quad 0.75]$. This procedure can be used when solving a two-state problem with a $2 \times 2$ transition matrix.

It is sometimes important to know whether a Markov process will eventually reach equilibrium. In Examples 5 and 6 we found the steady-state matrix. It happens to be true in those cases that we will eventually reach the steady-state matrix after a sequence of trials, regardless of the initial state matrix. Although this is not true for all transition matrices, there is a rather reasonable property that ensures that a Markov process will reach equilibrium. We call transition matrices with this property **regular**. A regular Markov process will eventually reach a steady state, and its transition matrix has the following property:

**DEFINITION**
**Regular Matrix**

A transition matrix $T$ of a Markov process is called **regular** if some power of $T$ has only positive entries.

A regular transition matrix is useful because it defines a Markov process that eventually reaches a steady state.

**Example 7**

$$T = \begin{bmatrix} 0.3 & 0.7 \\ 0.25 & 0.75 \end{bmatrix}$$

is regular because its first power contains all positive entries.

$$T = \begin{bmatrix} 0 & 1 \\ 0.6 & 0.4 \end{bmatrix}$$

is regular because

$$\begin{bmatrix} 0 & 1 \\ 0.6 & 0.4 \end{bmatrix}^2 = \begin{bmatrix} 0.6 & 0.4 \\ 0.24 & 0.76 \end{bmatrix}$$

has all positive entries.

**Example 8**  Find the steady-state matrix of the regular transition matrix

$$T = \begin{bmatrix} 0 & 0.5 & 0.5 \\ 0.5 & 0.5 & 0 \\ 0.5 & 0 & 0.5 \end{bmatrix}$$

(It is regular.)

**Solution**
The condition

$$\begin{bmatrix} x & y & z \end{bmatrix} \begin{bmatrix} 0 & 0.5 & 0.5 \\ 0.5 & 0.5 & 0 \\ 0.5 & 0 & 0.5 \end{bmatrix} = \begin{bmatrix} x & y & z \end{bmatrix}$$

with $x + y + z = 1$ yields a system of four equations whose augmented matrix is

$$\begin{bmatrix} 1 & 1 & 1 & 1 \\ -1 & 0.5 & 0.5 & 0 \\ 0.5 & -0.5 & 0 & 0 \\ 0.5 & 0 & -0.5 & 0 \end{bmatrix}$$

(Be sure that you can get this matrix.)
We will not show all the row operations that lead to the solution, but the final matrix is

$$\begin{bmatrix} 1 & 0 & 0 & \frac{1}{3} \\ 0 & 1 & 0 & \frac{1}{3} \\ 0 & 0 & 1 & \frac{1}{3} \\ 0 & 0 & 0 & 0 \end{bmatrix}$$

so the steady-state matrix is $\begin{bmatrix} \frac{1}{3} & \frac{1}{3} & \frac{1}{3} \end{bmatrix}$.

## 7.7 EXERCISES

Access end-of-section exercises online at **www.webassign.net**

## IMPORTANT TERMS

**7.1**
Experiment
Outcome
Trial
Sample Space
Sample Point
Event
Simple Event
Simple Outcome
Probability Assignment
Empirical Probability
Relative Frequency

**7.2**
Equally Likely
Successes
Failures
Random Selection
Random Outcome

**7.3**
Compound Event
Union
Intersection
Complement
Mutually Exclusive Events
Disjoint Events

**7.4**
Conditional Probability
Reduced Sample Space
Multiplication Rule

**7.5**
Independent Events
Dependent Events
Multiplication Rule for
    Independent Events
Pigeonhole Principle

**7.6**
Bayes' Rule

**7.7**
State
Present State
Next State
Transition Matrix
Probability-State Matrix
Initial State Matrix
Markov Chain
Steady State
Equilibrium
Steady-State Matrix
Regular Matrix

## IMPORTANT CONCEPTS

**Experiment**      An activity or phenomenon under consideration.

**Outcome**      An observable result of an experiment.

**Trial**      An observation or repetition of an experiment.

**Sample Space**      The set of all possible outcomes of an experiment.

**Probability**      A numerical estimate of the likelihood of a specific random outcome occurring.

**Event**      A subset of a sample space.

**Properties of Probability**      Each simple outcome is assigned a probability.

The probability of an event is the sum of the probabilities of all simple outcomes making up the event.

Each probability, $p$, has a value $0 \leq p \leq 1$

The probability of the sample space is 1.

| | |
|---|---|
| **Equally Likely Events** | Events that have equal probabilities. |
| **Compound Event** | An event formed from a union, intersection, or complement of events. |
| **Conditional Probability** | The probability of $E$ given $F$, denoted $P(E \mid F)$, has occurred. |

$$P(E \mid F) = \frac{P(E \cap F)}{P(F)}, P(F) \neq 0$$

| | |
|---|---|
| **Multiplication Rule** | $P(E \text{ and } F) = P(E \cap F) = P(E)P(F \mid E) = P(F)P(E \mid F)$ |
| **Independent Events** | $E$ and $F$ are independent if $P(E \mid F) = P(E)$ and $P(F \mid E) = P(F)$ |
| **Multiplication Rule for Independent Events** | $P(E \text{ and } F) = P(E \cap F) = P(E)P(F)$ |
| **Union of Independent Events** | When $E$ and $F$ are independent, $P(E \cup F) = P(E) + P(F) - P(E)P(F)$. |
| **Bayes' Rule** | $E_1$ and $E_2$ are mutually exclusive events whose union is the sample space, $S$, and $F$ is an event in $S$ and $P(F) \neq 0$ |

$$P(E_1 \mid F) = \frac{P(E_1 \cap F)}{P(F)} = \frac{P(E_1 \cap F)}{P(E_1 \cap F) + P(E_2 \cap F)}$$

$$= \frac{P(E_1)P(F \mid E_1)}{P(E_1)P(F \mid E_1) + P(E_2)P(F \mid E_2)}$$

| | |
|---|---|
| **State** | A category, situation, outcome, or position that a process can occupy at any given time. The states are disjoint and cover all possibilities. |
| **Markov Chain** | A sequence of independent trials with a finite number of discrete states. With each trial, the experiment can move from its present state to any other state or remain in the same state. The probability of going from one state to another depends only on the present state. |
| **Equilibrium or Steady State** | A trial in the sequence results in no change in the present state. |
| **Transition Matrix** | A matrix that represents the transition from a present state to the next state in a Markov Chain. It is a square matrix with entries in the interval zero through 1. The entries in a row add to 1. |

## 7  REVIEW EXERCISES

**1.** An experiment has six possible outcomes with the following probabilities:

$$P_1 = 0.02, \quad P_2 = 0.3, \quad P_3 = 0.1$$
$$P_4 = 0.0, \quad P_5 = 0.2, \quad P_6 = 0.3$$

Is this a valid probability assignment?

**2.** An experiment has the sample space $S = \{a, b, c, d\}$. Find the probability of each simple outcome in $S$ if

$$P(a) = P(b), \quad P(c) = 2P(b), \quad P(d) = 3P(c)$$

**3.** A refreshment stand kept a tally of the number of soft drinks sold. One day its records showed the following:

| Soft drinks | |
|---|---|
| **Size** | **Number sold** |
| Small | 94 |
| Medium | 146 |
| Large | 120 |

Find the probability that a person selected at random will buy a medium-sized soft drink.

**4.** Three people are selected at random from a group of five men and two women.

  **(a)** Find the probability that all three selected are men.

  **(b)** Find the probability that two men and one woman are selected.

  **(c)** Find the probability that all three selected are women.

**5.** In a class of 30 students, 10 participate in sports, 12 participate in band, and 5 participate in both. If a student is selected at random, find the probability that the student participates in sports or band.

**6.** In a group of 30 schoolchildren, 15 are eight-year-olds, 12 are nine-year-olds, and 3 are ten-year-olds. Of the eight-year-olds, 10 are boys; of the nine-year-olds, 5 are boys; and of the ten-year-olds, 2 are boys. One child is selected at random from the group. Find the probability that the child is

  **(a)** an eight-year-old.    **(b)** a boy.

  **(c)** a nine-year-old.     **(d)** a twelve-year-old.

  **(e)** a nine-year-old girl.  **(f)** a ten-year-old girl.

**7.** Mark and Melanie are two of the ten students who volunteer to tutor children after school. Four students are selected to tutor at South Elementary. Find the probability that Mark and Melanie are among the four selected.

**8.** A die is rolled. Find the probability that an even number or a number greater than 4 will be rolled.

**9.** A single card is picked from a deck of 52 playing cards. Find the probability that it will be a king or a spade.

**10.** A coin and a die are tossed. Find the probability of throwing a

  **(a)** head and a number less than 3.

  **(b)** tail and an even number.

  **(c)** head and a 6.

**11.** A coin is tossed five times. Find the probability that all five tosses will land heads up.

**12.** A card is selected from a deck of 52 playing cards. Find the probability that it will be a

  **(a)** red card or a 10.    **(b)** face card or a spade.

  **(c)** face card or a 10.

**13.** An elementary school teacher has a collection of mathematics review questions, including 10 addition, 8 subtraction, and 15 multiplication. A computer randomly selects problems for a student.

  **(a)** Find the probability that the first problem selected is an addition or subtraction problem.

  **(b)** Find the probability that the first two are subtraction or multiplication problems.

**14.** A card is selected at random from a deck of bridge cards. Find the probability that it is not

  **(a)** an ace.          **(b)** a face card.

**15.** A bargain table has 40 books; 10 are romance, 10 are biographies, 10 are crafts, and 10 are historical fiction. If 2 books are selected at random, what is the probability that they are

  **(a)** the same kind?

  **(b)** different kinds?

**16.** A load of lumber contains 40 pieces of birch and 50 pieces of pine. Of the lumber, 5 pieces of birch and 3 pieces of pine are warped. Let $F$, $G$, and $H$ be the events of selecting birch, pine, and a warped piece of wood, respectively. Compute and interpret the following probabilities.

  **(a)** $P(F)$, $P(G)$, $P(H)$    **(b)** $P(F \cap H)$

  **(c)** $P(F \cup H)$           **(d)** $P(F' \cup H)$

  **(e)** $P\big((F' \cup H)'\big)$

**17.** A card is picked from a deck of 52 playing cards. Let $F$ be the event of selecting an even-numbered card and let $G$ be that of selecting a 10. Are $G$ and $F$ independent events?

**18.** A card is picked from a deck of 52 playing cards. Let $R$ be the event of selecting a red card and let $Q$ be the event of selecting a Queen. Are $R$ and $Q$ independent?

**19.** A die is tossed four times. Find the probability of obtaining

  **(a)** 1, 2, 3, 4, in that order.

  **(b)** 1, 2, 3, 4, in any order.

  **(c)** two even numbers, then a 5, then a number less than 3.

**20.** Two dice are rolled. Find the probability that the

  **(a)** sum of the numbers on the dice is 6.

  **(b)** same number is obtained on each die.

21. A student has four examinations to take. She has determined that the probability of her passing the mathematics examination is 0.8; English, 0.5; history, 0.3; and chemistry, 0.7. Assuming independence of examinations, find the probability of her passing

    (a) mathematics, history, and English but failing chemistry.

    (b) mathematics and chemistry but failing history and English.

    (c) all four subjects.

22. A study of juvenile delinquents shows that 60% come from low-income families ($LI$), 45% come from broken homes ($BH$), and 35% come from both ($LI \cap BH$). A juvenile delinquent is selected at random.

    (a) Find the probability that the juvenile is not from a low-income family.

    (b) Find the probability that the juvenile comes from a broken home or a low-income family.

    (c) Find the probability that the juvenile comes from a low-income family, given that the juvenile comes from a broken home.

    (d) Are $LI$ and $BH$ independent?

    (e) Are $LI$ and $BH$ mutually exclusive?

23. A study of the adult population in a midwestern state found the following information on drinking habits:

    |            | Men | Women |
    |------------|-----|-------|
    | Abstain    | 20% | 40%   |
    | Infrequent | 10% | 20%   |
    | Moderate   | 50% | 35%   |
    | Heavy      | 20% | 5%    |

    The adult population of the state is 55% female and 45% male. An individual selected at random is found to be

    (a) an abstainer. Find the probability that the person is male.

    (b) a heavy drinker. Find the probability that the individual is female.

24. A stock analyst classifies stocks as either blue chip ($BC$) or not ($NBC$). The analyst also classifies stock by whether it goes up ($UP$), remains unchanged ($UC$), or goes down ($D$) at the end of a day's trading. One percent of the stocks are blue chip. The analyst summarizes the performance of stocks as follows:

    |     | Probability of | | |
    |-----|------|------|------|
    |     | *UP* | *UC* | *D*  |
    | *BC*  | 0.45 | 0.35 | 0.20 |
    | *NBC* | 0.35 | 0.25 | 0.40 |

    (a) Show this information with a tree diagram.

    (b) A customer selects a stock at random and asks the analyst to buy. Find the probability that the stock is a blue chip stock that goes up the next day.

25. A two-digit number is to be constructed at random from the numbers 1, 2, 3, 4, 5, and 6, repetition not allowed. Find the probability of getting

    (a) the number 33.       (b) the number 35.

    (c) a number the sum of whose digits is 10.

    (d) a number whose first digit is greater than its second.

    (e) a number less than 24.

26. A finite mathematics professor observed that 90% of the students who do the homework regularly pass the course. He also observed that only 20% of those who do not do the homework regularly pass the course. One semester, he estimated that 70% of the students did the homework regularly. Given a student who passed the course, find the probability that the student did the homework regularly.

27. A sociology class is composed of 10 juniors, 34 seniors, and 6 graduate students. Two juniors, 8 seniors, and 3 graduate students received an A in the course. A student is selected at random and is found to have received an A. What is the probability that the student is a junior?

28. In a certain population, 5% of the men are color-blind and 3% of the women are color-blind. The population is made up of 55% men and 45% women. If a person chosen at random is color-blind, what is the probability that the person is a man?

**29.** Two National Merit finalists and three semifinalists are seated in a row of five chairs on the stage. In how many ways can they be seated if the

(a) finalists are in the first two chairs and the semifinalists are seated in the last three?

(b) finalists and semifinalists alternate seats?

**30.** The history department hosts a distinguished visiting historian. Two students are randomly selected from the history majors—6 seniors, 5 juniors, and 3 sophomores—to join lunch with the scholar.

(a) Find the probability that the first student selected is a senior and the second is a sophomore.

(b) Find the probability that both are juniors.

**31.** Ten cards are numbered 1 through 10. A card is drawn. It is then replaced in the deck, and the cards are shuffled. A second card is drawn. Find the probability that

(a) the first card is less than 3 and the second greater than 7.

(b) both cards are less than 4.

**32.** A college has 1650 female students and 1460 males. The financial aid office reports that 35% of the males and 40% of the females receive financial aid. A student is selected at random. Find the probability that the student

(a) is female.      (b) receives financial aid.

(c) receives financial aid, given that the student is male.

(d) is female, given that the student receives financial aid.

**33.** A mathematics placement exam is scored high, middle, and low. The performance of these students in calculus is summarized as follows:

| | Score | | | |
|---|---|---|---|---|
| **Grade** | **High** | **Middle** | **Low** | **Total** |
| C or above | 98 | 124 | 3 | 225 |
| Below C | 12 | 118 | 65 | 195 |
| Total | 110 | 242 | 68 | 420 |

One of the 420 students is selected at random. Find the probability that the student

(a) makes a grade of C or above.

(b) scored low on the placement exam.

(c) made a grade below C, given that the student scored middle on the placement exam.

(d) scored high on the placement exam, given that the student made C or above.

**34.** A mathematics department compares SAT mathematics scores to performance in calculus. The findings are summarized in the following table:

| | SAT score | | | |
|---|---|---|---|---|
| **Grade** | **Below 550** | **550–650** | **Above 650** | **Total** |
| A or B | 1 | 55 | 28 | 84 |
| C | 21 | 69 | 25 | 115 |
| Below C | 43 | 56 | 2 | 101 |
| Total | 65 | 180 | 55 | 300 |

A student is selected at random.

(a) Find the probability that the student scored above 650.

(b) Find the probability that the student made an A or a B.

(c) Find the probability that the student scored above 650 and made an A or a B.

(d) Find the probability that the student's SAT score was in the 550–650 range.

(e) Find the probability that the student made a C.

(f) Show that "SAT above 650" and "grade of A or B" are dependent.

(g) Show that "SAT is 550–650" and "grade of C" are independent.

**35.** Reports on 192 accidents showed the following relationship between injuries and using a seat belt:

| | **Injuries** | **No injuries** |
|---|---|---|
| Seat belt not used | 66 | 28 |
| Seat belt used | 14 | 84 |

Determine whether the events "seat belt not used" and "injuries" are dependent or independent.

**36.** A professor gave two forms of an exam to an economics class. The grades by exam form are given in the following table:

| Grade | Exam | |
| --- | --- | --- |
| | Form I | Form II |
| A or B | 15 | 20 |
| Below B | 24 | 32 |

Show that the events "student took Form I" and "student made an A or a B" are independent.

**37.** A survey of working couples in Davidson County revealed the following information: The probability that the husband is happy with his job is 0.72. The probability that the wife is happy with her job is 0.55. The probability that both are happy with their jobs is 0.35.

A working couple is selected at random. Find the probability that

(a) at least one is happy with his or her job.

(b) neither is happy with his or her job.

# STATISTICS

When the president of the United States submits an annual budget to Congress in the trillion-dollar range, many taxpayers ask, "Where is all that money going?" But few of them wish to be handed a detailed budget a foot thick. They want the information summarized in a few broad categories such as defense, Social Security, education, agriculture, interest on the debt, and so on. Meanwhile, the president might want to know how the voters react to specific budget items such as defense and Social Security budgets. To poll all voters regarding their opinion of, for instance, the defense budget is impractical. However, the president can obtain valuable, although incomplete, information on voter opinion by a sample opinion poll.

"How did your class do on the finite exam?" and "How were the grades on the test?" are questions often asked by faculty and students after an exam. No one expects a response that includes a list of all the students' grades. Faculty want a response that indicates how well the class understood the material, and the students want to compare their performance with others in the class. On other occasions, a faculty member may wish to know if the students understood the concepts appropriate to a specific problem.

In these cases, the relevant information from all of the tests needs to be summarized in a form that conveys the desired information without going into great detail.

The exam summary and the president's budget summary provide examples of applying descriptive statistics. **Descriptive statistics** summarize data and describe their more relevant features.

The sample poll, on the other hand, falls into the category of inferential statistics. **Inferential statistics** make generalizations or draw conclusions from representative information. This chapter presents some of the methods used in descriptive and inferential statistics. Sections 8.1 through 8.3 deal with descriptive statistics, and Sections 8.4 through 8.8 deal with inferential statistics.

The discipline of statistics provides methods to collect data, organize them in a meaningful way, and interpret and report conclusions. Because numerous disciplines and organizations depend on statistical analyses, a variety of statistical methods exist. We find statistical specialists in economics, social sciences, sciences, business, medicine, engineering, governmental agencies, and education. To collect, interpret, and report statistical data often requires massive efforts. For example, accurate estimates of unemployment or incidence of crime may be rather easy to determine in a remote county in North Dakota but difficult to determine for the whole country. This chapter is a study of some useful statistical methods, mostly dealing with much simpler situations than those that occur routinely in the larger scheme of things.

## 8.1   FREQUENCY DISTRIBUTIONS

- Frequency Table
- Stem-and-Leaf Plots
- Construction of a Frequency Table

- Visual Representations of Frequency Distributions
- Histogram
- Pie Chart

### Frequency Table

Opinion polls and population studies use random samples to obtain information. Quite often it helps to organize the information in a tabular or visual form. One tabular form may be obtained by grouping similar observations into **categories** (or **classes**).

For example, the mathematics department gives a departmental final exam in calculus. The highest possible score on the exam is 160 points. A complete listing of all exam scores helps the faculty to decide on a grading curve, but a summary like the following table may give a better picture of the students' overall performance on the exam:

| Score | Number of students making score |
|-------|:-------------------------------:|
| 0–20 | 0 |
| 21–40 | 18 |
| 41–60 | 36 |
| 61–80 | 83 |
| 81–100 | 110 |
| 101–120 | 121 |
| 121–140 | 73 |
| 141–160 | 16 |
| Total | 457 |

We call this kind of summary a **frequency table** or **frequency distribution**. We call the number of observations in a category the **frequency** of that category. This frequency table gives the number of students for each 20-point interval of grades. A **range**, or **interval**, of numbers like 101 to 120 or 61 to 80, determines each category.

The frequency table of grades on the calculus exam does not give complete information. It does not tell us the highest or lowest score made. In fact, it does not tell the number of students who made any particular score. It does give general information about overall performance on the exam.

In some summaries, the categories might not be numerical. For example, we may summarize the students' majors by subject at a university with a frequency table like the following:

| Major | Number of students |
|-------|--------------------|
| Science | 429 |
| Arts | 132 |
| Languages | 41 |
| Social sciences | 631 |
| Engineering | 344 |
| Total | 1577 |

This example of **qualitative data** identifies the data by nonnumeric categories. The cases where data are represented by numerical values are **quantitative data**.

**Example 1** A mathematics quiz consists of five questions. The professor summarizes the performance of the class of 75 students with the following frequency distribution. Each quiz question determines a different category.

| Question | Number of correct answers (frequency) |
|----------|---------------------------------------|
| 1 | 36 |
| 2 | 41 |
| 3 | 22 |
| 4 | 54 |
| 5 | 30 |

**Example 2** A survey of students reveals that they spent the following amounts of money on books for three courses during a semester:

| | | | | | |
|------|------|------|------|------|------|
| $ 78 | $123 | $136 | $162 | $ 96 | $145 |
| $115 | $183 | $150 | $110 | $191 | $ 88 |
| $157 | $137 | $122 | $172 | $165 | $119 |
| $105 | $127 | $148 | $170 | $131 | $118 |

Make a frequency table to summarize the students' book expenses.

**Solution**

As the smallest amount is $78 and the largest amount is $191, the numbers cover a range of $191 - 78 = 113$. Let's form five categories of intervals of equal length. Our first estimate of the interval length is $\frac{113}{5} = 22.6$. We round this to 25 and start the first interval at 75, giving the intervals 75–99, 100–124, 125–149, 150–174, and 175–199. Next, we can make a table that places each number in the appropriate category:

| Interval | Numbers in the interval |
|----------|-------------------------|
| 75–99 | 78, 96, 88 |
| 100–124 | 123, 115, 110, 122, 119, 105, 118 |
| 125–149 | 136, 145, 137, 127, 148, 131 |
| 150–174 | 162, 150, 157, 172, 165, 170 |
| 175–199 | 183, 191 |

**NOTE**

The choice of the number of categories and the dollar range for each category could be made in several different ways.

From this table we have the count of each category and obtain the following frequency table:

| Book expenses | Number of students (frequency) |
|---------------|-------------------------------|
| $75–99 | 3 |
| $100–124 | 7 |
| $125–149 | 6 |
| $150–174 | 6 |
| $175–199 | 2 |

## Stem-and-Leaf Plots

To organize data into more orderly categories, a **stem-and-leaf plot** can be used. We do so by breaking the scores into two parts—the *stem,* consisting of the first one or two digits, and the *leaf,* consisting of the other digits.

**Example 3**  Make a stem-and-leaf plot of the following scores: 21, 13, 17, 24, 48, 7, 31, 46, 44, 39, 9, 15, 10, 41, 46, 33, and 24. The second digit becomes a leaf that we list next to its stem. This process gives a list of second digits corresponding to a stem.

**Solution**

We use the first digits 0, 1, 2, 3, 4 for the stems, which will divide the data into intervals 0–9, 10–19, 20–29, 30–39, and 40–49.

| Stem | Leaves |
|------|--------|
| 0 | 79 |
| 1 | 3750 |
| 2 | 144 |
| 3 | 193 |
| 4 | 86416 |

We can now easily count the frequency of each category.

Note that the digits for the leaves are not in order. They can be listed as they occur in the list.

**Example 4** Make a stem-and-leaf plot of the following scores: 3, 2, 6, 12, 14, 0, 11, 8, 2, 5, 7, 6, 6. For the categories, use the intervals 0–4, 5–9, and 10–14.

**Solution**

| Stem | Leaves |
|------|--------|
| 0 | 3202 |
| 0 | 685766 |
| 1 | 241 |

Note that the stem 0 occurs twice, the first time for leaves 0 through 4 and the second time for leaves 5 through 9.

**Example 5** Make a stem-and-leaf plot of the following data: 10.1, 9.3, 9.7, 11.4, 12.3, 10.8, 10.7, 10.3, 11.7, and 11.9.

**Solution**
We can use the first two digits for a stem.

| Stem | Leaves |
|------|--------|
| 9 | 37 |
| 10 | 1873 |
| 11 | 479 |
| 12 | 3 |

The stem-and-leaf plot has the advantage that the original data can be reconstructed from the plot.

## Construction of a Frequency Table

Here are some suggestions for setting up a frequency table.

**Construction of a Frequency Table**

Construct a frequency table in three steps, as follows:

**Step 1.** Choose the categories by which the data will be grouped, for example, the calculus grades in the 81–100 range. Generally, the categories are determined by intervals of equal length.

*(continued)*

**Step 2.** Place each piece of data in the appropriate category; for example, sort the calculus grades and place them in the appropriate category.

**Step 3.** Count the data in each category. This gives the frequency of the category; for example, perhaps 73 students scored in the 121–140 range.

The classification decisions in step 1 make the other two steps mechanical. The determination of categories can be a two-step decision. First, determine the number of categories (e.g., eight) and then the range of values that each category covers (e.g., 61–80, 81–100). No magic formula exists to make these decisions. They depend on the nature of the data and the message you wish to convey. However, some generally accepted rules of thumb might help. Just remember, exceptions are appropriate at times.

Hints on Setting up Category Intervals

1. **Number of categories.** Usually from 5 to 15. More categories might be unwieldy, and fewer categories might not distinguish between important features. Be sensible: Don't use five categories to summarize a two-category situation, such as the male–female breakdown in enrollment. Generally use a larger number of categories for larger amounts of data and fewer categories for smaller amounts of data.

2. **Range of each category.** A good guideline for choosing interval length and bounds is the following:

   **(a)** Find the difference between the largest and smallest observation and divide it by the number of intervals.

   **(b)** Adjust the length obtained to a relatively simple number.

   **(c)** Select a number less than, or equal to, the smallest observation for the lower bound of the first category. Drop down to, say, the first multiple of 5 or 10, or whatever number is appropriate for the data.

   In summarizing calculus grades on a 160-point test, the interval 80–99 range might be used as the range of one classification. However, a range of 81–100 could be used just as well.

   **(d)** Be sure to leave no gaps between categories when they might include some data. Don't use intervals like $2.00 to $4.00 and $5.00 to $7.00 when $4.75 is a valid data point.

   **(e)** Place each piece of data in only one category. The category designations 300–400 and 400–500 leave it unclear where to place 400. Category designations like 300–399 and 400–499 clearly indicate where to place 400.

   **(f)** Use category intervals that make sense for the situation. To summarize traffic speeds for a traffic study, use intervals like 31–35 and 36–40 rather than 33–38 and 39–42. Although intervals of equal length are generally preferred, sometimes different lengths make sense. For example, we might summarize performance on a test with intervals like 0–59, 60–69, 70–79, 80–89, and 90–100 because they represent letter grades.

## Visual Representations of Frequency Distributions

Because a picture conveys a more forceful message than a column of numbers, a visual presentation of a frequency table sometimes provides a better understanding of the data. We will study two common visual methods: the histogram and the pie chart. In each case, the graph shows information obtained from a frequency table.

### Histogram

A **histogram** is a bar chart in which each bar represents a category and its height represents the frequency of that category. Figure 8–1 shows the histogram of Example 1.

Mark the categories on a horizontal scale and the frequencies on a vertical scale. The bars are of equal width and are centered above the point that designates the category. In this case, a single number forms the category interval. The bars should be of equal width because two bars with the same height and different widths have different areas. This gives an impression of different frequencies. Thus, the *area*, not just the height of the bar, customarily represents the frequency when using different bar widths. Sometimes a space is left between bars when using a discrete variable. For a category defined by an interval (such as 1–5, 6–10), locate the bar between the end points of the interval.

The categories in the first two examples are **discrete**; that is, the values in one category are separated from those in another category by a "gap." A summary of the number of correct answers on a quiz uses the possible values 0, 1, 2, 3, 4, and 5 as categories. There is a jump from 3 to 4; a value of 2.65 is not valid. On the other hand, GPA scores are not discrete. Any GPA from 0 through 4.0 is valid. No jump occurs from one category to another. In such a case, the data are said to be **continuous**. For two different values of continuous data, all values between them are permissible values of the data. For example, 3.1 and 3.2 gives valid values of GPA, as well as any number between, such as 3.135 and 3.1999.

When representing continuous data by a histogram, a range of values such as 0–0.49, 0.50–1.0, and so on, determine the categories. The next example illustrates the use of a histogram with continuous data.

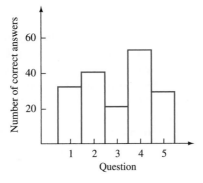

**FIGURE 8–1**    Summary of correct answers on a mathematics quiz.

**Example 6** The university registrar selects 100 transcripts at random and records the GPA for each, where all GPAs are rounded to two decimals. The frequency distribution follows:

| GPA | Frequency |
|---|---|
| 0–0.49 | 5 |
| 0.5–0.99 | 9 |
| 1.00–1.49 | 17 |
| 1.50–1.99 | 10 |
| 2.00–2.49 | 18 |
| 2.50–2.99 | 22 |
| 3.00–3.49 | 11 |
| 3.50–4.00 | 8 |

The histogram representing this information appears in Figure 8–2. Note that we have rotated the histogram so that the bars are horizontal. We did this because the category labels were too long to fit under a bar.

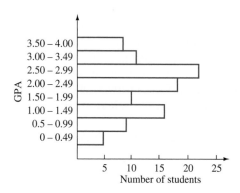

**FIGURE 8–2**

Sometimes we summarize a frequency table with a histogram using relative frequency instead of frequency for the vertical scale. **Relative frequency** counts the fractional part of the data that belong to a category. Compute relative frequency by dividing a frequency by the total number in the distribution. For example, the relative frequency of the 1.00–1.49 category of Example 6 is $\frac{17}{100}$, because 17 is the frequency of the category and 100 is the total of all frequencies. Relative frequency can also be stated as a percentage. If the relative frequency of a category is 0.17, then 17% of the scores belong to the category.

**Example 7** A question on an economics exam has five possible responses: A, B, C, D, and E. The number of students who gave each response follows:

| Response | Frequency |
|----------|-----------|
| A | 6 |
| B | 14 |
| C | 8 |
| D | 22 |
| E | 10 |

Draw a histogram that shows the relative frequency of each response.

**Solution**

Sixty students answered the question, so the relative frequency is the number responding divided by 60, the total of frequencies.

| Response | Relative Frequency |
|----------|--------------------|
| A | $\frac{6}{60} = 0.10$ |
| B | $\frac{14}{60} = 0.23$ |
| C | $\frac{8}{60} = 0.13$ |
| D | $\frac{22}{60} = 0.37$ |
| E | $\frac{10}{60} = 0.17$ |

The histogram is shown in Figure 8–3.

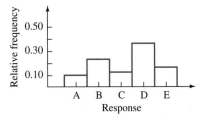

**FIGURE 8–3** Relative frequency of responses to a question.

A histogram sometimes conveys the erroneous impression that a natural break occurs between categories. When representing continuous data (such as the heights of 18-year-old males) it may be difficult to tell in which category a measurement should be placed. For example, if 5'9″ divides two categories, a smooth curve avoids the impression that a person slightly under 5'9″ is distinctly shorter than a person who is 5'9″. A smooth curve conveys the impression of continuous data better than a histogram. You can sketch a smooth curve based on a histogram by drawing it through the midpoints at the top of the bars. (See Figure 8–4.)

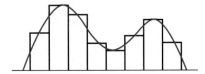

**FIGURE 8–4** A curve that smooths the histogram.

**Example 8**    A frequency table of the heights of male high school seniors in Ponca City is the following:

| Height (inches) | Frequency |
|---|---|
| 61–62.9 | 10 |
| 63–64.9 | 51 |
| 65–66.9 | 115 |
| 67–68.9 | 200 |
| 69–70.9 | 240 |
| 71–72.9 | 195 |
| 73–74.9 | 104 |
| 75–76.9 | 42 |
| 77–78.9 | 15 |

Draw a histogram and a smooth curve representing the data. (See Figure 8–5.)

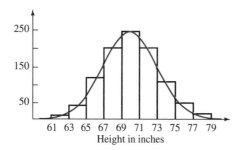

**FIGURE 8–5**

Note that Figure 8–5 identifies the boundaries of the bars with 61, 63, 65, 67, 69, and so on. Although this is a simple way to designate category boundaries, you cannot tell, just by looking at the graph, if the score 69, for example, lies in the 67–69 or the 69–71 category. We use the convention that a boundary point lies in the bar to its right. Thus, 69 is in the 69–71 category, which is consistent with the frequency table information.

If one category has boundaries $a$ and $b$, and the next category has boundaries $b$ and $c$, $a < b < c$, then point $b$ is placed in the category to the right with boundaries $b$ and $c$.

## Pie Chart

The second visual representation of data, the **pie chart**, emphasizes the proportion of data that falls into each category. You sometimes see a pie chart in the newspaper that represents the division of a budget into parts. You frequently see the parts reported as percentages of the total. You may obtain the percentage of the data that fall into each category from a frequency table.

**Example 9**  Jim Dandy has $200 for spending money this month. He carefully prepares the following budget:

| Category | Amount |
|---|---|
| Dates | $70 |
| Books and records | $20 |
| Laundry | $10 |
| Bicycle repairs | $48 |
| Miscellaneous | $52 |

Construct a pie chart that represents this budget.

**Solution**
First, compute the amount in each category as a percentage of the total by dividing the amount in the category by the total, $200.

| Category | Percentage of total amount |
|---|---|
| Dates | 35 |
| Books and records | 10 |
| Laundry | 5 |
| Bicycle repairs | 24 |
| Miscellaneous | 26 |

We "cut the pie" into pieces that have areas in the same proportion as the percentages representing the categories. (See Figure 8–6.) A glance at this pie chart tells us the relative share of each category. The angle at the center of each slice determines the size of the slice.

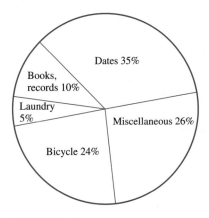

**FIGURE 8–6**

For the purpose of this book, you can sketch a pie chart by estimating the size of slices. If a more accurate drawing is desired, multiply each percentage by 360° to obtain the angle at the center of the slice. Then, use a protractor to mark off the required angle.

For example, the bicycle repairs category in Figure 8–6 accounts for 24% of the budget. Use the angle $0.24 \times 360° = 86°$ for this category.

**Example 10**   An advertising firm asked 150 children their favorite flavor of ice cream. Here is a frequency table of their findings:

| Flavor of ice cream | Number who favor |
|---|---|
| Vanilla | 60 |
| Chocolate | 33 |
| Strawberry | 18 |
| Peach | 24 |
| Other | 15 |
| Total | 150 |

Summarize this information with a pie chart.

**Solution**

Compute the percentages of each category and the size of the angle to be used.

| Flavor of ice cream | Percentage | Angle in degrees |
|---|---|---|
| Vanilla | 40 (from $\frac{60}{150}$) | $144° = 0.40 \times 360°$ |
| Chocolate | 22 (from $\frac{33}{150}$) | $79° = 0.22 \times 360°$ |
| Strawberry | 12 (from $\frac{18}{150}$) | $43° = 0.12 \times 360°$ |
| Peach | 16 (from $\frac{24}{150}$) | $58° = 0.16 \times 360°$ |
| Other | 10 (from $\frac{15}{150}$) | $36° = 0.10 \times 360°$ |

(See Figure 8–7.)

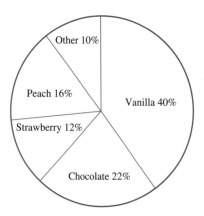

**FIGURE 8–7**

You might wonder why we have two ways to represent a frequency distribution. The histogram and pie chart give visual representations of the same information. The histogram shows the size of each category in reports on monthly sales, annual gross national product growth for several years, or enrollment in accounting courses. Use the pie chart when you wish to show the proportion of data that falls in each category in cases such as a breakdown of students by home state, a summary of family incomes, and a percentage breakdown of letter grades in a course.

## 8.1 EXERCISES

Access end-of-section exercises online at **www.webassign.net**

---

 ### Using Your TI Graphing Calculator

### Histograms

The graph of a histogram may be obtained by the following steps:

1. **Enter data**. Enter the scores in the list L1 and their frequencies in L2.

2. **Set the horizontal and vertical scales**. Set Xmin, Xmax, Xscl, Ymin, Ymax, and Yscl using WINDOW in the same way it is used to set the screen for graphing functions.

   Set Xmin and Xmax so that all scores will be in the interval (Xmin, Xmax). This interval will be the *x*-axis of the graph. Xscl is the width of a bar on the histogram and determines the interval length of each category. Ymin and Ymax determine the range of the frequencies.

3. **Define the histogram**. Press STAT PLOT <1:PLOT1> ENTER. You will see a screen similar to the following:

   Enter a 1 for Freq if the scores are simply a list in L1.

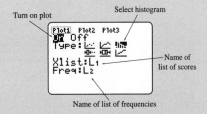

   On that screen select, as shown in the figure, <ON>, histogram for Type, L1 for Xlist (the scores), and L2 for frequencies. When you are finding the histogram for a list of scores, enter the number 1 instead of L2 for Freq. Press ENTER after each selection.

4. **Display the histogram**. Press GRAPH.

*(continued)*

## Example

Draw a histogram with three categories for the data summarized in the frequency table.

| Score | Frequency |
|:-----:|:---------:|
| 3 | 2 |
| 4 | 3 |
| 6 | 7 |
| 8 | 2 |
| 9 | 4 |
| 11 | 5 |
| 13 | 2 |

We set the range of the scores as 0 through 15 and the length of each category as 5. The frequency range is 0 through 15. This gives the following window settings:

and the window that defines the plot is

The lists L1 and L2 are

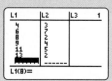

giving the histogram

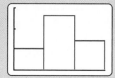

 **Using Excel**

## Histograms and Pie Charts

Both the histogram and the pie chart give a visual representation of data collected in categories. The histogram gives a picture of the number in each category, and the pie chart gives a picture of the percentage in each category. In Excel, both use the spreadsheet cells listing the *number* in each category.

## Example

Draw the histogram and the pie chart for data with four categories where the numbers in each category are 3, 4, 7, and 2.

### Solution

Let's call the categories A, B, C, and D and enter their frequencies in cells A2:A5. Because the procedure to draw the histogram and the pie chart have steps in common, we make just one list while showing which steps differ.

- Enter the categories in A2:A5 and the corresponding frequencies in cells B2:B5.
- Select the cells A2:B5.
- Click on the **Insert** tab. Move to the **Charts** group. Select **Column** for a histogram, with the first option shown. Select **Pie** for a pie chart with the first option shown.

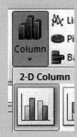 

- Click in the **Layout** tab under **Chart Tools**.
- For the histogram, click on **Data Labels** and then on the **Outside End** to place data labels on top of the bars.
- For the pie chart, click on **Data Labels > More Data Label Options**. Check the boxes for **Value** and **Percentage**, and **Outside End** for position of labels. Close the dialog box.

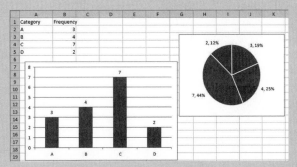

## Counting Frequencies in a Category

When you have a list of scores or a frequency table, you may need to count the number of scores for each category to use Excel. We illustrate how to do it.

### Given a List, Count the Number of Scores in Each Category

### Example

For the scores 1, 8, 6, 3, 4, 9, 5, 5, 3, 2, and 5, find the number in each of three categories 1–3, 4–6, and 7–9.

### Solution

Enter these data in A2:A12. We will place the count for category 1–3 in B2, the count for category 4–6 in B3, and the count for 7–9 in B4 using the COUNTIF function to first count the scores less than or equal to ($\leq$) to 3.

In cell B2, enter =COUNTIF(A2:A12,"<=3").

*(continued)*

**NOTE**

The symbol <= is used for ≤. A2:A12 is the range of the list.

For the frequency of category 4–6, enter into cell B3: =COUNTIF(A2:A12,"<=6")-B2.

**NOTE**

COUNTIF only counts scores less than (or less than or equal) a number, 6 in this case, so we count scores ≤ 6 and subtract the number ≤ 3.

For the frequency of category 7–9, enter into cell B4: =COUNTIF(A2:A12,"<=9")-B3-B2.

To create the pie chart or histogram, enter the text "1-3" in A14, "4-6" in A15, and "7-9" in A16. Copy the values from B2:B4 to B14:B16, respectively. You must "Paste by value" if using the *Paste* command. Select A14:B16 and create the pie chart or histogram.

## Given a Frequency Table, Count the Number in Each Category

### Example

Count the frequencies of the categories 1–6, 7–12, 13–18, and 19–24.

| Score | Frequency |
|-------|-----------|
| 2 | 3 |
| 5 | 4 |
| 6 | 8 |
| 7 | 2 |
| 11 | 3 |
| 17 | 5 |
| 18 | 2 |
| 21 | 3 |

### Solution

Enter the scores in cells A2:A9 and the corresponding frequencies in B2:B9. The category counts will be in cells C2:C5. To count the scores in the 1–6 category, we use the SUMIF function which searches A2:A9 for scores in the 1–6 category and adds their frequencies from B2:B9.

In C2, enter =SUMIF(A2:A9,"<=6",B2:B9).
In C3, enter =SUMIF(A2:A9,"<=12",B2:B9)-C2.
In C4, enter =SUMIF(A2:A9,"<=18",B2:B9)-C3-C2.
In C5, enter =SUMIF(A2:A9,"<=24",B2:B9)-C4-C3-C2.

**NOTE**

The function in C3 counts the scores ≤ 12 and subtracts the number ≤ 6.

Now you can use cells C2:C5 to create the pie chart or histogram.

## Exercises

**1.** Draw the histogram and pie chart for the following:

| Category | Frequency |
|----------|-----------|
| Adult male | 6 |
| Adult female | 9 |
| Children | 7 |

**2.** Draw the histogram and pie chart for the following:

| Category | Frequency |
|----------|-----------|
| A | 10 |
| B | 13 |
| C | 8 |
| D | 18 |

**3.** Draw the histogram for the following, using the categories 1–4, 5–8, 9–12:

$$3, 6, 8, 1, 2, 11, 9, 7, 4, 2, 10, 7$$

**4.** Draw the pie chart for the following, using the categories 1–5, 6–10, 11–15, 16–20:

$$8, 4, 19, 3, 11, 17, 20, 1, 5, 6, 13, 9, 16, 13, 7$$

---

## 8.2 MEASURES OF CENTRAL TENDENCY

- The Mean
- The Median
- The Mode
- Which Measure of Central Tendency Is Best?

We have used histograms and pie charts to summarize a set of data. These devices sometimes make it easier to understand the data.

At times, however, we want to be more concise in reporting information, so that comparisons can be easily made or so that some important aspect can be described. You often hear questions like:

"What was the class average on the exam?"

"What kind of gas mileage do you get on your car?"

"What happened to the price of homes from 1977 to 1984?"

You generally expect a response to questions like these to be a single number that is somehow "typical" or at the "center" of the exam grades, distance a car is driven on a certain amount of gas, or price of homes. We will study three ways to obtain such a "central number," which we call a **measure of central tendency**.

We associate a measure of central tendency with a **population**, a collection of objects, such as a collection of exams, a collection of cars, or a collection of homes. Each member of the population has a number associated with it, like a grade on an exam, gas mileage of a car, or the price of a home. We call these numbers **data**, and we find the measure of central tendency of those numbers. We often use subscripted variables to enumerate a list of numbers. Thus, $x_1$, denotes the first value, $x_2$ denotes the second value, and $x_n$ denotes the $n$th value.

A **sample** is a subcollection of a population. Five exam papers form a sample from the population of 37 students in my finite mathematics class. A study of obesity among teenagers could not find data on the weight and height of *all* teenagers. So we obtain estimates of teenagers' weight and height from representative samples of teenagers. Sometimes it is appropriate to find a measure of central tendency of a population, and other times that of a sample. The measures discussed in this chapter can be applied to both populations and samples.

## NOTE

The symbol $\mu$ (pronounced *mu*) is the standard notation for the mean of a population. For the mean of a sample, you will usually see the symbol $\bar{x}$ (*x* bar).

## The Mean

When we talk about "averages" like test averages, the average price of gasoline, or a basketball player's scoring average, we usually refer to one particular measure of central tendency, the arithmetic **mean**. To compute the mean of a set of numbers, simply add the numbers and divide by how many numbers were used.

**DEFINITION**
**Mean**

The population **mean** of $n$ numbers $x_1, x_2, \ldots, x_n$ is denoted by $\mu$ (the Greek letter *mu*) and is computed as follows:

$$\mu = \frac{x_1 + x_2 + \cdots + x_n}{n}$$

The **sample mean**, denoted by $\bar{x}$, (*x* bar) is

$$\bar{x} = \frac{x_1 + x_2 + \cdots + x_n}{n}$$

Notice that you calculate the population mean and the sample mean exactly the same. The notation $\mu$ and $\bar{x}$ indicate whether the data came from a population or from a sample.

Common terms for the mean are the **average** or the **arithmetic average**.

The Convenience Chain wants to compare sales in its 56 stores during July with sales a year ago, when it had 49 stores. A comparison of total sales for each July may be misleading because the number of stores differs. Sales can be down from a year ago in each of the 49 stores, but total sales can still be up because there are seven additional stores. The mean sales of all stores in each year should better indicate whether sales are improving.

**Example 1**     The mean of the test grades 82, 75, 96, 74 is

$$\frac{82 + 75 + 96 + 74}{4} = \frac{327}{4} = 81.75$$

**Example 2**     Find the mean of the annual salaries $25,000, $14,000, $18,000, $14,000, $20,000, $14,000, $18,000, and $14,000.

**Solution**
Add the salaries:

$$
\begin{array}{r}
25{,}000 \\
14{,}000 \\
18{,}000 \\
14{,}000 \\
20{,}000 \\
14{,}000 \\
18{,}000 \\
\underline{14{,}000} \\
137{,}000
\end{array}
$$

Divide this total by 8, the number of salaries, to obtain the mean:

$$
\text{Mean} = \frac{137{,}000}{8} = \$17{,}125
$$

This mean can be written in this more compact form:

$$
\frac{25{,}000 \; + \; 4 \times 14{,}000 \; + \; 2 \times 18{,}000 \; + \; 20{,}000}{8}
$$

where 14,000 has a frequency of 4, where 18,000 has a frequency of 2, and 25,000 and 20,000 each have a frequency of 1. The divisor, 8, is the sum of the frequencies. This form is useful in cases like the following, where the scores are summarized in a frequency table.

**Example 3**   Scores are summarized in the following frequency table:

| Score ($X$) | Frequency ($f$) |
|:-----------:|:---------------:|
| 3 | 2 |
| 4 | 1 |
| 5 | 8 |
| 6 | 4 |
|   | 15 |

The mean is given by

$$
\text{Mean} = \frac{2 \times 3 \; + \; 1 \times 4 \; + \; 8 \times 5 \; + \; 4 \times 6}{15} = \frac{74}{15} = 4.93
$$

The general formula for the mean of a frequency distribution is as follows:

**Formula for the Mean of a Frequency Distribution**

Given the scores $x_1, x_2, \ldots, x_k$, which occur with frequency $f_1, f_2, \ldots, f_k$, respectively, the mean is

$$\text{Mean} = \frac{f_1 x_1 + f_2 x_2 + \cdots + f_k x_k}{n}$$

where $n = f_1 + f_2 + \cdots + f_k$, the sum of frequencies.

The next example deals with **grouped data**. Scores are combined into categories, and the frequency of that category is given.

**Example 4**     Professor Tuff gave a 20-question quiz. She summarized the class performance with the following frequency table:

| Number of correct answers | Number of students |
|:---:|:---:|
| 0–5 | 8 |
| 6–10 | 14 |
| 11–15 | 23 |
| 16–20 | 10 |

Estimate the class mean for these grouped data.

**Solution**
We do not have specific values of the data, only the number in the indicated category. To obtain an *estimate* of the mean, we use the midpoint of each category as the representative value and compute the mean by

$$\text{Mean} = \frac{8 \times 2.5 + 14 \times 8.0 + 23 \times 13.0 + 10 \times 18.0}{55}$$

$$= 11.11$$

where 55 is the sum of the frequencies. (*Note:* The midpoint of an interval lies halfway between the left endpoint and the right endpoint of the interval. It can be found by taking the mean of the left endpoint and the right endpoint.)

**Formula for the Estimated Mean of Grouped Data**

If data are grouped in intervals with $m_1, m_2, \ldots, m_k$, the midpoints of each category interval, and $f_1, f_2, \ldots, f_k$ the frequency of each interval, respectively, then, an estimate of the mean is

$$\text{Mean} = \frac{f_1 m_1 + f_2 m_2 + \cdots + f_k m_k}{n}$$

where $n = f_1 + f_2 + \cdots + f_k$, the sum of frequencies.

Note that the formula for the mean of grouped data looks much like the formula for the mean of a frequency distribution. Why is it repeated? Be sure to notice the difference. First, we can only estimate the mean of grouped data. Second, the $m$'s used in the formula for grouped data are midpoints of the categories. Thus, one number is selected to represent a *range* of numbers in the category. In a frequency distribution, the $x$'s represent the actual scores.

The mean is a useful measure of central tendency because

1. It is familiar to most people.
2. It is easy to compute.
3. It can be computed for any set of numerical data.
4. Each set has just one mean.

**Example 5**   The salaries of the Acme Manufacturing Company are

| | |
|---|---|
| President | $300,000 |
| Vice president | $ 40,000 |
| Production workers | $ 25,000 |

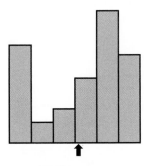

The company has 15 production workers. They complain to the president that company salaries are too low. Mr. President responds that the average company salary is about $42,060. He maintains that this is a good salary. The production workers and the vice president remain unimpressed with this information, because not a single one of them makes this much money. Although the president computed the mean correctly, he failed to mention that a single salary—his—was so large that the mean was in no way typical of all the salaries. This illustrates one of the disadvantages of the mean as a number that summarizes data. One or two **extreme values** can shift the mean, making it a poor representative of the data.

**FIGURE 8–8**   The mean is where the histogram balances.

When you summarize a set of data with a histogram, you can visualize the mean in the following way. Think of the histogram as being constructed from material of uniform weight, a thickness of plastic or cardboard. If you try to balance the histogram on a point, you will need to position it at the mean. (See Figure 8–8.)

## The Median

Another measure of central tendency, this one not so easily affected by a few extreme values, is the **median**. When listing data from an experiment according to size, people tend to focus on the middle of the list. Thus, the median is a useful measure of central tendency.

**DEFINITION**
**Median**

The **median** is the middle number after the data have been arranged in order. For an odd number of data items a single number lies in the middle, this forms the median. For an even number of data items, two numbers lie in the middle, and their mean gives the median.

Basically, the median divides the data into two equal parts. One part contains the lower half of the data, and the other part contains the upper half of the data.

**Example 6** The median of the numbers 3, 5, 8, 13, 19, 22, and 37 is the middle number, 13. Three terms lie below 13, and three lie above.

**Example 7** The median of the numbers 1, 5, 8, 11, 14, and 27 is the mean of the two middle numbers, 8 and 11. The median is

$$\frac{8 + 11}{2} = 9.5$$

Three terms lie below 9.5, and three lie above.

**Example 8** Find the median of the set of numbers 8, 5, 2, 17, 28, 4, 3, and 2.

**Solution**
First, the numbers must be placed in either ascending or descending order: 2, 2, 3, 4, 5, 8, 17, 28. As there are eight numbers, an even number, there is no middle number. We find the mean of the two middle numbers, 4 and 5, to obtain 4.5 as the median.

**Example 9** The set of numbers 2, 5, 9, 10, and 15 has a mean of 8.2 and a median of 9. If 15 is replaced by 140, the mean changes to 33.2, but the median remains 9. A change in one score of a set may make a significant change in the mean and yet leave the median unchanged.

The median is often used to report income, price of homes, and SAT scores.

## The Mode

The **mode** occurs less often as a measure of central tendency. It indicates which observation or observations dominate the data because of the frequency of their occurrence.

**DEFINITION**
**Mode**
The **mode** of a set of data is the value that occurs the largest number of times. If more than one value occurs this largest number of times, those values are also modes. When no value occurs more than once, we say that there is no mode. Thus, a distribution may have one mode, several modes, or no mode.

**Example 10** The mode of the numbers 1, 3, 2, 5, 4, 3, 2, 6, 8, 2, and 9 is 2 because 2 appears more often than any other value.

**Example 11**  The set of numbers 2, 4, 8, 3, 2, 5, 3, 6, 4, 3, and 2 has two modes, 2 and 3, because they both appear three times, more than any other value. The set of numbers 2, 5, 17, 3, and 4 has no mode because each number appears just once.

The mode provides useful information about the most frequently occurring categories. The clothing store does not want to know that the mean size of men's shirts is 15.289. However, the store likes to know when it sells more size $15\frac{1}{2}$ shirts than any other size. The mode best represents summaries where the most frequent response is desired, such as the most popular brand of coffee.

## Which Measure of Central Tendency Is Best?

With three different measures of central tendency, you probably are curious about which is best. The answer is, "It all depends." It all depends on the nature of the data and the information you wish to summarize.

The mean provides a good summary for values that represent magnitudes, like exam scores and price of shoes, if extreme values do not distort the mean. The mean is the best measure when equal distances between scores represent equal differences between the things being measured. For example, the difference between $15 and $20 is $5; the same amount of money is the difference between $85 and $90.

The median is a positional average. It is best used when ranking people or things. In a ranking, an increase or decrease by a fixed amount might not represent the same amount of change at one end of the scale as it does at the other. The difference between the number-one ranked tennis player and the number-two ranked player at Wimbledon may be small indeed. The difference between the tenth- and eleventh-ranked players may be significantly greater. In contests, in student standings in class, and in taste tests, numbers are assigned for ranking purposes. However, this does not imply that the people or things ranked all differ by equal amounts. In such cases, the median better measures central tendency.

The mode is best when summarizing dress sizes or the brands of bread preferred by families. The information desired indicates the most typical category, the one that occurs most frequently.

**Example 12**  In Example 5, we used the following data on salaries:

| | |
|---|---|
| President | $300,000 |
| Vice president | $ 40,000 |
| 15 production workers | $ 25,000 each |

We saw that the mean of these salaries, $42,060, was a poor representation of salaries because 16 of the 17 people had lower salaries than the mean. This is an instance of **skewed data**, in which relatively few values of the data distort the location of the central tendency. In cases like this, the median, $25,000, is a better measure of central tendency.

## 8.2    EXERCISES

Access end-of-section exercises online at **www.webassign.net**     WebAssign

---

## Using Your TI Graphing Calculator

### Calculating the Mean and Median

The TI graphing calculator has functions in the LIST menu to find the mean of a list or a frequency table and to find the median of a list. Six lists in the memory (L1, L2, . . . , L6) may be used.

### Example 1

Find the mean and median of the scores 5, 2, 8, 3, 7, 1.

**Solution**
Use STAT <4:Clrlist> L1,L2 ENTER to clear lists L1 and L2.
Enter the scores in the list L1.
For the mean, use LIST <MATH> <3:mean(> ENTER L1 ENTER gives mean = 4.33.
For the median, use LIST <MATH> <4:median(> ENTER L1 ENTER gives median = 4.

### Example 2

Find the mean of the frequency table.

| Score | Frequency |
|-------|-----------|
| 3 | 2 |
| 7 | 4 |
| 8 | 1 |

**Solution**
Enter the scores in L1 and the frequencies in L2. Use LIST <MATH> <3:mean(> ENTER L1,L2 ENTER to obtain mean = 6.

### Exercises

1. Find the mean and median of the following scores: 3, 4, 7, 14, 2, 8, 7.
2. Find the mean and median of the following scores: 5, 4, 8, 3, 9, 11, 13, 2.
3. Find the mean and median of the following scores: 7.2, 5.6, 2.1, 8.6, 6.9, 12.1, 16.5, 10.4

**4.** Find the mean of the scores summarized by the following frequency table:

| Score | Frequency |
|-------|-----------|
| 4 | 5 |
| 7 | 3 |
| 8 | 1 |
| 10 | 7 |
| 13 | 4 |

**5.** Find the mean of the scores summarized by the following frequency table:

| Score | Frequency |
|-------|-----------|
| 1 | 5 |
| 2 | 7 |
| 3 | 6 |
| 4 | 2 |

## Using Excel

### Calculate the Mean and Median

You can find the mean and median of a list of numbers using Excel. For example, if you have a list of 10 numbers, enter the numbers in cells, say, A1:A10, and enter the formula = AVERAGE(A1:A10) in a cell where you want the mean. Enter = MEDIAN(A1:A10) in the cell where you want the median.

### Exercises

**1.** Find the mean and median of the following scores: 3, 4, 7, 14, 2, 8, 7.

**2.** Find the mean and median of the following scores: 5, 4, 8, 3, 9, 11, 13, 2.

**3.** Find the mean and median of the following scores: 7.2, 5.6, 2.1, 8.6, 6.9, 12.1, 16.5, 10.4

## 8.3   MEASURES OF DISPERSION: RANGE, VARIANCE, AND STANDARD DEVIATION

- Range
- Variance and Standard Deviation
- Measurements of Position
- Application
- Five-Point Summary and Box Plot

A score often has little meaning unless it is compared with other scores. We have used the mean as one comparison. If you know your test score and the class average, you can compare how far you are above or below the class average. However, the average alone does not tell you how many in the class had grades closer to the average than you.

If two bowlers have the same average, do they have the same ability? If two students have the same average in a course, did they learn the same amount? Although the mean gives a rather simple representation of a set of data, sometimes more information is needed about how the scores are clustered about the mean in order to make valid comparisons. Let's use students' class averages to illustrate **measures of dispersion**.

| Student A | Student B |
|:---:|:---:|
| 80 | 65 |
| 87 | 92 |
| 82 | 95 |
| 92 | 75 |
| 84 | 98 |

Each student's mean is 85. However, student A is more consistent. Student B's scores vary more widely. The mean does not distinguish between these two sets of data. As this example shows, two data sets may have the same mean, but in one set the values may be clustered close to the mean, and in the other set the values may be widely scattered. To say something about the amount of clustering as well as the average, we need more information.

## Range

One way to measure the dispersion of a set of scores is to find the **range**, the distance between the largest and smallest scores. The range is 12 for student A and 33 for student B, so the range suggests that A's scores are clustered closer to the mean; the mean thereby represents A's scores better than B's scores. For grouped data, like a frequency table or histogram, the range is the difference between the smaller boundary of the lowest category and the larger boundary of the highest category.

Here are two frequency tables:

| TABLE I | |
|---|---|
| **Category** | **Frequency** |
| 1–2 | 10 |
| 3–4 | 5 |
| 5–6 | 3 |
| 7–8 | 6 |
| 9–10 | 16 |

| TABLE II | |
|---|---|
| **Category** | **Frequency** |
| 1–2 | 1 |
| 3–4 | 8 |
| 5–6 | 11 |
| 7–8 | 17 |
| 9–10 | 3 |

Each has an estimated mean of 6.15 (using the grouped mean data), and a range of 9. Clearly, the scores in Table II are clustered near the mean, whereas they tend to be more extreme in Table I. The range gives information only about the extreme scores and gives no information on their cluster near the mean.

The **range**, the difference between the largest and smallest scores, provides a simple measure of dispersion, but sometimes it helps to know whether the numbers are scattered rather uniformly throughout the range or whether most of them are clustered close together and a few are near the extremes of the range. For example, suppose a company manufactures ball bearings for an automobile company. The automobile company specifies that the bearings should be 0.35 inch in diameter. However, the automobile company and the bearing manufacturer both know that it is impossible to consistently make bearings that are *exactly* 0.35 inch in diameter. Slight variations in the material, limitations on the precision of equipment, and human error will create deviations from the desired diameter. So, the automobile company specifies that the bearings must be $0.35 \pm 0.001$ inch in diameter. The diameter may deviate as much as 0.001 inch from the desired diameter; that is, the acceptable range of diameters is 0.349 to 0.351. If the manufacturer produces a batch of bearings with all diameters in the range 0.3495 to 0.3508 inch, then there is no problem, because all of them are acceptable. However, if the diameters range from 0.347 to 0.353, there may or may not be a problem. Generally, they expect a few unacceptable bearings. If 90% of the bearings are unacceptable, then major problems exist in the manufacturing process. If fewer than one half of 1% are unacceptable, then the process may be considered satisfactory.

In this situation, a measure of how a batch of bearings varies from the desired diameter, 0.35 inch, provides more useful information than the largest and smallest diameters. With this situation in mind, we now look at two other useful indicators of variation.

## Variance and Standard Deviation

A widely used measure of dispersion is the **standard deviation**. A related indicator, the **variance**, is another. These indicators measure the degree to which the scores tend to cluster about a central value, in this case the mean. The variance and standard deviation give measures of the distance of observations from the mean. They are large if the observations tend to be far from the mean and small if they tend to be near the mean. Even when all scores remain within a certain range, the variance and standard deviation will increase or decrease as a score moves away from or closer to the mean.

Because the computation of the standard deviation is more complicated than that of the mean and the range, we will use the example of student A and student B to go through the steps to compute the variance and standard deviation.

**Step 1.** Determine the **deviation** of each score from the mean. Compute these deviations by subtracting the mean from each score. The following computations show the deviations for student A and student B using the mean of 85 in each case. Note that the deviations give the distance of the score from the mean a positive value for scores greater than the mean and a negative value for those less than the mean. You should also note that the deviations add to zero for both students. This always holds. The deviations will *always* add to zero. For this reason, we cannot accumulate the deviations to measure the overall deviation from the mean.

**Computation of Standard Deviation**

| | Student A | | | Student B | |
|---|---|---|---|---|---|
| Grade | Deviation | Squared deviation | Grade | Deviation | Squared deviation |
| 80 | $80 - 85 = -5$ | 25 | 65 | $65 - 85 = -20$ | 400 |
| 87 | $87 - 85 = 2$ | 4 | 92 | $92 - 85 = 7$ | 49 |
| 82 | $82 - 85 = -3$ | 9 | 95 | $95 - 85 = 10$ | 100 |
| 92 | $92 - 85 = 7$ | 49 | 75 | $75 - 85 = -10$ | 100 |
| 84 | $84 - 85 = -1$ | 1 | 98 | $98 - 85 = 13$ | 169 |
| 425 | 0 | 88 | 425 | 0 | 818 |

**Step 2.** Square each of the deviations. These appear under the heading "squared deviation."
This process of squaring the deviations allows us to accumulate a sum that is large when the scores tend to be far away from the mean. To adjust for cases in which two sets of data have different numbers of scores, we find a mean as in the next step.

**Step 3.** Find the mean of the squared deviations. For student A, the sum of the squared deviations is 88, and their mean is $\frac{88}{5}$, which equals 17.6. For student B, the sum of the squared deviations is 818, and their mean is $\frac{818}{5}$, which equals 163.6. The number 17.6 gives the **variance** for student A, and 163.6 gives the **variance** for student B.

**Step 4.** Find the square root of the means just obtained. For student A, we have $\sqrt{17.6} = 4.20$. For student B, we have $\sqrt{163.6} = 12.79$. These numbers are **standard deviations**. By tradition, the symbol $\sigma$ (sigma) denotes standard deviation, and $\sigma^2$ denotes variance. Thus, for student A, $\sigma = 4.20$, and for student B, $\sigma = 12.79$. The standard deviation measures the spread of the values about their mean. Student B has the larger standard deviation, so her grades are more widely scattered. The grades of student A cluster closer to the mean.

A formal statement of the formula for variance and standard deviation is the following:

**Formula for Population Variance and Standard Deviation**

Given the $n$ numbers $x_1, x_2, \ldots, x_n$ whose mean is $\mu$, the **population variance,** denoted $\sigma^2$, and **population standard deviation, $\sigma$,** of these numbers is given by:

$$\sigma^2 = \frac{(x_1 - \mu)^2 + (x_2 - \mu)^2 + \cdots + (x_n - \mu)^2}{n}$$

$$\sigma = \sqrt{\sigma^2}$$

Both the variance and the standard deviation measure the dispersion of data. The variance is measured in the **square** of the units of the original data. The standard deviation is measured in the units of the data, so we usually prefer it as a measure of dispersion.

No practical way exists to enable Congress to survey the total population of the United States regarding proposed legislation on the death penalty. However, valuable information can be obtained from a well-planned, representative sample of the population. From the sample, reasonable estimates can be made regarding the population as a whole. We have already seen that the population mean (when possible) and the sample mean, as an estimate of the population mean, use the same formula. Statisticians have found that using $n - 1$ as the divisor for the sample variance and standard deviation gives a better estimate than using $n$. Consequently, the sample variance and standard deviation use the following formulas:

**DEFINITION**
**Formula for Sample Variance and Standard Deviation**

Given the $n$ numbers $x_1, x_2, \ldots, x_n$ whose mean is $\bar{x}$, the **sample variance,** denoted by $s^2$, and **sample standard deviation,** $s$, of these numbers is given by

$$s^2 = \frac{(x_1 - \bar{x})^2 + (x_2 - \bar{x})^2 + \cdots + (x_n - \bar{x})^2}{n - 1}$$

$$s = \sqrt{s^2}$$

Unless otherwise stated, variance and standard deviation refer to *population* variance and standard deviation.

**Example 1**   Compute the mean and standard deviation of the numbers 8, 18, 7, and 10.

**Solution**

The mean is

$$\frac{8 + 18 + 7 + 10}{4} = \frac{43}{4} = 10.75$$

| Scores $(x)$ | Deviation $(x - \mu)$ | Squared deviation $((x - \mu)^2)$ |
|:---:|:---:|:---:|
| 8 | $8 - 10.75 = -2.75$ | 7.56 (rounded) |
| 18 | $18 - 10.75 = 7.25$ | 52.56 |
| 7 | $7 - 10.75 = -3.75$ | 14.06 |
| 10 | $10 - 10.75 = -0.75$ | 0.56 |
| | 0 | 74.74 |

$$\sigma = \sqrt{\frac{74.74}{4}} = \sqrt{18.69} = 4.32$$

We remind you that it is no accident that the deviations add to zero in each of the cases shown. Thus, the sum of deviations gives no information about the dispersion of scores. We therefore use a more complicated procedure, standard deviation, to determine dispersion. Let's summarize the procedure for obtaining standard deviation:

**Procedure for Computing Standard Deviation**

**Step 1.**   Compute the mean of the scores.

**Step 2.**   Subtract the mean from each value to obtain each deviation.

**Step 3.**   Square each deviation.

**Step 4.**   Find the mean of the squared deviations, using $n$ for population data and $n - 1$ for sample data. This gives the variance.

**Step 5.**   Take the square root of the variance. This gives the standard deviation.

**Example 2**   **(a)** Find the population standard deviation of the scores 8, 10, 19, 23, 28, 31, 32, and 41.
   **(b)** Find the sample standard deviation.

**Solution**

**(a) Step 1.**   Find the mean of the scores:

$$\mu = \frac{8 + 10 + 19 + 23 + 28 + 31 + 32 + 41}{8} = \frac{192}{8} = 24$$

**Step 2.**   Compute the deviation from the mean:

**Step 3.**  Square each deviation:

| $x$ | $x - \mu$ | $(x - \mu)^2$ |
|---|---|---|
| 8 | $8 - 24 = -16$ | $(-16)^2 = 256$ |
| 10 | $10 - 24 = -14$ | $(-14)^2 = 196$ |
| 19 | $19 - 24 = -5$ | $(-5)^2 = 25$ |
| 23 | $23 - 24 = -1$ | $(-1)^2 = 1$ |
| 28 | $28 - 24 = 4$ | $4^2 = 16$ |
| 31 | $31 - 24 = 7$ | $7^2 = 49$ |
| 32 | $32 - 24 = 8$ | $8^2 = 64$ |
| 41 | $41 - 24 = \underline{17}$ | $17^2 = \underline{289}$ |
|  | 0 | 896 |

**Step 4.**  Find the mean of the squared deviations:

$$\text{Variance} = \frac{896}{8} = 112$$

**Step 5.**  Take the square root of the result in step 4.

$$\sigma = \sqrt{112} = 10.58$$

**(b)**  Steps 1 through 3 are the same as those above.

**Step 4.**          $\text{Variance} = \dfrac{896}{7} = 128$

**Step 5.**          $s = \sqrt{128} = 11.31$

The standard deviation of a frequency distribution can be computed similar to the preceding process by keeping in mind the number of times a score is repeated.

**Example 3**   Find the mean and standard deviation for the following frequency distribution:

| Score | Frequency |
|---|---|
| 10 | 8 |
| 15 | 3 |
| 16 | 13 |
| 20 | 6 |

**Solution**

The total of the frequencies is 30, so we have 30 scores. First, compute the mean:

$$\mu = \frac{8 \times 10 + 3 \times 15 + 13 \times 16 + 6 \times 20}{30} = \frac{453}{30} = 15.1$$

Next, compute the variance, using the squares of deviations the number of times the corresponding score occurs:

$$= \frac{8 \times (10 - 15.1)^2 + 3 \times (15 - 15.1)^2 + 13 \times (16 - 15.1)^2 + 6 \times (20 - 15.1)^2}{30}$$

$$= \frac{8(5.1)^2 + 3(0.1)^2 + 13(0.9)^2 + 6(4.9)^2}{30} = \frac{362.7}{30} = 12.09$$

**NOTE**

We would use a divisor of 29 for the sample standard deviation.

The population standard deviation is $\sigma = \sqrt{12.09} = 3.48$.

You have estimated the mean of grouped data (Example 4, Section 8.2), and we use such a mean in the next example to estimate the standard deviation of grouped data.

**Example 4**  Estimate the standard deviation for the following grouped data:

| Score | Frequency |
|---|---|
| 0–5 | 6 |
| 6–10 | 3 |
| 11–19 | 13 |
| 20–25 | 8 |

**Solution**

Use the midpoint of each category to compute the mean. Because we know none of the scores specifically, we use the midpoint of a category as an estimate for each score in the category. By doing so, we convert the grouped data to the following frequency table:

| Score (midpoint) | Frequency |
|---|---|
| 2.5 | 6 |
| 8.0 | 3 |
| 15.0 | 13 |
| 22.5 | 8 |

**NOTE**

We would use a divisor of 29 for the sample standard deviation.

We compute the standard deviation of this frequency table to obtain the estimate of the standard deviation of the grouped data.

$$\text{Mean} = \frac{2.5 \times 6 + 8.0 \times 3 + 15.0 \times 13 + 22.5 \times 8}{30} = 13.8$$

The deviations are computed by using the mean and the midpoints of each category.

| Deviation | Squared deviation |
|---|---|
| $2.5 - 13.8 = -11.3$ | $(-11.3)^2 = 127.69$ |
| $8.0 - 13.8 = -5.8$ | $(-5.8)^2 = 33.64$ |
| $15.0 - 13.8 = 1.2$ | $(1.2)^2 = 1.44$ |
| $22.5 - 13.8 = 8.7$ | $(8.7)^2 = 75.69$ |

To obtain the variance, we need to use each squared deviation multiplied by the frequency of the corresponding category:

$$\begin{aligned}
\text{Variance} &= \frac{127.69 \times 6 + 33.64 \times 3 + 1.44 \times 13 + 75.69 \times 8}{30} \\
&= \frac{1491.30}{30} \\
&= 49.71
\end{aligned}$$

$$\text{Standard deviation} = \sqrt{49.71} = 7.05$$

*Note:* When asked to find a standard deviation, find the population standard deviation unless instructed to find the sample standard deviation.

You do not expect to compute a standard deviation during your daily activities. Yet on an intuitive level, all of us are interested in standard deviation. For example, you can recall a trip to the grocery store when you selected the shortest checkout line, only to find that it took longer than someone who got in a longer line at the same time. Let's look at the situation from the store's viewpoint. The store's managers are interested in customer satisfaction, so they work to reduce waiting time. They proudly announce that, on the average, they check out a customer in 3.5 minutes. The customer might agree that 3.5 minutes is good service, but an occasional long wait might make the customer question the store's claim. Basically, the customer wants the waiting time to have a small standard deviation. A small standard deviation indicates that you expect the waiting time to be near 3.5 minutes per customer.

A restaurant uses a different system of waiting lines. Customers do not choose a table and line up beside it. All customers wait in a single line and are directed to a table as it becomes available—and for a good reason. Diners do not want someone standing over them waiting for them to finish eating.

If we ignore the psychology of lining up at someone's table, which of these two systems is better? Is it better to let the customer choose one of several lines or to have a single line and allow a customer to go to a station as it becomes available? Both systems are reasonable in a bank. One bank may use a system where the customer chooses the teller, and a line forms at each teller. Another bank may use a system where the customers wait in a single line, and the customer at the head of the line goes to the next available teller. Both systems have the same average waiting time. Thus, from the bank's viewpoint, both systems allow it to be equally efficient. But it also turns out that the single-line concept tends to *decrease the standard deviation* and thereby reduce the extremes in waiting times. More customers will be served near the average waiting time than under the multiple-line system. Thus, from the customer's viewpoint, the single-line concept is better because it helps reduce an occasional long delay. As the bank's efficiency remains the same in both cases, management would do well to adopt the single-line concept.

## Measurements of Position

Scores are generally meaningless by themselves. "We scored ten runs in our softball game" gives no information about the outcome of the game unless the score of the opponent is known. A grade of 85 is not very impressive if everyone else scored 100.

The mean, median, and mode give a central point of reference for a score. You can compare your score to the mean or median and determine on which side of the central point your score lies.

We now look at other measures of position that prove useful.

One familiar measurement of position is the **rank**. A student ranks 15th in a class of 119; a runner places 38th in a field of 420 in the 10-kilometer run; a girl is the fourth runner-up in a science fair. Your rank simply gives the position of your score relative to all other scores. It gives no information on the location of the central point or on how closely scores cluster around it.

If you took the SAT test, you received a numerical score and a **percentile** score, another positional score. A score at the 84th percentile states that 84% of people taking the test scored the same or lower and 16% scored higher. A score at the 50th percentile is a median score: 50% scored higher and 50% scored the same or lower. The percentile score is used as a positional score when a large number of scores is involved. In addition to the 50th percentile (the median), the 25th and 75th percentiles give information on the dispersion of data.

**Example 5** (a) A golfer is ranked 22nd in a field of 114 players. What is his percentile?

(b) A golfer is at the 75th percentile in a field of 68. What is her rank?

**Solution**

(a) For a golfer ranked 22nd, there are $114 - 22 = 92$ players ranked lower. Counting the golfer himself, there are 93 ranked the same or lower. The percentile is $\frac{93}{114} \times 100\% = 81.6\%$, which we round down to 81st percentile.

(b) As 75% are ranked the same or lower, there are $0.75 \times 68 = 51$ players ranked the same or lower and 17 ranked higher. We say that she ranked 18th.

Another standard score is the *z*-**score**. It is more sophisticated than rank or percentile because it uses a central point of the scores (the mean) and a measure of dispersion (the standard deviation). It is useful in comparing scores when two different reference groups are involved. The *z*-score is

$$z = \frac{\text{score} - \text{mean}}{\text{standard deviation}}$$

**Example 6** If $\mu = 85$ and $\sigma = 5$, then the *z*-score for a raw score of 92 is

$$z = \frac{92 - 85}{5} = \frac{7}{5} = 1.4$$

For a score of 70,

$$z = \frac{70 - 85}{5} = \frac{-15}{5} = -3$$

A negative *z*-score indicates that the score is below the mean. $z = 1.4$ indicates that the score is 1.4 standard deviations above the mean. A *z*-score of $-3.0$ indicates the score lies 3 standard deviations *below* the mean. A *z*-score of 0 corresponds to the mean.

## Application

**Example 7** A student took both the SAT and ACT tests and made a 480 on the SAT verbal portion and a 22 on the ACT verbal. Which was the better score? (*Note:* You need to know that for the SAT verbal test, $\mu = 431$ and $\sigma = 111$; and for the ACT verbal, $\mu = 17.8$ and $\sigma = 5.5$.)

**Solution**
The *z*-score for each test is

$$\text{SAT } z = \frac{480 - 431}{111} = \frac{49}{111} = 0.44$$

$$\text{ACT } z = \frac{22 - 17.8}{5.5} = \frac{4.2}{5.5} = 0.76$$

The higher $z$-score, 0.76, indicates that the student performed better on the ACT, because the ACT score was more standard units above the mean.

## Five-Point Summary and Box Plot

We have used range and standard deviation to help understand the dispersion of data. Sometimes, we may want more detail about the dispersion. One method that provides such detail is the **five-point summary** and its visual representation, the **box plot**, also called the **box-and-whisker plot**.

We have already discussed three of the five points used to describe data:

*Minimum:* The smallest value in the data.

*Median:* The value in the middle of the data (when they are arranged in order).

*Maximum:* The largest value in the data.

Two other points that give information on the dispersion of data are the **first quartile**, $Q_1$, and the **third quartile, $Q_3$**.

**DEFINITION**
**Quartiles**

The *first quartile,* denoted $Q_1$, is the median of the values less than the median of the data (median itself excluded). It separates the lower 25% of the data from the upper 75%.

The *third quartile,* denoted $Q_3$, is the median of the values greater than the median of the data (median itself excluded). It separates the upper 25% of the data from the lower 75%.

The *second quartile,* denoted $Q_2$, is the median of all the data.

**Example 8**    The five-point summary of 2, 3, 6, 7, 9, 13, 17, 19, and 22 is:

Minimum:    2

$Q_1$:    4.5

Median:    9

$Q_3$:    18

Maximum:    22

This can be summarized as {2, 4.5, 9, 18, 22}. We now draw a figure, or box plot, that represents this five-point summary.

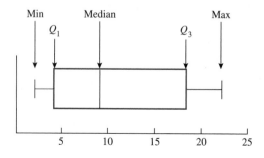

**NOTE**

The box contains the *middle* 50% of the data.

Note the construction of the box plot. The ends of the box are $Q_1$ and $Q_3$, with the location of the median shown in the box. Attach whiskers to each end of the box. The whisker on the left extends to the minimum value, and the whisker on the right extends to the maximum value.

---

## 8.3   EXERCISES

Access end-of-section exercises online at **www.webassign.net**

---

## Using Your TI Graphing Calculator

In Section 8.2, we used functions from the LIST menu to find the mean and median of data. The STAT menu has functions that calculate the measures of dispersion in this section. The mean and median are also calculated, so you have two routines that calculate mean and median.

### Calculating the Range

The range of a list of scores in L1 can be found using the **max** and **min** functions in the LIST menu. Here's how.
LIST <MATH> <2:max(> L1 − LIST <MATH> <1:min(> L1 ENTER

### Calculating the Mean and Standard Deviation

The TI graphing calculator has a routine in the STAT menu for calculating the mean and standard deviation for a list of numbers or for numbers summarized in a frequency table. The TI graphing calculator has six lists in the memory—L1, L2, ..., L6—that may be used.

Use with STAT <4:Clrlist> L1, L2 ENTER to clear lists L1 and L2.

Use one list—say L1—to enter the list of numbers and use a second list—say L2—to list the corresponding frequencies if you have a frequency table.
STAT <EDIT> will show a screen of the lists.

Let's illustrate with the following frequency table:

| $x$ | Frequency |
|-----|-----------|
| 2   | 3         |
| 5   | 2         |
| 6   | 4         |
| 9   | 1         |

*(continued)*

With the numbers $x$ in L1 and the frequencies in L2,

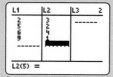

To find the mean and standard deviation of the list L1 only, select $\boxed{\text{STAT}}$ <CALC> <1 : 1-Var Stats> L1 $\boxed{\text{ENTER}}$

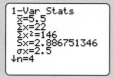 

The mean is shown as $x$-bar = 5.5 (the $x$ with a bar over it), $\sigma x$ = 2.5 is the population standard deviation, and $Sx$ = 2.8867 is the sample standard deviation. Scrolling down the screen, we see the five-point summary, minX = 2, $Q_1$ = 3.5, Median = 5.5, $Q_3$ = 7.5, and maxX = 9.

To find the mean and standard deviation of the frequency table with numbers in L1 and their frequency in L2, select $\boxed{\text{STAT}}$ <CALC> <1 : 1-Var Stats> L1, L2 $\boxed{\text{ENTER}}$

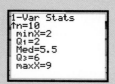

The mean is shown as $x$-bar = 4.9 (the $x$ with a bar over it); $\sigma x$ = 2.1656 is the population standard deviation, and $Sx$ = 2.2827 is the sample standard deviation. Scrolling down the screen, we see the five-point summary, minX = 2, $Q_1$ = 2, Median = 5.5, $Q_3$ = 6, and maxX = 9.

## Exercises

Find the mean and standard deviation (population and sample) for each of the following sets of data:

**1.** 3, 5, 1, 6, 8, 2

**2.** 3.2, 4.1, 6.3, 4.4, 7.5, 5.2

**3.**

| $x$ | Frequency |
|-----|-----------|
| 4   | 2         |
| 5   | 1         |
| 8   | 3         |
| 9   | 1         |
| 11  | 4         |

**4.**

| $x$ | Frequency |
|-----|-----------|
| 5.5 | 3         |
| 6.1 | 2         |
| 8.7 | 1         |
| 9.3 | 3         |
| 7.2 | 1         |

## Box Plot

You can draw a box plot of data using your TI graphing calculator. Refer back to "Using Your TI Graphing Calculator" in Section 8.1. It describes how to use your TI graphing calculator to draw a histogram. To draw a box plot, follow Steps 1 through 4 for drawing a histogram but substitute *box plot* for *histogram.* Be sure you select the box plot symbol for Type in Step 3. It is the next to last Type symbol.

## Exercises

1. Draw the box plot for the following scores: 3, 6, 2, 9, 4, 5, 13, 14, 10.
2. Draw the box plot for the scores summarized by the following frequency table:

| Score | Frequency |
|-------|-----------|
| 4 | 2 |
| 7 | 1 |
| 8 | 4 |
| 10 | 7 |
| 12 | 5 |
| 13 | 3 |

 # Using Excel

## Measures of Dispersion

Excel has routines for finding measures of dispersion for a list of scores. Let's illustrate assuming the data 2, 6, 9, 1, 7, 4, 3, 7, 5, 6 are stored in cells A1:A10.

## Range

In the cell where you want the range stored, enter the formula =MAX(A1:A10)-MIN(A1:A10), which gives 8.

## Standard Deviation

*Sample Standard Deviation:* In the cell where you want the sample standard deviation stored, enter the formula =STDEV(A1:A10), which gives 2.4944.

*Population Standard Deviation*: In the cell where you want the population standard deviation stored, enter the formula =STDEVP(A1:A10), which gives 2.3664.

*(continued)*

## Exercises

1. Find the range, sample standard deviation, and population standard deviation for the following: 8, 14, 22, 6, 19, 13, 9, 22, 15, 17, 11.

2. Find the range, sample standard deviation, and population standard deviation for the following: 45, 47, 39, 49, 33, 54, 51, 59, 44, 38, 55, 58.

3. Find the range as well as the sample and population standard deviation for the following scores: 17, 22, 25, 33, 37, 34, 29, 41, 35, 48.

4. Find the range as well as the sample and population standard deviation for the following scores: 1, 5, 7, 3, 11, 15, 9, 18, 20, 11, 8, 22, 14

# 8.4 RANDOM VARIABLES AND PROBABILITY DISTRIBUTIONS OF DISCRETE RANDOM VARIABLES

- Random Variables
- Probability Distribution of a Discrete Random Variable

## Random Variables

We routinely use numbers to convey information, to help express an opinion or an emotion, to help evaluate a situation, and to clarify our thoughts. Think how many times you have used phrases such as, "On a scale from 1 to 10, what do you think of . . . ?"; "What is your GPA?"; "The chance of rain this weekend is 40%"; "Which team is first in the American League?"; "How many A's did Professor X give?"; and "What did you make on your math test?"

All of these statements use numbers in some form. Notice that the numbers are not the heart of the statement, they represent a key facet of the message. When he says she is an 8 and she says he is a 3, the numbers do not become the emotions evoked, but rather indicate something about the *strength* of the emotions. An examination produces an exam paper, but we assign a grade to the paper to indicate something about the quality of the paper. A trip to the grocery store results in bags of grocery items, but we usually associate a number—the cost—with the purchase.

We associate numbers with outcomes of activities and phenomena because we can compare and analyze numbers. To summarize grades with a GPA is easier than averaging letter grades. The auto driver and the highway patrolman can more nearly agree on the meaning of "drive at 65 mph or less" than they can on "drive at a safe and reasonable speed."

This idea of assigning a number to outcomes of an experiment (activity or process) gives us an important and useful statistical concept. We call this assignment of numbers a **random variable**.

**DEFINITION**
**Random Variable**

A **random variable** is a rule that assigns a number to each outcome of an experiment.

We will usually denote the random variable by a capital letter such as $X$ or $Y$. Let's look at some examples of random variables.

**Example 1**   If a coin is tossed twice, we have four possible outcomes: HH, HT, TH, and TT. We can assign a number to each outcome by simply giving the number of heads that appear in each case. The values of the random variable $X$ are assigned as follows:

| Outcome | $X$ |
|---------|-----|
| HH | 2 |
| HT | 1 |
| TH | 1 |
| TT | 0 |

**Example 2**   A true–false quiz consists of 15 questions. A random variable may be defined by assigning the total number of correct answers to each quiz. In this case, the random variable ranges from 0 through 15.

**Example 3**   The hardware store has a box of super-special items. Selecting an item from the box defines an experiment with each of the items as an outcome. Two possible ways to define a random variable are to

**(a)** assign the original price to each item (outcome) or

**(b)** assign the sale price to each item.

**Example 4**   **(a)** Each judge of an Olympic diving contest assigns a number from 1 to 10 to each outcome (dive). We can think of each judge's assignment as a way of defining a random variable.

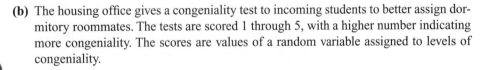

**(b)** The housing office gives a congeniality test to incoming students to better assign dormitory roommates. The tests are scored 1 through 5, with a higher number indicating more congeniality. The scores are values of a random variable assigned to levels of congeniality.

**(c)** An airline baggage claim office records the number of bags claimed by each passenger. We can view the number of bags as a random variable associated with each passenger.

**Example 5**   Three people are selected from a group of five men and four women. The number of women selected defines a random variable $X$. List the possible values of $X$ and the number of different outcomes that can be associated with each value.

**Solution**

*X* can assume the value 0, 1, 2, or 3. The number of different outcomes associated with each value is the following:

| *X* | Outcome | Number of possible outcomes |
|---|---|---|
| 0 | Three men | $C(5, 3) = 10$ |
| 1 | One woman, two men | $C(4, 1)C(5, 2) = 40$ |
| 2 | Two women, one man | $C(4, 2)C(5, 1) = 30$ |
| 3 | Three women | $C(4, 3) = 4$ |

In one sense, you may be quite arbitrary in the way you define a random variable for an experiment. It depends on what is most useful for the problem at hand. Here are several examples:

| Experiment | Random variable, *X* |
|---|---|
| A survey of cars entering a mall parking lot | Number of passengers in a car: $1, 2, \ldots, 8$ |
| Rolling a pair of dice | Sum of numbers that turn up: $2, 3, \ldots, 12$ |
| Tossing a coin three times | Number of times tails occurs |
| Selecting a sample of five tires from an assembly line | Number of defective tires in the sample |
| Measuring the height of a student selected at random | Observed height |
| Finding the average life of a brand X tire | Number of miles driven |
| Selecting a box of Crunchies cereal | Weight of the box of cereal |
| Checking the fuel economy of a hybrid car | Distance *X* the car travels on a gallon of gas |

We call the first four random variables **discrete variables**, because the values assigned come from a set of distinct numbers and the values between are not permitted as outcomes. For example, the number of auto passengers can take on only the values 1, 2, 3, and so on. The number 2.63 is not a valid assignment for the number of passengers. On the other hand, the last four examples are **continuous variables**. Assuming an accurate measuring device, 5 feet 4.274 inches is a valid height. There are no distinct gaps that must be excluded as a valid height. Similarly, even though one might expect a car to travel about 23 miles on a gallon of gas, 22.64 miles cannot be excluded as a valid distance. Any distance in a reasonable interval is a valid possibility.

## Probability Distribution of a Discrete Random Variable

We now merge two concepts that we studied earlier. We have used the concept of probability to give a measure of the likelihood of a certain outcome or outcomes occurring in an experiment (Section 7.1). At the beginning of this section, we introduced the concept of a random variable, the assigning of a number to each outcome of an experiment. The value assigned to the random variable depends on the outcome that actually occurs. Some outcomes may have a small likelihood of occurring, whereas others may be quite likely to occur. Thus, the associated values of the random variable usually have different levels of likelihood.

We sometimes ask about the chance of observing a certain value of a random variable $X$ (that is, find the *probability* that a certain value occurs). It is no surprise that such a probability closely relates to the probabilities of the outcomes. Rather than look at the probability of just one value of $X$, we now focus on the probability of each of the possible values of $X$. We call this assignment of a probability to each value of a discrete random variable a **probability distribution**. Let's look at a simple example.

**Example 6**  A children's game has 15 cards with 5 colored red, 3 colored black, and 7 colored green. The cards are shuffled, and a child selects 2 cards. If the cards are different colors, no points are given. Five points are given if both cards are green, 10 points are given if both cards are red, and 15 points are given if both cards are black. Because we assign a number (the points) to each outcome (cards drawn), this determines a random variable $X$ with outcomes and numbers related as follows:

| Outcome | Value of $X$ (points) |
|---|---|
| Red and black | 0 |
| Red and green | 0 |
| Red and red | 10 |
| Black and black | 15 |
| Black and green | 0 |
| Green and green | 5 |

The probability of drawing two greens is $\dfrac{C(7, 2)}{C(15, 2)} = 0.200$

of two reds $\dfrac{C(5, 2)}{C(15, 2)} = 0.095$

of two blacks $\dfrac{C(3, 2)}{C(15, 2)} = 0.029$

All other outcomes will be different colors, so the probability of different colors is $1 - 0.200 - 0.095 - 0.029 = 0.676$. Now we can give the probability of each value of $X$ and thereby obtain the probability distribution of $X$.

| $X$ | $P(X)$ |
|---|---|
| 0 | 0.676 |
| 5 | 0.200 |
| 10 | 0.095 |
| 15 | 0.029 |

> **DEFINITION**
> **Probability Distribution**
>
> If a discrete random variable has the values
>
> $$x_1, x_2, x_3, \ldots, x_k$$
>
> then a **probability distribution** $P(x)$ is a rule that assigns a probability $P(x_i)$ to each value $x_i$. More specifically,
>
> **(a)** $0 \le P(x_i) \le 1$ for each $x_i$,
> **(b)** $P(x_1) + P(x_2) + \cdots + P(x_k) = 1$

**Example 7** A coin is tossed twice. Define a random variable as the number of times that heads appears. Compute the probability of zero, one, or two heads in the usual manner to get

$$P(0) = P(TT) = \frac{1}{4}$$

$$P(1) = P(HT \text{ or } TH) = \frac{1}{2}$$

$$P(2) = P(HH) = \frac{1}{4}$$

> **NOTE**
>
> The width of a bar of a probability distribution is one unit; therefore, the *area* of the rectangular bar equals the probability of that value of X.

We can represent a probability distribution graphically using a histogram. (See Figure 8–9.)

Form a category for each value of the random variable and center the rectangle over the category mark. We typically assign the probability, $P(x)$, of a value, $x$, of the random variable by the area of the rectangle. When $x$ has only integral values, as in this case, the width of the rectangle is 1, and the height of the rectangle equals the probability, $P(x)$.

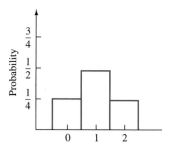

**FIGURE 8–9** Probability graph of the number of heads in two tosses of a coin.

**Example 8** An experiment randomly selects two people from a group of five men and four women. A random variable $X$ is the number of women selected. Find the probability distribution of $X$.

**Solution**
The values of $X$ range over the set $\{0, 1, 2\}$, because 0, 1, or 2 women can be selected. The probability of each value is computed in the usual manner.

$$P(0) = \frac{C(5, 2)}{C(9, 2)} = \frac{10}{36} \qquad \text{(probability both are men)}$$

$$P(1) = \frac{C(5, 1) \times C(4, 1)}{C(9, 2)} = \frac{20}{36} \qquad \text{(probability of one man and one woman)}$$

$$P(2) = \frac{C(4, 2)}{C(9, 2)} = \frac{6}{36} \qquad \text{(probability of two women)}$$

The probability distribution and its graph are shown in Figure 8–10.

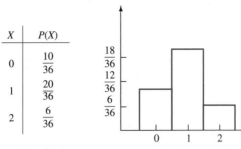

| $X$ | $P(X)$ |
|---|---|
| 0 | $\dfrac{10}{36}$ |
| 1 | $\dfrac{20}{36}$ |
| 2 | $\dfrac{6}{36}$ |

(a) Probability distribution

(b) Graph of probability distribution

**FIGURE 8–10**

**Example 9**   Five red cards are numbered 1 through 5, and five black cards are numbered 1 through 5. Cards are drawn, without replacement, until a card is drawn that matches the color of the first card. Let the number of cards drawn be the random variable $X$. Find the probability distribution of $X$.

**Solution**

The possible values of $X$ are 2, 3, 4, 5, 6, and 7, because at least two cards must be drawn in order to have a match. There are five cards that differ from the first color, so the sixth card is the last time a card different from the first can be drawn. The seventh card must then be a match. To compute probabilities, designate the color of the first card by C and the different color by D. Then the possible outcomes of the sequence of colors are the following:

$$X = 2: \text{CC}$$
$$X = 3: \text{CDC}$$
$$X = 4: \text{CDDC}$$
$$X = 5: \text{CDDDC}$$
$$X = 6: \text{CDDDDC}$$
$$X = 7: \text{CDDDDDC}$$

We now compute the probability of each sequence.

Note that it doesn't matter which color is drawn first, so the probability the first card is some color equals 1.

$$\text{CC:} \quad P(X = 2) = 1 \times \frac{4}{9} = \frac{4}{9}$$

$$\text{CDC:} \quad P(X = 3) = 1 \times \frac{5}{9} \times \frac{4}{8} = \frac{5}{18}$$

$$\text{CDDC:} \quad P(X = 4) = 1 \times \frac{5}{9} \times \frac{4}{8} \times \frac{4}{7} = \frac{10}{63}$$

$$\text{CDDDC:} \quad P(X = 5) = 1 \times \frac{5}{9} \times \frac{4}{8} \times \frac{3}{7} \times \frac{4}{6} = \frac{5}{63}$$

$$\text{CDDDDC:} \quad P(X = 6) = 1 \times \frac{5}{9} \times \frac{4}{8} \times \frac{3}{7} \times \frac{2}{6} \times \frac{4}{5} = \frac{2}{63}$$

$$\text{CDDDDDC:} \quad P(X = 7) = 1 \times \frac{5}{9} \times \frac{4}{8} \times \frac{3}{7} \times \frac{2}{6} \times \frac{1}{5} \times \frac{4}{4} = \frac{1}{126}$$

## 8.4    EXERCISES

Access end-of-section exercises online at **www.webassign.net**

## 8.5    EXPECTED VALUE OF A RANDOM VARIABLE

- Expected Value
- Variance and Standard Deviation of a Random Variable

If we are honest, most of us will admit to having fantasized about getting rich quick. Some will act on the fantasy by buying lottery tickets, by gambling, or by purchasing merchandise so that we can win a magazine's sweepstakes award. Do you ever ask yourself how much you are willing to pay for the *chance* to become rich quick? Would you pay $1 for the chance to win $1 million? Millions of people would and do by purchasing lottery tickets. Would you pay $5,000 for a chance to win $1 million? Far fewer will. In this section, we study one way to judge the worth of a chance to win money, in large or small amounts, and to determine the expected long-term performance of an activity. We study a measure called **expected value**. We analyze a hypothetical activity to develop the concept of expected value.

### Expected Value

Each day a student puts $1.00 in a vending machine to buy an $0.85 candy bar. She observes three possible outcomes:

1. She gets a candy bar and $0.15 change. This happens 80% of the time.
2. She gets a candy bar and no change. This happens 16% of the time.
3. She gets a candy bar and the machine returns her $1.00. This happens 4% of the time.

Over a period of time, what is the average cost of a candy bar?

The three possible costs, $0.85, $1.00, and $0.00 have a mean of $0.617. This is *not* the average cost, because it assumes that $0.85, $1.00, and $0.00 occur equally often, whereas, in fact, $0.85 occurs more often than the other two. If the machine behaved this way for 500 purchases, we would expect the following:

| Cost (X) | Frequency (per 500) | Probability (P[X]) |
|---|---|---|
| $0.85 | 400 | 0.80 |
| $1.00 | 80 | 0.16 |
| $0.00 | 20 | 0.04 |
|  | 500 | 1.00 |

This presents the information as a frequency table, so we can compute the mean as we have before with a frequency table (Section 8.3). The mean is

$$\frac{400 \times 0.85 + 80 \times 1.00 + 20 \times 0}{500} = \frac{340 + 80 + 0}{500} = 0.84$$

so the average cost is $0.84.

Let's write the mean in another way:

$$\frac{400 \times 0.85 + 80 \times 1.00 + 20 \times 0}{500} = \frac{400}{500} \times 0.85 + \frac{80}{500} \times 1.00 + \frac{20}{500} \times 0$$
$$= 0.80 \times 0.85 + 0.16 \times 1.00 + 0.04 \times 0$$

This last expression is simply the sum obtained by adding each cost of a candy bar times the probability that cost occurred; that is,

$$P(0.85) \times 0.85 + P(1.00) \times 1.00 + P(0) \times 0$$

This illustrates the procedure used to compute the mean when each value occurs with a specified probability. We call this mean the **expected value**; it represents the long-term mean of numerous trials. Although a few trials likely would not average to the expected value, a larger and larger number of trials will tend to give a mean closer to the expected value.

We emphasize that the term *expected value* does not mean the value we expect in an everyday sense. In the candy bar example, the amount paid for a candy bar is $0.00, $0.85, or $1.00, whereas the expected value is $0.84. As $0.84 is never the amount paid, you never expect to pay that. However, $0.84 is the average amount paid over a large number of purchases. We call this long-term average the expected value.

We now make a formal definition of expected value in terms of a random variable and probability distribution.

**DEFINITION**
**Expected Value**

If $X$ is a random variable with values $x_1, x_2, \ldots, x_n$ and corresponding probabilities $p_1, p_2, \ldots, p_n$, then the **expected value** of $X$, $E(X)$, is

$$E(X) = p_1 x_1 + \cdots + p_n x_n$$

**Example 1**  Find the expected value of $X$, where the values of $X$ and their corresponding probabilities are given by the following table:

| $x_i$ | 2 | 5 | 9 | 24 |
|-------|-----|-----|-----|-----|
| $p_i$ | 0.4 | 0.2 | 0.3 | 0.1 |

**Solution**

$$E(X) = 0.4 \times 2 + 0.2 \times 5 + 0.3 \times 9 + 0.1 \times 24$$
$$= 0.8 + 1.0 + 2.7 + 2.4$$
$$= 6.9$$

**Example 2**    **(a)** A contestant tosses a coin and receives $5 if heads appears and $1 if tails appears. What is the expected value of a trial?

**(b)** A contestant receives $4 if a coin turns up heads and pays $3 if it turns up tails. What is the expected value?

**(c)** A contestant receives $5 if a coin turns up heads and pays $5 when the coin turns up tails. What is the expected value?

**Solution**

**(a)** The probability of receiving $5 is $\frac{1}{2}$ (the probability of tossing a head) and the probability of receiving $1 is $\frac{1}{2}$ (the probability of tossing a tail). Then,

$$E(X) = \frac{1}{2}(5) + \frac{1}{2}(1) = \$3$$

**(b)** Likewise, the probability of receiving $4 is $\frac{1}{2}$, and the probability of paying $3 is $\frac{1}{2}$. So,

$$E(X) = \frac{1}{2}(4) + \frac{1}{2}(-3) = \$2.00 - \$1.50 = \$0.50$$

**(c)** Because the probability of receiving $5 is $\frac{1}{2}$ and the probability of paying $5 is $\frac{1}{2}$, the expected value is

$$\frac{1}{2}(5) + \frac{1}{2}(-5) = \$0$$

We call a game **fair** whenever the expected value is zero. Thus, the game in part (c) is fair and the games in parts (a) and (b) are not fair. In a fair game, a player's expected winnings and losses are equal.

**Example 3**    An IRS study shows that 60% of all income tax returns audited have no errors; 6% have errors that cause overpayments averaging $25; 20% have minor errors that cause underpayments averaging $35; 13% have more serious errors averaging $500 underpayments; and 1% have flagrant errors averaging $7000 underpayment. If the IRS selects returns at random:

**(a)** What is the average amount per return owed to the IRS, that is, what is the expected value of a return selected at random?

**(b)** How much should the IRS expect to collect if one million returns are audited at random?

**(c)** If the budget for the Audit Department is $15 million, how many returns must they examine to collect enough to cover their budget expenses?

**Solution**

**(a)** Because an underpayment error eventually results in additional money paid to the IRS and a taxpayer overpayment is money paid back by the IRS, we use positive values for underpayment and negative for overpayment. The probability in each case is the fraction of returns with that type of error.

$$E(X) = 0.60(0) + 0.06(-25) + 0.20(35) + 0.13(500) + 0.01(7000)$$
$$= 140.5$$

Thus, the IRS expects to collect $140.50 for each return selected.

**(b)** If one million returns are selected at random, they may expect to collect 140.50 × 1 million = $140.5 million.

**(c)** As the average amount they expect to collect is $140.50 per return audited, they must examine

$$15{,}000{,}000/140.50 = 106{,}762 \text{ returns}$$

The next example illustrates that the expected value need not be an amount of money. It can be any "payoff" associated with each outcome.

**Example 4**  A tray of electronic components contains nine good components and three defective components. If two components are selected at random, what is the expected number of defective components?

**Solution**

Let the random variable $X$ be the number of defective components selected. $X$ can have the value 0, 1, or 2. We need the probability of each of those numbers:

$$P(0) = \text{probability of no defective (both good)}$$

$$= \frac{C(9, 2)}{C(12, 2)} = \frac{36}{66} = \frac{12}{22}$$

$$P(1) = \text{probability of one good and one defective}$$

$$= \frac{C(9, 1)C(3, 1)}{C(12, 2)} = \frac{27}{66} = \frac{9}{22}$$

$$P(2) = \text{probability of two defective}$$

$$= \frac{C(3, 2)}{C(12, 2)} = \frac{3}{66} = \frac{1}{22}$$

The expected value is

$$E(X) = \frac{12}{22}(0) + \frac{9}{22}(1) + \frac{1}{22}(2) = \frac{11}{22} = \frac{1}{2}$$

so the expected number of components is $\frac{1}{2}$. Clearly, you don't expect to get half of a component. The value $\frac{1}{2}$ simply says that if a large number of selections are made, you will *average* one half each time. You expect to get no defectives a little less than half the time and either one or two the rest of the time, but the average will be one half.

## Variance and Standard Deviation of a Random Variable

The expected value represents a "central tendency." As illustrated by the vending machine example at the beginning of this section, the expected value is a long-term average, or mean, of the values of a random variable, taking the probability of their occurrence into consideration. Think back on the mean and median; we shouldn't be surprised that we sometimes need the dispersion, or spread, of a random variable to give more information. We can find the **variance of a random variable**, and it too measures the dispersion, or spread, of a random variable from the mean (expected value). The greater dispersion gives a greater variance. The histogram of the probability distribution shown in Figure 8–11a has a smaller variance than that shown in Figure 8–11b.

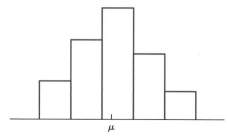

These scores are clustered closer
to the mean than those in (b).

(a)

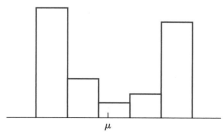

These scores have greater
dispersion than those in (a).

(b)

**FIGURE 8–11**

Let's use the candy bar example to illustrate variance and standard deviation. The information is given in the form of a frequency distribution:

| Amount paid | Frequency |
|:-----------:|:---------:|
| $0.85 | 400 |
| $1.00 | 80 |
| $0.00 | 20 |
|  | 500 |

We compute the standard deviation using the method shown in Example 3 of Section 8.3 for frequency distributions:

$$\text{Variance} = \frac{400(0.85 - 0.84)^2 + 80(1.00 - 0.84)^2 + 20(0.00 - 0.84)^2}{500}$$

We rewrite this expression as

$$\text{Variance} = \frac{400}{500}(0.85 - 8.84)^2 + \frac{80}{500}(1.00 - 0.84)^2 + \frac{20}{500}(0.00 - 0.84)^2$$
$$= 0.80\,(0.85 - 0.84)^2 + 0.16\,(1.00 - 0.84)^2 + 0.04\,(0.00 - 0.84)^2$$

This last expression takes the form

$$\text{Variance} = P(0.85)(0.85 - \mu)^2 + P(1.00)(1.00 - \mu)^2 + P(0)(0 - \mu)^2$$

where $\mu$ is 0.84. Let's complete the computations for variance and standard deviation:

$$\text{Variance} = 0.80(0.01)^2 + 0.16(0.16)^2 + 0.04(-0.84)^2$$
$$= 0.00008 + 0.004096 + 0.028224 = 0.0324$$
$$\text{Variance} = \$0.0324, \text{ about } \$0.03$$
$$\text{Standard deviation} = \sigma = \sqrt{0.0324} = 0.18, \text{ that is, } \$0.18$$

Let's give the formal definition of the variance and standard deviation of a random variable and then give an example showing the computation. We denote the variance by $\sigma^2(X)$ and let the Greek letter $\mu$ (mu) represent the mean (expected value).

**DEFINITION**
**Variance, Standard Deviation**

If $X$ is a random variable with values $x_1, x_2, \ldots, x_n$, corresponding probabilities $p_1$, $p_2, \ldots, p_n$, and expected value $E(X) = \mu$, then

$$\textbf{Variance} = \sigma^2(X) = p_1(x_1 - \mu)^2 + p_2(x_2 - \mu)^2 + \cdots + p_n(x_n - \mu)^2$$

$$\textbf{Standard deviation} = \sigma(X) = \sqrt{\text{variance}}$$

**Example 5**   Find the variance and standard deviation for the random variable defined by the following table:

| $x_i$ | 4 | 7 | 10 | 8 |
|-------|-----|-----|-----|-----|
| $p_i$ | 0.2 | 0.2 | 0.5 | 0.1 |

**Solution**
Set up the computations in the following way:

| $x_i$ | $p_i$ | $p_i x_i$ | $x_i - \mu$ | $(x_i - \mu)^2$ | $p_i(x_i - \mu)^2$ |
|-------|-------|-----------|-------------|-----------------|---------------------|
| 4 | 0.2 | 0.8 | −4 | 16 | 3.2 |
| 7 | 0.2 | 1.4 | −1 | 1 | 0.2 |
| 10 | 0.5 | 5.0 | 2 | 4 | 2.0 |
| 8 | 0.1 | 0.8 | 0 | 0 | 0 |
|   |   | $\mu = 8.0$ |  |  | $\sigma^2(X) = 5.4$ |

The variance of $\sigma^2(X) = 5.4$. So the standard deviation is

$$\sigma(X) = \sqrt{5.4} = 2.32$$

## 8.5   EXERCISES

Access end-of-section exercises online at **www.webassign.net**    Web**Assign**

## 8.6   BERNOULLI EXPERIMENTS AND BINOMIAL DISTRIBUTION

- Bernoulli Trials
- Probability of a Bernoulli Experiment
- Justification of the Bernoulli Experiment Formula
- Binomial Distribution

### Bernoulli Trials

We have studied the probability of an event occurring in an experiment. We now look at a sequence of experiments to determine probabilities related to the sequence. The experiments involved are not arbitrary; they have specific characteristics like those found in the problems "What is the probability that heads appears seven times in ten tosses of a coin?" or "If you guess at the answers of 15 multiple-choice questions, what are your chances of a passing grade?"

These problems illustrate a certain type of probability problem, **Bernoulli trials**. Such problems involve **repeated trials** of an experiment with only two possible outcomes: heads or tails, right or wrong, yes or no, and so on. We classify the *two outcomes* as **success** or **failure**.

To classify an experiment as a Bernoulli trial experiment, several properties must hold.

---

**Bernoulli Experiment**   **Bernoulli Experiment with *n* Trials**

1. The experiment is repeated a fixed number of times (*n* times).

2. Each trial has only two possible outcomes: success and failure. The possible outcomes are exactly the same for each trial.

3. The probability of success remains the same for each trial. (We use $p$ for the probability of success and $q = 1 - p$ for the probability of failure.)

4. The trials are independent. (The outcome of one trial has no influence on later trials.)

5. We are interested in the total number of successes, not the order in which they occur.

   There may be $0, 1, 2, 3, \ldots,$ or $n$ successes in $n$ trials.

---

**Example 1** **(a)** We count the number of times heads occurs when a coin is tossed eight times. Each toss of the coin is a trial, so there are eight repeated trials ($n = 8$). We consider the outcome *heads* a success and *tails* a failure. The probability of success (heads) on each trial is $p = \frac{1}{2}$, and the probability of failure (tails) is $1 - \frac{1}{2} = \frac{1}{2}$. This is an example of a Bernoulli trial.

**(b)** A student guesses at all the answers on a 10-question multiple-choice quiz (four choices of an answer on each question). This fulfills the properties of a Bernoulli trial because:

**1.** Each guess is a trial ($n = 10$).

**2.** There are two possible outcomes: correct and incorrect.

**3.** The probability of a correct answer is $\frac{1}{4}$ ($p = \frac{1}{4}$) and the probability of an incorrect answer is $\frac{3}{4}$ ($q = \frac{3}{4}$), on each trial.

**4.** The guesses are independent because guessing an answer on one question gives no information on other questions.

**(c)** Suppose eight cards are drawn from a deck with none replaced. We are interested in the number of spades drawn. This is *not* a Bernoulli trial because the trials (selecting a card) are not independent (the first card drawn affects the possible choices of the second card; consequently, the conditional probability of drawing a spade changes each time a card is removed).

**(d)** Suppose a card is drawn from a deck, the card is noted and placed back in the deck, and the deck is shuffled. Repeat this eight times. We count the number of times a spade is drawn.

This experiment is a Bernoulli trial because each trial is the same, the trials are independent, and the probability of obtaining a spade remains the same for each trial.

Technically, each trial has 52 outcomes, each card in the deck. However, we reduce the outcomes to the two possible outcomes "success" or "failure" when we collect all spades into the event defined as success and all other cards into the event defined as failure. Then, $p = \frac{13}{52}$ and $q = \frac{39}{52}$.

The following example illustrates how a tree diagram may be used to count the number of successes for a small number of trials.

**Example 2** A coin is tossed three times. What is the probability of exactly two heads in the three tosses?

**Solution**
Look at the paths of Figure 8–12 that begin at 0 and terminate at the end of a branch. There are a total of eight paths, and three of them contain exactly two heads. The probability of terminating at the end of those branches is $\frac{1}{8}$ for each one. (Because the probability of each branch is $\frac{1}{2}$, the probability of taking a sequence of three paths is $\frac{1}{2} \times \frac{1}{2} \times \frac{1}{2} = \frac{1}{8}$.) So, the probability of exactly two heads in three tosses of a coin is $\frac{3}{8}$.

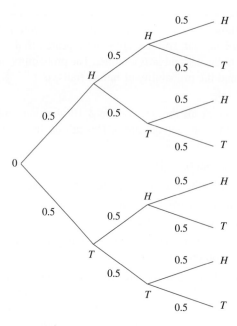

**FIGURE 8-12**

## Probability of a Bernoulli Experiment

The tree diagram of a Bernoulli trial problem can become unwieldy with a large number of trials, so we use an algebraic formula for computing the probability of a specified number of successes. Let's start with a simple example.

**Example 3**    A quiz has five multiple-choice questions with four possible answers to each. A student wildly guesses the answers. What is the probability that he guesses exactly three correctly?

**Solution**

As a Bernoulli trial problem, $n = 5, p = \frac{1}{4}$, and $q = \frac{3}{4}$. We are about to give a formula that calculates the desired probability, but you need to be aware that you have no reason, at this point, to know why it is true. That will be explained later. We want you to understand the quantities used in the formula so that you can more easily follow the justification. Now for the formula. The probability of exactly three correct answers is

$$P(3 \text{ correct}) = C(5, 3)\left(\frac{1}{4}\right)^3\left(\frac{3}{4}\right)^2$$

$$= 10\left(\frac{1}{64}\right)\left(\frac{9}{16}\right)$$

$$= 0.088 \quad \text{(rounded to three decimals)}$$

Let's make some observations about the computation, because they will hold for the general formula.

1. In $C(5, 3)$, 5 is the number of trials, and 3 is the number of successes. (In general, in $C(n, x)$, $n$ is the number of trials, and $x$ is the number of successes.)

2. In $\left(\frac{1}{4}\right)^3$, $\frac{1}{4}$ is the probability of success in a single trial, and 3 is the number of successes in the five trials. (In general, in $p^x$, $p$ is the probability of success in a single trial, and $x$ is the number of successes in $n$ trials.)

3. In $\left(\frac{3}{4}\right)^2$, $\frac{3}{4}$ is the probability of failure in a single trial, and 2 is the number of failures in five trials. It may seem trivial, but note that $2 = 5 - 3$; the number of failures equals the number of trials minus the number of successes. (In general, in $(1 - p)^{n-x}$, $1 - p$ is the probability of failure in a single trial, and $n - x$ is the number of failures in $n$ trials.)

We now give you the general formula.

---

**Probability of a Bernoulli Experiment**

Given a Bernoulli experiment and

$n$ independent repeated trials,

$p$ is the probability of success in a single trial,

$q = 1 - p$ is the probability of failure in a single trial,

$x$ is the number of successes ($0 \leq x \leq n$).

Then, the probability of $x$ successes in $n$ trials is

$$P(x \text{ successes in } n \text{ trials}) = C(n, x)p^x q^{n-x}$$

$P(k$ successes in $n$ trials) may be written $P(X = k)$.

---

**Example 4**   A single die is rolled three times. Find the probability that a 5 turns up exactly twice.

**Solution**

$$n = 3, \quad x = 2, \quad p = \frac{1}{6}, \quad q = \frac{5}{6}$$

so

$$P(X = 2) = C(3, 2)\left(\frac{1}{6}\right)^2\left(\frac{5}{6}\right)^1$$

$$= 3\left(\frac{1}{36}\right)\left(\frac{5}{6}\right)$$

$$= 0.0694$$

Note that in the tree diagram (Figure 8–13) there are three branches with exactly two 5's. This number $C(3, 2)$, gives the number of sequences two 5's can occur in three rolls of the die.

The probability of termination at the end of one such branch is

$$\left(\frac{1}{6}\right)^2\left(\frac{5}{6}\right)$$

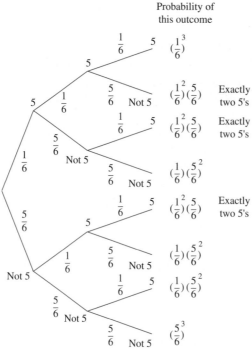

Probability of
this outcome

**FIGURE 8–13**   Exactly two 5's occur three ways
in three tosses of a die.

The notation $P(X = 2)$ is used to indicate the probability of two successes.

$$\text{Thus, } P(X = 2) = C(3, 2)\left(\frac{1}{6}\right)^2\left(\frac{5}{6}\right)$$

**Example 5**   A coin is tossed 10 times. What is the probability that heads occurs 6 times?

**Solution**
In this case, $n = 10$, $x = 6$, $p = \frac{1}{2}$, and $q = \frac{1}{2}$.

$$P(X = 6) = C(10, 6)\left(\frac{1}{2}\right)^6\left(\frac{1}{2}\right)^4$$

$$= 210\left(\frac{1}{64}\right)\left(\frac{1}{16}\right)$$

$$= 0.205$$

## Justification of the Bernoulli Experiment Formula

Now let's go back to Example 3 and explain how to obtain the expression used to compute
the probability of a Bernoulli trial.

Recall the problem: A multiple-choice quiz has five questions with four possible answers to each, one of which is correct. A student guesses the answers. What is the probability that three of the five are correct?

Let's look at how the student succeeds in answering three of the five questions. There are several ways, and the exact number plays a key role in the solution. First, let's list some ways. We will list the sequence of successes (correct answers) and failures (incorrect answers). One sequence is

$$SSSFF$$

which indicates that the first three were correct and the last two were incorrect. Some other sequences of three correct and two incorrect are

$$SSFSF$$

$$FSSSF$$

and so on. Rather than list all the ways in which three correct and two incorrect answers can be given, let's compute the number.

The basic procedure for forming a sequence of three S's and two F's amounts to selecting the three questions with correct answers. The other two answers are automatically incorrect.

In how many different ways can three questions be selected from the five questions? You should recognize this as $C(5, 3)$. As $C(5, 3) = 10$, there are 10 possible sequences of three S's and two F's. The student succeeds in passing if his sequence of guesses is any one of the 10 sequences.

It helps that all 10 sequences have exactly the same probabilities. Note that the probability of SSSFF is

$$\left(\frac{1}{4}\right)\left(\frac{1}{4}\right)\left(\frac{1}{4}\right)\left(\frac{3}{4}\right)\left(\frac{3}{4}\right)$$

by the Multiplication Rule. Also, the probability of SSFSF is

$$\left(\frac{1}{4}\right)\left(\frac{1}{4}\right)\left(\frac{3}{4}\right)\left(\frac{1}{4}\right)\left(\frac{3}{4}\right)$$

Both of these are simply

$$\left(\frac{1}{4}\right)^3\left(\frac{3}{4}\right)^2$$

each written in different order. In fact, the probability of any sequence of three S's and two F's will contain $\frac{1}{4}$ three times and $\frac{3}{4}$ twice, which gives $\left(\frac{1}{4}\right)^3 \left(\frac{3}{4}\right)^2$.

When we add up the probabilities of the 10 sequences, we are adding $\left(\frac{1}{4}\right)^3 \left(\frac{3}{4}\right)^2$ 10 times, which is

$$10\left(\frac{1}{4}\right)^3\left(\frac{3}{4}\right)^2$$

We obtained this probability in Example 3. A similar situation holds in general for other values of $n, x, p,$ and $q$.

**Example 6**   Plantation Foods has found that 25% of those hired to load trucks will work more than one week. Find the probability that three of four new hires will work more than one week.

**Solution**
This experiment has four repeated trials. A hire that works more than one week is considered a success.

$$n = 4, \quad x = 3, \quad p = 0.25, \quad q = 0.75$$
$$P(X = 3) = C(4, 3)(0.25)^3(0.75)^1$$
$$= 4(0.0156)(0.75) = 0.0469$$

The probability that three of the four new hires will work more than one week is about 0.047.

**Example 7**   Professor Purdue gives a multiple-choice quiz with five questions. Each question has four possible answers. A student guesses all answers. What is the probability that she passes the test if at least three correct answers are needed to pass?

**Solution**
In this case, $n = 5, p = \frac{1}{4}$, and $q = \frac{3}{4}$. The student passes if she gets three, four, or five correct answers. We must find the probability of each of these outcomes and add them:

$$P(X = 3) = C(5, 3)(0.25)^3(0.75)^2$$
$$= 10(0.015625)(0.5625)$$
$$= 0.0879$$
$$P(X = 4) = C(5, 4)(0.25)^4(0.75)^1$$
$$= 5(0.0039062)(0.75)$$
$$= 0.0146$$
$$P(X = 5) = C(5, 5)(0.25)^5(0.75)^0$$
$$= 1(0.0009765)(1)$$
$$= 0.00098 \quad \text{(rounded)}$$

Then the probability of three or more correct, which we indicate by $P(X \geq 3)$, is $P(X \geq 3) = 0.0879 + 0.0146 + 0.00098 = 0.10348$.

## Binomial Distribution

We can use Bernoulli trials to form an important probability distribution. We call this distribution a **binomial distribution**. Define it in the following way.

**DEFINITION**
**Binomial Distribution**   For a sequence of Bernoulli experiments of $n$ repeated trials, define a random variable $X$ as the number of successes in $n$ trials. For each value of $x$, $0 \leq x \leq n$, find the probability of $x$ successes in $n$ trials. The probability distribution obtained is the **binomial distribution**.

**Example 8**   Form the binomial distribution for the experiment of rolling a die three times and counting the times a 4 appears.

**Solution**

The random variable $X$ takes on the values 0, 1, 2, and 3, the possible number of successes in three trials. The probability of each value occurring is computed by using binomial trials with $p = \frac{1}{6}$ and $q = \frac{5}{6}$. ($\frac{1}{6}$ is the probability of rolling a 4 in a single trial.)

| $X$ | $P(X)$ |
|---|---|
| 0 | $C(3, 0)\left(\frac{5}{6}\right)^3 = 0.5787$ |
| 1 | $C(3, 1)\left(\frac{1}{6}\right)\left(\frac{5}{6}\right)^2 = 0.3472$ |
| 2 | $C(3, 2)\left(\frac{1}{6}\right)^2\left(\frac{5}{6}\right) = 0.0694$ |
| 3 | $C(3, 3)\left(\frac{1}{6}\right)^3 = 0.0046$ |

The binomial distribution got its name from the binomial $(p + q)^n$. Each term of the expansion of the binomial gives one of the probabilities in the binomial distribution. Note that

$$(p + q)^3 = p^3 + 3p^2q + 3pq^2 + q^3$$

which can be written

$$C(3, 3)p^3q^0 + C(3, 2)p^2q + C(3, 1)pq^2 + C(3, 0)p^0q^3$$

where each term represents the probability of 3, 2, 1, or 0 successes, respectively, in a binomial experiment with three trials.

**Example 9**   Find the binomial distribution for $n = 4$ and $p = 0.3$.

**Solution**

The random variable $X$ may take on the values 0, 1, 2, 3, and 4. The probability distribution is the following.

| $X$ | $P(X)$ |
|---|---|
| 0 | $C(4, 0)(0.7)^4 = 0.2401$ |
| 1 | $C(4, 1)(0.3)(0.7)^3 = 0.4116$ |
| 2 | $C(4, 2)(0.3)^2(0.7)^2 = 0.2646$ |
| 3 | $C(4, 3)(0.3)^3(0.7) = 0.0756$ |
| 4 | $C(4, 4)(0.3)^4 = 0.0081$ |

**Example 10**   Form the binomial distribution of the experiment of tossing a coin six times and counting the number of heads.

**Solution**

The random variable $X$ takes on the values 0, 1, 2, 3, 4, 5, and 6, the possible number of successes in six tosses. Both $p$ and $q$ are $\frac{1}{2}$. The values of $X$ and the corresponding probabilities computed using binomial trials are the following:

| $X$ | $P(X)$ |
|---|---|
| 0 | $C(6, 0)(\frac{1}{2})^6 = 0.0156$ |
| 1 | $C(6, 1)(\frac{1}{2})^1(\frac{1}{2})^5 = 0.0938$ |
| 2 | $C(6, 2)(\frac{1}{2})^2(\frac{1}{2})^4 = 0.2344$ |
| 3 | $C(6, 3)(\frac{1}{2})^3(\frac{1}{2})^3 = 0.3125$ |
| 4 | $C(6, 4)(\frac{1}{2})^4(\frac{1}{2})^2 = 0.2344$ |
| 5 | $C(6, 5)(\frac{1}{2})^5(\frac{1}{2}) = 0.0938$ |
| 6 | $C(6, 6)(\frac{1}{2})^6 = 0.0156$ |

The histogram of the distribution is given in Figure 8–14.

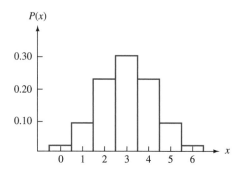

**FIGURE 8–14**  Probability distribution of the number of heads in six tosses of a coin.

---

## 8.6    EXERCISES

Access end-of-section exercises online at **www.webassign.net**      ENHANCED **WebAssign**

---

## Using Your TI Graphing Calculator

### Binomial Probability Distribution

We give you two programs that help with the binomial distribution. The first, **BIOD,** calculates all the probabilities of a binomial distribution. The second, **BIOH,** gives you the histogram of a binomial distribution.

**BIOD**

| | |
|---|---|
| : Lbl 1 | : C → [A](I, 2) |
| : Prompt P, N | : I+1 → I |
| : {N+1, 2} → dim ([A]) | : If X = N |
| : 0 → X | : Goto 3 |
| : 1 → I | : X+1 → X |
| : Lbl 2 | : Goto 2 |
| : N nCr X → C | : Lbl 3 |
| : C*(P^X)*(1-P)^(N-X) → C | : Disp "X PR" |
| : round (C, 4) → C | : Pause [A] |
| : X → [A](I, 1) | : Goto 1 |
| | : End |

The BIOD program calculates the binomial probability distribution

$$P(X = x) = C(N, x)p^x(1 - p)^{N-x} \qquad (\text{for } x = 0, 1, 2, \ldots, N)$$

Start the program with PRGM <BIOD> ENTER When requested, enter the probability of success ($p$) and the number of trials ($N$) and press ENTER . We illustrate with $p = 0.4$ and $N = 5$. The beginning screen is

The screen showing the distribution with the value of $x$ in the first column and the corresponding probabilities in the second column is

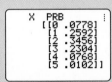

After the probability distribution appears, press ENTER to go to the start of the program.

## Exercises

1. Find the binomial probability distribution for $p = 0.35$ and $n = 4$.
2. Find the binomial probability distribution for $p = 0.20$ and $n = 6$.
3. Find the binomial probability distribution for $p = 0.70$ and $n = 5$.
4. Find the binomial probability distribution for $p = 0.5$ and $n = 6$.

## Histogram of a Probability Distribution

This program calculates the binomial probability distribution

$$P(X = x) = C(N, x)p^x(1 - p)^{N-x} \qquad (\text{for } x = 0, 1, 2, \ldots, N)$$

and draws its histogram. **Warning:** Turn off or clear all $y=$ functions before running the program so that they don't appear on the screen.

*(continued)*

**BIOH**

: Lbl 1                                    : N nCr X → C
: Prompt P, N                        : C*(P^X)*(1-P)^(N-X) → C
: 1 → I                                   : 1000*C → C
: 0 → X                                  : round (C, 0) → L2 (I)
: ClrDraw                             : I-1 → L1 (I)
: ClrList L1, L2                     : I+1 → I
: 0 → Xmin                           : If X=N
: N+1 → Xmax                      : Goto 3
: 1 → Xscl                             : X+1 → X
: 0 → Ymin                           : Goto 2
: 800 → Ymax                       : Lbl 3
: 100 → Yscl                         : Plot 1(Histogram, L1, L2)
: N+1 → dim (L1)                 : DispGraph
: N+1 → dim (L2)                 : Pause
: Lbl 2                                   : Goto 1
                                               : End

*Note:* Xmin, Xmax, Ymin, and so on are found under VARS . <1 : Window> Histogram is found under STAT PLOT <TYPE>.

Start the program with PRGM <BIOH> ENTER . When requested, enter the probability of success ($p$) and the number of trials ($N$) and press ENTER . We illustrate with $p = 0.4$ and $N = 5$.

The beginning and final screens are

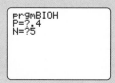
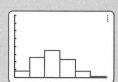

From left to right, each bar represents the probability for $x = 0, 1, 2, \ldots, n$, respectively. The tick marks on the vertical scale represent a probability of $0.1, 0.2, 0.3, \ldots, 0.8$. After the histogram is displayed, press ENTER to go to the beginning of the program.

## Exercises

1. Draw the histogram of the binomial probability distribution for $p = 0.35$ and $n = 4$.
2. Draw the histogram of the binomial probability distribution for $p = 0.20$ and $n = 6$.
3. Draw the histogram of the binomial probability distribution for $p = 0.70$ and $n = 5$.
4. Draw the histogram of the binomial probability distribution for $p = 0.5$ and $n = 6$.

# 8.7  NORMAL DISTRIBUTION

- Normal Curve
- Area Under a Normal Curve
- The z-Score
- Approximating the Binomial Distribution

- Mean and Standard Deviation of a Binomial Distribution
- Using the Normal Curve to Approximate a Binomial Distribution
- Application

In Section 8.4, we used histograms to give a graphic representation of a probability distribution of **discrete** data, that is, the values of the observations had space between that contained no data values. Counting the number of heads in five tosses of a coin gives discrete data because two different values must be at least one unit apart. When we deal with **continuous data**, the values can be arbitrarily close together. All numbers between two different values of the data are permissible values of data. For example, where we measure the heights of male college freshmen, two men could be of different but nearly the same height. Thus, assuming a precise measuring device, any height between 5′9″ and 5′10″ is a possible height for some male.

## Normal Curve

One of the most important probability distributions of data is the **normal distribution**. The normal distribution gives a valid representation of the distribution of many populations, such as IQ of 18-year-olds, heights and weights of 12-year-old girls, and scores on standardized tests such as SAT scores. Just as we used histograms in Section 8.4 to graph discrete probability distributions, we use a **normal curve** to graph a normal distribution. Although normal curves vary in size and shape, they all have "bell shapes" similar to that shown in Figure 8–15. The horizontal baseline represents the values of the data, and the height of the curve tells something about the probability of the value.

**FIGURE 8–15**    The graph of a normal curve.

A histogram can be used to represent continuous data, and in many cases a smooth curve drawn through the tops of the bars gives a normal curve. Studies indicate that the height of 18-year-old males has a mean of 68 inches and a standard deviation of 3 inches, and the normal curve gives a good representation of the heights. Figure 8–16 shows a histogram of heights and a bell-shaped curve obtained by smoothing the histogram.

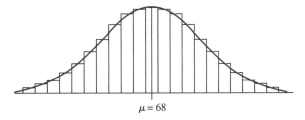

$\mu = 68$

**FIGURE 8–16**    Smooth histogram of heights of 18-year-old males.

The normal curve forms a smooth curve that peaks at the mean and is perfectly symmetric about the mean. This symmetry means that one half of the area under the curve lies to the left of the mean and the other half lies to the right of the mean. We will use areas under the normal curve to represent probabilities, so the total area is taken as 1, representing the probability of the sample space. The normal curve has the property that the mean and standard deviation of the data determine the shape of the normal curve. Figure 8–17 shows three normal distributions with mean = 0 and three different standard deviations.

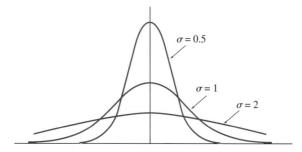

**FIGURE 8–17**    Three normal curves with the same mean, but different standard deviations.

Note that the curves approach the baseline but never touch it. This indicates that fewer and fewer observations are found as you move away from the mean. The normal curve with the smaller standard deviation has a sharper peak, and a larger standard deviation gives a flatter curve.

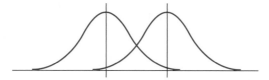

**FIGURE 8–18**    Two normal curves with different means and equal standard deviations.

Figure 8–18 illustrates that two normal distributions with the same standard deviation but different means give normal curves that are the same size and shape, but with the curves located in different positions.

When dealing with continuous data, like heights of 18-year-old males, remember that few if any males are exactly 70 inches tall. When we say a fellow is 70 inches tall, we usually mean that he is "close to" 70 inches. Depending on how accurately we attempt to measure his height, we may mean he is within $\frac{1}{2}$ inch or $\frac{1}{4}$ inch of 70 inches. When we ask for the probability of a randomly chosen 18-year-old male who is approximately 70 inches tall, we can be more specific and ask for the probability that his height lies between 69.5 and 70.5 inches.

## Area Under a Normal Curve

The ability to find the area under a portion of the normal curve is important because it gives the probability that an observation lies in a specified category or interval. For example, suppose that a normal distribution has a mean of 85. (See Figure 8–19.) The area under the curve between 90 and 95 represents the probability that the observation lies between 90 and 95.

Notice that the area under the curve between 85 and 90 is larger than the area between 90 and 95. This indicates a larger probability of values between 85 and 90.

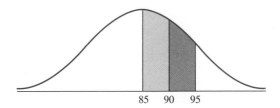

**FIGURE 8–19**

---

**The Probability for a Normal Distribution**

The probability that a value of $X$ lies between two values, $x_1$ and $x_2$, is determined by and equals the fraction of the area under the normal curve that lies between $x_1$ and $x_2$.

We use the notation $P(x_1 \leq X \leq x_2)$ to denote the probability that $X$ lies between $x_1$ and $x_2$.

---

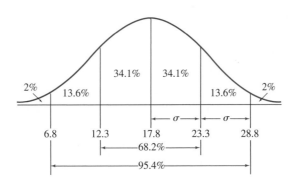

**FIGURE 8–20**    Normal curve with $\mu = 17.8$ and $\sigma = 5.5$.

The normal curve has the unusual property of being completely determined by the mean and standard deviation. This suggests that we need to deal with a different normal curve for each value of the mean and standard deviation. Fortunately, that is not the case. The normal curve can be standardized in a way that allows us to find the area under a portion of a normal curve from one table of values. The distance from the mean, measured in standard deviations, determines the area under the curve in that interval. For example, for all normal curves, about 68% of the scores will lie within 1 standard deviation of the mean, 34% on one side and 34% on the other. Approximately 95% of the scores will lie within 2 standard deviations, and more than 99% will lie within 3 standard deviations. (See Figure 8–20.) Let's apply this idea to the following example.

---

**Example 1**    One year, the ACT English test had a mean $\mu = 17.8$ and a standard deviation $\sigma = 5.5$, and the scores formed a normal distribution.

(a) The scores 12.3 and 23.3 are both 1 standard deviation from the mean, so 34% of the area under the normal curve lies between 12.3 and 17.8, and 34% of the area lies between 17.8 and 23.3. Thus, the probability that an ACT English score lies between 12.3 and 17.8 is 0.34. Similarly, the probability that the score lies between 17.8 and 23.3 is 0.34. (See Figure 8–20.)

**(b)** About 95% of the area lies between 6.8 and 28.8, that is, within 2 standard deviations of the mean, so the probability that an ACT English score lies between 6.8 and 28.8 is 0.95.

**(c)** Because more than 99% of the area lies within 3 standard deviations of the mean (a distance of 16.5 or less from the mean), you expect less than 1% of the area to lie more than 3 standard deviations away. So, the probability is less than 0.01 that an ACT English score is greater than 34.3 or less than 1.3.

---

**NOTE**

The *standard normal curve* is the graph of the function

$$f(x) = \frac{e^{-x^2/2}}{\sqrt{2\pi}}$$

It happens that we can answer questions about normal curves in general by referring to a specific normal curve, the **standard normal curve**. The standard normal curve has mean $\mu = 0$ and $\sigma = 1$.

In the standard normal curve, the letter $z$ is used for the variable, and because $\mu = 0$ and $\sigma = 1$, $z$ also represents the number of standard deviations between the observation and the mean. Thus, $z = 1.6$ indicates 1.6 standard deviations between $z$ and the mean, 0.

Based on the properties of a normal curve that we have already discussed, we can draw some conclusions about the probabilities associated with a standard normal distribution.

---

**Example 2**   A standard normal distribution has a mean of zero and a standard deviation of 1.

**(a)** The probability that a value of $z$ lies between 0 and 1 is about 0.34, and the probability that a value of $z$ lies between $-1$ and 1 (within 1 standard deviation of the mean) is about 0.68.

**(b)** The probability that a value of $z$ lies between $z = -2$ and $z = 2$ (within $2\sigma$ of the mean) is about 0.95.

**(c)** The probability that a value of $z$ lies between $z = -3$ and $z = 3$ (within 3 standard deviations of the mean) is about 0.99.

---

### The z-Score

We have already discussed the probability that an observation lies within 1, 2, or 3 standard deviations of the mean in a normal distribution. How do we determine the number of observations that lie within 1.25, 0.63, or 2.50 standard deviations of the mean? Tables such as the **standard normal table** found inside the back cover of the book enable you to find the area between the mean and an observation. The key lies in the ability to locate the observation according to the number of standard deviations it lies from the mean. Traditionally, the **z-score** represents the number of standard deviations between an observation and the mean.

Whatever scale we use for $x$ on a normal curve, we can associate a value of $z$ with each value of $x$.

Here's how we determine a value of $z$ that corresponds to a given value of $x$. As $z$ represents the number of standard deviations between the mean and the score $x$, find $z$ by

$$z = \frac{\text{difference between the value of } x \text{ and the mean}}{\text{standard deviation}}$$

$$= \frac{\text{value of } x - \text{mean}}{\text{standard deviation}} = \frac{x - \mu}{\sigma}$$

We use $z$ to find the area under the normal curve between two scores. To do so, we use the standard normal table found inside the back cover of the book. The table does not give area between any two scores; that would require a prohibitively lengthy, cumbersome table. You must understand that the table gives area *between the mean and a z-score* for selected *z-scores*. In that table the numbers in the column labeled $A$ represent the area under the normal curve between the mean and the corresponding $z$-score.

**Example 3**   Compute $z$ for each of the following values of the mean, the standard deviation, and a given value of $x$:

**(a)** Mean = 25, standard deviation = 2, $x = 31$
**(b)** Mean = 25, standard deviation = 2, $x = 19$
**(c)** Mean = 7.5, standard deviation = 1.2, $x = 10.5$
**(d)** Mean = 16.85, standard deviation = 2.1, $x = 14.12$

**Solution**
**(a)** Using the formula

$$z = \frac{x - \text{mean}}{\text{standard deviation}}$$

we obtain

$$z = \frac{31 - 25}{2} = \frac{6}{2} = 3$$

**(b)** $z = \dfrac{19 - 25}{2} = \dfrac{-6}{2} = -3$

A negative value of $z$ indicates that $x$ is less than the mean.

**(c)** $z = \dfrac{10.5 - 7.5}{1.2} = \dfrac{3}{1.2} = 2.5$

**(d)** $z = \dfrac{14.12 - 16.85}{2.1} = \dfrac{-2.73}{2.1} = -1.3$

**Example 4**   Use the standard normal table (inside the back cover) to find the fraction of area, $A$, between the mean and

**(a)** $z = 0.75$                     **(b)** $z = -0.75$
**(c)** $z = 2.58$                     **(d)** $z = -1.92$

**Solution**
**(a)** The value for $A$ that corresponds to $z = 0.75$ is 0.2734, the fraction of the area between the mean and $z = 0.75$. (See Figure 8–21.)

**(b)** Because the curve is symmetric about the mean, the area between the mean and $z = -0.75$ equals the area between the mean and $z = 0.75$, which is 0.2734. (See Figure 8–22.)

**(c)** From the table at $z = 2.58$, we find $A = 0.4951$.

**(d)** We use $z = 1.92$ to find $A = 0.4726$, which equals $A$ for $z = -1.92$.

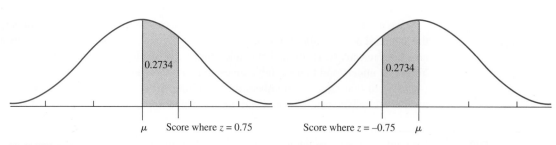

FIGURE 8–21                         FIGURE 8–22

In Section 7.1 on an introduction to probability, we noted that relative frequency is a probability distribution that we call empirical probability. At times, we find it appropriate to estimate the relative frequency with which a value lies in an interval by finding the probability that the value lies in the interval.

For a normal distribution, the same procedure is used to find

**(a)** the probability that $X$ is between $a$ and $b$, $P(a \leq X \leq b)$.

**(b)** an estimate of the fraction of values of $X$ that lie between $a$ and $b$. (Fraction $= P(a \leq X \leq b)$.)

**(c)** an estimate of the percentage of values of $X$ that lie between $a$ and $b$. (Percentage $= 100 \cdot P(a \leq X \leq b)$.)

**Example 5** Estimate the fraction of scores between $z = 1.15$ and $z = -1.15$ under the normal curve. (We may also say that we find the area *within* 1.15 standard deviations of the mean.)

**Solution**

We obtain this area by combining the area between the mean and $z = 1.15$ with the area between the mean and $z = -1.15$, because the table gives only the area between the mean and a score, not between two scores. For the area between the mean and $z = 1.15$, look for the area corresponding to $z = 1.15$. It is 0.3749. The area between the mean and $z = -1.15$ is the same, 0.3749, so the total area is $0.3749 + 0.3749 = 0.7498$. (See Figure 8–23.)

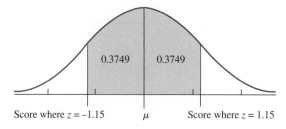

FIGURE 8–23

**Example 6** Find the area under the normal curve between $z = -0.46$ and $z = 2.32$.

**Solution**

The point where $z = -0.46$ lies below the mean, so the area between the mean and where $z = -0.46$ is 0.1772 (look up $A$ for $z = 0.46$). The point where $z = 2.32$ lies above the mean, and the area between the mean and where $z = 2.32$ is 0.4898. As the two points lie on opposite sides of the mean, add the two areas found, $0.1772 + 0.4898 = 0.6670$, to find the total area. (See Figure 8–24.)

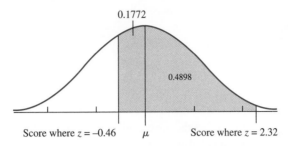

0.1772

0.4898

Score where $z = -0.46$     $\mu$     Score where $z = 2.32$

**FIGURE 8–24**

**Example 7**     Find the probability that a score lies to the right of $z = 0.80$.

**Solution**
Figure 8–25 shows the desired area. The standard normal table gives $A = 0.2881$ for $z = 0.80$. However, this area lies *below* $z = 0.80$ and above the mean, and we want to know the area *above* $z$. Remember that all the area above the mean is 0.5000 of the area under the curve. Therefore, the area *above* $z = 0.80$ is $0.5000 - 0.2881 = 0.2119$ of the total area. Thus, $P(Z > 0.80) = 0.2119$.

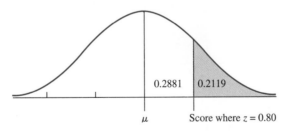

0.2881    0.2119

$\mu$     Score where $z = 0.80$

**FIGURE 8–25**

**Example 8**     Estimate the fraction of scores that are more than 1.40 standard deviations away from the mean.

**Solution**
This asks for the area to the right of the score where $z = 1.40$ and the area to the left of the score where $z = -1.40$. (See Figure 8–26.) The area between the mean and a score where $z = 1.40$ is 0.4192, so the area to the right of the score is $0.5000 - 0.4192 = 0.0808$. By symmetry, the area to the left of the score where $z = -1.40$ is also 0.0808. The total area more than 1.40 standard deviations away from the mean is 0.1616, so about 0.1616 of the scores are more than 1.40 standard deviations away from the mean.

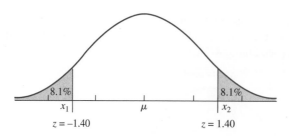

**FIGURE 8–26**

Generally, you will be given the mean and scores, not values of $z$. The following examples show how to find the area between two scores.

**Example 9**    A normal distribution has a mean of 30 and a standard deviation of 7. Find the probability that the value of $x$ is between 30 and 42 $\left(P(30 \leq X \leq 42)\right)$.

**Solution**
To use the standard normal table, we must find the $z$-score that corresponds to 42. It is

$$z = \frac{42 - 30}{7} = \frac{12}{7} = 1.71$$

From the standard normal table, we obtain $A = 0.4564$ when $z = 1.71$. Thus, $P(30 \leq X \leq 42) = 0.4564$. (See Figure 8–27.)

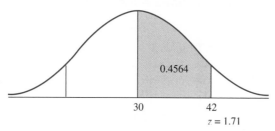

**FIGURE 8–27**

When two values of $x$ lie on opposite sides of the mean and are different distances from the mean, we add areas to find the area between the values.

**Example 10**    A normal distribution has a mean $\mu = 50$ and a standard deviation $\sigma = 6$. Estimate the percentage of scores between 47 and 58.

**Solution**
As the mean lies between 47 and 58, we need to find the area under the curve in two steps; that is, we need to find the area between the mean and each score.

**1.** The area between 47 and 50:

$$z_1 = \frac{47 - 50}{6} = \frac{-3}{6} = -0.50$$

From the standard normal table, using that the area between 0 and –0.50 equals the area between 0 and 0.50, this area is $A = 0.1915$.

2.  The area between 50 and 58:

$$z_2 = \frac{58 - 50}{6} = \frac{8}{6} = 1.33$$
$$A = 0.4082$$

The total area between 47 and 58 is $0.1915 + 0.4082 = 0.5997$, so 0.5997, or 59.97%, of the scores lie between 47 and 58. (See Figure 8–28.)

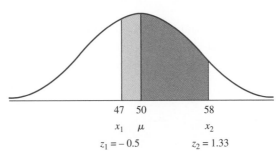

47   50          58
$x_1$  $\mu$          $x_2$
$z_1 = -0.5$     $z_2 = 1.33$

**FIGURE 8–28**

Example 10 illustrates the fundamental concept used to find the probability that a value of $x$ lies in a specified interval of a normal distribution—say, $X$ lies between $x_1$ and $x_2$ $(x_1 \leq X \leq x_2)$.

For the interval $x_1 \leq X \leq x_2$,

$$P(x_1 \leq X \leq x_2) = P(z_1 \leq Z \leq z_2)$$

where $z_1$, $Z$, and $z_2$ are the $z$-values corresponding to $x_1$, $X$, and $x_2$, respectively.
$P(z_1 \leq Z \leq z_2)$ is found using the standard normal table.

Sometimes we want to find an area between two values of $x$ that lie on the same side of the mean.

**Example 11**  The Welding Program Department at Paul's Valley Technical School gives an exit test to evaluate the students' skills and knowledge of procedures.

The design of the test gives scores reasonably close to a normal distribution with a mean of $\mu = 100$ and a standard deviation $\sigma = 8$. A student is selected at random. Find the probability that the student will score between 110 and 120.

**Solution**
Because we always measure areas from the mean to a score, we can find the area between 100 and 110 and the area between 100 and 120. To find the area between 110 and 120, subtract the area between 100 and 110 from the area between 100 and 120.

For $x_1 = 110$,

$$z_1 = \frac{110 - 100}{8} = \frac{10}{8} = 1.25$$

and $A_1 = 0.3944$.
For $x_2 = 120$,

$$z_2 = \frac{120 - 100}{8} = \frac{20}{8} = 2.5$$

and $A_2 = 0.4938$.

The area between 110 and 120 is then $0.4938 - 0.3944 = 0.0994$, so the probability that a student scores between 110 and 120 $(P(110 \le X \le 120))$ is 0.0994. (See Figure 8–29.)

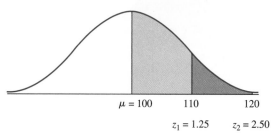

$$\mu = 100 \qquad 110 \qquad 120$$
$$z_1 = 1.25 \qquad z_2 = 2.50$$

**FIGURE 8–29**

**Example 12**  Students at Flatland University spend an average of 24.3 hours per week on homework, with a standard deviation of 1.4 hours. Assume a normal distribution.

**(a)** Estimate the percentage of the students who spend more than 28 hours per week on homework.

**(b)** What is the probability that a student spends more than 28 hours per week on homework?

**Solution**
**(a)** The value of $z$ corresponding to 28 hours is

$$z = \frac{28 - 24.3}{1.4} = \frac{3.7}{1.4} = 2.64$$

From the standard normal table, we have $A = 0.4959$ when $z = 2.64$. The value $A = 0.4959$ represents the area from the mean, 24.3, to 28 ($z = 2.64$). All of the area under the curve to the right of the mean is one half of the total area. The area to the right of $z = 2.64$ is $0.5000 - 0.4959 = 0.0041$. Therefore, about 0.41% of the students study more than 28 hours.

**(b)** The probability that a student studies more than 28 hours is the area that is to the right of 28—that is, 0.0041. (See Figure 8–30.)

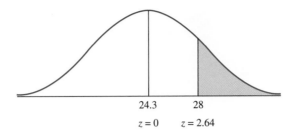

24.3          28

$z = 0$       $z = 2.64$

**FIGURE 8–30**

**Example 13**   (a)  Find the value of $z$ such that an estimated 4% of the scores lie to the right of $z$.

(b)  Find the value of $z$ such that 0.04 is the probability that a score lies to the right of $z$.

**Solution**

(a)  If 4% of the scores lie to the right of $z$, then the other 46% of the scores to the right of the mean lie between the mean and $z$. (Remember that 50% of the scores lie to the right of the mean.) Look for $A = 0.4600$ in the standard normal table. It occurs at $z = 1.75$. This is the desired value of $z$.

(b)  This also occurs at $z = 1.75$. (See Figure 8–31.)

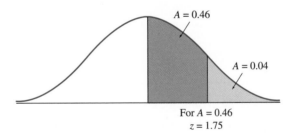

$A = 0.46$

$A = 0.04$

For $A = 0.46$
$z = 1.75$

**FIGURE 8–31**

These examples ask for the probability that a score lies in a certain interval (such as $P(50 \leq X \leq 60)$) or greater than a certain score (such as $P(X \geq 75)$). A natural question is "How do you find the probability of a certain score, such as $P(X = 65)$?" The answer is "The probability is zero." In general, $P(X = c) = 0$ for any value $c$ in a normal distribution. For example, this states that the probability of randomly selecting an 18-year-old male who is *exactly* 5 feet 10 inches tall is zero.

Because of the symmetry of the normal curve, the standard normal table gives $A$ only for positive values of $z$. For negative values of $z$, when the score lies to the left of the mean, simply use the value of $A$ for the corresponding positive $z$. Keep in mind that each $z$-value determines an area *from the mean* to the $z$ position.

We can find a variety of areas by adding or subtracting areas given from the table. (See Figure 8–32.)

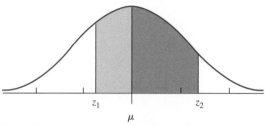

(a) Add the areas for $z_1$ and $z_2$ to get the area between $z_1$ and $z_2$.

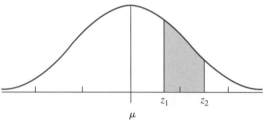

(b) Subtract $z_1$ area from $z_2$ area
to get the area between $z_1$ and $z_2$.

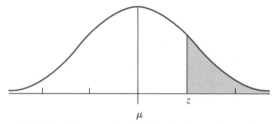

(c) Subtract the $z$ area from 0.500 to get the area beyond $z$.

**FIGURE 8–32**

**Procedure to Determine $P(c \leq X \leq d)$ of a Normal Distribution**

**Step 1.** Determine the $z$-value for $x = c$ and $x = d$ and call them $z_1$ and $z_2$, respectively.

**Step 2.** From the standard normal table, determine $A$ corresponding to $z_1$ and to $z_2$.

**Step 3.**

**(a)** If $c$ and $d$ are on opposite sides of the mean ($z_1$ and $z_2$ have opposite signs), add the values of $A$ corresponding to $z_1$ and $z_2$.

**(b)** If $c$ and $d$ are on the same side of the mean ($z_1$ and $z_2$ have the same signs), subtract the smaller value of $A$ from the larger value.

**Procedure to Determine $P(X < c)$ or $P(X > d)$ of a Normal Distribution**

**Step 1.** Determine the $z$-value corresponding to $c$ or $d$.

**Step 2.** From the standard normal table, determine the value of $A$ corresponding to $z$ of step 1.

**Step 3.**

**(a)** $P(X < c)$: Find the area below $c$. If $c$ lies to the right of the mean ($z$ is positive), add $A$ to 0.5000. If $c$ lies to the left of the mean ($z$ is negative), subtract $A$ from 0.5000.

**(b)** $P(X > d)$: Find the area above $d$. If $d$ lies to the right of the mean ($z$ is positive), subtract $A$ from 0.5000. If $d$ lies to the left of the mean ($z$ is negative), add $A$ to 0.5000.

**Example 14**   Several hundred thousand junior high students take a standardized test. The mean is 100, and the standard deviation is 10. For a student selected at random, find the probability that the student scores in the 114 to 120 interval.

**Solution**

This may be answered by using the properties of the normal curve, as it represents standardized test scores well.

The values of $z$ corresponding to 114 and 120 are $z = 1.4$ and $z = 2.0$, respectively. For $z = 1.4$, $A = 0.4192$; for $z = 2.0$, $A = 0.4773$. So, the area between 114 and 120 is $0.4773 - 0.4192 = 0.0581$. The probability that the student's score lies in the 114 to 120 interval is 0.0581.

**Example 15**   The Quality Cola Bottling Company sells its Quality Cola in the standard size can, 355 milliliters (ml). The manager does not expect every can to contain *exactly* 355 ml of cola but would like to be consistently close. Working with the quality-control manager, she agrees that the quantity can be expected to vary as a normal distribution, but the cans should have a mean of 355 ml, and at least 95% should vary from 355 ml by no more than 5 ml. What value of the standard deviation does this require?

**Solution**

The company wants 95% of the values to lie between 350 and 360 ml ($355 \pm 5$); that is, 47.5% lie between 355 and 360 ($A = 0.475$), while the other 47.5% lie between 350 and 355 ($A = 0.475$). From the standard normal table, we find for $A = 0.475$, $z = 1.96$, so $z = -1.96$ at 350 and $z = 1.96$ at 360.

As $5 = 1.96\sigma$,

$$\sigma = \frac{5}{1.96} = 2.55$$

**Summary of Properties of a Normal Curve**

1. All normal curves have the same general bell shape.

2. The curve is symmetric with respect to a vertical line that passes through the peak of the curve.

3. The vertical line through the peak occurs where the mean, median, and mode coincide.

4. The area under any normal curve is always 1.

5. The mean and standard deviation completely determine a normal curve. For the same mean, a smaller standard deviation gives a taller and narrower peak. A larger standard deviation gives a flatter curve.

6. The area to the right of the mean is 0.5; the area to the left of the mean is 0.5.

7. About 68.26% of the area under a normal curve is enclosed in the interval formed by the score 1 standard deviation to the left of the mean and the score 1 standard deviation to the right of the mean.

*(continued)*

8. If a random variable $X$ has a normal probability distribution, the probability that a score lies between $x_1$ and $x_2$ is the area under the normal curve between $x_1$ and $x_2$.

9. The probability that a score is less than $x_1$ equals the probability that a score is less than, or equal to $x_1$; that is,

$$P(X < x_1) = P(X \le x_1)$$

10. $P(X = c) = 0$.

## Approximating the Binomial Distribution

**Example 16**   The probability that a new drug will cure a certain blood disease is 0.7. If it is administered to 100 patients with the disease, what is the probability that 60 of them will be cured?

**Solution**
You should set up

$$P(X = 60) = C(100, 60)(0.7)^{60}(0.3)^{40}$$

with little difficulty. You then may find it tedious to compute the probability and even more tedious to form the binomial distribution of the experiment. Some calculators have functions that make these computations relatively easy.

We typically use the normal distribution for continuous data and the binomial distribution for discrete data, but the normal distribution can sometimes be used to estimate binomial probabilities.

We do not calculate $P(X = 60)$ in the preceding example because the normal distribution provides a means to avoid this wearisome computation. It can be used to estimate the binomial distribution for large values of $n$. We will soon show how to do this. First, we need to find the mean and standard deviation of a binomial distribution.

## Mean and Standard Deviation of a Binomial Distribution

Recall from Section 8.5 that the computation of the mean (or expected value), the variance, and the standard deviation of a probability distribution can be time-consuming for a random variable with many values.

In Example 9 of Section 8.6, we used the binomial distribution with $n = 4$ and $p = 0.3$. The expected value of the random variable $X$ is

$$E(X) = 0.2401(0) + 0.4116(1) + 0.2646(2) + 0.0756(3) + 0.0081(4)$$
$$= 0 + 0.4116 + 0.5292 + 0.2268 + 0.0324$$
$$= 1.20$$

The significance of this result is that $1.20 = 4(0.3)$, which is $np$ for this example. This result holds for all binomial distributions, $E(X) = np$. Likewise, the variance and standard deviation of a binomial distribution can be expressed in rather simple terms of $n$, $p$, and $q$, as follows.

**DEFINITION**

Mean, Variance, and Standard Deviation of a Binomial Distribution

Let $X$ be the random variable for a binomial distribution with $n$ repeated trials, with $p$ the probability of success, $q$ the probability of failure, and

$$P(X = x) = C(n, x)\, p^x (1 - p)^{n - x}$$

Then, the **mean** (expected value), **variance**, and **standard deviation** of $X$ are given by

Mean: $\quad \mu = np$

Variance: $\quad \sigma^2(X) = np(1 - p) = npq$

Standard deviation: $\quad \sigma(X) = \sqrt{np(1 - p)} = \sqrt{npq}$

**Example 17**　(a) For the binomial distribution with $n = 20$, $p = 0.35$:

$$\text{Mean} = \mu = 20(0.35) = 7$$
$$\sigma^2(X) = 20(0.35)(0.65) = 4.55$$
$$\sigma(X) = \sqrt{4.55} = 2.133$$

(b) If $n = 160$, $p = 0.21$,

$$\mu = 160(0.21) = 33.6$$
$$\sigma^2(X) = 160(0.21)(0.79) = 26.544$$
$$\sigma(X) = \sqrt{26.544} = 5.152$$

## Using the Normal Curve to Approximate a Binomial Distribution

To compute some binomial probabilities, such as the probability of 40 to 65 successes in 200 trials, can be tedious and subject to mistakes. Fortunately, the normal curve can often be used to obtain a satisfactory **estimate of a binomial probability**.

To illustrate how to use the normal curve to estimate binomial probabilities, we use the following simple example.

**Example 18**　Use the normal curve to estimate the probability of

(a) three heads in six tosses of a coin.

(b) three or four heads in six tosses of a coin.

**Solution**

(a) We first point out that the binomial distribution of this problem, and its histogram, are given in Example 10 of Section 8.6. From it, for example, we see that $P(X = 3) = 0.3125$.

For the binomial distribution with $n = 6$ and $p = 0.5$, the mean and standard deviation are

$$\mu = 6(0.5) = 3$$
$$\sigma = \sqrt{6(0.5)(0.5)} = \sqrt{1.5} = 1.225$$

Figure 8–33 shows a normal curve with $\mu = 3$ and $\sigma = 1.225$ superimposed on the binomial distribution for $n = 6$ and $p = 0.5$. Note that a portion of the histogram lies above the curve and that some space under the curve is not filled by the rectangles. It appears that if the portions of the rectangle outside were moved into the empty spaces below the curve, then the area under the curve would pretty well be filled. This states that the total area enclosed by the histogram is "close" to the area under the normal curve.

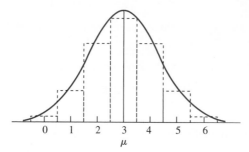

**FIGURE 8–33**     A normal curve superimposed on a binomial distribution.

Let's make another observation. Each bar in the histogram is of width 1, and its height is equal to the probability it represents.

The bar representing $P(X = 3)$ is centered at 3 and is therefore located between 2.5 and 3.5. The area of the bar ($1 \times 0.3125$) can also be used to represent the probability that $X = 3$. We mention this because it holds the key to using the normal distribution. To find $P(X = 3)$, find the area under the normal curve between 2.5 and 3.5. (See Figure 8–34.)

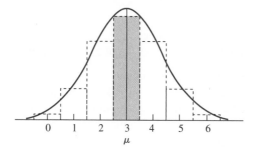

**FIGURE 8–34**     To estimate $P(X = 3)$ in a binomial distribution, find the area under the normal curve between $x = 2.5$ and $x = 3.5$.

Here's how we use the normal curve to estimate $P(X = 3)$. Use $\mu = 3$, $\sigma = 1.225$, $x_1 = 2.5$, and $x_2 = 3.5$. Then,

$$z_1 = \frac{2.5 - 3}{1.225} = \frac{-0.5}{1.225} = -0.41$$

$$z_2 = \frac{3.5 - 3}{1.225} = \frac{0.5}{1.225} = 0.41$$

The standard normal table shows that the area under the normal curve between the mean and $z = 0.41$ is $A = 0.1591$. Then, the area between $z_1 = -0.41$ and $z_2 = 0.41$ is $0.1591 + 0.1591 = 0.3182$. This estimates $P(X = 3)$ as $0.3182$. This compares to the actual probability of $0.3125 \left(C(6, 3)(0.5)^3(0.5)^3\right)$. Notice that we approximated the binomial probability $P(X = 3)$ by the normal probability $P(2.5 \le X \le 3.5)$.

**(b)** To find the probability of three or four heads in six tosses of a coin, $P(3 \le X \le 4)$, we find the total area of the bars for three and four. This amounts to finding the area in the histogram between 2.5 and 4.5. The normal approximation is the area between

$$z_1 = \frac{2.5 - 3}{1.225} = -0.41 \quad \text{and} \quad z_2 = \frac{4.5 - 3}{1.225} = 1.22$$

We find the areas determined by

$$z_1 = -0.41 \quad \text{and} \quad z_2 = 1.22$$

The corresponding areas are

$$A_1 = 0.1591 \quad \text{and} \quad A_2 = 0.3888$$

so the desired probability estimate is $0.1591 + 0.3888 = 0.5479$. The actual probability, from Example 10, Section 8.6, is $0.3125 + 0.2344 = 0.5469$. Note that we approximated the binomial probability $P(3 \le X \le 4)$ by the normal probability $P(2.5 \le X \le 4.5)$.

At this point, we must confess that the normal approximation worked rather well in the preceding example because $p$ and $q$ were both $\frac{1}{2}$. Had we used $p = \frac{1}{6}$ and $q = \frac{5}{6}$, the normal approximation would not have worked very well. However, if $n$ is 50 instead of 6, the normal approximation gives reasonable results. So when is it reasonable to use the normal distribution to approximate the binomial distribution? The answer is, "It depends." It depends on the values of $n$ and $p$. Several rules of thumb are used by statisticians to judge when the normal approximation is reasonable. We use the following:

> The normal distribution provides a good estimate of the binomial distribution when both $np$ and $nq$ are greater than or equal to 5.

Figures 8–35, 8–36, and 8–37 show the binomial distribution for $p = 0.60$ and $n = 3, 7$, and 15. Notice that the distribution for $n = 15$ more closely resembles a normal distribution than those for $n = 3$ or $n = 7$. We observe that, for $n = 3$, $np = 1.8$ and $nq = 1.2$, both smaller than the desired value of 5 for a good normal approximation. Also, for $n = 7$, $np = 4.2$ and $nq = 2.8$, but for $n = 15$, $np = 9.0$ and $nq = 6.0$. Thus, by our rule, the normal approximation is a reasonable approximation of the binomial distribution for $n = 15$ and $p = 0.6$.

$P(x)$

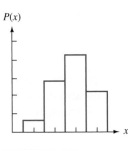

**FIGURE 8–35**
Binomial distribution for $p = 0.6$ and $n = 3$.

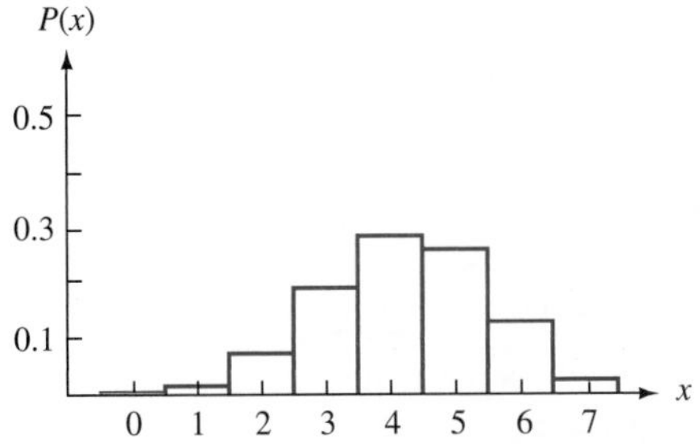

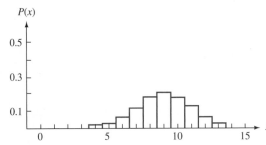

**FIGURE 8–36**   Binomial distribution for $p = 0.6$ and $n = 7$.

**FIGURE 8–37**   Binomial distribution for $p = 0.6$ and $n = 15$.

**Example 19**     For a binomial distribution with $p = 0.6$, find the smallest value of $n$ so that the normal distribution gives a reasonable approximation to the binomial distribution.

**Solution**
For the normal distribution to be a reasonable approximation, both $np \geq 5$ and $nq \geq 5$.
For $np \geq 5$,

$$n(0.6) \geq 5 \quad \text{and} \quad n \geq \frac{5}{0.6} = 8.33$$

For $nq \geq 5$,

$$n(0.4) \geq 5 \quad \text{and} \quad n \geq \frac{5}{0.4} = 12.5$$

Thus, $n \geq 12.5$, so the smallest integer value of $n$ is 13.

In the next three examples, we estimate some binomial probabilities, using the normal distribution. In each of the three examples, we use the binomial distribution with

$$n = 14, \quad p = 0.4, \quad q = 0.6$$

On the basis of these values, we have

$$\mu = 14(0.4) = 5.6 \quad \text{and} \quad \sigma = \sqrt{14(0.4)(0.6)} = 1.83$$

These values will be used in the three examples.

**Example 20**     The probability that a gasoline additive increases gasoline mileage in a car is 0.4. In a test conducted by an automotive class at the Rocky Mountain Technical Institute, the additive was used on 14 cars selected at random. Use the normal curve to estimate

**(a)** the probability that gasoline mileage improves in 3 to 7 cars ($P(3 \leq X \leq 7)$).

**(b)** the probability that gasoline mileage improves in more than 3 cars and fewer than 7 cars ($P(3 < X < 7)$).

**Solution**
**(a)** Figure 8–38 shows the graph of the binomial probability with the bars representing $3 \leq X \leq 7$ shaded and a normal curve superimposed. It shows that the area under the normal curve between $X = 2.5$ and 7.5 approximates the desired probability. To find the area, we need the corresponding $z$-scores.

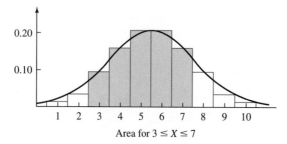

Area for $3 \leq X \leq 7$

**FIGURE 8–38**

Using $\mu = 5.6$ and $\sigma = 1.83$ we have the following.
At $X = 2.5$,

$$z = \frac{2.5 - 5.6}{1.83} = -1.69 \quad \text{and} \quad A = 0.4545$$

(from the standard normal table).
At $X = 7.5$,

$$z = \frac{7.5 - 5.6}{1.83} = 1.04 \quad \text{and} \quad A = 0.3508$$

The area under the normal curve between 2.5 and 7.5 is then $0.4545 + 0.3508 = 0.8053$, so $P(3 \le X \le 7) = 0.8053$.

**(b)** To find $P(3 < X < 7)$, we need to find the area under the normal curve between 3.5 and 6.5, which includes the bars for 4, 5, and 6. (See Figure 8–39.)
At $X = 3.5$,

$$z = \frac{3.5 - 5.6}{1.83} = -1.15 \quad \text{and} \quad A = 0.3749$$

At $X = 6.5$,

$$z = \frac{6.5 - 5.6}{1.83} = 0.49 \quad \text{and} \quad A = 0.1879$$

The area under the normal curve between $X = 3.5$ and $X = 6.5$ is then $0.3749 + 0.1879 = 0.5628$, so $P(3 < X < 7) = 0.5628$.

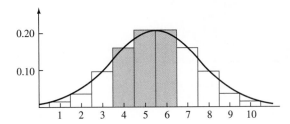

**FIGURE 8–39**

**Example 21** Using the gasoline additive data in Example 20 ($n = 14$, $p = 0.4$), use the normal curve to estimate the probability that gasoline mileage improves in more than three cars $(P(X > 3))$.

**Solution**
To approximate $P(X > 3)$, we need to find the area under the normal curve to the right of 3.5. (See Figure 8–40.) From Example 20, we have at $X = 3.5$, $z = -1.15$, and the area from $X = 3.5$ to the mean, 5.6, is 0.3749. As the area to the right of the mean is 0.5000, $P(X > 3) = 0.3749 + 0.5000 = 0.8749$.

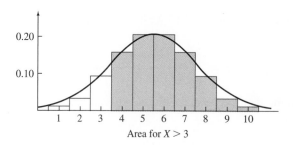

Area for $X > 3$

**FIGURE 8–40**

**Example 22**    Using the gasoline additive data in Example 20 ($n = 14$, $p = 0.4$), use the normal curve to estimate the probability that gasoline mileage improves in seven or fewer cars $(P(X \leq 7))$.

**Solution**
To approximate $P(X \leq 7)$, we need to find the area under the normal curve to the left of 7.5. (See Figure 8–41.) From Example 20, we have at $X = 7.5$, $z = 1.04$, and the area from the mean, 5.6, to $X = 7.5$ is 0.3508. As the area to the left of the mean is 0.5000, $P(X \leq 7) = 0.3508 + 0.5000 = 0.8508$.

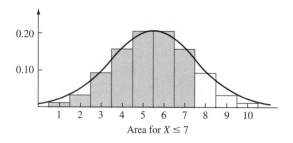

Area for $X \leq 7$

**FIGURE 8–41**

In summary, estimate a binomial probability with a normal distribution as follows:

**Procedure for Estimating a Binomial Probability**

1. If $np$ and $nq$ are both greater than or equal to 5, you may assume that the normal distribution provides a good estimate.
2. Compute $\mu = np$ and $\sigma = \sqrt{npq}$.
3. To estimate $P(X = c)$, find the area under the normal curve between $c - 0.5$ and $c + 0.5$.

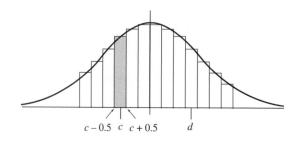

$c - 0.5$  $c$  $c + 0.5$      $d$

**4.** To estimate $P(c \le X \le d)$, $c < d$, find the area under the normal curve between $c - 0.5$ and $d + 0.5$.

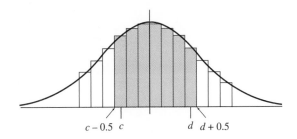

**5.** To estimate $P(c < X < d)$, $c < d$, find the area under the normal curve between $c + 0.5$ and $d - 0.5$.

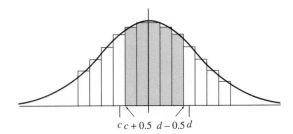

**6.** To estimate $P(X > c)$, find the area under the normal curve to the right of $c + 0.5$.

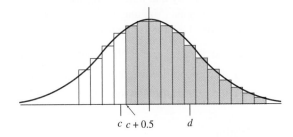

**7.** To estimate $P(X \ge c)$, find the area under the normal curve to the right of $c - 0.5$.

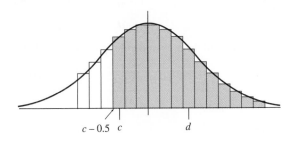

*(continued)*

**8.** To estimate $P(X < c)$, find the area under the normal curve to the left of $c - 0.5$.

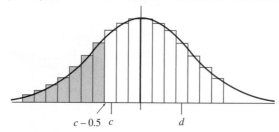

**9.** To estimate $P(X \leq c)$, find the area under the normal curve to the left of $c + 0.5$.

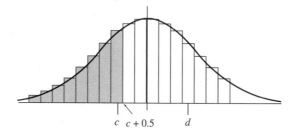

## Application

**Example 23**    The probability that a new drug will cure a certain blood disease is 0.7. It is administered to 100 patients. (You may recognize this as the problem we set up but did not solve in Example 16.) Use a normal curve to estimate

**(a)** the probability that 60 of them will be cured.

**(b)** the probability that 60 to 75 of them will be cured.

**(c)** the probability that more than 75 will be cured.

### Solution
Because $n = 100$, $p = 0.7$, and $q = 0.3$, we know that $np = 70$ and $nq = 30$. As both values are greater than 5, a normal curve provides a good estimate, and

$$\mu = 100(0.7) = 70$$
$$\sigma = \sqrt{100(0.7)(0.3)} = \sqrt{21} = 4.583$$

**(a)** Find the area under the normal curve between 59.5 and 60.5—that is, between

$$z_1 = \frac{59.5 - 70}{4.583} = \frac{-10.5}{4.583} = -2.29$$

and

$$z_2 = \frac{60.5 - 70}{4.583} = \frac{-9.5}{4.583} = -2.07$$

The corresponding areas are $A_1 = 0.4890$ and $A_2 = 0.4808$. As the $z$-scores lie on the same side of the mean, we must subtract areas, $0.4890 - 0.4808 = 0.0082$. So, $P(X = 60) = 0.0082$.

**(b)** To estimate $P(60 \le X \le 75)$, we find the area between 59.5 and 75.5. The corresponding $z$-values and areas are

$$z_1 = \frac{59.5 - 70}{4.583} = -2.29 \quad \text{and} \quad z_2 = \frac{75.5 - 70}{4.583} = 1.20$$

$A_1 = 0.4890$, and $A_2 = 0.3849$. As the scores lie on opposite sides of the mean, we add areas to obtain the probability:

$$P(60 \le X \le 75) = 0.4890 + 0.3849 = 0.8739$$

**(c)** To estimate the probability that more than 75 patients will be cured, $P(X > 75)$, find the area under the normal curve that lies to the right of 75.5. (*Note:* To find the probability of 75 or more, $P(X \ge 75)$, find the area to the right of 74.5.)

For a score of 75.5, $z = 1.20$ and $A = 0.3849$. (See part (b).) Because we want the area above $z = 1.20$, we need to subtract $0.5000 - 0.3849 = 0.1151$ to get $P(X > 75) = 0.1151$.

---

## 8.7 EXERCISES

### Using Your TI Graphing Calculator

### Area Under the Normal Curve

The standard normal table enables you to find the area under a normal curve for limited values of $z$. We have a program for the TI graphing calculator that finds the area for other $z$-values. This program calculates the area under the normal curve between two scores.

**NORML**

```
: Lbl 1                          : Input U
: Disp "MEAN"                    : (L-M)/S → W
: Input M                        : (U-M)/S → Z
: Disp "STD DEV"                 : fnInt((e^(-X²/2))/√ (2π),X,W,Z) → A
: Input S                        : Disp "AREA="
: Disp "LOW LIMIT"               : Pause A
: Input L                        : Goto 1
: Disp "UP LIMIT"                : End
```

*(continued)*

## NOTE

fnInt is found in the [MATH] <MATH> menu.

Input: The mean and standard deviation of the normal distribution, the scores that define the lower and upper limits of the area. If the area above a specified score is desired, use a score that is 3.5 standard deviations above the mean for the upper limit. Similarly, a score 3.5 standard deviations below the mean is used as the lower limit when the area below a specified score is desired.

Output: The area under the normal curve bounded by the given upper and lower limit scores.

We give two illustrations.

**(a)** A data set has a mean of 26 and a standard deviation of 1.5. Find the area under the normal curve between 23 and 30. The first screen shows the mean, standard deviation, and limits entered. The second screen shows the area between limits, $A = 0.9734$.

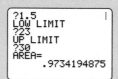

**(b)** For the data set in (a), find the area above 24. Notice in the first screen that the upper limit is 32 because it is more than 3.5 standard deviations above the mean, 26. ($3.5 \times 1.5 = 5.25$, and 32 is 6 above the mean.) The second screen shows the area under the normal curve above 24, $A = 0.9087$.

## NOTE

This program may give slightly different answers than those obtained using the standard normal table because $z$ in the table is rounded to two decimals and $A$ is rounded to four decimals. The program carries more decimal places in its calculations.

## Exercises

**1.** A data set has a mean of 48 and a standard deviation of 3. Find the area under the normal curve between 46 and 53.

**2.** A data set has a mean of 120 and a standard deviation of 10. Find the area under the normal curve between 125 and 135.

**3.** A data set has a mean of 65 and a standard deviation of 4.2. Find the area under the normal curve above 62.

**4.** A data set has a mean of 76 and a standard deviation of 5. Find the area under the normal curve below 73.5.

## Using Excel

Excel has a function that finds the area under the normal curve that is *below* a score $x$. This differs from the standard normal table we have used, that finds the area between the mean and the score. Here's how to use it.

Store the value of $x$ in A2, the mean in C2, and the standard deviation in D2. Let's put the area under the normal curve below $x$ in cell A3 by entering the formula =NORMDIST(A2,C2,D2,1) in A3. For example, if we enter $x = 35$ in A2, mean = 30 in C2, and standard deviation = 5 in D2, then the formula returns 0.84134474 in A3.

To find the area between two scores, we enter the smaller score, say 23, in A2, the larger score, say 35, in B2, the mean, 30, in C2, and the standard deviation, 5, in D2. We obtain the area between the two scores in A4 by subtracting the area below the smaller score from the area below the larger score. In A4, enter the formula =NORMDIST(B2,C2,D2,1)-NORMDIST(A2,C2,D2,1). In this case, it returns the area 0.76058803.

To find the area above the score in A2, use the formula =1-NORMDIST(A2,C2,D2,1).

### Exercises

In the following exercises, find the area (to 4 decimals) under the normal curve with the given mean and standard deviation.

**1.** Mean = 45, standard deviation = 3. Find the area below $x = 49$.

**2.** Mean = 45, standard deviation = 3. Find the area below 40.

**3.** Mean = 130, standard deviation = 8. Find the area between 125 and 136.

**4.** Mean = 130, standard deviation = 8. Find the area between 120 and 128.

**5.** Mean = 130, standard deviation = 8. Find the area between 134 and 138.

**6.** Mean = 92, standard deviation = 6. Find the area above 95.

**7.** Mean = 92, standard deviation = 6. Find the area above 85.

## 8.8  ESTIMATING BOUNDS ON A PROPORTION

- Confidence Intervals
- Standard Error of a Proportion
- Computing Error Bounds for a Proportion
- Quality Control

Mr. Alexander Quality, President of Quality Cola Company, grew tired of Quality Cola. He wanted a cola with more zest and a new taste. Like his father before him, he had vowed never to drink his competitor's cola. Thus, he must develop a new quality cola or resign himself to the traditional taste. He discussed the problem with departmental heads. The head of the research division agreed they could develop a new formula, but it would cost thousands of dollars. The chief accountant insisted that they should recover the development cost and make a profit. The marketing manager hesitated to put a new product on the market unless she was confident that it would succeed.

Mr. Quality, an astute executive, agreed that his managers had valid points, so he instructed them to develop a new formula, find out if the public liked it, and if so, pour money into advertising it. After weeks of work, the research division developed a formula that both they and Mr. Quality liked. Now the marketing manager wanted to know if the public liked it. She quickly determined that it was quite unrealistic and prohibitive in cost to give everyone in the country a taste test. So, she asked the company statistician to help her. She told the statistician that she was confident that the new cola would be successful if 40% or more of the population liked it. The statistician outlined the following plan:

1. Select, by a random means, 500 people throughout the country.
2. Give each one a taste test.
3. Find the proportion of the sample that like the new cola.
4. Use the sample proportion as an estimate of the proportion of the total population that like the new cola.

It took the statistician several weeks to select and survey the sample. When the information was in and tabulated, it showed that 43% of the people in the sample test liked the new cola. At first the marketing manager was elated. Enough people liked the new product to make it successful. Then, she had second thoughts. What about the millions of people who did not participate in the taste test? They were the ones who would determine the success of the new cola, so she called the statistician.

"Can I depend on 43% of everyone liking the new cola? Perhaps you just happened to pick the few people who like it."

"I cannot guarantee that precisely 43% of the general public will like it. I told you this was an estimate."

"How good is the estimate? If the estimate is off 2 or 3 percentage points, we are O.K. If the proportion for the entire population is actually only 20%, we are in real trouble. Can you put some bounds on how much the estimate might be in error?"

We interrupt this saga to give some background of how this analysis works. The market analysis in this story involves a population and a sample. The **population** consists of all the people who are potential customers, millions of them, perhaps. The **sample** consists of the 500 people selected for the taste test. The results of the taste test showed that the **proportion** of the sample who liked Quality Cola was 43%. We denote the **sample proportion** by $\bar{p}$ and $\bar{p} = 0.43$. (In this case, we use the decimal form rather than the percentages.)

For the entire population, some proportion will like Quality Cola, and we denote that **population proportion** with $p$. For all practical purposes, the population proportion, $p$, is impossible to determine even though the marketing manager would dearly love to know it.

Let's use the Quality Cola story to describe an important relationship between the proportion, $p$, of an arbitrary population and the proportion, $\bar{p}$, of a sample taken from the population.

The proportion of the population who like Quality Cola is unknown, but let's suppose that somehow we know it is $p = 0.45$ (45% like Quality Cola). In such a case, the sample proportion $\bar{p} = 0.43$ yields a reasonably good estimate.

However, if we took other random samples of 500 people, we would not expect to find the same proportion who like Quality Cola. We might find $\bar{p} = 48\%$ in one sample, 39% in another, 41% in another, or maybe even as much as 65% in a sample.

Now for an interesting, and useful, relationship. Take a large number of random samples and in each case record the percentage who like Quality Cola. This gives a data set of sample proportions. As a set of numbers they have a mean and standard deviation.

For a large enough sample size, and a large number of samples, the proportions obtained resemble a normal distribution. An amazing result occurs if we could compute *all* possible sample proportions. That mean equals the population proportion, $p$. Let's illustrate this with a simple example.

**Example 1**  A deck of cards contains a large number of cards (thousands or maybe even millions of cards). Each card in the deck is numbered with a 1, 2, 3, or 4. Each of the numbers appears on one fourth of the cards.

The population in this example is the deck of cards. The proportion of cards that contain a 1 is $p = \frac{1}{4}$. Now suppose someone who doesn't know the make-up of the deck wants to find the proportion of cards that contain a 1. They proceed by randomly selecting two cards from the deck and noting the proportion of 1's, that is, they select a sample of two cards and note $\bar{p}$ for the sample. The cards are replaced and other random samples are taken. For each sample taken, three kinds of outcomes are possible:

**(a)** Neither card is a 1 ($\bar{p} = 0$).

**(b)** One card is a 1 and the other card is not ($\bar{p} = \frac{1}{2}$).

**(c)** Both cards are 1's ($\bar{p} = 1$).

After a number of samples, a list of numbers consisting of 0's, $\frac{1}{2}$'s, and 1's is obtained. The mean of these numbers estimates the population proportion. If we did the impractical and took *all* possible samples of two cards, the theory indicates that the mean of the sample proportions would equal the population proportion. Rather than trying the impractical, let's find the mean of the proportions from the *expected value* of finding a 1. We need the probability of the three possible values of $\bar{p}$:

**(a)** $\bar{p} = 0$ when neither card is a 1. The probability that a single card drawn is not a 1 is $\frac{3}{4}$. Thus, the probability the first card is not a 1, and the second card is not a 1 is $\left(\frac{3}{4}\right)\left(\frac{3}{4}\right) = \frac{9}{16}$.

**(b)** $\bar{p} = \frac{1}{2}$ when one card is a 1 and the other isn't. This occurs when the first card is a 1 and the second isn't, or the first card isn't a 1 and the second is. The probability of this occurring is $\frac{1}{4}\left(\frac{3}{4}\right) + \frac{3}{4}\left(\frac{1}{4}\right) = \frac{6}{16}$.

**(c)** $\bar{p} = 1$ when both cards are 1's. The probability of this occurring is $\frac{1}{4}\left(\frac{1}{4}\right) = \frac{1}{16}$.

We have $P(\bar{p} = 0) = \frac{9}{16}, P(\bar{p} = \frac{1}{2}) = \frac{6}{16}$, and $P(\bar{p} = 1) = \frac{1}{16}$. So, the expected value of $\bar{p}$ is

$$0\left(\frac{9}{16}\right) + \frac{1}{2}\left(\frac{6}{16}\right) + 1\left(\frac{1}{16}\right) = \frac{4}{16} = \frac{1}{4}$$

Thus, the mean of *all* sample proportions equals the population proportion $p = \frac{1}{4}$.

Note that *none* of the sample proportions could possibly equal the population proportion, but the mean of the sample proportions does.

It is unrealistic to take all possible samples of a large population, so one possible strategy is similar to that taken at a nearby university. Here's the situation.

**Example 2**  The dean of students received a student petition requesting a change in the date of spring break. The dean had no desire to consider a change unless there was strong student support beyond the relatively small number who signed the petition.

The dean approached Professor Turner, a statistics professor, for help. Professor Turner had just taught the chapter on sampling to the class, so it seemed like a good project to assign to the students. Each of the 15 students was to make a random sample of 60 students and determine the proportion of the sample that favored a change in spring break.

The dean was somewhat dismayed when the sample proportions were all different. Which one should be used?

Because the mean of all sample proportions equals the population proportion, it seems reasonable that the mean of the 15 samples should provide a better estimate of the population proportion than any sample alone. So, Professor Turner recommended that the dean use the mean of the 15 sample proportions as the estimate of the population proportion.

The procedure of taking several samples and averaging the $\bar{p}$'s is not the method generally used because it is often a major effort to find just one random sample. So, we go to the other extreme and take just one sample. We then use the proportion obtained as the estimate for the population proportion. When only one sample is used, we cannot take a small sample like Example 1 where it was impossible to obtain a sample proportion near the population proportion. The sample size must be large enough to allow a sample proportion close to the population proportion. Realizing the sample proportion likely differs from the population proportion, it is important to know how close the sample proportion is to the population proportion. The fact that the sample proportions form a normal distribution helps answer that.

We remind ourselves that we are dealing with

1. A population of which a proportion, $p$, has a characteristic of interest (such as liking Quality Cola).

2. Random samples of the population are taken and the proportion, $\bar{p}$, of the sample that have the characteristic of interest is obtained. The set of possible $\bar{p}$'s has a mean equal to the population proportion $p$.

3. For a large sample size, the distribution of the $\bar{p}$'s is approximately normal.

Because the $\bar{p}$'s form a normal distribution, they tend to be clustered fairly close to the population proportion, $p$.

## Confidence Intervals

In the Quality Cola example, the marketing manager would feel comfortable if she knew the sample proportion was within 3% of the actual, but unknown, population proportion. (See Figure 8–42a.)

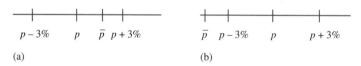

(a)                                      (b)

**FIGURE 8–42**    The Quality Cola marketing manager would like $\bar{p}$ to be in the interval $p - 3\%$ to $p + 3\%$ [as in part (a)], not outside the interval [as in part (b)].

**FIGURE 8–43**    A confidence interval for $p$.

Because the population proportion, $p$, is unknown and unattainable, we use a sample proportion, $\bar{p}$, as an estimate of the population proportion, even though $\bar{p}$ rarely turns out to be $p$. Still, we expect the sample proportion $\bar{p}$ to be close to $p$ in the sense that $\bar{p}$ is in some interval centered at $p$, say in the interval from $p - E$ to $p + E$. We call $E$ the **maximum error of the proportion**. (In the Quality Cola example, $E = 3\%$.) The longer this interval, the more confident we are that it contains $p$. (See Figure 8–43.)

When we state how close $p$ is to $\bar{p}$, we consider the interval $\bar{p} - E$ to $\bar{p} + E$ and discuss the probability that $p$ lies in this interval.

It is reasonable to expect a larger value of $E$, and thus a longer interval, to be more likely to contain $p$ than a shorter interval. We say we are more *confident* the longer interval contains the population proportion $p$. We would like to do more than say we are more or less confident. We want to measure the confidence. This leads to the idea of a confidence interval and confidence level. We call the degree of confidence a **confidence level**, and we express it as a percentage. When we say that we are 100% confident that $p$ lies in the interval from $\bar{p} - E$ to $\bar{p} + E$, we are saying we are certain. When we say that we are 90% confident $p$ lies in the interval $\bar{p} - E$ to $\bar{p} + E, (\bar{p} - E \leq p \leq \bar{p} + E)$, we are saying that if we take a sample, the probability that $p$ for the population lies in the confidence interval is 0.90.

| | |
|---|---|
| **DEFINITION**<br>Confidence Interval,<br>Confidence Level | Let $\bar{p}$ be the proportion of any random sample from a population and let $E$ be a positive number.<br><br>    The numbers between $\bar{p} - E$ and $\bar{p} + E$ form an interval called the **confidence interval** for the population proportion.<br>    The **confidence level** associated with this interval is the probability $P(\bar{p} - E \leq p \leq \bar{p} + E)$ expressed as a percentage. |

We now proceed to discuss how we determine $E$.

## Standard Error of a Proportion

We introduce the term **standard error**, abbreviated S.E. We use this term when referring to the standard deviation of the $\bar{p}$'s, and it is rather simple to compute:

$$\text{S.E.} = \sqrt{\frac{p(1 - p)}{n}}$$

where $n$ is the sample size and $p$ is the mean of the $\bar{p}$'s and is also the population proportion.

Unfortunately, we don't know the value of $p$, which is the very thing we are attempting to find. Because we are going to need S.E. to estimate a confidence interval, we use

$$\text{S.E.} \approx \sqrt{\frac{\bar{p}(1 - \bar{p})}{n}}$$

which is considered a reasonable estimate.

| | |
|---|---|
| **DEFINITION**<br>Estimate of<br>Standard Error | $$\text{S.E.} \approx \sqrt{\frac{\bar{p}(1 - \bar{p})}{n}}$$ |

The properties of a normal distribution tell us, for example, that about 68% of the $\bar{p}$'s lie within 1 S.E. of the $p$. (See Figure 8–44.)

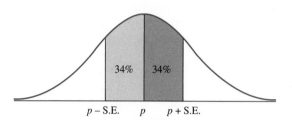

$$p - \text{S.E.} \qquad p \qquad p + \text{S.E.}$$

**FIGURE 8–44**   Because $\bar{p}$ forms a normal distribution, about 68% of the $\bar{p}$'s lie within 1 S.E. of the mean of the $\bar{p}$'s (mean = $p$).

We also say that the probability that $\bar{p}$ lies within 1 S.E. of $p$ is 0.68, and consequently, the probability that $p$ lies within 1 S.E. of $\bar{p}$ is 0.68. The interval $\bar{p} - \text{S.E.}$ to $\bar{p} + \text{S.E.}$ is a 68% confidence interval.

We are primarily interested in three confidence levels: 90%, 95%, and 99%, with the 95% confidence level being the one most used in practice.

## Computing Error Bounds for a Proportion

Let's look at how we find a 95% confidence interval. We want to find the interval $(\bar{p} - E, \bar{p} + E)$ so that the probability that $p$ lies in that interval is 0.95; that is, $P(\bar{p} - E \le p \le \bar{p} + E) = 0.95$. As $\bar{p}$ forms a normal distribution, we are seeking the area under the normal curve such that 95% of the area is between $\bar{p} - E$ and $\bar{p} + E$, 47.5% to the left of $\bar{p}$ and 47.5% to the right of $\bar{p}$.

From the standard normal table, $z = 1.96$ for $A = 0.475$, so we conclude that $E$ lies 1.96 S.E. from the mean; that is, $E = 1.96 \,(\text{S.E.})$. (See Figure 8–45.) To move to the right of $p$ by 1.96 S.E. gives the boundary point $p + 1.96$ S.E. Similarly, the left boundary point is $p - 1.96$ S.E.

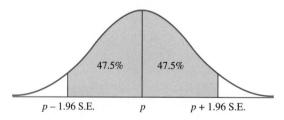

$$p - 1.96 \text{ S.E.} \qquad p \qquad p + 1.96 \text{ S.E.}$$

**FIGURE 8–45**   95% of all $\bar{p}$'s lie within 1.96 S.E.'s of the population proportion, $p$.

The procedure for estimating $p$ by using the proportion of a random sample, $\bar{p}$, is justified by a fundamental theorem in statistics known as the Central Limit Theorem.

**DEFINITION**
**Central Limit Theorem**

Let $\bar{p}$ represent the proportion of a sample of size $n$ that has a certain property and let $p$ represent the proportion of the entire population that has the same property. When the sample size is large ($n\bar{p} \geq 5$ and $n(1 - \bar{p}) \geq 5$), then a normal distribution is a good representation of the distribution of $\bar{p}$ with

Mean of the $\bar{p}$'s $= \mu_{\bar{p}} = p =$ the population mean

Standard error of the $\bar{p}$'s = S.E. $= \sqrt{\dfrac{p(1 - p)}{n}}$

An acceptable estimate for S.E. is

$$\text{S.E.} \approx \sqrt{\dfrac{\bar{p}(1 - \bar{p})}{n}}$$

The Central Limit Theorem allows us to use the properties of a normal distribution in the study of a distribution that is not normal. Even if a population does not have normal distribution properties, the sample proportions, $\bar{p}$, are normally distributed. When we use sufficiently large samples, generally over 50, we can make inferences about the population based on the properties of a normal distribution.

Now let's go back to the Quality Cola story and find the 95% confidence interval for the sample of 500 that gave $\bar{p} = 0.43$.

**Example 3**   $n = 500$   and   $\bar{p} = 0.43$

Find the 95% confidence interval.

**Solution**

$$\text{S.E.} = \sqrt{\dfrac{0.43(0.57)}{500}} = 0.0221$$

Then the maximum error $E = 1.96(0.0221) = 0.043$. The 95% confidence interval is from $0.43 - 0.043 = 0.387$ to $0.43 + 0.043 = 0.473$.

Thus, we say $0.387 \leq p \leq 0.473$ with 95% confidence. The Quality Cola marketing manager should be pleased because the lower end of the confidence interval is very nearly the 40% favorable response she desired.

In general, we use $E = z \times$ S.E., where $z$ depends on the confidence level desired. For a 90% confidence level, find $z$ in the standard normal table that corresponds to $A = 0.45$ (one-half of 0.90). You will find $z = 1.65$. For a 99% confidence level, use $z = 2.58$ because it corresponds to $A = 0.495$ (one-half of 0.99).

**Procedure for Computing Error Bounds for a Proportion**

Let $n$ be the sample size and $\bar{p}$ the proportion of the sample that respond favorably.

1. Decide on the confidence level to be used and write it as a decimal $c$. The most common values of $c$ are 0.90., 0.95, and 0.99, with 0.95 the most used.

2. Compute $A = \frac{c}{2}$. This corresponds to the area under the standard normal curve between $z = 0$ and the $z$-score that corresponds to an area equal to $A$. ($A$ is the same as that found in the normal distribution table.)

3. Find the value of $z$ in the standard normal table that corresponds to $A$.

$$\text{For } A = 0.45 \; (c = 0.90), z = 1.65$$
$$\text{For } A = 0.475 \; (c = 0.95), z = 1.96$$
$$\text{For } A = 0.495 \; (c = 0.99), z = 2.58$$

4. Compute the standard error estimate S.E. $\approx \sqrt{\dfrac{\bar{p}(1 - \bar{p})}{n}}$.

5. Compute the maximum error $E = z \times \text{S.E.} = z \times \sqrt{\dfrac{\bar{p}(1 - \bar{p})}{n}}$.

6. Compute the upper and lower bounds

$$\bar{p} + E \quad \text{and} \quad \bar{p} - E$$

7. Then $c$, the confidence level, is the probability that the proportion of the *total* population, $p$, lies in the interval

$$\bar{p} - E \leq p \leq \bar{p} + E$$

**Example 4**

Compute the error bounds for the proportion $\bar{p} = 0.55$ obtained from a sample of size $n = 120$. Use the 95% confidence level.

**Solution**

The steps for the procedure give

1. $n = 120$, $\bar{p} = 0.55$, and $c = 0.95$.

2. $A = \frac{0.95}{2} = 0.475$.

3. The value of $z$ that corresponds to $A = 0.475$ is $z = 1.96$.

4. S.E. $= \sqrt{\dfrac{(0.55)(0.45)}{120}} = \sqrt{0.0020625} = 0.045$.

5. $E = z \times \text{S.E.} = 1.96 \times 0.045 = 0.0882$.

6. The upper and lower bounds of the proportion are

$$0.55 + 0.0882 = 0.6382$$
$$0.55 - 0.0882 = 0.4618$$

7. The 95% confidence interval is $0.4618 < p < 0.6382$.

**Example 5**   A marketing class made a random selection of 150 shoppers at a shopping mall to participate in a taste test of different brands of coffee. They found that 54 shoppers preferred brand X. Find, at the 95% confidence level, the error bounds of the proportion of shoppers who prefer brand X.

**Solution**
For this problem, $n = 150, \bar{p} = \frac{54}{150} = 0.36$, and $c = 0.95$. Then,

$$A = \frac{0.95}{2} = 0.475$$

$z = 1.96$ corresponds to $A = 0.475$ in the standard normal table

$$\text{S.E.} = \sqrt{\frac{(0.36)(0.64)}{150}} = \sqrt{0.001536} = 0.0392$$

$$E = 1.96(0.0392) = 0.0768$$

The bounds are

$$0.36 + 0.0768 = 0.4368 \qquad \text{and} \qquad 0.36 - 0.0768 = 0.2832$$

giving the 95% confidence interval $0.2832 < p < 0.4368$.
   The marketing class is 95% confident that between 28.32% and 43.68% of all shoppers prefer brand X.

It should not escape your attention that the error bounds in this example give a rather large confidence interval. One way to maintain the same confidence level and reduce the size of the confidence interval is to increase the sample size.

**Example 6**   Suppose that the sample in Example 5 was $n = 500$ in size but the proportion remained the same. Compute the error bounds of the proportion.

**Solution**
Now, $n = 500, \bar{p} = 0.36$, and $c = 0.95$. We still have $z = 1.96$, but

$$\text{S.E.} = \sqrt{\frac{(0.36)(0.64)}{500}} = \sqrt{0.0004608} = 0.0215$$

Then, $E = 1.96(0.0215) = 0.0421$, and the upper and lower bounds are

$$0.36 + 0.0421 = 0.4021 \qquad \text{and} \qquad 0.36 - 0.0421 = 0.3179$$

giving the confidence interval $0.3179 < p < 0.4021$.

Note that this confidence interval is smaller, so the sample proportion gives a better estimate of the total population. This is generally true; a larger sample size reduces the maximum error. The sample size that will keep the maximum error to a specified level can be determined. (See Example 9.)

**Example 7**     A random sample of 200 people shows that 46 of them use No-Plaque toothpaste. Based on a 99% confidence level, estimate the proportion of the general population that uses the toothpaste.

**Solution**
For this sample,

$$n = 200$$

$$\bar{p} = \frac{46}{200} = 0.23$$

$$c = 0.99$$

$$A = 0.495$$

$$z = 2.58$$

$$\text{S.E.} = \sqrt{\frac{(0.23)(0.77)}{200}} = 0.02976$$

$$E = 2.58(0.02976) = 0.0768$$

Then, the interval that contains the proportion of the general population is from $0.23 - 0.0768 = 0.1532$ to $0.23 + 0.0768 = 0.3068$. We conclude, with 99% confidence, that about 15% to 31% of the population use No-Plaque toothpaste. Note that the higher confidence level, 99% in this case, requires a larger confidence interval.

**Example 8**     A random sample of 25 shoppers showed that 24% shopped at Cox's Department Store. Find a 90% confidence interval of the proportion of the general population that shop there.

**Solution**
$n = 25$, $\bar{p} = 0.24$, $c = 0.90$, $A = 0.45$, $z = 1.65$, S.E. $= \sqrt{0.007296} = 0.0854$, and $E = 0.1409$. (Be sure you check these computations.) So, the upper and lower bounds are

$$0.24 + 0.1409 = 0.3809 \qquad \text{and} \qquad 0.24 - 0.1409 = 0.0991$$

So, the 90% confidence interval is $0.0991 < p < 0.3809$. Notice that this small sample yields a wide confidence interval.

**Example 9**     KWTX television station wants an estimate of the proportion of the population that watches its late movie. The station wants the estimate correct within 5% at the 95% confidence level. How big a sample should it select?

**Solution**
Basically, the television station wants a maximum error of 5%, written 0.05 in our computations, at the 95% confidence level.
     Look at the computations to obtain $E$. We have

$$E = z\sqrt{\frac{\bar{p}(1 - \bar{p})}{n}}$$

We are given $E = 0.05$, and we know that $z = 1.96$ for the 95% confidence level. We need to find $n$ so that

$$0.05 = 1.96\sqrt{\frac{\bar{p}(1 - \bar{p})}{n}}$$

We face a dilemma. We need to know the value $\bar{p}$ so we can solve for $n$. However, we find $\bar{p}$ from the sample. Thus, it appears that we need $\bar{p}$ before we know the size of the sample, and we need the sample to find $\bar{p}$. We have a way out of this vicious circle. It can be shown that the largest possible value of $\bar{p}(1 - \bar{p})$ is 0.25 and occurs when $\bar{p} = 0.5$. So, if we use 0.25 for $\bar{p}(1 - \bar{p})$, the value of $E$ may be a little too high and the resulting confidence interval a little larger than necessary, but we have erred on the safe side. We proceed using 0.25. Then,

$$0.05 = 1.96\sqrt{\frac{0.25}{n}}$$

Squaring both sides, we get

$$0.0025 = (1.96)^2 \frac{0.25}{n} = \frac{0.9604}{n}$$

Then,

$$0.0025n = 0.9604$$

$$n = \frac{0.9604}{0.0025} = 384.16$$

A sample size of 385 will be sufficient to provide the desired maximum error.

## Quality Control

Manufacturing companies want to maintain the quality of their products. Defective items can occur because of a random glitch or because a machine may become worn or out of adjustment or a new employee may not understand the procedures. A manufacturer usually recognizes that some defective items caused by random glitches will occur, but if the defective items become excessive, production is halted to take corrective measures.

A quality-control manager monitors the production process for defective items by taking random samples and using the proportion of defective items in the sample to estimate the proportion of defective items overall. As with opinion polls, the sample proportion provides only an estimate, and error bounds can be quite useful. Let's look at a simple example.

**Example 10**    The quality-control policy of Arita China Company specifies that production must continue as long as the quality-control manager is 95% confident that defects occur in less than 1% of the pieces produced. Otherwise, production is halted, and the equipment is adjusted and calibrated. The quality-control manager periodically takes a random sample of 450 pieces of china and carefully examines them for defects. On February 15, she found two defective pieces in the sample of 450. Should she allow production to continue?

**Solution**

The sample estimate of the proportion of defective pieces is

$$\bar{p} = \frac{2}{450} = 0.0044 = 0.44\%$$

$$\text{S.E.} = \sqrt{\frac{0.0044(0.9956)}{450}} = 0.00312$$

giving the error bounds

$$0.0044 + 1.96(0.00312) = 0.0105 \quad \text{and} \quad 0.0044 - 1.96(0.00312) = -0.00172$$

So, with 95% confidence the population proportion of defective items is in the interval $0 < x < 0.0105$. (0 was used because the proportion is never negative.)

The accepted interval for proportion of defective items is $0 < x < 0.01$. As the computed interval $0 < x < 0.0105$ is so close to the desired interval, the quality-control manager should not halt production.

Had the computed interval been $0 < x < 0.0195$, the quality-control manager would have been justified in requesting that production be halted to calibrate the equipment.

## 8.8    EXERCISES

Access end-of-section exercises online at **www.webassign.net**

## IMPORTANT TERMS

**8.1**
Descriptive Statistics
Inferential Statistics
Categories
Frequency Table
Frequency Distribution
Qualitative Data
Quantitative Data
Histogram
Discrete Data
Continuous Data
Relative Frequency
Pie Chart
Stem-and-Leaf Plot

**8.2**
Measure of Central Tendency
Population Sample
Mean
Average (Arithmetic Average)
Grouped Data
Median

Mode
Skewed Data

**8.3**
Measures of Dispersion
Range
Standard Deviation
Variance
Deviation
Squared Deviation
Rank
Percentile
z-Score
Quartile
Box Plot
Five-Point Summary

**8.4**
Random Variable
Discrete Variable
Continuous Variable
Probability Distribution

**8.5**
Expected Value
Variance of a Random Variable
Standard Deviation of a
   Random Variable
Fair Game

**8.6**
Bernoulli Trials
Repeated Trials
Bernoulli Experiment
Binomial Distribution

**8.7**
Discrete Data
Continuous Data
Normal Distribution
Normal Curve
Standard Normal Curve
z-Score
Mean of a Binomial Distribution
Variance of a Binomial
   Distribution

Standard Deviation of a Binomial
   Distribution
Estimate of a Binomial
   Probability

**8.8**
Sample Proportion
Population Proportion
Maximum Error of the
   Proportion

Confidence Interval
Confidence Level
Standard Error
Error Bounds

 **8**

## IMPORTANT CONCEPTS

| | |
|---|---|
| **Frequency Table** | A tabular summary of information that shows categories of the information and the frequency with which observations fall in each category. |
| **Histogram** | A visual representation, a bar chart, of a frequency table. Each category is represented by a bar and the height of the bar denotes the frequency of the category. |
| **Pie Chart** | A pie-shaped chart that represents the proportion of scores in each category. The proportion of scores that lies in a category is represented by a "slice of the pie." |
| **Measure of Central Tendency** | A number that represents a center of a list of scores. Three measures of central tendency are mean, median, and mode. |
| **Mean** | The mean of a set of scores is the sum of the scores divided by the number of scores. |
| **Median** | The median of a set of scores is the middle score after the scores are arranged in descending or ascending order. For an even number of scores the median is the mean of the two middle scores. |
| **Mode** | The score that occurs most often. There may be one, more than one, or no modes. |
| **Standard Deviation** | A measure of the degree to which scores tend to cluster around the mean. (See Section 8.3.) |
| **First Quartile** | Separates the lower 25% of the data from the upper 75%. |
| **Second Quartile** | The median of the data. |
| **Third Quartile** | Separates the upper 25% of the data from the lower 75%. |
| **Five-Point Summary** | Five points used to help understand the dispersion of data: minimum value, maximum value, median, first quartile, and third quartile. |
| **Random Variable** | A rule that assigns a number to each outcome of an experiment. |
| **Probability Distribution of a Random Variable** | A rule that assigns a probability to each value of the random variable. |
| **Expected Value** | The long term mean, $E(X)$, of numerous trials of a random variable where $E(X) = p_1 x_1 + p_2 x_2 + \cdots + p_n x_n$ |
| **Bernoulli Experiment** | An experiment that is repeated a fixed number of times. Each trial has exactly two outcomes and the outcomes are the same for each trial. The trials are independent and the probability of each outcome remains the same for each trial. |

| | |
|---|---|
| **Bernoulli Experiment Probability** | The probability of $x$ successes in $n$ trials is $C(n, x)\, p^x\, q^{n-x}$. |
| **Normal Curve (Distribution)** | A bell-shaped curve that represents many data distributions that arise naturally. The mean and median lie at the center of the data and the curve is symmetric around the mean. The shape of the curve is determined by the mean and standard deviation of the data. |
| **z-Score** | In a normal distribution the $z$-score is the number of standard deviations that a score lies from the mean. |
| **Mean and Standard Deviation of a Binomial Distribution** | Mean $= np$ and standard deviation $= \sqrt{npq}$. |
| **Sample** | A subset of a population under study. |
| **Population Proportion** | The proportion of a population that gives the response of interest. |
| **Sample Proportion** | The proportion of a sample that gives the response of interest. |
| **Confidence Interval** | When $\bar{p}$ is the proportion of a random sample and $E$ is a positive number, the numbers between $\bar{p} - E$ and $\bar{p} + E$ form a confidence interval for the population. |
| **Confidence Level** | The probability, expressed as a percent, that the population proportion lies in the confidence interval. The most common confidence level is 95%. |
| **Standard Error Estimate** | $\text{S.E.} = \sqrt{\dfrac{\bar{p}\,(1 - \bar{p})}{n}}$. |
| **Maximum Error** | $E = z \times \text{S.E.}$ |

# 8  REVIEW EXERCISES

**1.** Draw a histogram and a pie chart based on the following frequency table:

| New accounts opened | Frequency |
|---|---|
| Monday | 17 |
| Tuesday | 31 |
| Wednesday | 20 |
| Thursday | 14 |
| Friday | 8 |

**2.** Find the mean of 4, 6, –5, 12, 3, 2, and 9.

**3.** Find the median of
   (a) 8, 12, 3, 5, 6, 3, and 9.
   (b) 4, 9, 16, 12, 3, 22, 1, and 95.
   (c) 3, –2, 6, 1, 4, and –3.

**4.** Find the mean for the following quiz data:

| Score on quiz | Frequency |
|---|---|
| 0 | 2 |
| 1 | 3 |
| 2 | 6 |
| 3 | 9 |
| 4 | 2 |
| 5 | 4 |

**5.** Estimate the mean for the number of passengers in cars arriving at a play.

| Number of passengers in a car | Frequency |
|---|---|
| 1–2 | 54 |
| 3–4 | 32 |
| 5–6 | 12 |

**6.** Find the mean, median, and mode for the numbers 2, 8, 4, 3, 2, 9, 6, 2, and 7.

**7.** A shopper paid a total of $90.22 for his purchases. The mean price was $3.47. How many items did he purchase?

**8.** Find the variance and standard deviation for the numbers 8, 18, 10, 16, 3, and 11.

**9.** A professor selects three students each day. Let $X$ be the random variable that represents the number who completed their homework. Give all possible values of $X$.

**10.** Professor Delgado asked her students to evaluate her teaching at the end of the semester. Their response to the statement "The tests were a good measure of my knowledge" were marked on a scale of 1 to 5 as follows:

| Outcome | Random variable $X$ | Responses |
|---|---|---|
| Strongly agree | 1 | 10 |
| Agree | 2 | 35 |
| Neutral | 3 | 30 |
| Disagree | 4 | 20 |
| Strongly disagree | 5 | 15 |

Find the probability distribution of $X$ and sketch its graph.

**11.** A store has a special on bread with a five-loaf limit. The probability distribution for sales is listed in the following table. Find the average number of loaves per customer.

| Number of loaves per customer | Probability |
|---|---|
| 0 | 0.05 |
| 1 | 0.20 |
| 2 | 0.15 |
| 3 | 0.20 |
| 4 | 0.25 |
| 5 | 0.15 |

**12.** Find the expected value and variance for the following probability distribution:

| $X$ | $P(X)$ |
|---|---|
| 1 | 0.14 |
| 2 | 0.06 |
| 3 | 0.22 |
| 4 | 0.15 |
| 5 | 0.36 |
| 6 | 0.07 |

**13.** An instructor summarized his student ratings (scale of 1 to 5) in the following probability distribution:

| Response | $X$ | Probability |
|---|---|---|
| Excellent | 1 | 0.20 |
| Good | 2 | 0.32 |
| Average | 3 | 0.21 |
| Fair | 4 | 0.15 |
| Poor | 5 | 0.12 |

What is his expected average rating?

**14.** A normal distribution has a mean of 80 and a standard deviation of 6. Find the fraction of scores

(a) between 80 and 88. (b) between 70 and 84.

(c) greater than 90.

**15.** A distributor averages sales of 350 mopeds per month with a standard deviation of 25. Assume that sales follow a normal distribution. What is the probability that sales will exceed 400 during the next month?

**16.** A construction company contracts to build an apartment complex. The total construction time follows a normal distribution with an average time of 120 days and a standard deviation of 15 days.

(a) The company will suffer a penalty if construction is not completed within 140 days. What is the probability that it will be assessed the penalty?

(b) The company will be given a nice bonus if it completes construction in less than 112 days. What is the probability that it will receive the bonus?

(c) What is the probability that the construction will be completed in 115 to 130 days?

**17.** Find the binomial distribution for $n = 4$ and $p = 0.25$.

**18.** A binomial distribution has $n = 22$ and $p = 0.35$. Find the mean and standard deviation of the distribution.

**19.** The probability that a new drug will cure a certain disease is 0.65. If it is administered to 80 patients, estimate the probability that it will cure

(a) more than 50 patients.

(b) 65 patients.

(c) more than 55 and fewer than 60 patients.

**20.** For a set of scores, the mean $= 240$ and $\sigma = 10$. Find

(a) $z$ for the score 252.    (b) $z$ for the score 230.

(c) the score corresponding to $z = -2.3$

**21.** In a cross-country ski race, a skier came in 43rd in a field of 216 skiers. What was her percentile?

**22.** A brother and sister took two different standardized tests. He scored 114 on a test that had a mean of 100 and a standard deviation of 18. She scored 85 on a test that had a mean of 72 and a standard deviation of 12. Which one had the better score?

**23.** The housing office measures the congeniality of incoming students with a test scored 1–5. A higher score indicates a higher level of congeniality. A summary of 800 tests is the following:

| $X$ (score) | Number receiving score |
|---|---|
| 1 | 20 |
| 2 | 160 |
| 3 | 370 |
| 4 | 215 |
| 5 | 35 |

On the basis of this information, determine a probability distribution of $X$.

**24.** A number is drawn at random from $\{1, 2, 3, 4, 5, 6, 7\}$. A random variable $X$ has the value $X = 0$ when the number drawn is even and $X = 1$ when the number is odd. Find the probability distribution of $X$.

**25.** Three papers in a creative writing class were graded A, and five were graded B. The professor selects three of them to read to the class. Let the random variable, $X$, be the number of A papers selected. List the possible values of $X$ and give the number of possible outcomes that can be associated with each value.

**26.** A chemistry teacher gives a challenge problem for each meeting of a class of 19 students. A random variable, $X$, is the number of students who work the challenge problem. What are the possible values of $X$?

**27.** A die is rolled, and the player receives, in dollars, the number rolled on the die or $3, whichever is smaller. Let $X$ be the number of dollars received. Determine the number of different ways in which each can occur.

**28.** A store ran a special on six-pack cartons of cola, with a maximum of four allowed. A cola representative recorded the number purchased by each customer. The results are summarized as follows:

| Number of cartons | Frequency |
|---|---|
| 0 | 85 |
| 1 | 146 |
| 2 | 268 |
| 3 | 204 |
| 4 | 122 |

Find the mean number of cartons purchased.

**29.** Estimate the mean score of the following grouped data:

| Score | Frequency |
|---|---|
| 0–6 | 8 |
| 7–10 | 13 |
| 11–14 | 6 |
| 15–20 | 15 |

**30.** Roy bought four textbooks at a mean price of $164.60, and Rhonda bought three textbooks at a mean price of $159.70. Find the mean price of the seven books.

**31.** Compute the standard error for the given sample sizes and proportions:

(a) $n = 60, \bar{p} = 0.35$  (b) $n = 700, \bar{p} = 0.64$

(c) $n = 950, \bar{p} = 0.40$

**32.** Compute the error bounds for the following:

(a) $n = 50, \bar{p} = 0.45$, 95% confidence level

(b) $n = 100, \bar{p} = 0.30$, 99% confidence level

**33.** A survey of 30 individuals revealed that 22 watched Monday Night Football. Find, at the 95% confidence level, the error bounds of the proportion who watch Monday Night Football.

**34.** A manufacturer wants an estimate of the proportion of customers who will respond favorably to its new product. The manufacturer wants the estimate to be correct within 3% at the 95% confidence level. How big a sample should it select?

**35.** The U.S. Department of Labor Statistics estimated that 59.2% of the female population over age 15 were in the labor force. A TV station randomly selects seven females over age 15 to interview for a documentary on female experiences in the work place. Find the probability that three are not in the labor force.

**36.** The U.S. Department of Transportation estimated that 17.1% of licensed drivers in 2008 were in the 20–29 age group. The highway patrol conducts a check of valid safety stickers by randomly stopping 75 vehicles. Use the normal distribution to estimate the probability that fewer than twenty of the drivers were in the 20–29 age group.

**37.** In 2010, the National Highway Safety Administration found that 78% of Idaho residents use seat belts. Six Idaho residents are randomly selected. Find the probability that four or five of them use seat belts.

**38.** The U.S. Census Bureau reports on the educational attainment of the American population. The percentage of the population attaining a college degree or more is given (for selected years) in the following table.

| Year | A college degree or more |
|------|--------------------------|
| 1970 | 10.7 |
| 1980 | 16.2 |
| 1990 | 21.3 |
| 2000 | 25.6 |
| 2010 | 29.9 |

Show this information with a histogram.

**39.** The National Endowment for the Arts found that 56.6% of the American adult population had read at least one book in the last year. For a random selection of 300 adults, use the normal distribution to estimate the probability that

**(a)** from 150 to 180 had read a book.

**(b)** more than 175 had read a book.

**40.** The seven most expensive cars made in 2011 were:

| Make | Price (in millions) |
|------|---------------------|
| Bugatti Super Sports | 2.400 |
| Pagani Roadster | 1.850 |
| Lamborghini | 1.600 |
| McLaren F1 | 0.970 |
| Ferrari Enzo | 0.670 |
| Pagani Zonda | 0.667 |
| SSC Ultimate Aero | 0.654 |

Find the mean and median prices of these cars.

**41.** The U.S. Centers for Disease Control and Prevention found that 51% of high-school females played on a sports team. Find the probability that 5 females in a random selection of 10 high-school females played on a sports team.

# GAME THEORY

People of all ages like to play games. The games may range from simple children's games (hide and seek), to games of luck (roulette), to games requiring special skills (baseball) or strategies (chess). In some games, each opponent tries to anticipate the other's actions and act accordingly. For example, the defense in a football game may anticipate a pass and set up a blitz while the offense, in turn, tries to anticipate the defense's strategy. The study of game theory is not restricted to our usual concept of a game. It analyzes strategies used when the goals of "competing" parties conflict. Such a situation could arise when labor and management meet to discuss a contract, when airlines vie for a certain route, or when supermarkets compete against one another.

## 9.1    TWO-PERSON GAMES

* Strictly Determined Games

This chapter focuses on a simple situation with exactly two sides, or **players**, involved. These two players compete for a **payoff** that one player pays to the other. Let's begin with a simple example using a coin-matching game.

Two friends, Rob and Chad, play the following game. Each one has a coin, and each decides which side to turn up. They show the coins simultaneously and make payments according to the following.

If both show heads, Rob pays Chad 50¢. If both show tails, Rob pays Chad 25¢. If a head and a tail show, Chad pays Rob 35¢. The following figure shows these payoffs for each possible outcome:

|  |  | Chad | |
|---|---|---|---|
|  |  | **Heads** | **Tails** |
| **Rob** | **Heads** | Rob pays 50¢ <br> Chad gets 50¢ | Rob gets 35¢ <br> Chad pays 35¢ |
|  | **Tails** | Rob gets 35¢ <br> Chad pays 35¢ | Rob pays 25¢ <br> Chad gets 25¢ |

We call this game a **two-person game** because exactly two people participate. Notice that the amount won by a player is exactly the same as the amount lost by the opponent. Whenever this is the case, we call the game a **zero-sum game**. We can represent the payoffs in a two-person, zero-sum game by a **payoff matrix**. The payoffs indicated by the figure can be simplified to the following matrix:

$$
\begin{array}{cc}
 & \textbf{Chad} \\
\textbf{Rob}\begin{array}{c} \text{Heads} \\ \text{Tails} \end{array} &
\begin{array}{cc} \text{Heads} & \text{Tails} \end{array} \\
 & \begin{bmatrix} -50¢ & 35¢ \\ 35¢ & -25¢ \end{bmatrix}
\end{array}
$$

The matrix shows all possible payoffs for the way heads and tails are paired. An entry represents the amount Rob receives for that pair. Consequently, a negative entry indicates that Rob pays Chad.

We call a player's plan of action against the opponent a **strategy**. Game theory attempts to determine the best strategy so that each player will maximize his payoff. We make the assumption that each player knows all strategies available to himself and to his opponent, but each player selects a strategy without the opponent knowing the strategy selected. In this coin-matching game, the choice of a strategy simply amounts to selecting heads or tails. Note that when Rob selects the strategy *heads,* he has selected the first row of the payoff matrix. Chad's selection of a strategy is equivalent to selecting a column of the matrix.

**Example 1**    Rob and Chad match quarters. When the coins match, Rob receives 25¢. When the coins differ, Chad receives 25¢. The payoff matrix is

$$
\begin{array}{cc}
 & \textbf{Chad} \\
 & \begin{array}{cc} H & T \end{array} \\
\textbf{Rob} \quad \begin{array}{c} H \\ T \end{array} & \begin{bmatrix} 25¢ & -25¢ \\ -25¢ & 25¢ \end{bmatrix}
\end{array}
$$

**Example 2**    The payoff matrix for a game is

$$
\begin{array}{cc}
 & C \\
 & \begin{array}{cc} c_1 & c_2 \end{array} \\
R \quad \begin{array}{c} r_1 \\ r_2 \end{array} & \begin{bmatrix} 14 & -3 \\ -6 & -5 \end{bmatrix}
\end{array}
$$

(a) What are the possible payoffs to $R$ if strategy $r_1$ is selected?

(b) What are the possible payoffs to $C$ if strategy $c_2$ is selected?

(c) What is the payoff when $R$ selects $r_2$ and $C$ selects $c_1$?

**Solution**

(a) $R$ receives 14 if $C$ selects $c_1$ and $R$ pays 3 if $C$ selects $c_2$.

(b) $C$ receives 3 or 5, depending on whether $R$ selects $r_1$ or $r_2$.

(c) $C$ receives 6, and $R$ pays 6.

## Strictly Determined Games

We begin our analyses of the best strategy with some simple games that have fixed strategies.

In a competitive situation, we might ask if a strategy exists that will improve our chances of winning or that will increase our payoff. Some simple games have a fixed strategy or strategies that result in the best payoff possible when the competitors both have knowledge of the payoff associated with each strategy. Other games have no fixed best strategy. We first analyze the best strategy for games that have fixed strategies. We call them **strictly determined games**.

Two players, $R$ and $C$, play a game in which $R$ can take two alternative actions (strategies), called $r_1$ and $r_2$, and $C$ can take two actions, $c_1$ and $c_2$. Assume that each player chooses a strategy with no foreknowledge of the other's likely strategy. Because there are two possible strategies for $R$ and two for $C$, a $2 \times 2$ payoff matrix represents the outcome, or payoff, corresponding to each pair of strategies selected. Here is an example of a payoff matrix:

$$
\begin{array}{cc}
 & C \\
 & \begin{array}{cc} c_1 & c_2 \end{array} \\
R \quad \begin{array}{c} r_1 \\ r_2 \end{array} & \begin{bmatrix} 4 & -9 \\ 6 & 8 \end{bmatrix}
\end{array}
$$

Following the convention that the entries represent the payoff to the row player, $R$, we interpret the matrix as follows: If $R$ chooses strategy $r_1$ and $C$ chooses $c_1$, then $C$ pays 4 to $R$. If $R$ chooses strategy $r_1$ and $C$ chooses $c_2$, then $R$ pays 9 to $C$, and so on. Note that the sum of the amounts won by $R$ and $C$ is zero regardless of strategies selected. For example, if $R$ selects $r_2$, and $C$ selects $c_1$, then $R$ wins 6 and $C$ loses 6 (or wins $-6$), so the total is zero. Therefore, this is called a *zero-sum game*.

We emphasize that a matrix entry represents an amount paid by player $C$ to player $R$. To indicate the situation where $C$ receives payment from $R$, enter a negative amount.

$R$ and $C$ each wish to gain as much as they can (or lose as little as possible). Which strategies should they select? Let's analyze the situation. First, observe that for $R$ to gain as much as possible, $R$'s strategy attempts to select the *maximum* entry. As a gain for $C$ is represented by a negative number, $C$ attempts to select the *minimum* entry.

$R$ should select $r_2$, because $R$ then stands to gain the most, 6 or 8, regardless of $C$'s strategy.

At first glance, it appears that $C$'s strategy should be $c_2$, because that is $C$'s only chance of winning. However, we assume that both parties know all possible strategies, so $C$ knows that $R$'s best strategy is $r_2$. Thus, $C$ expects $R$ to choose $r_2$, so $C$ can choose only between losing 8 or losing 6. $C$ should select strategy $c_1$, because $C$ then risks giving the least away. With the strategies $r_2$ and $c_1$, $C$ pays 6 to $R$. We call this entry of the matrix the **value of the game**; its location in the matrix is called a **saddle point**; and the pair of strategies leading to the value is called a **solution**. This game has the value 6 with the saddle point located at $(2, 1)$, and the strategies $r_2$ and $c_1$ form the solution.

Let's use this example and analyze the general approach that determines whether a solution exists in a two-person game and how we can find a solution, if a solution exists.

Several courses of action may be available to both players, and there need not be the same number for each.

The strategies available to $R$ and $C$ in a two-person game can be represented by a **payoff matrix** $A$:

$$
R \quad
\begin{array}{c}
\\ r_1 \\ r_2 \\ \vdots \\ r_m
\end{array}
\overset{\displaystyle C}{
\overset{\begin{array}{cccc} c_1 & c_2 & \cdots & c_n \end{array}}{
\begin{bmatrix}
a_{11} & a_{12} & \cdots & a_{1n} \\
a_{21} & a_{22} & \cdots & a_{2n} \\
\vdots & \vdots & & \vdots \\
a_{m1} & a_{m2} & \cdots & a_{mn}
\end{bmatrix}}} = A
$$

Note that the number of strategies available to $R$ is $m$ and the number of strategies available to $C$ is $n$. The entry $a_{ij}$ represents the payoff that $R$ receives when $R$ adopts strategy $r_i$ and $C$ adopts $c_j$. The following gives the conditions of the game:

**Underlying Assumptions of a Strictly Determined Game**

1. Each player aims to choose the strategy that will enable the player to obtain as large a payoff as possible (or to lose as little as possible).

2. It is assumed that neither player has any prior knowledge of what strategy the other will adopt.

3. Each player makes a choice of strategy under the assumption that the opponent is an intelligent person adopting an equally rational approach to the game.

Let's first approach the game from the viewpoint of $R$ who is trying to maximize winnings. $R$ scans the rows of $A$ trying to decide which strategy, $r_1, r_2, \ldots, r_m$, to adopt. For any given strategy (row), what is the least possible payoff $R$ can expect? It is the *minimum* entry of that row. In the example with

$$
R \begin{array}{c} \\ r_1 \\ r_2 \end{array} \overset{\overset{\textstyle C}{\overline{\phantom{xxxxxx}}}}{\overset{\begin{array}{cc} c_1 & c_2 \end{array}}{\begin{bmatrix} 4 & -9 \\ 6 & 8 \end{bmatrix}}}
$$

as the payoff matrix, the minimum row entries are $-9$ for row 1 and 6 for row 2.

$R$ is aware of playing an intelligent opponent who aims to hold $R$ to a small payoff. When $R$ chooses a row, $R$ is guaranteed at least the minimum entry of that row. $R$ then selects the row containing the largest of these minima. This guarantees $R$ that payoff, and gives the largest payoff that $R$ can hope for against $C$'s best counterstrategy.

We now look at the situation from the viewpoint of $C$, who wants the minimum payoff because a gain for $C$ is represented by a negative number. $C$ scans the columns of $A$ trying to decide which strategy, $c_1, c_2, \ldots, c_n$, to adopt. $C$ marks the maximum element in each column, $C$'s least possible gain (or largest possible loss) for that strategy. In the example given, this is 6 in column 1 and 8 in column 2. $C$ then selects the strategy that has the smallest of those maxima. The smallest of the maxima represents $C$'s greatest gain if it is negative and $C$'s smallest loss if it is positive. This is the least payoff that $C$ need make against $R$'s strategy.

Thus, the approaches that $R$ and $C$ should adopt are the following:

| | |
|---|---|
| **Strategy for a Two-Person Zero-Sum Game** | $R$ marks off the minimum element in each row, and selects the row that has the largest of these minima. |
| | $C$ marks off the maximum element in each column, and selects the column that has the smallest of these maxima. |

If the largest of the row minima occurs in the same location as the smallest of the column maxima, that payoff is the **value** of the game.

The two strategies leading to the value, the **solution**, will be the ones that should be adopted.

When the largest row minimum and the smallest column maximum are the same, the game is said to be **strictly determined**. The location of this element is the **saddle point** of the game. (There are games that are not strictly determined. We discuss strategies for such games in the following section.)

In the example, $R$ selects the largest of the row minima, 6. $C$ selects the smallest of the column maxima, 6. Because these are the same, the game is strictly determined, and the (2, 1) location is the saddle point of the game. This game is strictly determined because each player should always play the same strategy. If $R$ wants to maximize average earnings, $R$ should always play $r_2$. If $C$ wants to minimize average payoff, $C$ should always play $c_1$. It is in this sense that the game is strictly determined.

**Example 3**    The following payoff matrix shows the strategies of a game played by players $R$ and $C$, with $R$ having three strategies and $C$ having two strategies:

$$
\begin{array}{cc}
 & C \\
 & \begin{array}{cc} c_1 & c_2 \end{array} \\
R \begin{array}{c} r_1 \\ r_2 \\ r_3 \end{array} & \begin{bmatrix} 1 & 2 \\ 3 & 4 \\ 7 & 5 \end{bmatrix}
\end{array}
$$

Determine the value of the game, if it exists.

**Solution**

First, analyze $R$'s strategies by selecting the minimum entry from each row and write it on the right of the matrix. Next, analyze $C$'s strategies by finding the maximum entry in each column and write it below the column. Then select the largest of the minima and the smallest of the maxima.

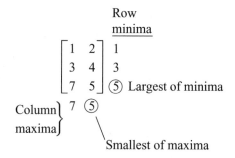

Observe that the smallest of the column maxima and the largest of the row minima are the same and both occur in the (3, 2) location. Thus, the game is strictly determined with value 5. The saddle point is (3, 2) and the strategies $r_3$ for $R$ and $c_2$ for $C$ form the solution.

When the players adopt these strategies, as they should, player $R$ will receive 5 from player $C$.

Perhaps this game seems unfair because $C$ has no chance of winning. The best $C$ can do is to lose 5 each time the game is played. Unfortunately, a game like this may correspond to a real situation. Sometimes a business manager may have several options for the business, but economic conditions are such that each option results in a financial loss. The best strategy tries to minimize losses.

What happens when one player chooses another strategy? Suppose $C$ adopts $c_1$ in the hope of reducing losses to 1. If $R$ keeps to strategy $r_3$, then $C$ loses 7, rather than the loss of 5 by choosing $c_2$.

The following example illustrates the possibility for a two-person game to have more than one saddle point.

**Example 4**  Look at the analysis of the following two-person game:

$$
\begin{array}{c}
& & \multicolumn{3}{c}{C} & \text{Row} \\
& & c_1 & c_2 & c_3 & \text{minima} \\
& r_1 & 3 & 4 & 11 & 3 \\
R & r_2 & -5 & 2 & -3 & -5 \\
& r_3 & 6 & 6 & 9 & ⑥ \text{ Max.} \\
& \text{Column} & ⑥ & ⑥ & 11 & \\
& \text{maxima} & \text{Min.} & \text{Min.} & &
\end{array}
$$

Each 6 in this matrix occurs at a saddle point, $(3, 1)$ and $(3, 2)$. (Verify this.) The best strategy for $R$ is $r_3$, whereas $C$ has the option of either strategy $c_1$ or $c_2$. In either case, the payoff is 6. When more than one saddle point occurs, the entries in the matrix at the saddle points are the same, being the value of the game.

A game may have no saddle point, and thereby is not strictly determined. The next example illustrates this type of situation.

**Example 5**  We have the payoff matrix of a two-person game:

$$
\begin{array}{c}
& & \multicolumn{2}{c}{C} & \text{Row} \\
& & c_1 & c_2 & \text{minima} \\
R & r_1 & 3 & -2 & ⊖② \text{ Max.} \\
& r_2 & -5 & 4 & -5 \\
& \text{Column} & ③ & 4 & \\
& \text{maxima} & \text{Min.} & &
\end{array}
$$

For a game to be strictly determined, the largest of the row minima and the smallest of the column maxima must be the same and occur in the same location in the matrix. That does not occur in this game because the largest of the row minima is $-2$ and the smallest of the column maxima is 3, and they appear in different locations. Thus, there is no saddle point.

The next section discusses strategies each player should take in such a game.

**Example 6**  Two major discount stores, A-Mart and B-Mart, compete for the business of the same customer base. A-Mart has 55% of the business and B-Mart 45%. Both companies are considering building new superstores to increase their market share. If both build, or neither builds, they expect their market share to remain the same. If A-Mart builds and B-Mart doesn't, A-Mart's share increases to 65%. If B-Mart builds and A-Mart doesn't, then A-Mart's share drops to 50%. Determine which strategy, to build or not build, each company should take.

**Solution**

Set up a payoff matrix showing A-Mart's share in all possible decisions. Let $B$ represent the decision to build a superstore and $NB$ the decision not to build.

$$
\begin{array}{c}
\textbf{B-Mart} \\
\begin{array}{cc}
B & NB
\end{array}
\end{array}
$$

$$
\textbf{A-Mart} \quad
\begin{array}{c}
B \\
NB
\end{array}
\begin{bmatrix}
55 & 65 \\
50 & 55
\end{bmatrix}
$$

The row minima and the column maxima follow:

|  |  | **B-Mart** | | Row |
|---|---|---|---|---|
|  |  | $B$ | $NB$ | minima |
| **A-Mart** | $B$ | 55 | 65 | ⑤⑤ Max. |
|  | $NB$ | 50 | 55 | 50 |
| Column maxima |  | ⑤⑤ | 65 |  |
|  |  | Min. |  |  |

Each store should adopt the build strategy in which case A-Mart retains 55% market share and B-Mart retains 45%.

---

Access end-of-section exercises online at **www.webassign.net**       ENHANCED WebAssign

---

## 9.2 MIXED-STRATEGY GAMES

- Games That Are Not Strictly Determined
- Fair Games
- Should I Have the Operation?

### Games That Are Not Strictly Determined

In the preceding section, we analyzed strictly determined two-person games in which each person had one strategy that was best under the conditions of the game. We also saw that some games are not strictly determined.

The following game is not strictly determined because the largest of the row minima is 10 and the smallest column maxima is 16.

|  |  | **C** | | Row |
|---|---|---|---|---|
|  |  | $c_1$ | $c_2$ | minima |
| $R$ | $r_1$ | 10 | 16 | 10 |
|  | $r_2$ | 17 | 8 | 8 |
| Column maxima |  | 17 | 16 |  |

Thus, the game has no saddle point. Note that $C$'s best strategy is $c_1$ if $R$ plays $r_1$ and is $c_2$ if $R$ plays $r_2$. The best strategy for $R$ is $r_1$ when $C$ plays $c_2$ and is $r_2$ when $C$ plays $c_1$. Which strategy should each player adopt?

Games are often played a number of times, not just once. For example, a football team has a variety of offensive strategies (plays) available and the opposing team has several defensive strategies. Although the offensive team may have a play that is successful more consistently than others, they do not choose it every time because the opponents would then use the defensive strategy that works best against that play. The best offensive strategy uses a variety of plays in a way that maximizes overall efforts. Similarly, the defensive team uses a variety of strategies that they hope will keep the offensive gains to a minimum. In the simple example

$$
\begin{array}{c}
C \\
\begin{array}{cc}
c_1 & c_2
\end{array} \\
R \quad
\begin{array}{c}
r_1 \\
r_2
\end{array}
\begin{bmatrix}
-17 & 16 \\
10 & 8
\end{bmatrix}
\end{array}
$$

players must conceal their choice of strategy or else the other players can select the strategy that is most beneficial to them. If $R$ realizes that $C$ selects strategy $c_2$, then $R$ will select $r_1$ to maximize payoff. If $C$ selects $c_1$, then $R$ knows to select $r_2$.

If each player keeps the strategy secret, with a mixture of strategies, then the best strategy for each selects a strategy in a random manner but with a frequency that provides the greatest benefit. Thus, $R$ should select $r_1$ part of the time in a random manner and select $r_2$ the rest of the time. In a similar manner, $C$ selects $c_1$ part of the time and $c_2$ the rest of the time.

Player $R$ would like to adopt a sequence of strategies that would maximize $R$'s long-term average payoff from $C$, and $C$ would like to minimize the long-term average payoff to $R$. How often should $R$ adopt $r_1$ and how often should $C$ adopt $c_1$ for the most beneficial payoff to each? We use probability theory to control the randomness of each player's strategy.

We use the following payoff matrix for a general two-person game:

$$
\begin{array}{c}
C \\
\begin{array}{cc}
c_1 & c_2
\end{array} \\
R \quad
\begin{array}{c}
r_1 \\
r_2
\end{array}
\begin{bmatrix}
a_{11} & a_{12} \\
a_{21} & a_{22}
\end{bmatrix}
\end{array}
$$

We use the notation $p_1$ for the **probability** that $R$ will adopt strategy $r_1$ and $p_2$ for the probability that $R$ will adopt strategy $r_2$. For player $C$, we use $q_1$ for the probability that $C$ will adopt strategy $c_1$, and $q_2$ represents the probability that $c_2$ will be adopted. In practice, over a sequence of a number of games, player $R$ uses the strategy $r_1$ the fraction $p_1$ of the time and uses $r_2$ the fraction $p_2$ of the time. Similarly, $C$ adopts $c_1$ the fraction $q_1$ of the time and adopts $c_2$ the fraction $q_2$ of the time. Whenever the players adopt strategies in a random manner with the fraction of times indicated, then the probability of the payoff $a_{11}$ is $p_1 q_1$—that is, the probability that $R$ selects $r_1$ and $C$ selects $c_1$. Similarly, the probability of payoff $a_{12}$ is $p_1 q_2$, the probability of $a_{21}$ is $p_2 q_1$, and the probability of $a_{22}$ is $p_2 q_2$.

With each possible pair of strategies the following probabilities and pay-offs hold:

| Strategies | $r_1 c_1$ | $r_1 c_2$ | $r_2 c_1$ | $r_2 c_2$ |
|---|---|---|---|---|
| Probability | $p_1 q_1$ | $p_1 q_2$ | $p_2 q_1$ | $p_2 q_2$ |
| Payoff | $a_{11}$ | $a_{12}$ | $a_{21}$ | $a_{22}$ |

When each player randomly selects a strategy according to these probabilities, the average long-term payoff is the expected value of the payoff. (Note that this is the expected value as defined in Section 8.5.) We call this the **expected payoff** of the game, and it takes the form

$$\text{Expected payoff} = p_1 q_1 a_{11} + p_1 q_2 a_{12} + p_2 q_1 a_{21} + p_2 q_2 a_{22}$$

The expected payoff denotes the average payoff to $R$ when the game is played a large number of times.

We can represent the strategy probability with the matrices

$$P = [p_1 \ \ p_2] \quad \text{and} \quad Q = \begin{bmatrix} q_1 \\ q_2 \end{bmatrix}$$

and denote the payoff matrix by $A$ and the expected payoff associated with these probabilities by $E(P, Q)$. Then, in matrix form $E(P, Q)$ becomes $E(P, Q) = PAQ$.

We call the matrices $P$ and $Q$ the **strategies adopted** by $R$ and $C$, respectively.

We verify this by performing the matrix multiplication

$$[p_1 \ \ p_2] \begin{bmatrix} a_{11} & a_{12} \\ a_{21} & a_{22} \end{bmatrix} \begin{bmatrix} q_1 \\ q_2 \end{bmatrix} = [p_1 a_{11} + p_2 a_{21} \ \ p_1 a_{12} + p_2 a_{22}] \begin{bmatrix} q_1 \\ q_2 \end{bmatrix}$$
$$= (p_1 a_{11} + p_2 a_{21})q_1 + (p_1 a_{12} + p_2 a_{22})q_2$$
$$= a_{11} p_1 q_1 + a_{21} p_2 q_1 + a_{12} p_1 q_2 + a_{22} p_2 q_2$$

which is the expected payoff.

**Example 1**   Now for a payoff matrix of a mixed strategy game:

$$\begin{array}{c} \qquad\qquad C \\ \qquad\quad \begin{array}{cc} c_1 & c_2 \end{array} \\ R \ \ \begin{array}{c} r_1 \\ r_2 \end{array} \begin{bmatrix} 15 & 75 \\ 45 & -30 \end{bmatrix} \end{array}$$

Let's compute the expected payoff using the strategy $\begin{bmatrix} \frac{2}{3} & \frac{1}{3} \end{bmatrix}$ for $R$ and the strategy $\begin{bmatrix} \frac{2}{5} & \frac{3}{5} \end{bmatrix}$ for $C$. This means that $\frac{2}{3}$ of the time $R$ will adopt strategy $r_1$ and $\frac{1}{3}$ of the time will adopt strategy $r_2$, whereas $\frac{2}{5}$ of the time $C$ will adopt strategy $c_1$ and $\frac{3}{5}$ of the time $C$ will adopt strategy $c_2$

$$E(P, Q) = \begin{bmatrix} \frac{2}{3} & \frac{1}{3} \end{bmatrix} \begin{bmatrix} 15 & 75 \\ 45 & -30 \end{bmatrix} \begin{bmatrix} \frac{2}{5} \\ \frac{3}{5} \end{bmatrix} = [25 \ \ 40] \begin{bmatrix} \frac{2}{5} \\ \frac{3}{5} \end{bmatrix} = [34]$$

$R$ can expect an average payoff of 34 if the game is played a large number of times provided that the players select strategies in a random manner with the relative frequency given by $P$ and $Q$. If the players choose to vary the relative frequency of their strategies, then the expected payoff will change.

We have seen that a strictly determined game has a solution that provides an optimal strategy for both players. In the case of a mixed-strategy game, no such single strategy exists. However, player $R$ would like a strategy $P$ that gives the best payoff against player $C$'s best strategy. We call such a strategy the **optimal strategy for player $R$**.

Likewise, $C$ wants to adopt a strategy $Q$ that yields the least payoff to $R$ against $R$'s best strategy. We call it the **optimal strategy for player $C$**.

Both players in a mixed-strategy game have the option of selecting their strategy probabilities. Because different strategies generally yield different expected payoffs, a player would like to know how to determine the optimal strategy. Mathematicians have determined how to find optimal strategies from the $2 \times 2$ payoff matrix of a two-person game.

**DEFINITION**
**Optimal Strategy**

For the payoff matrix

$$\begin{array}{c} & C \\ R & \begin{bmatrix} a_{11} & a_{12} \\ a_{21} & a_{22} \end{bmatrix} \end{array}$$

the optimal strategy for player $R$ is $P = [p_1 \quad p_2]$ where

$$p_1 = \frac{a_{22} - a_{21}}{a_{11} + a_{22} - a_{12} - a_{21}} \quad \text{and} \quad p_2 = 1 - p_1$$

The optimal strategy for player $C$ is $Q = [q_1 \quad q_2]$ where

$$q_1 = \frac{a_{22} - a_{12}}{a_{11} + a_{22} - a_{12} - a_{21}} \quad \text{and} \quad q_2 = 1 - q_1$$

The expected payoff for the optimal strategies is

$$E = PAQ = \frac{a_{11}a_{22} - a_{12}a_{21}}{a_{11} + a_{22} - a_{12} - a_{21}}$$

The values of $p_1$, $p_2$, $q_1$, and $q_2$ are meaningless when $a_{11} + a_{22} - a_{12} - a_{21} = 0$. However, this never occurs for a game not strictly determined.

**Example 2**

Now let's find the optimal strategies, and the value associated with the strategies, for the game with payoff matrix

$$\begin{array}{c} & C \\ R & \begin{bmatrix} 15 & 75 \\ 45 & -30 \end{bmatrix} \end{array}$$

**Solution**

$$p_1 = \frac{-30 - 45}{15 - 30 - 75 - 45} = \frac{-75}{-135} = \frac{5}{9}, \quad p_2 = \frac{4}{9}$$

$$q_1 = \frac{-30 - 75}{-135} = \frac{7}{9}, \quad q_2 = \frac{2}{9}$$

The optimal strategy for $R$ is $\begin{bmatrix} \frac{5}{9} & \frac{4}{9} \end{bmatrix}$ and the optimal strategy for $C$ is $\begin{bmatrix} \frac{7}{9} & \frac{2}{9} \end{bmatrix}$. These strategies yield the expected payoff

$$E = \frac{15(-30) - 75(45)}{-135} = \frac{-3825}{-135} = 28.3$$

For a large number of games, $R$ can expect, on the average, to receive about 28.3 from $C$ provided that both players adopt their optimal strategies.

**Example 3**   A two-person game with payoff matrix

$$C$$
$$R \begin{bmatrix} 4 & 6 \\ 2 & -5 \end{bmatrix}$$

is strictly determined with value 4. $R$'s strategy is $r_1$, and $C$'s strategy is $c_1$, which we can represent in matrix form as $P = [1 \quad 0]$ for $R$ and $Q = [1 \quad 0]$ for $C$.

If we use the formulas for optimal strategies and expected payoff, we obtain for $R$ the strategy $P = \begin{bmatrix} \frac{7}{9} & \frac{2}{9} \end{bmatrix}$ and for $C$ the strategy $Q = \begin{bmatrix} \frac{11}{9} & \frac{-2}{9} \end{bmatrix}$ with $E = \frac{32}{9}$.

This gives an erroneous result because neither $\frac{11}{9}$ nor $\frac{-2}{9}$ is a valid probability and because $E = 4$ is the optimal value, not $\frac{32}{9}$.

**CAUTION**

If a game is strictly determined, the preceding method of computing the value is not correct.

This example emphasizes that a mixed-strategy analysis does not apply to a strictly determined game.

In some instances, a payoff matrix can be simplified to a smaller matrix. Here is an example.

**Example 4**   In this payoff matrix each player has the option of three strategies. The game is not strictly determined.

$$C$$

|   | $c_1$ | $c_2$ | $c_3$ |
|---|---|---|---|
| $r_1$ | 50 | 20 | $-10$ |
| $R \quad r_2$ | 0 | 15 | 5 |
| $r_3$ | $-15$ | 10 | $-5$ |

Notice that each entry in row 2 is greater than the corresponding entry in row 3. Thus, $R$ should never adopt strategy $r_3$ because, regardless of the strategy adopted by $C$, the strategy $r_2$ yields a better value for $R$. As $C$ wants to minimize the payoff to $R$, $C$ should never se-lect strategy $c_2$ because strategy $c_3$ yields a smaller payoff to $R$ regardless of the strategy $R$ adopts. In a game like this, we say that strategy $r_2$ **dominates** strategy $r_3$, strategy $c_3$ **dominates** strategy $c_2$, and strategies $r_3$ and $c_2$ can be removed from consideration.

**DEFINITION**
**Dominant Row**
**or Column**

Row $i$ of a payoff matrix **dominates** row $j$ if every entry of row $i$ is greater than or equal to the corresponding entry of row $j$.

Column $i$ of a payoff matrix **dominates** column $j$ if every entry of column $i$ is less than or equal to the corresponding entry of column $j$.

Whenever row $i$ dominates row $j$, then row $j$ may be removed from the payoff matrix without affecting the analysis.

Whenever column $i$ dominates column $j$, then column $j$ may be removed from the payoff matrix without affecting the analysis.

Because row 2 dominates row 3 and column 3 dominates column 2 of the payoff matrix of the example, we can remove row 3 and column 2 to obtain the payoff matrix

$$A = \begin{bmatrix} 50 & -10 \\ 0 & 5 \end{bmatrix}$$

From this $2 \times 2$ matrix, we can now find the optimal strategy for $R$, $P = [p_1 \; p_2 \; 0]$. The zero occurs because $r_3$ will never be used.

$$p_1 = \frac{5 - 0}{50 + 5 - (-10) - 0} = \frac{5}{65} = \frac{1}{13} \quad \text{and} \quad p_2 = \frac{12}{13}$$

Thus, $P = \left[ \frac{1}{13} \; \frac{12}{13} \; 0 \right]$.

For $C$, the optimal strategy is $Q = [q_1 \; 0 \; q_3]$. $q_2 = 0$ because $c_2$ is never used.

$$q_1 = \frac{5 - (-10)}{65} = \frac{15}{65} = \frac{3}{13} \quad \text{and} \quad q_3 = \frac{10}{13}$$

Thus, $Q = \left[ \frac{3}{13} \; 0 \; \frac{10}{13} \right]$.

The expected payoff is

$$\frac{50(5) - (-10)(0)}{65} = \frac{250}{65} = 3.85$$

Thus, in the long term, $R$ will receive an average value of 3.85 from $C$.

### Fair Games

A game not strictly determined is called a **fair game** when its value is zero. This means that the long-term average gain of each player is zero; that is, the losses of each player equal his gains, so his net gain is zero.

**Example 5**   The game

$$\begin{bmatrix} 3 & -1 \\ 2 & 4 \end{bmatrix}$$

is not strictly determined, and its value is

$$\frac{3(4) - 2(-1)}{3 + 4 + 1 - 2} = \frac{14}{6}$$

so it is not a *fair* game.

The game

$$\begin{bmatrix} 2 & -4 \\ -3 & 6 \end{bmatrix}$$

is not strictly determined, and its value is

$$\frac{2(6) - (-3)(-4)}{2 + 6 + 4 + 3} = \frac{0}{15} = 0$$

so this game is fair.

Generally, a strictly determined game is not fair.

## Should I Have the Operation?

Medical diagnoses have an element of uncertainty. For two people with the same symptoms, one may have the disease, and the other may not. Because of the uncertainty involved, further tests or a second opinion is often recommended. Some medical procedures involve nontrivial risks. Should a patient undergo surgery when, historically, there is a relatively small probability of a malignant tumor?

Let's look at the decision to have surgery or not as a game between the patient and nature. To illustrate, we use the example in which the expected years of life remaining for the patient are

1.  3 years if the disease is present and no surgery is performed.
2.  15 years if the disease is present and surgery is performed.
3.  30 years if no disease is present and no surgery is performed.
4.  25 years if no disease is present and surgery is performed.

Thinking of this as a two-person game, nature has the options of disease ($D$) and no disease ($ND$), whereas the patient has the options of surgery ($S$) or no surgery ($NS$).

The matrix representing these options is

$$
\begin{array}{c}
 & \text{Nature} \\
 & \begin{array}{cc} D & ND \end{array} \\
\text{Patient} \begin{array}{c} S \\ NS \end{array} & \begin{bmatrix} 15 & 25 \\ 3 & 30 \end{bmatrix}
\end{array}
$$

where the patient's strategy is $P = [p_1 \ \ p_2]$ and nature's strategy is $Q = [q_1 \ \ q_2]$. If nature's strategy is $[1 \ \ 0]$, the disease is present; then the patient's strategy is clearly $[1 \ \ 0]$, have surgery. However, because the patient knows only the probability of nature's options, we ask how the probability should influence the patient's decision. Suppose the probability that the disease is present is 0.20. Then, $Q = [0.20 \ \ 0.80]$. If the patient opts for surgery, then $P = [1 \ \ 0]$ and the expected number of years remaining is

$$[1 \ \ 0]\begin{bmatrix} 15 & 25 \\ 3 & 30 \end{bmatrix}\begin{bmatrix} 0.20 \\ 0.80 \end{bmatrix} = [15 \ \ 25]\begin{bmatrix} 0.20 \\ 0.80 \end{bmatrix} = 23$$

If the patient opts for no surgery, then the expected survival time is

$$[0 \ 1]\begin{bmatrix} 15 & 25 \\ 3 & 30 \end{bmatrix}\begin{bmatrix} 0.20 \\ 0.80 \end{bmatrix} = [3 \ 30]\begin{bmatrix} 0.20 \\ 0.80 \end{bmatrix} = 24.6$$

In this case, no surgery appears to be a slightly better option.

The following example, using the preceding payoff matrix, illustrates how we can estimate when surgery is a better option and when it is not.

**Example 6**  The payoff matrix for nature's options of disease or no disease and a patient's options of surgery or no surgery is

$$\begin{array}{cc} & \text{Nature} \\ & \begin{array}{cc} D & ND \end{array} \\ \text{Patient} \begin{array}{c} S \\ NS \end{array} & \begin{bmatrix} 15 & 25 \\ 3 & 30 \end{bmatrix} \end{array}$$

For what values of $Q$, the probability of disease, is surgery a better option?

**Solution**
We determine if surgery is the better option by considering the expected survival time for both the surgery and no surgery options and using $Q = [q_1 \quad 1 - q_1]$.

*For surgery:*

$$\text{Expected survival} = [1 \ 0]\begin{bmatrix} 15 & 25 \\ 3 & 30 \end{bmatrix}\begin{bmatrix} q_1 \\ 1 - q_1 \end{bmatrix}$$

$$= [15 \ 25]\begin{bmatrix} q_1 \\ 1 - q_1 \end{bmatrix}$$
$$= 15q_1 + 25(1 - q_1)$$
$$= 25 - 10q_1$$

*For no surgery:*

$$\text{Expected survival} = [0 \ 1]\begin{bmatrix} 15 & 25 \\ 3 & 30 \end{bmatrix}\begin{bmatrix} q_1 \\ 1 - q_1 \end{bmatrix} = [3 \ 30]\begin{bmatrix} q_1 \\ 1 - q_1 \end{bmatrix}$$
$$= 3q_1 + 30(1 - q_1)$$
$$= 30 - 27q_1$$

Surgery is the better option when

$$25 - 10q_1 > 30 - 27q_1$$
$$17q_1 > 5$$
$$q_1 > \frac{5}{17} = 0.294$$

Thus, surgery seems the better option when the probability is greater than 0.294 that the disease is present.

You might question the use of game theory to help make decisions about medical treatment because of these reasons:

1. Game theory assumes both players are intelligent people who know all strategies available to both players and who will adopt a rational approach to the game.

2. Data on life expectancies and the probability of a successful treatment give information that may be reliable for large groups of people but doesn't predict the outcome for a particular individual very well. When we are at risk, we tend to want to be treated and treated aggressively.

Although nature is not an intelligent person, nature tends to "behave" in a random manner, often with known probability. We can play a mixed-strategy game with nature *as if* nature were an intelligent person who decided to choose options randomly with a known probability. The medical researcher might find such a game useful in determining research actions.

## 9.2    EXERCISES

Access end-of-section exercises online at **www.webassign.net**

## IMPORTANT TERMS

**9.1**
Payoff
Two-Person Game
Strategy
Strictly Determined Game
Zero-Sum Game
Value of the Game

Saddle Point
Solution
Payoff Matrix

**9.2**
Mixed-Strategy Game
Probability of a Strategy

Expected Payoff
Optimal Strategy
Dominant Row or Column
Fair Game

## IMPORTANT CONCEPTS

**Strictly Determined Game**

1. Each player aims to choose the strategy that results in the largest payoff.
2. Neither player knows what strategy the other will adopt.
3. Each player assumes the opponent will adopt an equally rational approach to the game.

**Strategy for a Two-Person Zero-Sum Game**

$R$ marks the minimum element in each row and selects the largest. $C$ marks the maximum element in each column and selects the smallest. If the selections of $R$ and $C$ occur in the same location, that is the value of the game and those selections form the strategy of each.

**Mixed-Strategy Game**

A game that is not strictly determined. Each player uses a mixture of strategies.

**Optimal Strategy**

For a mixed-strategy game with payoff matrix $R \begin{array}{c} \phantom{a} \\ \begin{bmatrix} a_{11} & a_{12} \\ a_{21} & a_{22} \end{bmatrix} \end{array}$, the optimal

strategies are: For $R$, $p_1 = \dfrac{a_{22} - a_{21}}{a_{11} + a_{22} - a_{12} - a_{21}}$ and $p_2 = 1 - p_1$.

For $C$, $q_1 = \dfrac{a_{22} - a_{12}}{a_{11} + a_{22} - a_{12} - a_{21}}$ and $q_2 = 1 - q_1$.

This expected payoff is $\dfrac{a_{11}a_{22} - a_{12}a_{21}}{a_{11} + a_{22} - a_{12} - a_{21}}$.

## REVIEW EXERCISES

1. The following represent payoff matrices for two-person games. For each one, determine whether the game is strictly determined. If a game is strictly determined, find the saddle point and value.

(a) $\begin{bmatrix} 5 & -1 \\ 2 & 4 \end{bmatrix}$

(b) $\begin{bmatrix} 1 & 3 & 9 \\ 7 & 4 & 8 \\ -5 & 3 & 4 \end{bmatrix}$

(c) $\begin{bmatrix} 140 & 210 \\ 300 & 275 \end{bmatrix}$

(d) $\begin{bmatrix} -6 & 2 & 9 & 1 \\ 5 & -4 & 0 & 2 \\ 4 & 2 & 8 & 3 \end{bmatrix}$

2. $R$ and $C$ are competing retail stores. Each has two options for an advertising campaign. The results of the campaign are expected to result in the following payoff matrix:

$$R \begin{array}{c} \quad\quad C \\ \begin{bmatrix} 40 & 35 \\ -25 & 30 \end{bmatrix} \end{array}$$

The entries represent the amount in sales that $R$ will gain from $C$. What strategy should each adopt?

3. The following are payoff matrices that describe the payoff (in yards gained) of possible strategies of the offense and defense in a football game. Determine the solutions, if they exist.

(a) $\begin{bmatrix} 3 & 15 & -4 \\ 2 & -1 & 6 \\ 4 & 2 & 7 \end{bmatrix}$

(b) $\begin{bmatrix} -5 & 8 & 7 \\ 4 & 5 & 25 \\ 3 & -2 & 2 \end{bmatrix}$

4. Two players $X$ and $Y$ play the following game. $X$ gives $Y$ \$5 before the game starts. They each write 1, 2, or 3 on a slip of paper. Then, $Y$ gives $X$ an

amount determined by the numbers written. The following payoff matrix gives the amount paid to $X$ in each possible case:

$$X \begin{array}{c} \quad\quad\quad Y \\ \begin{array}{ccc} 1 & 2 & 3 \end{array} \\ \begin{array}{c} 1 \\ 2 \\ 3 \end{array} \begin{bmatrix} 2 & 3 & 7 \\ 4 & 0 & 6 \\ 5 & 4 & 9 \end{bmatrix} \end{array}$$

What strategy should each player use?

5. Determine the expected payoff of the following games:

(a) The strategy of $X$ is [0.3  0.7]; the strategy of $Y$ is [0.6  0.4]; the payoff matrix is

$$X \begin{array}{c} \quad\quad Y \\ \begin{bmatrix} 5 & 9 \\ 11 & 2 \end{bmatrix} \end{array}$$

(b) The strategy of $X$ is [0.5  0.5]; the strategy of $Y$ is [0.1  0.9]; the payoff matrix is

$$X \begin{array}{c} \quad\quad Y \\ \begin{bmatrix} -2 & 6 \\ 3 & 9 \end{bmatrix} \end{array}$$

(c) The strategy of $X$ is [0.1  0.4  0.5]; the strategy of $Y$ is [0.2  0.2  0.6]; the payoff matrix is

$$X \begin{array}{c} \quad\quad Y \\ \begin{bmatrix} -3 & 2 & 1 \\ 4 & -2 & 5 \\ 3 & 1 & 2 \end{bmatrix} \end{array}$$

**6.** Determine the optimal strategy and its resulting value for each of the following games:

(a) $\begin{bmatrix} 2 & 9 \\ 6 & 3 \end{bmatrix}$

(b) $\begin{bmatrix} -2 & 5 \\ 8 & 4 \end{bmatrix}$

(c) $\begin{bmatrix} 3 & 5 & 7 \\ 1 & 4 & 6 \\ 6 & 7 & 5 \end{bmatrix}$

**7.** A truck-garden farmer raises vegetables in the field and in a greenhouse. The weather determines where they should be planted. The following payoff matrix describes the strategies available:

$$\begin{array}{c c} & \begin{array}{c c} \text{Wet} & \text{Dry} \end{array} \\ \begin{array}{c} \text{Field} \\ \text{Greenhouse} \end{array} & \begin{bmatrix} 250 & 140 \\ 175 & 210 \end{bmatrix} \end{array}$$

What percentage of the crop should be planted in the field and what percentage in the greenhouse?

# LOGIC

Communication is an important activity of the human race, but we often find it difficult. We sometimes have difficulty expressing our thoughts. We make statements that another person interprets in a way we did not intend. To be sure that no question arises about the meaning, we sometimes ask lawyers to draw up a document to convey the precise intention of the information in the document. Even so, a lawsuit may arise when two parties disagree on the meaning and intent of the document.

Because problems in clear and precise communication do exist, mathematicians have sought to study these problems in order to clarify them and make some areas of communication more precise. In this chapter, we introduce you to this area of mathematics, logic.

We emphasize that we use and perform logical arguments. Many discussions, arguments, or debates begin with statements of opinion or facts followed by a line of reasoning that leads to a, hopefully desired, conclusion. A conclusion may, or may not, be valid. An argument based on erroneous "facts" or faulty reasoning can lead to an invalid conclusion.

We will study basic statements and logical arguments that we represent with letters and symbols to provide simple and concise forms of logical statements and arguments. The symbolic form allows us to better understand the relations and differences of various logical statements and to describe valid forms of arguments.

## 10.1 STATEMENTS

- Notation
- Compound Statements
- Conjunction
- Disjunction
- Negation

In logic, we study sentences and relationships between certain kinds of sentences. In fact, we limit the study to just a portion of sentences used in our daily speech. If we examine the sentences that we use in daily life, we find they can be classified into several categories, which include the following:

1. Open your book to page 93.        (a command)
2. Did you enjoy the concert?        (a question)
3. What an exciting game!        (an exclamation)
4. The sun rises in the east.        (a true declarative statement)
5. Three plus two is nine.        (a false declarative statement)
6. That was a good movie.        (an ambiguous sentence)

The last sentence is ambiguous because there may be no agreement on what makes a movie good.

We restrict our study of logic to unambiguous declarative sentences that can be classified as true or false but not both. We call such sentences **statements**. Only sentences 4 and 5 above are statements.

**Example 1**    Classify each of the following sentences as a statement or not a statement.

(a) When is your next class?
(b) George Washington was the first president of the United States.
(c) Andrew Jackson was a great president.
(d) In 2018, February will have 29 days.
(e) That was a hard test!
(f) Three plus five is eight.

**Solution**

(a) This is a question, not a declarative sentence, so it is not a statement.
(b) This is a true declarative sentence, so it is a statement.
(c) This is an ambiguous sentence, because there is no uniform understanding of the meaning of a "great" president. Therefore, it is not a statement.
(d) This is a false declarative sentence, so it is a statement.
(e) This is not a statement because it is an exclamation, not a declarative sentence. It is also ambiguous.
(f) This is a true declarative sentence, so it is a statement.

## Notation

We use the letters $p, q, r, \ldots$ to denote statements. Thus, we might use $p$ to represent a specific statement such as "Mathematics is required for my degree." In many cases, $p$ or $q$ is used to represent a general or unspecified statement, just as John Doe or Jane Doe is used to represent a general or unspecified person. For example, we might let $p$ represent an arbitrary statement taken from *The Story of My Life* by Helen Keller.

Because we deal only with statements that can be classified as "true" or "false," we can assign a **truth value** to a statement $p$. We use $T$ to represent the value "true" and $F$ to represent the value "false" (as you do on a true–false quiz).

## Compound Statements

We all frequently use statements in daily conversation that can be constructed by combining two or more simple statements. We call these **compound statements**. Compound statements can be long, complicated, and difficult to understand. Their meaning depends on the meaning of the component statements and how these components are put together into a single statement. We will look at some compound statements and how to analyze them.

## Conjunction

We first study a compound statement such as "Ted is taking art, and Susan is taking history" that can be formed by connecting the statement "Ted is taking art" to the statement "Susan is taking history" using the word "and." If we use the $p, q$ notation, this becomes:

> Let $p$ be the statement "Ted is taking art."
>
> Let $q$ be the statement "Susan is taking history."
>
> Then "$p$ and $q$" represents the statement "Ted is taking art, and Susan is taking history."

We use the notation "$p \wedge q$" to denote "$p$ and $q$." A statement of the form $p \wedge q$ is called a **conjunction**.

When we form a statement by combining statements that we know to be true or false, can we decide on the truth of the compound statement? The answer is yes. We determine the truth value of a conjunction by the truth value of the statements making up the conjunction. The conjunction "Ted is taking art, and Susan is taking history" is true when both statements "Ted is taking art" and "Susan is taking history" are true. If either one or both statements are false, then the conjunction is false. In general, we can summarize such a situation with a **truth table**. When we form a compound statement from two statements, the two statements could be both true, both false, or one true and one false. The truth table below gives the truth value of $p \wedge q$ using the four possible combinations of truth values for $p$ and for $q$:

| $p$ | $q$ | $p \wedge q$ |
|-----|-----|--------------|
| $T$ | $T$ | $T$ |
| $T$ | $F$ | $F$ |
| $F$ | $T$ | $F$ |
| $F$ | $F$ | $F$ |

Note that $p \wedge q$ has the truth value T only when both $p$ and $q$ are true.

**Example 2**    The statement "2012 was a leap year, and the Fourth of July is a national holiday in the United States" is true because both statements "2012 was a leap year" and "The Fourth of July is a national holiday in the United States" are true. The statement "2012 was a leap year, and February has 30 days" is false because the statement "February has 30 days" is false.

## Disjunction

Two statements can be connected with the word "or" to form a **disjunction**. The statement "We advertise in the Sunday paper, or we buy time on TV" is the disjunction of the statement "We advertise in the Sunday paper" and the statement "We buy time on TV." The notation $p \vee q$ represents the disjunction of statement $p$ with statement $q$. We may read $p \vee q$ as "$p$ or $q$."

**Example 3**    Identify each of the following statements as conjunction, disjunction, or neither:

(a) The toast is burned, and the eggs are cold.

(b) The cafeteria opens for lunch at 11:00 A.M.

(c) Students in English 102 are required to analyze two poems, or they are required to analyze two short stories.

**Solution**

(a) Conjunction      (b) Neither      (c) Disjunction

Again, the truth value of $p \vee q$ is determined by the truth values of $p$ and of $q$. The statement $p \vee q$ is true when one or both of $p$ and $q$ are true; otherwise, it is false. The following truth table summarizes the truth value of $p \vee q$:

| $p$ | $q$ | $p \vee q$ |
|-----|-----|-----------|
| $T$ | $T$ | $T$ |
| $T$ | $F$ | $T$ |
| $F$ | $T$ | $T$ |
| $F$ | $F$ | $F$ |

One common use of the connective $p$ or $q$ is interpreted as "either $p$ or $q$ but not both." We call this **exclusive or**. Unless otherwise specified, mathematicians use $p$ or $q$ to mean "either $p$ or $q$ or both." This is called the **inclusive or**.

**Example 4**    Determine the truth value of each of the following statements:

(a) June has 30 days, or September has 31 days.

(b) There are 24 hours in a day, or there are seven days in a week.

(c) Thomas Jefferson was the first U.S. president, or there are 65 states in the United States.

**Solution**

(a) This is true because the statement "June has 30 days" is true.

(b) This is true because both statements forming the disjunction are true.

**(c)** This statement is false because both statements "Thomas Jefferson was the first U.S. president" and "there are 65 states in the United States" are false.

## Negation

Sometimes we want to make a statement that means the opposite of a given statement. We call such a statement the **negation** of the given statement. We can write the negation of statement $p$ by writing "It is not true that $p$" (denoted by $\sim p$, which is read "not $p$"). The negation of "March is a summer month" is "It is not true that March is a summer month." You would probably express this rather awkward sentence as "March is not a summer month." We define the negation of a true statement as false and the negation of a false statement as true. The following table summarizes the truth values of $p$ and $\sim p$ for the possible truth values of $p$:

| $p$ | $\sim p$ |
|-----|----------|
| T | F |
| F | T |

**Example 5**  Write the negation of the following statements:

**(a)** Thanksgiving is in November.    **(b)** $2 + 2 = 7$

**(c)** My car won't start.

Solution

**(a)** Thanksgiving is not in November.    **(b)** $2 + 2 \neq 7$

**(c)** My car will start.

**Example 6**  Let $p$ be the statement "Susan's dog is a poodle" and let $q$ be the statement "Jake has a black cat." Write the following statements:

**(a)** $p \vee q$     **(b)** $\sim q$     **(c)** $p \wedge (\sim q)$

Solution

**(a)** Susan's dog is a poodle, or Jake has a black cat.

**(b)** Jake does not have a black cat.

**(c)** Susan's dog is a poodle, and Jake does not have a black cat.

**Example 7**  Determine the truth value of each of the following:

**(a)** February has 30 days, or Franklin Roosevelt was not the first president of the United States.

**(b)** A quart is not larger than a liter, and water does not run uphill.

**Solution**

(a) This statement is of the form $p \vee (\sim q)$, where $p$ is false and $q$ is false (so $\sim q$ is true). Therefore, the statement is true.

(b) This statement is of the form $(\sim p) \wedge (\sim q)$, where $p$ is false ($\sim p$ is true) and $q$ is false ($\sim q$ is true). As both $\sim p$ and $\sim q$ are true, the statement is true.

**Example 8**   Find the truth value of $(p \vee \sim q) \wedge q$ for all possible truth values of $p$ and $q$.

**Solution**
The following table shows all values of $p$ and $q$ and the corresponding values of the parts making up the given statement.

| $p$ | $q$ | $p \vee \sim q$ | $(p \vee \sim q) \wedge q$ |
|---|---|---|---|
| $T$ | $T$ | $T$ | $T$ |
| $T$ | $F$ | $T$ | $F$ |
| $F$ | $T$ | $F$ | $F$ |
| $F$ | $F$ | $T$ | $F$ |

In the next example, you are to find truth values of a statement with variables $p$, $q$, and $r$. You can construct the table of all possible values of $p$, $q$, and $r$ as follows:

1. List all possible values of $p$ and $q$ and for each case enter $T$ under the $r$ column.

2. Again list all possible values of $p$ and $q$ under the part of the table just formed and for each case enter $F$ in the $r$ column. Going from a table with two variables (four rows) to one with three variables (eight rows), double the number of rows in the table. Each time you add another variable, double the number of rows needed to list all possible values of the variables.

**Example 9**   Find the truth value of $p \vee (q \wedge r)$ for all possible truth values of $p$, $q$, and $r$.

**Solution**
We list all possible combinations of $T$ and $F$ for the statements and fill in the following truth table:

| $p$ | $q$ | $r$ | $q \wedge r$ | $p \vee (q \wedge r)$ |
|---|---|---|---|---|
| $T$ | $T$ | $T$ | $T$ | $T$ |
| $T$ | $F$ | $T$ | $F$ | $T$ |
| $F$ | $T$ | $T$ | $T$ | $T$ |
| $F$ | $F$ | $T$ | $F$ | $F$ |
| $T$ | $T$ | $F$ | $F$ | $T$ |
| $T$ | $F$ | $F$ | $F$ | $T$ |
| $F$ | $T$ | $F$ | $F$ | $F$ |
| $F$ | $F$ | $F$ | $F$ | $F$ |

## 10.1    EXERCISES

Access end-of-section exercises online at **www.webassign.net**

## 10.2    CONDITIONAL STATEMENTS

- Conditional
- Converse, Inverse, and Contrapositive
  of a Conditional
- Biconditional

### Conditional

We sometimes make sentences that contain a condition rather than an outright assertion. For example, the sentence "If the Sun shines, I will cut the grass" contains a condition (If the Sun shines) regarding the cutting of grass. We call such sentences **conditional** or **implication statements**.

| DEFINITION | |
|---|---|
| **DEFINITION**<br>Conditional (Implication)<br>Statement | A **conditional (implication) statement** is a statement of the form<br> "If $p$, then $q$." <br>We denote it by $p \rightarrow q$.<br>In an implication, $p \rightarrow q$, we call $p$ the **hypothesis** and $q$ the **conclusion**. |

**Example 1**    Let $p$ be the statement "I study four hours" and $q$ the statement "I can make an A on the exam."

**(a)** Write the statement $p \rightarrow q$.    **(b)** Write the statement $q \rightarrow p$.

**(c)** Write the statement with $p$ as the hypothesis and $q$ as the conclusion.

Solution

**(a)** If I study four hours, then I can make an A on the exam.

**(b)** If I can make an A on the exam, then I study four hours.

**(c)** If I study four hours, then I can make an A on the exam.

As the statements (a) and (c) in the solution are the same statements, this illustrates that the notation in part (a) describes the same statement as does part (c).

We usually think that the truth value of the hypothesis of a conditional has some effect on the conclusion. For example, the amount of time spent studying has some effect on the exam grade. However, we can construct implications where the hypothesis and conclusion have no relationship. The sentence "If June has 30 days, then Humpty Dumpty sat on the wall" is a true conditional even though the number of days in June has nothing to do with Humpty Dumpty. Although such a conditional might appear to be somewhat nonsensical, and you might not use this form in conversation, we still want to assign it a truth value. The following table shows how the truth values of $p$ and $q$ determine the truth value of $p \rightarrow q$:

| $p$ | $q$ | $p \rightarrow q$ |
|-----|-----|-------------------|
| $T$ | $T$ | $T$ |
| $T$ | $F$ | $F$ |
| $F$ | $T$ | $T$ |
| $F$ | $F$ | $T$ |

To assign $T$ as the truth value of $p \rightarrow q$ in the last two lines may not seem natural. You might prefer a value of $F$, or you may think it doesn't make much sense to put either $T$ or $F$. Granted, we most often use a conditional statement in situations where the first two lines apply. However, in logic we insist that statements be either true or false, so we must assign values to the last two cases. The truth table gives the definition accepted by mathematicians.

Even so, we do not arbitrarily assign true values. The following examples help us to understand this.

Your professor tells you, "If you make an A on the final exam, I will give you an A in the course." You make an A on the final exam, and the professor gives you a B in the course. Isn't your conclusion that the professor made a false statement? (Or in less polite terms, the professor lied.) Thus, assigning the value $F$ to $p \rightarrow q$ when $p$ is $T$ and $q$ is $F$ is consistent with our usual interpretation.

Now let's consider the case for a false hypothesis.

Dick has just received a driver's license, and his parents impose some rules to help him become a responsible driver. One of the rules is "If you speed, you will be grounded." This rule is of the form $p \rightarrow q$, a conditional. When Dick returns one evening, he is asked, "Did you speed?" He had not sped and truthfully answered, "No." His parents do not ground Dick. Thus, $p$ and $q$ both have the truth value $F$. Even though both parts of the rule are false, the rule has not been violated, so we should assign the value $T$ to the conditional in this case.

Another evening Dick returns quite late and he again is asked "Did you speed?" His truthful answer was, "No." Dick was grounded because he stayed out past the time to be home. Here $p$ has the value $F$ and $q$ the value $T$. Has the rule been violated? The rule does not state that speeding is the only reason for grounding. Thus, the rule has not been violated, and we should assign the value $T$ to the rule.

**Example 2**    Determine the truth of the following conditionals:

(a) If $2 \times 3 = 6$, then $15 - 4 = 11$.

(b) If $3 + 3 = 6$, then $2 = 1$.

(c) If there are 8 days in a week, then June has 30 days.

(d) If there are 8 days in a week, then August has 79 days.

**Solution**

(a) True because the hypothesis and conclusion are true

(b) False because the hypothesis is true and the conclusion is false

(c) True because the hypothesis is false and the conclusion is true

(d) True because the hypothesis and conclusion are false

Note that a false hypothesis always makes a conditional true, regardless of whether or not we have a true conclusion. A true conclusion always makes a conditional true.

## Converse, Inverse, and Contrapositive of a Conditional

We have just discussed how to form a conditional statement from two given statements. Actually, more than one conditional can be formed. For example, the statements "My car has a flat tire" and "I want to ride to the game with you" can be used to form the conditional "If my car has a flat tire, then I want to ride to the game with you." They can also be combined to form the conditional "If I want to ride to the game with you, then my car has a flat tire." Two other conditionals, which are commonly formed by using the negation of the given statements, are "If my car does not have a flat tire, then I do not want to ride to the game with you" and "If I do not want to ride to the game with you, then my car does not have a flat tire." Because the four conditionals of these forms occur frequently, we give them names: the **original conditional**, the **converse**, the **inverse**, and the **contrapositive**.

| **DEFINITION** | From a given conditional "If $p$ then $q$" we can form the following conditionals: |
|---|---|
| Converse, Inverse, Contrapositive | **Converse:** "If $q$, then $p$." $(q \rightarrow p)$ |
| | **Inverse:** "If not $p$, then not $q$." $(\sim p \rightarrow \sim q)$ |
| | **Contrapositive:** "If not $q$, then not $p$." $(\sim q \rightarrow \sim p)$ |

**Example 3**  Write the converse, inverse, and contrapositive of "If it does not rain, then we want to go on a picnic."

**Solution**

*Converse:* "If we want to go on a picnic, then it does not rain."

*Inverse:* "If it does rain, then we do not want to go on a picnic."

*Contrapositive:* "If we do not want to go on a picnic, then it does rain."

Now let's compare the truth tables of a conditional and its converse, inverse, and contrapositive. We can summarize them in one table as follows:

| $p$ | $q$ | $p \rightarrow q$ | $q \rightarrow p$ | $\sim p \rightarrow \sim q$ | $\sim q \rightarrow \sim p$ |
|---|---|---|---|---|---|
| T | T | T | T | T | T |
| T | F | F | T | T | F |
| F | T | T | F | F | T |
| F | F | T | T | T | T |

Note that $p \rightarrow q$ and $q \rightarrow p$ have different truth values when $p$ is true and $q$ is false or vice versa. This indicates that a conditional and its converse cannot be interchanged because a converse might not be true even though the conditional is. It is a common error for some people to think that a conditional and its converse have exactly the same meaning. You certainly would not accept that the statement "If I live in Chicago, then I live in Illinois" has the same meaning as the statement "If I live in Illinois, then I live in Chicago."

As a matter of curiosity, you may want to listen for instances when a politician, a letter to the editor writer, or an acquaintance uses the converse, in error, instead of the implication.

Also note that $p \rightarrow q$ (a statement) and $\sim q \rightarrow \sim p$ (its contrapositive) have exactly the same truth values in all cases. This means that one statement can be substituted for the other. Similarly, $q \rightarrow p$ (the converse of $p \rightarrow q$) and $\sim p \rightarrow \sim q$ (the inverse of $p \rightarrow q$) also have the same truth values. For example, the statements "If I go home, I will have some home cooking" and "If I do not have some home cooking, then I do not go home" have the same meaning. Also the statements "If I have home cooking, then I go home" and "If I do not go home, then I do not have home cooking" have the same meaning.

## Biconditional

Another fundamental compound statement is the **biconditional** statement, denoted $p \leftrightarrow q$, and read "$p$ if and only if $q$." You can also think of the biconditional, $p \leftrightarrow q$, as $(p \rightarrow q) \wedge (q \rightarrow p)$. We develop the truth table for the biconditional as follows:

| $p$ | $q$ | $p \rightarrow q$ | $q \rightarrow p$ | $(p \rightarrow q) \wedge (q \rightarrow p)$ | $p \leftrightarrow q$ |
|---|---|---|---|---|---|
| $T$ | $T$ | $T$ | $T$ | $T$ | $T$ |
| $T$ | $F$ | $F$ | $T$ | $F$ | $F$ |
| $F$ | $T$ | $T$ | $F$ | $F$ | $F$ |
| $F$ | $F$ | $T$ | $T$ | $T$ | $T$ |

We summarize this table with the following:

| $p$ | $q$ | $p \leftrightarrow q$ |
|---|---|---|
| $T$ | $T$ | $T$ |
| $T$ | $F$ | $F$ |
| $F$ | $T$ | $F$ |
| $F$ | $F$ | $T$ |

Note that $p \leftrightarrow q$ is true exactly when $p$ and $q$ have the same truth values.

**Example 4** Determine the truth value of the following biconditionals:

(a) $2 + 2 = 4$ if and only if $5 \times 2 = 10$.

(b) The English alphabet contains 36 letters if and only if Mexico is north of New York.

(c) $x + x = 2x$ if and only if $2 = 1$.

### Solution

(a) The biconditional is true because both $2 + 2 = 4$ and $5 \times 2 = 10$ are true.

(b) This is true because the component statements have the same truth value, false.

(c) This is false because the statements $x + x = 2x$ and $2 = 1$ have different truth values, true and false, respectively.

We can combine the basic compound and simple statements to form other compound statements. Using the truth tables of the basic compound statements, we can find the truth values of other compound statements.

**Example 5**    **(a)** Find the truth table for $\sim p \to (p \land q)$.

**(b)** Find the truth table for $(p \land q) \to r$.

**Solution**

**(a)** It is easier to analyze $\sim p \to (p \land q)$ if we include columns for $\sim p$ and $p \land q$ in the truth table:

| $p$ | $q$ | $\sim p$ | $p \land q$ | $\sim p \to (p \land q)$ |
|---|---|---|---|---|
| $T$ | $T$ | $F$ | $T$ | $T$ |
| $T$ | $F$ | $F$ | $F$ | $T$ |
| $F$ | $T$ | $T$ | $F$ | $F$ |
| $F$ | $F$ | $T$ | $F$ | $F$ |

**(b)** Because there are three statements $p$, $q$, and $r$, eight rows are required in the truth table to allow for all the possible combinations of the truth values of $p$, $q$, and $r$:

| $p$ | $q$ | $r$ | $p \land q$ | $(p \land q) \to r$ |
|---|---|---|---|---|
| $T$ | $T$ | $T$ | $T$ | $T$ |
| $T$ | $F$ | $T$ | $F$ | $T$ |
| $F$ | $T$ | $T$ | $F$ | $T$ |
| $F$ | $F$ | $T$ | $F$ | $T$ |
| $T$ | $T$ | $F$ | $T$ | $F$ |
| $T$ | $F$ | $F$ | $F$ | $T$ |
| $F$ | $T$ | $F$ | $F$ | $T$ |
| $F$ | $F$ | $F$ | $F$ | $T$ |

## **10.2**    **EXERCISES**

## 10.3     EQUIVALENT STATEMENTS

Two compound statements are **logically equivalent**, or **equivalent**, if they have exactly the same truth values. This means that a statement can be substituted for its equivalent because when one is true, the other is also true, and when one is false, the other is false. Two equivalent statements will have the same truth values for any choice of truth values of the component parts of the compound statements.

**Example 1**     Show that $p \rightarrow q$ and $\sim p \vee q$ are logically equivalent statements.

**Solution**
We form the truth table for each statement and see whether they are the same. We can form the table as follows:

| $p$ | $q$ | $\sim p$ | $p \rightarrow q$ | $\sim p \vee q$ |
|---|---|---|---|---|
| $T$ | $T$ | $F$ | $T$ | $T$ |
| $T$ | $F$ | $F$ | $F$ | $F$ |
| $F$ | $T$ | $T$ | $T$ | $T$ |
| $F$ | $F$ | $T$ | $T$ | $T$ |

Note that for each choice of $p$ and $q$ (each row), the truth values of $p \rightarrow q$ and $\sim p \vee q$ are the same. Therefore, the statements are equivalent.

**Example 2**     Show that $p \rightarrow q$ and $q \rightarrow p$ are not logically equivalent.

**Solution**
Form the truth table and compare truth values:

| $p$ | $q$ | $p \rightarrow q$ | $q \rightarrow p$ |
|---|---|---|---|
| $T$ | $T$ | $T$ | $T$ |
| $T$ | $F$ | $F$ | $T$ |
| $F$ | $T$ | $T$ | $F$ |
| $F$ | $F$ | $T$ | $T$ |

Note that the truth values of $p \rightarrow q$ and $q \rightarrow p$ differ in rows 2 and 3, so the statements are not equivalent.
    A difference of truth values in just one row prevents two statements from being equivalent.

**Example 3**     Determine whether the two following statements are equivalent:

(a) If you check the box, you cannot take the standard deduction.

(b) If you can take the standard deduction, you do not check the box.

**Solution**

Let $p$ represent the statement "You check the box" and let $q$ represent the statement "You can take the standard deduction."

Then statement (a) is $p \rightarrow \sim q$ and statement (b) is $q \rightarrow \sim p$. We form a truth table and compare the truth values of $p \rightarrow \sim q$ and $q \rightarrow \sim p$.

| $p$ | $q$ | $\sim p$ | $\sim q$ | $p \rightarrow \sim q$ | $q \rightarrow \sim p$ |
|---|---|---|---|---|---|
| T | T | F | F | F | F |
| T | F | F | T | T | T |
| F | T | T | F | T | T |
| F | F | T | T | T | T |

As $p \rightarrow \sim q$ and $q \rightarrow \sim p$ have identical truth values, they are equivalent.

---

## 10.3 EXERCISES

Access end-of-section exercises online at **www.webassign.net**

---

## 10.4 VALID ARGUMENTS

We all like to win an argument or prove a point. Unfortunately, false information or intimidation, not logic, sometimes wins arguments. Although we like to win arguments, it is more important that we draw a valid conclusion from information given. We would like a jury to reach a verdict supported by evidence. We do not like to see a person's reputation ruined because an invalid conclusion was drawn from information given. A business executive can harm the corporation with a decision not supported by the facts relevant to the decision.

Some arguments are generally recognized as invalid. For example, those who accept the premise "If you live in Omaha, then you live in Nebraska" will not accept the conclusion "If you live in Nebraska, then you live in Omaha." However, it appears that a significant number of people who accept the premise "If you are homosexual, then you will acquire AIDS" will accept the conclusion "If you acquire AIDS, then you are homosexual." Actually, this AIDS argument and the Omaha argument are of the same form. Each one has a premise of the form $p \rightarrow q$ and a conclusion of the form $q \rightarrow p$. We would like an argument to be valid or invalid based on whether or not the premises support the conclusion, not on which statements are used. We want logically sound **arguments** or **proofs**; we call them **valid arguments**.

An argument consists of a statement, called the **conclusion**, that follows from one or more statements, called the **premises**. If the conclusion follows logically from the premises, we say the argument is **valid**. You might ask what we mean by "the conclusion follows logically from the premises." Here is what we mean. Suppose an argument has two statements for its premises: call them $p$ and $q$. Call the conclusion $r$. The conclusion, $r$, follows logically from the premises, $p$ and $q$, if whenever all the premises, $p$ and $q$, are true, then the conclusion, $r$, is also true. An argument may have any number (one or more) premises.

An argument with premises $p_1, p_2, p_3, \ldots, p_n$ and a conclusion $r$ is **valid** if $r$ is true when all of $p_1, p_2, p_3, \ldots, p_n$ are true.

Here is a simple example of a valid argument.

**Example 1**

Premises:    If Jenny has a job, she saves money.
             Jenny has a job.
Conclusion:  Jenny saves money.

Let's put this in symbolic form. Write

$p$:  Jenny has a job.
$q$:  Jenny saves money.

Now the argument can be written in the notation

Premises:    $p \rightarrow q$
             $p$
Conclusion:  $q$

where the premises are written above the line and the conclusion below. The argument is valid if $q$ is true whenever both $p \rightarrow q$ and $p$ are true. Check the following truth table:

| $p$ | $q$ | $p \rightarrow q$ |
|-----|-----|-------------------|
| $T$ | $T$ | $T$ |
| $T$ | $F$ | $F$ |
| $F$ | $T$ | $T$ |
| $F$ | $F$ | $T$ |

Notice that when $p \rightarrow q$ and $p$ are both true (first line only), $q$ is also true, so the argument is valid.

The entries in the truth table would be the same if statements other than "Jenny has a job" and "Jenny saves money" are used for $p$ and $q$. Therefore, for any statements $p$ and $q$ the argument

Premises:    $p \rightarrow q$
             $p$
Conclusion:  $q$

is a valid argument. We call it the **Law of Detachment**.

For any pair of statements $p$ and $q$, the argument

Premises:    $p \rightarrow q$
             $p$
Conclusion:  $q$

is a valid argument.

**Example 2**   By the Law of Detachment, the following arguments are valid:

(a) Premises:     If my car has a flat tire, I will ride to the game with Sam.
                          My car has a flat tire.
                          ─────────────────────────────────────────────
      Conclusion:  I will ride to the game with Sam.

(b) Premises:     If I make an A on the final, I will make an A in the course.
                          I made an A on the final.
                          ─────────────────────────────────────────────
      Conclusion:  I will make an A in the course.

We can determine the validity of an argument in a way a little different from the definition by observing that if all the premises $p_1, p_2, p_3, \ldots, p_n$ are true, then

$$p_1 \wedge p_2 \wedge \cdots \wedge p_n \to r$$

is true only when $r$ is also true. If any one of $p_1, p_2, \ldots, p_n$ is false, then $p_1 \wedge p_2 \wedge \cdots \wedge p_n$ is false. In this case, the conditional $p_1 \wedge p_2 \wedge \cdots \wedge p_n \to r$ is true.

All of this makes for a situation where an argument with premises $p_1, p_2, \ldots, p_n$ and conclusion $r$ is valid provided the conditional $p_1 \wedge p_2 \wedge \cdots \wedge p_n \to r$ is true for all possible values of the premises and conclusion. (A statement that is always true is called a **tautology**.) This approach will be used to test the validity of arguments.

**THEOREM**

**Valid Argument**

An argument with premises $p_1, p_2, p_3, \ldots, p_n$ and a conclusion $r$ is valid, if

$$p_1 \wedge p_2 \wedge \cdots \wedge p_n \to r$$

is true for all possible truth values of $p_1, p_2, p_3, \ldots, p_n$ and $r$.

**Example 3**   Show that the following argument is valid:

Premises:     $p \wedge q$

                      $q$
                      ───────
Conclusion:  $p$

**Solution**
By the theorem above we can prove the argument valid by showing $(p \wedge q) \wedge q \to p$ is true for all possible values of $p$ and $q$. We check the truth table for

| $p$ | $q$ | $p \wedge q$ | $(p \wedge q) \wedge q$ | $(p \wedge q) \wedge q \to p$ |
|---|---|---|---|---|
| $T$ | $T$ | $T$ | $T$ | $T$ |
| $T$ | $F$ | $F$ | $F$ | $T$ |
| $F$ | $T$ | $F$ | $F$ | $T$ |
| $F$ | $F$ | $F$ | $F$ | $T$ |

The argument is valid because only $T$ appears in the last column.

**Example 4** Determine whether the following argument is valid:

$$\text{Premises:} \quad p \rightarrow q$$
$$\underline{\qquad q \rightarrow r}$$
$$\text{Conclusion:} \quad p \rightarrow r$$

We call this a **syllogism**.

**Solution**
Check $[(p \rightarrow q) \wedge (q \rightarrow r)] \rightarrow (p \rightarrow r)$:

| $p$ | $q$ | $r$ | $p \rightarrow q$ | $q \rightarrow r$ | $p \rightarrow r$ | $(p \rightarrow q) \wedge (q \rightarrow r)$ | $[(p \rightarrow q) \wedge (q \rightarrow r)] \rightarrow (p \rightarrow r)$ |
|---|---|---|---|---|---|---|---|
| T | T | T | T | T | T | T | T |
| T | F | T | F | T | T | F | T |
| F | T | T | T | T | T | T | T |
| F | F | T | T | T | T | T | T |
| T | T | F | T | F | F | F | T |
| T | F | F | F | T | F | F | T |
| F | T | F | T | F | T | F | T |
| F | F | F | T | T | T | T | T |

The argument is valid.

**Example 5** Determine whether the following argument is valid:

Premises: If Mike jogs daily, he keeps his weight under control.
If Mike keeps his weight under control, he can wear last year's clothes.

Conclusion: If Mike jogs daily, he can wear last year's clothes.

**Solution**
Write the arguments in symbolic form.
Let

$p$: Mike jogs daily.
$q$: Mike keeps his weight under control.
$r$: Mike can wear last year's clothes.

The argument is of the form

$$\text{Premises:} \quad p \rightarrow q$$
$$\underline{\qquad q \rightarrow r}$$
$$\text{Conclusion:} \quad p \rightarrow r$$

Example 4 shows that this argument (a syllogism) is valid.

**Example 6** Determine whether the following argument is valid:

Premises: Marci will mow the lawn, or Lee will mow the lawn.
Marci cannot mow the lawn.

Conclusion: Lee will mow the lawn.

**Solution**
Let

$p:$ Marci will mow the lawn.
$q:$ Lee will mow the lawn.

The argument is of the form

Premises: $p \lor q$

$\sim p$

Conclusion: $q$

Check the following truth table for $[(p \lor q) \land (\sim p)] \to q$:

| $p$ | $q$ | $p \lor q$ | $\sim p$ | $(p \lor q) \land (\sim p)$ | $[(p \lor q) \land (\sim p)] \to q$ |
|---|---|---|---|---|---|
| $T$ | $T$ | $T$ | $F$ | $F$ | $T$ |
| $T$ | $F$ | $T$ | $F$ | $F$ | $T$ |
| $F$ | $T$ | $T$ | $T$ | $T$ | $T$ |
| $F$ | $F$ | $F$ | $T$ | $F$ | $T$ |

The last column indicates that the argument is valid.

We call this a **disjunctive syllogism**.

**Example 7** Show that the following argument is not valid:

Premises: $p \to q$

$\sim p$

Conclusion: $\sim q$

**Solution**
Check $[(p \to q) \land (\sim p)] \to (\sim q)$:

| $p$ | $q$ | $p \to q$ | $\sim p$ | $\sim q$ | $(p \to q) \land (\sim p)$ | $[(p \to q) \land (\sim p)] \to \sim q$ |
|---|---|---|---|---|---|---|
| $T$ | $T$ | $T$ | $F$ | $F$ | $F$ | $T$ |
| $T$ | $F$ | $F$ | $F$ | $T$ | $F$ | $T$ |
| $F$ | $T$ | $T$ | $T$ | $F$ | $T$ | $F$ |
| $F$ | $F$ | $T$ | $T$ | $T$ | $T$ | $T$ |

The argument is not valid because an $F$ appears in the last column.

**Example 8**    Show that the following argument is valid:

$$p \rightarrow q$$
$$\dfrac{\sim q}{\sim p}$$

**Solution**

Here is the truth table:

| $p$ | $q$ | $p \rightarrow q$ | $\sim q$ | $(p \rightarrow q) \wedge \sim q$ | $[(p \rightarrow q) \wedge \sim q] \rightarrow \sim p$ |
|---|---|---|---|---|---|
| $T$ | $T$ | $T$ | $F$ | $F$ | $T$ |
| $T$ | $F$ | $F$ | $T$ | $F$ | $T$ |
| $F$ | $T$ | $T$ | $F$ | $F$ | $T$ |
| $F$ | $F$ | $T$ | $T$ | $T$ | $T$ |

The last column indicates that the argument is valid.

We call this **indirect reasoning**.

We now summarize the valid arguments discussed:

**Four Valid Arguments**

**Law of Detachment**

$$p \rightarrow q$$
$$\dfrac{p}{q}$$

**Syllogism**

$$p \rightarrow q$$
$$\dfrac{q \rightarrow r}{p \rightarrow r}$$

**Disjunctive syllogism**

$$p \vee q$$
$$\dfrac{\sim p}{q}$$

**Indirect reasoning**

$$p \rightarrow q$$
$$\dfrac{\sim q}{\sim p}$$

## 10.4    EXERCISES

Access end-of-section exercises online at **www.webassign.net**     **WebAssign**

## IMPORTANT TERMS

**10.1**
Statement
Truth Value
Compound Statement
Conjunction
Truth Table
Disjunction
Negation
Exclusive Or
Inclusive Or

**10.2**
Conditional (Implication)
Converse
Inverse
Contrapositive
Biconditional

**10.3**
Equivalent Statements

**10.4**
Valid Argument
Proof
Premise
Valid Conclusion
Law of Detachment
Tautology
Syllogism
Disjunctive Syllogism
Indirect Reasoning

## IMPORTANT CONCEPTS

**Conjuction**  A statement of the form "$p$ and $q$."

**Disjunction**  A statement of the form "$p$ or $q$."

**Conditional (Implication) Statement**  A statement of the form "If $p$, then $q$."

**Converse Statement**  The converse of "If $p$, then $q$" is "If $q$, then $p$."

**Inverse Statement**  The inverse of "If $p$, then $q$" is "If not $p$, then not $q$."

**Contrapositive Statement**  The contrapositive of "If $p$, then $q$" is "If not $q$, then not $p$."

**Biconditional Statement**  A statement of the form "$p$ if and only if $q$."

**Law of Detachment**  A valid argument of the form: Premises: $p \rightarrow q$
$$\dfrac{p}{}$$
Conclusion: $q$

**Syllogism**  A valid argument of the form: Premises: $p \rightarrow q$
$$\dfrac{q \rightarrow r}{}$$
Conclusion: $p \rightarrow r$

**Disjunctive Syllogism**  A valid argument of the form: Premises: $p \vee q$
$$\dfrac{\sim p}{}$$
Conclusion: $q$

**Indirect Reasoning**  A valid argument of the form: Premises: $p \rightarrow q$
$$\dfrac{\sim q}{}$$
Conclusion: $\sim p$

## REVIEW EXERCISES

**1.** Determine whether each of the following sentences is a statement.

   **(a)** It rained on March 30.

   **(b)** Did you see him make a hole-in-one?

   **(c)** Harriet's painting is the best work of art in the exhibit.

   **(d)** February 2010 had five Fridays.

2. Identify each of the following as a conjunction, disjunction, or neither.

   (a) The wind blew over the house plant, and rain soaked the carpet.

   (b) We will attend the Brazos River Festival, or we will drive out to see the wildflowers.

   (c) My antique car won't start today.

   (d) We plan to watch a documentary on TV or go to a concert.

3. Let $p$ be the statement "Rhonda is sick today" and let $q$ be the statement "Rhonda has a temperature." Write out the following statements.

   (a) $\sim p$        (b) $p \wedge q$

   (c) $\sim p \wedge \sim q$        (d) $p \vee q$

4. Convert the following statements into symbolic form using $p$ for "Angela is in biology lab today" and $q$ for "Pete has a history exam tomorrow."

   (a) Angela is in biology lab today, and Pete does not have a history exam tomorrow.

   (b) Angela is not in biology lab today, or Pete does not have a history exam tomorrow.

5. Determine whether the following statements are true or false.

   (a) Water freezes at 0°C, and water boils at 100°C.

   (b) Water freezes at 0°C, or George Washington was the first president of the United States.

   (c) The year 2011 was a leap year, and January has 31 days.

   (d) The year 2011 was not a leap year, or February has 30 days.

6. Write the negation of the following statements.

   (a) This sentence is false.

   (b) Dillard's operates a chain of department stores.

   (c) Halloween is not a national holiday.

7. Find the truth value of each of the following where $p$ and $q$ have truth values $T$ and $r$ has the truth value $F$:

   (a) $p \wedge (q \vee r)$        (b) $(p \wedge q) \vee (p \wedge r)$

   (c) $p \wedge q \wedge \sim r$        (d) $\sim[p \vee (q \wedge r)]$

8. Let $p$ be the statement "Monty has a laptop computer" and $q$ be the statement "Tanya has a graphing calculator."

   (a) Write the statement $p \rightarrow q$.

   (b) Write the statement $q \rightarrow p$.

9. Determine the truth of the following conditionals.

   (a) If 10 dimes make a dollar, then 10% of $3.50 is $0.35.

   (b) If New Year's Day is January 1, then Thanksgiving Day is in June.

10. Determine the truth value of the following conditionals:

    (a) If Florida is the northernmost state of the United States, then New York City is on the West Coast.

    (b) If William Shakespeare was a French novelist, then Franklin Roosevelt was president of the United States.

11. Write the inverse, converse, and contrapositive of the statement "If I turn my paper in late, then I will be penalized."

12. Determine the truth value of the following biconditionals:

    (a) Water freezes at 0°C if and only if water boils at 100°C.

    (b) Ten dimes make a dollar if and only if 10% of $5.00 is $0.65.

**Construct truth tables for Exercises 13 and 14.**

13. $\sim p \rightarrow (p \wedge q)$

14. $(p \wedge \sim q) \leftrightarrow (\sim p \vee q)$

15. Use truth tables to determine if $p \wedge (p \vee q)$ and $p$ are logically equivalent.

16. Use truth tables to determine if $p \wedge (q \vee r)$ and $(p \wedge q) \vee (p \wedge r)$ are logically equivalent.

17. Determine whether the following argument is valid:

    Premises: If you study logic, mathematics is easy. Mathematics is not easy.

    Conclusion: You did not study logic.

18. Determine whether the following argument is valid:

    Premises: If Sally studies music, then she is artistic. Sally is not artistic.

    Conclusion: Sally does not study music.

**19.** Determine whether the following argument is valid:

Premises:    If there is money in my account, then I will pay my rent.
If I pay my rent, then I will not be evicted.

Conclusion:  If there is money in my account, then I will not be evicted.

**20.** Determine whether the following argument is valid: If wheat prices are steady, then exports will rise. Wheat prices are steady. Therefore, exports will rise.

**21.** Ms. Lopez will request a transfer out of state, or she will start a business locally. Ms. Lopez did not request a transfer out of state. Therefore, she will start a business locally. Is this a valid argument?

# REVIEW TOPICS

This appendix contains basic algebra topics that are necessary for the materials in this book. You are encouraged to study those topics for which you need review and skip those topics with which you are familiar.

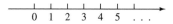

# PROPERTIES OF REAL NUMBERS

- The Real Number Line
- The Arithmetic of Real Numbers

## The Real Number Line

The most basic numbers in our study of mathematics are the **natural numbers,**

$$1, 2, 3, 4, \ldots$$

These are in fact the numbers that we use to count. They can be represented graphically as points on a line, as in Figure A–1.

$$
\begin{array}{ccccccc}
\mid & \mid & \mid & \mid & \mid & \mid \\
0 & 1 & 2 & 3 & 4 & 5 & \ldots
\end{array}
$$

**FIGURE A–1**

Begin with a starting point on the line, which we label 0, and a convenient unit scale. Starting at 0, mark off equal lengths to the right. These marks represent $1, 2, 3, \ldots$.

We can mark off lengths in the opposite direction from 0 and let them represent numbers, the negatives of the natural numbers (Figure A–2). This collection of numbers represented by the marks on the line is written

$$\ldots, -5, -4, -3, -2, -1, 0, 1, 2, 3, 4, 5, \ldots$$

and is called the set of **integers.** We call the number represented by the symbol 0 zero. It is neither positive nor negative.

$$
\begin{array}{ccccccccccc}
& \mid & \mid & \mid & \mid & \mid & \mid & \mid & \mid & \mid & \mid & \mid \\
\ldots & -5 & -4 & -3 & -2 & -1 & 0 & 1 & 2 & 3 & 4 & 5 & \ldots
\end{array}
$$

**FIGURE A–2**

Other numbers, the fractions, can be represented with points between the integers. In Figure A–3, point $A$, halfway between 1 and 2, represents $1\frac{1}{2}$; point $B$, one quarter of the way from 3 to 4, represents $3\frac{1}{4}$. Point $C$ is three quarters of the way from $-4$ to $-5$, so it represents $-4\frac{3}{4}$. These numbers together with the integers are called **rational numbers.**

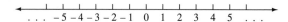

**FIGURE A–3**

All these points can be expressed in terms of finite or infinite decimals. Point $A$ is the point 1.5, $B$ the point 3.25, and $C$ the point $-4.75$. Every rational number can be expressed either as a finite decimal that terminates, such as $A$, $B$, and $C$ above, or as a decimal that repeats infinitely. For example, $5\frac{1}{3}$ is a rational number that can be written in decimal form as $5.333\ldots$; the 3 repeats endlessly.

There are, however, certain numbers called **irrational numbers** that do not have any pattern of repetition in their decimal form. One such number is $\sqrt{2}$. Its decimal form is 1.414213 . . . .

The set of all rational and irrational numbers is called the set of **real numbers.** One way to visualize the set of real numbers is to think of each point on the line as representing a real number.

## The Arithmetic of Real Numbers

There are four useful operations on the set of real numbers: addition, subtraction, multiplication, and division. The following tables summarize some rules that govern these operations:

| Rule | | Example |
| --- | --- | --- |
| Division by zero is not allowed. | | $\frac{5}{0}$ and $\frac{0}{0}$ have no meaning. |

| Rules of Operations | | Examples |
| --- | --- | --- |
| $a + b = b + a$ | Numbers can be added in either order. | $4 + 9 = 9 + 4$ |
| $ab = ba$ | Numbers can be multiplied in either order. | $3 \times 11 = 11 \times 3$ |
| $ab + ac = a(b + c)$ | A common number can be factored from each term in a sum. | $12a + 15b = 3(4a + 5b)$ |

| Rules of Signs | | Examples |
| --- | --- | --- |
| $-a = (-1)a$ <br> $-(-a) = a$ | The negative of a number $a$ is the product $(-1)a$. | $-12 = (-1)12$ <br> $-(-7) = 7$ |
| $(-a)b = a(-b) = -ab$ | The product of a positive and a negative number is a negative number. | $(-4)8 = -32$ <br> $5(-3) = -15$ |
| $(-a)(-b) = ab$ | The product of two negative numbers is a positive number. | $(-2)(-7) = 14$ |
| $(-a) + (-b) = -(a + b)$ <br> $-(a - b) = (-a) + b$ | The sum of two negative numbers is a negative number. | $(-4) + (-5) = -9$ <br> $-(8 - 3) = -8 + 3$ |
| $\dfrac{-a}{b} = \dfrac{a}{-b}$ <br><br> $= -\left(\dfrac{a}{b}\right)$ | Division using a negative number and a positive number is a negative number. | $\dfrac{-18}{3} = -6$ <br><br> $\dfrac{22}{(-11)} = -2$ |

| Arithmetic of Fractions | | Examples |
|---|---|---|
| $\dfrac{ac}{bc} = \dfrac{a}{b}$ | The value of a fraction is unchanged if both the numerator and the denominator are multiplied or divided by the same nonzero number, $c \neq 0$. | $\dfrac{10}{15} = \dfrac{2}{3}$ $\dfrac{7}{4} = \dfrac{14}{8}$ |
| $\dfrac{a}{b} + \dfrac{c}{b} = \dfrac{a+c}{b}$ | To add two fractions with the same denominators, add the numerators and keep the same denominator. | $\dfrac{5}{3} + \dfrac{2}{3} = \dfrac{7}{3}$ |
| $\dfrac{a}{b} + \dfrac{c}{d} = \dfrac{ad}{bd} + \dfrac{bc}{bd}$ $= \dfrac{ad+bc}{bd}$ $(b, d \neq 0)$ | To add two fractions with different denominators, convert them to fractions with the same denominators by multiplying the numerator and denominator of each fraction by the denominator of the other. | $\dfrac{3}{7} + \dfrac{2}{5}$ $= \dfrac{3(5)}{7(5)} + \dfrac{7(2)}{7(5)}$ $= \dfrac{3(5) + 7(2)}{7(5)}$ $= \dfrac{15 + 14}{35}$ $= \dfrac{29}{35}$ |
| $\dfrac{a}{b} \times \dfrac{c}{d} = \dfrac{ac}{bd}$ | To multiply two fractions, multiply their numerators and multiply their denominators. | $\dfrac{3}{4} \times \dfrac{6}{11} = \dfrac{18}{44} = \dfrac{9}{22}$ |
| $\dfrac{a}{b} \div \dfrac{c}{d} = \dfrac{a}{b} \times \dfrac{d}{c} = \dfrac{ad}{bc}$ Division may also be written in the form $\dfrac{\frac{a}{b}}{\frac{c}{d}} = \dfrac{a}{b} \div \dfrac{c}{d}$ $= \dfrac{a}{b} \times \dfrac{d}{c} = \dfrac{ad}{bc}$ $(b, c, d \neq 0)$ $\dfrac{a}{b} = \dfrac{c}{d}$ if and only if $ad = bc$ | To divide by a fraction, invert the divisor and multiply. | $\dfrac{2}{3} \div \dfrac{5}{8} = \dfrac{2}{3} \times \dfrac{8}{5}$ $= \dfrac{16}{15}$ $\dfrac{\frac{5}{3}}{\frac{4}{7}} = \dfrac{5}{3} \div \dfrac{4}{7}$ $= \dfrac{5}{3} \times \dfrac{7}{4}$ $= \dfrac{35}{12}$ $\dfrac{3}{7} = \dfrac{9}{21}$ because $3(21) = 7(9)$ $\dfrac{5}{9} \neq \dfrac{3}{4}$ because $5(4) \neq 9(3)$ |

## A.1     EXERCISES

**Evaluate the expressions in Exercises 1 through 49.**

**1.** $(-1)13$

**2.** $(-1)(-7)$

**3.** $-(-23)$

**4.** $(-10)(-4)$

**5.** $(-5)(6)$

**6.** $(-2)(-4)$

**7.** $5(-7)$

**8.** $(-6) + (-11)$

**9.** $-(7 - 2)$

**10.** $\dfrac{-10}{5}$

**11.** $\dfrac{21}{-3}$

**12.** $\dfrac{5 \times 3}{7 \times 3}$

**13.** $(-4) + (-6)$

**14.** $(-3)(-2)$

**15.** $(-4)2$

**16.** $\dfrac{4}{9} + \dfrac{2}{9}$

**17.** $\dfrac{5}{3} + \dfrac{4}{3}$

**18.** $\dfrac{4}{11} - \dfrac{2}{11}$

**19.** $\dfrac{12}{5} - \dfrac{3}{5}$

**20.** $\dfrac{6}{10} - \dfrac{13}{10}$

**21.** $\dfrac{2}{3} + \dfrac{3}{4}$

**22.** $\dfrac{5}{8} - \dfrac{1}{3}$

**23.** $\dfrac{5}{6} - \dfrac{7}{4}$

**24.** $\dfrac{5}{12} - \dfrac{1}{6}$

**25.** $\dfrac{2}{5} + \dfrac{1}{4}$

**26.** $(-3) + 6$

**27.** $\dfrac{4}{7} - \dfrac{3}{5}$

**28.** $\dfrac{2}{3} \times \dfrac{4}{5}$

**29.** $\dfrac{\frac{3}{4}}{\frac{9}{8}}$

**30.** $\dfrac{3}{8} + \dfrac{2}{5}$

**31.** $\dfrac{\frac{2}{7}}{\frac{4}{5}}$

**32.** $\left(\dfrac{4}{3}\right)\left(\dfrac{6}{7}\right)$

**33.** $\left(\dfrac{1}{3}\right)\left(\dfrac{1}{5}\right)$

**34.** $6 \times (-3)$

**35.** $\dfrac{2}{5} \times \dfrac{4}{3}$

**36.** $\left(-\dfrac{2}{3}\right)\left(\dfrac{1}{9}\right)$

**37.** $\left(-\dfrac{3}{5}\right)\left(-\dfrac{4}{7}\right)$

**38.** $\dfrac{4}{5} \div \dfrac{2}{15}$

**39.** $\dfrac{3}{11} + \dfrac{1}{3}$

**40.** $\dfrac{-\frac{4}{9}}{\frac{5}{2}}$

**41.** $\dfrac{5}{7} \div \dfrac{15}{28}$

**42.** $\dfrac{\frac{4}{9}}{\frac{16}{3}}$

**43.** $\dfrac{5}{8} \div \dfrac{1}{3}$

**44.** $\left(\dfrac{1}{2} - \dfrac{1}{3}\right)\left(\dfrac{5}{7}\right)$

**45.** $\left(\dfrac{3}{4} + \dfrac{1}{5}\right) \div \left(\dfrac{2}{9}\right)$

**46.** $5(4a + 2b)$

**47.** $-2(3a + 11b)$

**48.** $2(a - 3b)$

**49.** $-5(2a + 10b)$

---

## **A.2**    SOLVING LINEAR EQUATIONS

Numerous disciplines including science, technology, the social sciences, business, manufacturing, and government find mathematical techniques essential in day-to-day operations. They depend heavily on mathematical equations that describe conditions or relationships between quantities.

In an equation such as

$$4x - 5 = 7$$

the symbol $x$, called a **variable,** represents an arbitrary, an unspecified, or an unknown number just as John Doe and Jane Doe often denote an arbitrary, unspecified, or unknown person.

The equation $4x - 5 = 7$ may be true or false depending on the choice of the number $x$. If we substitute the number 3 for $x$ in

$$4x - 5 = 7$$

both sides become equal and we say that $x = 3$ is a **solution** of the equation. If 5 is substituted for $x$, then both sides are *not* equal, so $x = 5$ is *not* a solution.

It may help a sales representative to know that the expression $0.20x + 11$ describes the daily rental of a car, where $x$ represents mileage. Furthermore, the solution of the equation $0.20x + 11 = 40$ answers the question "You paid \$40 for car rental, how many miles did you drive?" Solutions of equations can sometimes help one to make a decision or gain needed information.

One basic procedure for solving an equation is to obtain a sequence of equivalent equations with the goal of isolating the variable on one side of the equation and the appropriate number on the other side.

The following two operations help to isolate the variable and find the solution.

1. The same number may be added to or subtracted from both sides of an equation.

2. Both sides of an equation may be multiplied or divided by a nonzero number.

Either of these operations yields another equation that is equivalent to the first, in other words, a second equation that has the same *solution* as the first.

**Example 1**   Solve the equation $3x + 4 = 19$.

**Solution**
We begin to isolate $x$ by subtracting 4 from both sides:

$$3x + 4 - 4 = 19 - 4$$
$$3x = 15$$

Next, divide both sides by 3:

$$\frac{3x}{3} = \frac{15}{3}$$
$$x = 5$$

is the solution. We can check our answer by substituting $x = 5$ into the original equation,

$$3(5) + 4 = 15 + 4 = 19$$

so the solution checks.

Now You Are Ready to Work Exercise 3

**Example 2**   Solve $4x - 2 = 2x + 12$.

**Solution**

$$4x - 2 = 2x + 12 \qquad \text{(First, add 2 to both sides.)}$$
$$4x - 2 + 2 = 2x + 12 + 2$$
$$4x = 2x + 14 \qquad \text{(Next, subtract } 2x \text{ from both sides.)}$$
$$4x - 2x = 2x + 14 - 2x$$
$$2x = 14 \qquad \text{(Now divide both sides by 2.)}$$
$$x = 7$$

*Check:* $4(7) - 2 = 28 - 2 = 26$ (left-hand side) and $2(7) + 12 = 14 + 12 = 26$ (right-hand side), so it checks.

Now You Are Ready to Work Exercise 7

**Example 3**   Solve $7x + 13 = 0$.

Solution

$$7x + 13 = 0 \qquad \text{(Subtract 13 from both sides.)}$$
$$7x = -13 \quad \text{(Divide both sides by 7.)}$$
$$x = -\frac{13}{7}$$

Now You Are Ready to Work Exercise 9

The previous examples all use **linear equations**.

**DEFINITION**
Linear Equation

A **linear equation in one variable**, $x$, is an equation that can be written in the form

$$ax + b = 0 \quad \text{where} \quad a \neq 0$$

A **linear equation in two variables** is an equation that can be written in the form

$$y = ax + b \quad \text{where} \quad a \neq 0$$

**Example 4**   Solve $\dfrac{3x - 5}{2} + \dfrac{x + 7}{3} = 8$.

Solution
We show two ways to solve this. First, use rules of fractions to combine the terms on the left-hand side:

$$\frac{3x - 5}{2} + \frac{x + 7}{3} = 8 \qquad \text{(Convert fractions to the same denominator.)}$$

$$\frac{3(3x - 5)}{6} + \frac{2(x + 7)}{6} = 8 \qquad \text{(Now add the fractions.)}$$

$$\frac{3(3x - 5) + 2(x + 7)}{6} = 8$$

$$\frac{9x - 15 + 2x + 14}{6} = 8$$

$$\frac{11x - 1}{6} = 8 \qquad \text{(Now multiply both sides by 6.)}$$

$$11x - 1 = 48$$
$$11x = 49$$
$$x = \frac{49}{11}$$

An alternative, and simpler, method is the following:

$$\frac{3x - 5}{2} + \frac{x + 7}{3} = 8$$

Multiply through by 6 (the product of the denominators):

$$3(3x - 5) + 2(x + 7) = 48$$
$$9x - 15 + 2x + 14 = 48$$
$$11x - 1 = 48$$
$$11x = 49$$
$$x = \frac{49}{11}$$

**Now You Are Ready to Work Exercise 13**

---

**Example 5**   Rent-A-Car charges $0.21 per mile plus $10 per day for car rental. Thus, the daily fee is represented by the equation

$$y = 0.21x + 10$$

where $x$ is the number of miles driven and $y$ is the daily fee.

**(a)** Determine the rental fee if the car is driven 165 miles during the day.

**(b)** Determine the rental fee if the car is driven 420 miles during the day.

**(c)** The rental fee for one day is $48.64. Find the number of miles driven.

**Solution**

**(a)** $x = 165$, so $y = 0.21(165) + 10 = 44.65$. The fee is $44.65.

**(b)** $x = 420$, so $y = 0.21(420) + 10 = 98.2$. The fee is $98.20.

**(c)** $y = 48.64$, so $x$ is the solution of the equation.

$$0.21x + 10 = 48.64$$
$$0.21x = 48.64 - 10$$
$$0.21x = 38.64$$
$$x = \frac{38.64}{0.21} = 184$$

The car was driven 184 miles.

**Now You Are Ready to Work Exercise 17**

---

## A.2   EXERCISES

**Determine which of the following values of $x$ are solutions to the equations in Exercises 1 and 2. Use $x = 1, 2, -3, 0, 4,$ and $-2$.**

**1.** $2x - 4 = -10$       **2.** $3x + 1 = x + 5$

**Solve the equations in Exercises 3 through 16.**

**3.** *(See Example 1)*     **4.** $-4x + 2 = 6$
$2x - 3 = 5$

**5.** $4x - 3 = 5$       **6.** $7x - 4 = 0$

**7.** *(See Example 2)*     **8.** $2x - 4 = -5x + 2$
$7x + 2 = 3x + 4$

**9.** *(See Example 3)*     **10.** $5 - x = 8 + 3x$
$12x + 21 = 0$

**11.** $3(x - 5) + 4(2x + 1) = 9$

**12.** $6(4x + 5) + 7 = 2$

**13.** *(See Example 4)*       **14.** $\dfrac{4x + 7}{6} + \dfrac{2 - 3x}{5} = 5$

$\dfrac{2x + 3}{3} + \dfrac{5x - 1}{4} = 2$

**15.** $\dfrac{12x + 4}{2x + 7} = 4$       **16.** $\dfrac{x + 1}{x - 1} = \dfrac{3}{4}$

**17.** *(See Example 5)* The U-Drive-It Rental Company charges $0.20 per mile plus $112 per week for car rental. The weekly rental fee for a car is represented by the linear equation

$$y = 0.20x + 112$$

where $x$ is the number of miles driven and $y$ is the weekly rental charge.

**(a)** Determine the rental fee if the car is driven 650 miles during the week.

**(b)** Determine the rental fee if the car is driven 1500 miles.

**(c)** The weekly rental fee is $302. How many miles were driven?

**18.** Joe Cool has a summer job selling real estate in a subdivision development. He receives a base salary of $100 per week plus $50 for each lot sold. Therefore, the equation

$$y = 50x + 100$$

represents his weekly income, where $x$ is the number of lots sold.

**(a)** What is his weekly income if he sells 7 lots?

**(b)** What is his weekly income if he sells 15 lots?

**(c)** If he receives $550 one week, how many lots did he sell?

**19.** A Girl Scout troop collects aluminum cans for a project. The recycling center weighs the cans in a container that weighs 8 pounds, so the scouts are paid according to the equation

$$y = 0.42(x - 8)$$

where $x$ is the weight in pounds given by the scale and $y$ is the payment in dollars.

**(a)** How much money do the Girl Scouts receive if the scale reads 42 pounds?

**(b)** How much do they receive if the scale reads 113 pounds?

**(c)** The scouts received $22.26 for one weekend's collection. What was the reading on the scale?

**20.** The tuition and fees paid by students at a local junior college are given by the equation

$$y = 27x + 85$$

where $x$ is the number of hours enrolled and $y$ is the total cost of tuition and fees ($85 fixed fees and $27 per hour tuition).

**(a)** How much does a student pay who is enrolled in 13 hours?

**(b)** A student who pays $517 is enrolled in how many hours?

---

## A.3  COORDINATE SYSTEMS

We have all seen a map, a house plan, or a wiring diagram that shows information recorded on a flat surface. Each of these uses some notation unique to the subject to convey the desired information. In mathematics, we often use a flat surface called a **plane** to draw figures and locate points. We place a reference system in the plane to record and communicate information accurately. The standard mathematical reference system consists of a horizontal and a vertical line (called **axes**). These two perpendicular axes form a **Cartesian,** or **rectangular, coordinate system.** They intersect at a point called the **origin.**

We name the horizontal axis the **x-axis,** and we name the vertical axis the **y-axis.** The origin is labeled $O$.

Two numbers are used to describe the location of a point in the plane, and they are recorded in the form $(x, y)$. For example, $x = 3$ and $y = 2$ for the point $(3, 2)$. The first number, 3, called the **x-coordinate,** or **abscissa,** represents the horizontal distance from the $y$-axis to the point. The second number, 2, called the **y-coordinate** or **ordinate,** represents the vertical distance measured from the $x$-axis to the point. The point $(3, 2)$ is shown as point $P$ in Figure A–4. Points located to the right of the $y$-axis have positive $x$-coordinates; those to the left have negative $x$-coordinates. The $y$-coordinate is positive for points located above the $x$-axis and negative for those located below.

Figure A–4 shows other examples of points in this coordinate system: $Q$ is the point $(-4, 3)$, and $R$ is the point $(-3, -2.5)$. The origin $O$ has coordinates $(0, 0)$.

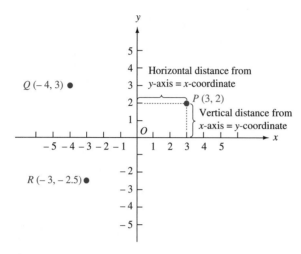

**FIGURE A–4**

Figure A–5 shows the points $(-3, 2)$, $(-4, -2)$, $(1, 1)$, and $(1, -2)$ plotted on the Cartesian coordinate system.

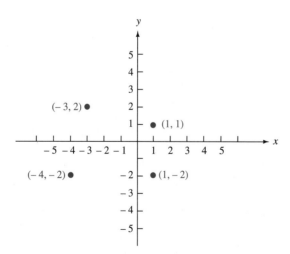

**FIGURE A–5**

The coordinate axes divide the plane into four parts called **quadrants.** The quadrants are labeled I, II, III, and IV as shown in Figure A–6. Point $A$ is located in the first quadrant, where $x$ and $y$ are both positive; $B$, in the second quadrant, where $x$ is negative and $y$ is positive; $C$ in the third quadrant, where both $x$ and $y$ are negative; and $D$ is in the fourth quadrant, where $x$ is positive and $y$ is negative. Points $A$ and $E$ lie in the same quadrant.

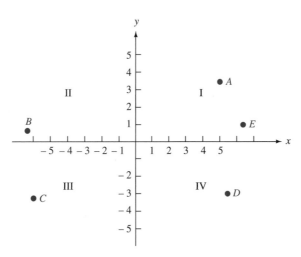

**FIGURE A–6**

René Descartes (1596–1650), a French philosopher–mathematician, invented the Cartesian coordinate system. His invention of the coordinate system is one of the outstanding ideas in the history of mathematics because it combined algebra and geometry in a way that enables us to use algebra to solve geometry problems and to use geometry to clarify algebraic concepts.

## A.3   EXERCISES

1. The following are the coordinates of points in a rectangular Cartesian coordinate system. Plot these points.

   $(-5, 4), (-2, -3), (-2, 4), (1, 5), (2, -5)$

2. What are the coordinates of the points $P$, $Q$, $R$, and $S$ in the coordinate system in Figure A–7?

3. Locate the following points on a Cartesian coordinate system:

   $(-2, 5), (3, -2), (0, 4), (-2, 0), (\frac{7}{2}, 2)$
   $(\frac{2}{3}, \frac{9}{4}), (-4, -2), (0, -5), (0, -2), (-6, -3)$

4. Give the coordinates of $A$, $B$, $C$, $D$, $E$, and $F$, in the coordinate system shown in Figure A–8.

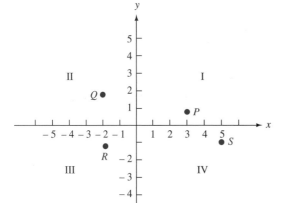

**FIGURE A–7**

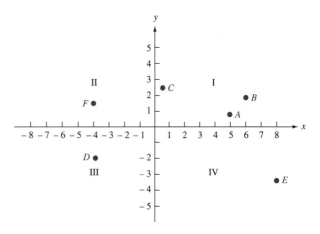

**FIGURE A–8**

**5.** Note that all points in the first quadrant have positive $x$-coordinates and positive $y$-coordinates. What are the characteristics of the points in:

  **(a)** the second quadrant?

  **(b)** the third quadrant?

  **(c)** the fourth quadrant?

**6.** For each case shown in Figure A–9, find the property the points have in common.

**7.** An old map gives these instructions to find a buried treasure: Start at giant oak tree. Go north 15 paces, then east 22 paces to a half-buried rock. The key to the treasure chest is buried at the spot that is 17 paces west and 13 paces north of the rock. From the place where the key is buried, go 32 paces west and 16 paces south to the location of the buried treasure. Use a coordinate system to represent the location of the oak tree, the rock, the key, and the treasure.

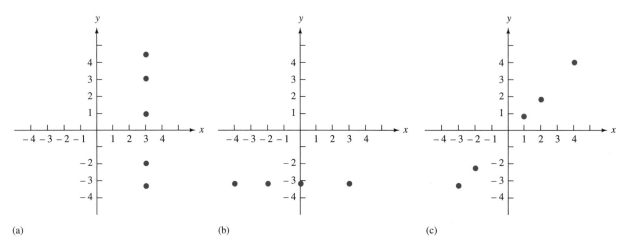

(a)                              (b)                              (c)

**FIGURE A–9**

---

## A.4     LINEAR INEQUALITIES AND INTERVAL NOTATION

- Solving Inequalities
- Interval Notation

We frequently use inequalities in our daily conversation. They may take the form "Which store has the lower price?" "Did you make a higher grade?" "Our team scored more points." "My expenses are greater than my income." Statements such as these basically state that one quantity is greater than another. Statements using the terms "greater than" or "less than" are called **inequalities.** Our goal is to solve inequalities. First, we give some terminology and notation.

The symbol $<$ means "less than," and $>$ means "greater than." Just remember that each of these symbols points to the smaller quantity. The notations $a > b$ and $b < a$ have exactly the same meaning. We may interpret the definition of $a > b$ in three equivalent ways. At times, one may be more useful than the other, so choose the most appropriate one.

**DEFINITION**    If $a$ and $b$ are real numbers, the following statements have the same meaning.
$a > b$
  **(a)** $a > b$ means that $a$ lies to the right of $b$ on a number line.
  **(b)** $a > b$ means that there is a positive number $p$ such that $a = b + p$.
  **(c)** $a > b$ means that $a - b$ is a positive number $p$.

The positive numbers lie to the right of zero on a number line, and the negative numbers lie to the left.

We will also use the symbols $<$ (less than), $\geq$ (greater than or equal to), and $\leq$ (less than or equal to).

| **DEFINITION** | $a < b$ means $b > a$. |
|---|---|
| $a < b, a \geq b, a \leq b$ | $a \geq b$ means $a = b$ or $a > b$. |
| | $a \leq b$ means $a = b$ or $a < b$. |

**Example 1**  The numbers 5, 8, 17, $-2$, $-3$, and $-15$ are plotted on a number line in Figure A–10.

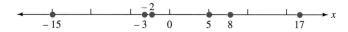

**FIGURE A–10**

Notice the following:

1. (a) 17 lies to the right of 5.    (b) $17 = 5 + 12$    (c) $17 - 5 = 12$
   Each of these three statements is equivalent to saying that $17 > 5$.
2. $8 > -3$ because $8 - (-3) = 8 + 3 = 11$ (by part (c) of the above definition).
3. $-2 > -15$ because $-2 = -15 + 13$ (by part (b) of the above definition, where $p = 13$).

Now You Are Ready to Work Exercise 1

## Solving Inequalities

By the **solution of an inequality** like

$$3x + 5 > 23$$

we mean the value or values of $x$ that make the statement true. The method for solving inequalities is similar to that for solving equations. We want to operate on an inequality in a way that gives an equivalent inequality but that enables us to determine the solution. Here are some simple examples of useful properties of inequalities.

**Example 2**

1. Because $18 > 4$, $18 + 6 > 4 + 6$; that is, $24 > 10$ (6 added to both sides).
2. Because $23 > -1$, $23 - 7 > -1 - 7$; that is, $16 > -8$ (7 subtracted from both sides).
3. Because $6 > 2$, $4(6) > 4(2)$; that is, $24 > 8$ (both sides multiplied by 4).
4. Because $10 > 3$, $-2(10) < -2(3)$; that is, $-20 < -6$ (both sides multiplied by $-2$).
5. Because $-15 > -21$, $\dfrac{-15}{3} > \dfrac{-21}{3}$; that is, $-5 > -7$. (Divide both sides by 3.)
6. Because $20 > 6$, $\dfrac{20}{-2} < \dfrac{6}{-2}$; that is, $-10 < -3$. (Divide both sides by $-2$.)

**CAUTION**

The inequality symbol reverses when we multiply each side by a negative number.

**CAUTION**

The inequality symbol reverses when dividing each side by a negative number.

These examples illustrate basic properties that are useful in solving inequalities.

**Properties of Inequalities**    For real numbers $a$, $b$, and $c$, the following are true:

1. Adding a number to both sides of an inequality leaves the direction of the inequality unchanged.

$$\text{If } a > b, \text{ then } a + c > b + c.$$

2. Subtraction of a number from both sides of an inequality leaves the direction of the inequality unchanged.

$$\text{If } a > b, \text{ then } a - c > b - c.$$

3. Multiply both sides of an inequality by a nonzero number:
   **(a)** If $a > b$ and $c$ is positive, then $ac > bc$.
   **(b)** If $a > b$ and $c$ is negative, then $ac < bc$. (Notice the change from $>$ to $<$.)

4. Divide both sides of an inequality by a nonzero number:

   **(a)** If $a > b$ and $c$ is positive, then $\dfrac{a}{c} > \dfrac{b}{c}$.

   **(b)** If $a > b$ and $c$ is negative, then $\dfrac{a}{c} < \dfrac{b}{c}$. (Notice the change from $>$ to $<$.)

(*Note:* All these properties hold if $>$ is replaced by $<$ and vice versa; and if $>$ is replaced by $\geq$ and $<$ is replaced by $\leq$.)

We use these properties to solve an inequality—that is, to find the values of $x$ that make the inequality true. In general, we proceed by finding equivalent inequalities that will eventually isolate $x$ on one side of the inequality and the appropriate number on the other side.

**Example 3**    Solve the inequality $3x + 5 > 14$.

**Solution**
Begin with the given inequality:

$$3x + 5 > 14 \quad \text{(Next, subtract 5 from each side (Property 2))}$$
$$3x > 9 \quad \text{(Now divide each side by 3 (Property 4))}$$
$$x > 3$$

Thus, all $x$ greater than 3 make the inequality true. This solution can be graphed on a number line as shown in Figure A–11. The empty circle indicates that $x = 3$ is omitted from the solution, and the heavy line indicates the values of $x$ included in the solution.

**FIGURE A–11**    $x > 3$.

**Now You Are Ready to Work Exercise 3**

**Example 4**   Solve the inequality $5x - 17 > 8x + 14$ and indicate the solution on a graph.

**Solution**
Start with the given inequality:

$$5x - 17 > 8x + 14 \quad \text{(Now add 17 to both sides (Property 1))}$$
$$5x > 8x + 31 \quad \text{(Now subtract } 8x \text{ from both sides (Property 2))}$$
$$-3x > 31 \quad \text{(Now divide both sides by } -3 \text{ (Property 4))}$$
$$x < -\frac{31}{3} \quad \text{(This reverses the inequality symbol)}$$

Thus, the solution consists of all $x$ to the left of $-\frac{31}{3}$. See Figure A–12.

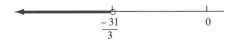

**FIGURE A–12**   $x < \dfrac{-31}{3}$.

Now You Are Ready to Work Exercise 9

**Example 5**   Solve and graph $2(x - 3) \le 3(x + 5) - 7$.

**Solution**

$$2(x - 3) \le 3(x + 5) - 7 \text{ (First perform the indicated multiplications)}$$
$$2x - 6 \le 3x + 15 - 7$$
$$2x - 6 \le 3x + 8 \qquad \text{(Now add 6 to both sides (Property 1))}$$
$$2x \le 3x + 14 \qquad \text{(Subtract } 3x \text{ from both sides (Property 2))}$$
$$-x \le 14 \qquad \text{(Multiply both sides by } -1 \text{ (Property 3))}$$
$$x \ge -14$$

Because the solution includes $-14$ and all numbers greater, the graph shows a solid circle at $-14$ (see Figure A–13).

**FIGURE A–13**   $x \ge -14$.

Now You Are Ready to Work Exercise 13

The next example illustrates a problem that involves two inequalities.

**Example 6**   Solve and graph $3 < 2x + 5 \le 13$.

**Solution**
This inequality means both $3 < 2x + 5$ *and* $2x + 5 \le 13$. Solve it in a manner similar to the preceding examples except that you try to isolate the $x$ in the middle.
Begin with the given inequality:

$$3 < 2x + 5 \le 13 \quad \text{(Subtract 5 from all parts of the inequality.)}$$
$$-2 < 2x \le 8 \qquad \text{(Divide each part by 2.)}$$
$$-1 < x \le 4$$

The solution consists of all numbers between −1 and 4, including 4 but not including −1. The graph of the solution (see Figure A–14) shows an empty circle at −1 because −1 is not a part of the solution. It shows a solid circle at 4 because 4 is a part of the solution. The solid line between −1 and 4 indicates that all numbers between −1 and 4 are included in the solution.

**FIGURE A–14**    $-1 < x \le 4$.

Now You Are Ready to Work Exercise 17

## Interval Notation

The solution of an inequality can be represented by yet another notation, the **interval notation.** Identify the portion of the number line that represents the solution of an inequality by its end points; brackets or parentheses indicate whether or not the end point is included in the solution. A parenthesis indicates that the end point is not included, and a bracket indicates that the end point is included. For example, the notation $(-1, 4]$ means $-1 < x \le 4$ and indicates the set of all numbers between −1 and 4 with −1 excluded and 4 included in the set. The notation $(-1, 4)$ means $-1 < x < 4$ and indicates that both −1 and 4 are excluded from the set. The notation $(-1, \infty)$ denotes the set of all numbers greater than $-1$, $x > -1$. The symbol $\infty$ denotes infinity and indicates that there is no upper bound to the interval.

Table A–1 shows the variations of the interval notation.

TABLE **A.1**

| Inequality Notation | | Interval Notation | | Graph of Interval |
|---|---|---|---|---|
| **General** | **Example** | **General** | **Example** | |
| $a < x < b$ | $-1 < x < 4$ | $(a, b)$ | $(-1, 4)$ | |
| $a \le x < b$ | $-1 \le x < 4$ | $[a, b)$ | $[-1, 4)$ | |
| $a < x \le b$ | $-1 < x \le 4$ | $(a, b]$ | $(-1, 4]$ | |
| $a \le x \le b$ | $-1 \le x \le 4$ | $[a, b]$ | $[-1, 4]$ | |
| $x < b$ | $x < 4$ | $(-\infty, b)$ | $(-\infty, 4)$ | |
| $x \le b$ | $x \le 4$ | $(-\infty, b]$ | $(-\infty, 4]$ | |
| $a < x$ | $-1 < x$ | $(a, \infty)$ | $(-1, \infty)$ | |
| $a \le x$ | $-1 \le x$ | $[a, \infty)$ | $[-1, \infty)$ | |

**Example 7**    Solve $1 \le 2(x - 5) + 3 < 5$.

**Solution**

$1 \le 2(x - 5) + 3 < 5$    (Multiply to remove parentheses.)
$1 \le 2x - 10 + 3 < 5$
$1 \le 2x - 7 < 5$    (Add 7 throughout.)
$8 \le 2x < 12$    (Divide through by 2.)
$4 \le x < 6$

The solution consists of all values of $x$ in the interval $[4, 6)$. The graph is shown in Figure A–15.

**FIGURE A–15** $4 \leq x < 6$.

Now You Are Ready to Work Exercise 25

**Example 8**  The total points on an exam given by Professor Passmore are 20 points plus 2.5 points for each correct answer. A total score in $[70, 80)$ is a C. If Scott made a C on the exam, how many questions did he answer correctly?

**Solution**
The score on an exam is given by $20 + 2.5x$, where $x$ is the number of correct answers. Thus the condition for a C is

$$70 \leq 20 + 2.5x < 80$$

Solve for $x$ to obtain the number of correct answers:

$$70 \leq 20 + 2.5x < 80$$
$$50 \leq 2.5x < 60$$
$$\frac{50}{2.5} \leq x < \frac{60}{2.5}$$
$$20 \leq x < 24$$

In this case, only whole numbers make sense, so Scott got $20, 21, 22$, or $23$ correct answers.

Now You Are Ready to Work Exercise 31

## A.4  EXERCISES

### Level 1

**1.** *(See Example 1)*  The following inequalities are of the form $a > b$. Verify the truth or falsity of each one by the property $a > b$ means $a - b$ is a positive number.

(a) $9 > 3$   (b) $4 > 0$   (c) $-5 > 0$

(d) $-3 > -15$   (e) $\dfrac{5}{6} > \dfrac{2}{3}$

**2.** Plot the numbers $10, 2, 5, -4, 3$, and $-2$ on a number line. Verify the truth or falsity of each of the following by the property $a > b$ if $a$ lies to the right of $b$.

(a) $10 > 5$   (b) $-4 > 2$   (c) $10 > 2$

(d) $5 > 10$   (e) $-4 > -2$   (f) $-2 > -4$

**Solve the inequalities in Exercises 3 though 8. State your solution using inequalities.**

**3.** *(See Example 3)*
$3x - 5 < x + 4$

**4.** $12 > 1 - 5x$

**5.** $5x - 22 \leq 7x + 10$

**6.** $13x - 5 \leq 7 - 4x$

**7.** $3(2x + 1) < 9x + 12$

**8.** $14 - 5x \geq 6x - 15$

**Solve the inequalities in Exercises 9 through 20. Graph the solution.**

**9.** *(See Example 4)*
$3x + 2 \leq 4x - 3$

**10.** $3x + 2 < 2x - 3$

**11.** $6x + 5 < 5x - 4$

**12.** $78 < 6 - 3x$

**13.** *(See Example 5)*    $3(x + 4) < 2(x - 3) + 14$

**14.** $4(x - 2) > 5(2x + 1)$

**15.** $3(2x + 1) < -1(3x - 10)$

**16.** $-2(3x + 4) > -3(1 - 6x) - 17$

**17.** *(See Example 6)*    $-16 < 3x + 5 < 22$

**18.** $124 > 5 - 2x \geq 68$

**19.** $14 < 3x + 8 < 32$

**20.** $-9 \leq 3(x + 2) - 15 < 27$

**Solve the inequalities in Exercises 21 through 26. Give the solution in interval form.**

**21.** $3x + 4 \leq 1$

**22.** $5x - 7 > 3$

**23.** $-7x + 4 \geq 2x + 3$

**24.** $-3x + 4 < 2x - 6$

**25.** *(See Example 7)*

$-45 < 4x + 7 \leq -10$

**26.** $16 > 2x - 10 \geq 4$

## Level 2

**Solve the inequalities in Exercises 27 through 30.**

**27.** $\dfrac{6x + 5}{-2} \geq \dfrac{4x - 3}{5}$

**28.** $\dfrac{2x - 5}{3} < \dfrac{x + 7}{4}$

**29.** $\dfrac{2}{3} < \dfrac{x + 5}{-4} \leq \dfrac{3}{2}$

**30.** $\dfrac{3}{4} < \dfrac{7x + 1}{6} < \dfrac{5}{2}$

## Level 3

**31.** *(See Example 8)*    Prof. Tuff computes a grade on a test by $35 + 5x$, where $x$ is the number of correct answers. A grade in the interval $[75, 90)$ is a B. If a student receives a B, how many correct answers were given?

**32.** A professor computes a grade on a test by $25 + 4x$ where $x$ is the number of correct answers. A grade in the interval $[70, 79]$ is a C. What number of correct answers can be obtained to receive a C?

**33.** On a final exam, any grade in the interval $[85, 100]$ was an A. The professor gave 3 points for each correct answer and then adjusted the grades by adding 25 points. If a student made an A, how many correct answers were given?

**34.** A sporting goods store runs a special on jogging shoes. The manager expects to make a profit if the number of pairs of shoes sold, $x$, satisfies $32x - 4230 > 2x + 480$. How many pairs of shoes need to be sold to make a profit?

## IMPORTANT TERMS

**A.1**

Natural Numbers

Integers

Rational Numbers

Irrational Numbers

Real Numbers

**A.2**

Variable

Solution

Linear Equation

**A.3**

Axes

Cartesian Coordinate System

Rectangular Coordinate System

Origin

$x$-Axis

$y$-Axis

Abscissa

Ordinate

Quadrants

**A.4**

Inequalities

$>, <, \geq, \leq$

Solution of an Inequality

Properties of Inequalities

Interval Notation

# USING A TI-83/84 GRAPHING CALCULATOR

**B**

The purpose of this appendix is to provide a summary of some of the key instructions that are useful in applying the graphing calculator to this course. This appendix is not intended to replace the instruction guidebook. It covers only the TI-83/84 families of graphing calculator. Students are free to use other graphing calculators and computers in working the Technology Explorations.

**NOTE**

A key may be used to make two or more selections. For example, the notation "TEST A" appears above the MATH key. When two names appear above a key, the one on the left is selected by the 2nd key, and the one on the right is selected by the ALPHA key. Press 2nd MATH and the TEST menu will appear. Press ALPHA MATH and the letter A will appear. When only one name appears above a key, use the 2nd key to select it. For example, INS appears above the DEL key and is accessed with 2nd DEL. INS is used to insert characters/ numbers in an exisisting string of characters/ numbers. The use of the 2nd and ALPHA with other keys will give the menus or letters indicated above the key.
    Thus, when you see a notation like TEST in this material, you are to press 2nd MATH.

# NOTATION

We will use the following notation:

Symbols enclosed in a rectangle, like

$$\boxed{A}, \boxed{5}, \boxed{Y=}, \boxed{MATRX}, \text{ and } \boxed{2nd}$$

refer to keys to be pressed.

Symbols enclosed with < >, like

$$<OPS>, <NUM>, \text{ and } <PROB>$$

refer to commands selected from a menu.

# ARITHMETIC OPERATIONS

Arithmetic calculations are done much like the calculations on a nongraphing calculator. Here are a few basic hints that might be helpful:

1. The $\boxed{ENTER}$ key plays the same role as "equals" in calculation.
2. The $\boxed{-}$ key denotes subtraction.
3. The $\boxed{(-)}$ key denotes "negative."
   To calculate $2 - 5$, you key $\boxed{2}\,\boxed{-}\,\boxed{5}$
   To calculate $-2 \cdot 8$, you key $\boxed{(-)}\,\boxed{2}\,\boxed{\times}\,\boxed{8}$
4. The $\boxed{\wedge}$ key is used to indicate that an exponent follows.

   $$3^4 \text{ is keyed as } \boxed{3}\,\boxed{\wedge}\,\boxed{4}$$

5. The parentheses keys are used in the usual manner to group calculations:

   $$3(5 + 8) \text{ is keyed } \boxed{3}\,\boxed{(}\,\boxed{5}\,\boxed{+}\,\boxed{8}\,\boxed{)}$$

# GRAPHING

The TI graphing calculator will graph functions written in the form $y = f(x)$. It will not graph functions written as $3x + 2y = 27$ or $x^2 + y^2 = 5$. To graph $3x + 2y = 27$, you must solve for $y$ and graph $y = -\frac{3}{2}x + \frac{27}{2}$ or $y = -1.5x + 13.5$ or $y = (27 - 3x)/2$.

    To enter the equation to be graphed, display the <y=> screen. It is obtained by pressing $\boxed{Y=}$.

    To enter an equation like $y = -2x + 3$, key the numbers and symbols of the equation. The "$x$" in the equation is entered with the key $\boxed{X, T, \theta, n}$. To initiate graphing press the $\boxed{GRAPH}$ key.

## The Range

From time to time, you will want to change the portion of the $x$-$y$ plane that shows on the screen. Here's how. Press $\boxed{\text{WINDOW}}$ and you will obtain a menu that includes something like

WINDOW

Xmin=$-10$

Xmax=10

Xscl=1

Ymin=$-10$

Ymax=10

Yscl=1

Yres=1

$x$Min and $x$Max specify the range of $x$-values that will show on the screen, $x$Scl specifies the spacing of the tick marks on the $x$-axis. $y$Min, $y$Max, and $y$Scl do the same for $y$.

## The Standard Screen

The screen that shows $x$ from $-10$ to 10 and $y$ from $-10$ to 10 is called the **standard screen.** For the standard screen, $x$Min and the rest of the range settings can be set automatically with $\boxed{\text{ZOOM}}$ <6:ZStandard>.

## The Square Window

The screen on the graphing calculator is 1.5 times as wide as high. Therefore, the standard window with ranges of $-10$ to 10 for both $x$ and $y$ results in a different scale for the two axes. A graph is actually distorted somewhat. To obtain a graph with no distortion, the axes must be scaled the same. This is done by selecting the **square window** option. Here's how to do it: $\boxed{\text{ZOOM}}$ <5:ZSquare>.

## The TRACE Function

The TRACE function has the useful feature that it moves the cursor along a curve and gives the coordinates of the point where the cursor is located.

**1.** To initiate the TRACE function press $\boxed{\text{TRACE}}$.

**2.** To trace, move the cursor toward the left with the $\boxed{<}$ key and to the right with the $\boxed{>}$ key. As you do so, the $x$- and $y$-coordinates show at the bottom of the screen.

# EVALUATING A FUNCTION

**1.** You can evaluate a function for a given value of $x$ by tracing the graph until the $x$-coordinate reaches the desired value. The corresponding function value shows as the $y$-coordinate.

The TRACE function is useful in locating a value of $x$ that corresponds to a given $y$-value. Trace the graph until the given value of $y$ shows as the $y$-coordinate and read the value of the $x$-coordinate.

Because the cursor moves in discrete steps when using TRACE, it may skip over a value of $x$ that you want to use in evaluating the function. By zooming in, you reduce the step size and thus can obtain values of $x$ closer to, if not actually equal to, the desired value.

2. Using **value** you can evaluate a function more accurately for a given value of $x$. To illustrate how, suppose you have entered $y1 = 2x^2 + 3$ in the $<y=>$ window. For a value of $x$—say, $x = 4.2$—you obtain $y1 = 2(4.2)^2 + 3$ with the sequence $\boxed{\text{CALC}}$ $<1:\text{value}>$ $\boxed{\text{ENTER}}$. When $<x=>$ shows on the screen, enter 4.2 $\boxed{\text{ENTER}}$. The value $y = 38.28$ will show on the screen.

### ZOOM

The ZOOM feature zooms in or out at the location of the cursor and gives a magnified or reduced view. The location of the cursor determines the center of the new area. The ranges of $x$ and $y$ are reduced by one fourth or enlarged by a factor of 4.

To zoom in and magnify a part of the area:

1. On the graph, move the cursor near the point you want to be the center of the new area and press $\boxed{\text{ENTER}}$.

2. Now zoom in by $\boxed{\text{ZOOM}}$ $<2:\text{Zoom In}>$ $\boxed{\text{ENTER}}$.

## FINDING THE INTERSECTION OF TWO GRAPHS

The TRACE command can be used to estimate the point of intersection of two graphs. Locate the cursor on one of the graphs and trace toward the point of intersection. You may zoom in to obtain a more accurate estimate.

Press the $\boxed{\vee}$ or $\boxed{\wedge}$ key to move the cursor from one graph to the other. As you move from one graph to the other, compare the $y$-coordinates of points on the graph to help determine the accuracy of the estimates.

You can locate the point of intersection of two graphs more accurately using the **Intersect** command. Graph the two functions with their intersection or intersections showing on the graph. Then follow these steps:

1. Select $\boxed{\text{CALC}}$ $<5:\text{intersect}>$. The screen will display $<\text{First curve?}>$. The equation of the curve selected will show in the upper left corner of the screen.

2. Press $\boxed{\vee}$ or $\boxed{\wedge}$, if necessary, to move the cursor to one of the curves, then press $\boxed{\text{ENTER}}$. The screen will display $<\text{Second curve?}>$.

3. Press $\boxed{\vee}$ or $\boxed{\wedge}$, if necessary, so that the cursor moves to the second curve. Press $\boxed{\text{ENTER}}$. If the display still shows $<\text{Second curve?}>$, press the other of $\boxed{\vee}$ or $\boxed{\wedge}$.

4. The screen will display $<\text{Guess?}>$. Move the cursor near the point of intersection and press $\boxed{\text{ENTER}}$. Then the screen will display the $x$- and $y$-coordinates of the point of intersection.

## CONSTRUCTING A TABLE

(See also Section 1.1.) The **Table** function allows you to specify a list of values of $x$ and then compute the corresponding values of a function, or functions, as defined in the $\boxed{\text{Y=}}$ menu. You may let the calculator generate a list of equally spaced values of $x$ or you may list them one by one.

## Setting Up the *x*-List

**(a)** Let the calculator generate the list.

- Select $\boxed{\text{TBLSET}}$ and on the screen that appears enter the starting value of *x,* say 5, at **TblStart.**
- Enter the amount by which you want to increment the values of *x,* say 3, at **ΔTbl.**
- Select **Auto** for **Indpnt.**

The calculator will generate a list for *x* beginning with 5 and increasing by 3 throughout the list.

**(b)** The user enters the *x* values.

- Select $\boxed{\text{TBLSET}}$ and on the screen that appears you can ignore **TblStart** and **ΔTbl.**
- Select **Ask** for **Indpnt.**

## Calculating the Formulas for *y*

Open the $\boxed{\text{Y=}}$ screen and enter the desired formula for Y1. If other formulas are needed, enter them in Y2, Y3, . . . .

To view the table, press $\boxed{\text{TABLE}}$ and the screen will show the table of *x* and *y* values. Only two columns of *y* values show. If you have entered three or more formulas for *y,* scroll to the right to view those columns. If you have entered the formulas for *y* before entering the values of *x,* the calculated values of *y* will apear as you enter the values of *x.*

The table shows as many as six digits of the numbers. If a value of *y* is more than six digits, select the number and the 12-digit form shows at the bottom of the screen.

# MATRICES

Ten matrices are allowed and, depending on memory available, may have up to 99 rows or columns.

## Matrix Names

The names of matrices may be obtained from the $\boxed{\text{MATRIX}}$ menu, [A] through [J].

## Entering a Matrix into Memory

To enter a matrix into memory, select $\boxed{\text{MATRIX}}$ <EDIT>, select the name you wish to use, then press $\boxed{\text{ENTER}}$. Enter the size of the matrix and the matrix entries, row by row. To return to the home screen, press $\boxed{\text{QUIT}}$.

## Displaying a Matrix

To display a matrix $A$, press $\boxed{\text{MATRIX}}$, select the name of the matrix, and press $\boxed{\text{ENTER}}$.

## Storing a Matrix

To store a matrix in a second matrix, display the name of the first matrix on the screen, press $\boxed{\text{STO>}}$, display the name of the second matrix on the screen, press $\boxed{\text{ENTER}}$.

## Performing Matrix Operations

Here are the commands to perform the basic matrix operations.

**Adding Matrices [A] and [B]**    [A] + [B] ENTER

**Multiplying Matrices [A] and [B]**    [A] × [B] ENTER or [A] [B] ENTER

**Multiplying All Entries of a Matrix [A] by Scalar such as 5**    5 × [A] ENTER or
5 [A] ENTER

**Finding the Inverse of a Matrix [A]**    [A] x⁻¹ ENTER

**Entering the Powers of a Matrix** $A$ **(Say,** $A^5$**)**    [A] ∧ 5 ENTER

**Using Row Operations to Reduce a Matrix.**    The row operations are used on an augmented matrix to solve a system of equations. Think through the steps used to manually solve a system. Use the corresponding row operation from the menu.

**Row Swap.**    This interchanges two rows of a matrix. Here's how to interchange rows 2 and 3 of a matrix named [B].

> MATRX <MATH> <C:rowSwap (> ENTER

The screen will display <rowSwap(>. Complete the command as <rowSwap ([B], 2, 3)> ENTER

**Multiplying All Entries in a Row by a Constant.**    To multiply row 1 of a matrix named [B] by 5:

> MATRX <MATH> <E:*row(> ENTER

Complete the command with <*row(5, [B], 1)> ENTER

**Adding Two Rows of a Matrix.**    This adds two rows and places the result in the second row named.
To add row 2 to row 3 of a matrix named [B] and to place the result in row 3:

> MATRX <MATH> <D:row+(> ENTER

Complete the command as <row+([B], 2, 3)>.

**Multiplying a Row by a Constant and Adding to Another Row.**    To multiply row 2 by −5 and add the result to row 4 and then replace row 4 with the result:

> MATRX <MATH> <F:*row+(> ENTER

When <*row+(> appears, complete the command with <*row+(−5, [B], 2, 4)>.

**A Sequence of Row Operations on a Matrix.**    If you perform a sequence of row operations on matrix [B] then, at any stage, the answer is the last operation performed on the *original* matrix. For example, the following sequence

$$*\text{row}(2, [B], 1)$$
$$\text{row}+([B], 1, 3)$$
$$*\text{row}+(-4, [B], 2, 3)$$

will give the same final answer as

$$*row+(-4, [B], 2, 3)$$

gives. Each operation in the sequence operates on the original matrix $B$ and saves the result in a temporary location, ANS, not in $B$.

On the other hand, the sequence

$$*row(2, [B], 1)$$
$$row+(ANS, 1, 3)$$
$$*row+(-4, ANS, 2, 3)$$

will give the answer that carries through each operation in the sequence with the final result stored in ANS.

A row operation on a matrix–say, [B]–does not modify [B] itself unless other action is taken. You can record the effect of a row operation on [B] by a store command.

The row operations can be carried through the sequence with the following commands:

$$*row(2, [B], 1) \boxed{STO>} [B]$$
$$row+([B], 1, 3) \boxed{STO>} [B]$$
$$*row+(-4, [B], 2, 3) \boxed{STO>} [B]$$

**Row Echelon Form of a Matrix.**   A system of equations can be solved by reducing the augmented matrix to the reduced echelon form through a sequence of row operations. This can be quite tedious for larger systems of equations.

The basic procedure of reducing a matrix to solve a system uses a series of row operations in an attempt to reduce the columns (except for the last one) to columns containing a single entry of 1 and all other entries 0.

**rref.**   The command <rref> reduces a matrix to its reduced echelon form. Here is the procedure for using <rref>:

Display <rref(> on the screen by: $\boxed{MATRX}$ <MATH> <B:rref(>. Then fill in the matrix name, say [A], giving <rref([A])>. Press $\boxed{ENTER}$ and the reduced echelon matrix will appear on the screen.

# STATISTICS

The data for statistical analyses are entered in a list, or lists, and the analysis is done on the lists.

## Entering Data Into a List

You may use up to six lists with names $L_1, L_2, \ldots, L_6$ and the names appear on the keyboard above the keys for $1, 2, \ldots, 6$.

## Clearing Lists

To clear the lists $L_1$ and $L_2$:

$\boxed{STAT}$ <EDIT> <4:CLRLST> $\boxed{ENTER}$ $\boxed{L1}$ $\boxed{,}$ $\boxed{L2}$ $\boxed{ENTER}$.

## Entering Data

To enter data in $L_1$ (scores) and $L_2$ (frequency): $\boxed{\text{STAT}}$ <EDIT> $\boxed{\text{ENTER}}$. Enter a score in $L_1$ and press $\boxed{\text{ENTER}}$. Repeat until all scores are entered. Press $\boxed{>}$ to move to $L_2$. Enter frequencies in a similar manner; then press $\boxed{\text{QUIT}}$.

## Statistical Calculations

The TI has two menus, the STAT and LIST menus, for calculating some of the measures of central tendency and dispersion. In both cases, the data are entered in list L1 for a single list of scores. For a frequency table, list the scores in L1 and the corresponding frequencies in L2.

## STAT Menu

To obtain the statistical calculations for a single list in L1 use

$\boxed{\text{STAT}}$ <CALC><1:1-Var stats> $\boxed{\text{ENTER}}$

The screen shows the values for:

| | |
|---|---|
| $\overline{x}$ | (Mean) |
| $Sx$ | (Sample standard deviation) |
| $\sigma x$ | (Population standard deviation) |
| $\min X$ | (Smallest score) |
| $Q_1$ | (First quartile) |
| Med | (Median) |
| $Q_3$ | (Third quartile) |
| $\max X$ | (Largest score) |

These values give you the five-point summary and both forms of the standard deviation. For a frequency table with scores in L1 and frequencies in L2 use

$\boxed{\text{STAT}}$ <CALC><1:1-Var stats> L1,L2 $\boxed{\text{ENTER}}$

to obtain the same list of values.

## LIST Menu

Using the $\boxed{\text{LIST}}$ <MATH> menu, you can obtain the mean, maximum, minimum, median, sample variance, and standard deviation from a single list in L1 or a frequency table in L1 and L2. Unlike the STAT menu, these values do not appear all at the same time. They must be obtained separately. For example, the mean is obtained with

$\boxed{\text{LIST}}$ <MATH><3:mean(L1)

for a single list and with

$\boxed{\text{LIST}}$ <MATH><3:mean(L1,L2)

for a frequency table. Although this is less convenient than using STAT, it has the advantage that the calculations can be used in other formulas. One example is the calculation of the range.

## Range

$\boxed{\text{LIST}}$ <MATH><2:max> L1-<1:min> L1 $\boxed{\text{ENTER}}$

## Permutations and Combinations

Find $P(9, 7)$, $C(9, 7)$, and 6!

$P(9, 7)$: $\boxed{9}$ $\boxed{\text{MATH}}$ <PRB> <2:nPr> $\boxed{\text{ENTER}}$ $\boxed{7}$ $\boxed{\text{ENTER}}$

$C(9, 7)$: $\boxed{9}$ $\boxed{\text{MATH}}$ <PRB> <3:nCr> $\boxed{\text{ENTER}}$ $\boxed{7}$ $\boxed{\text{ENTER}}$

6!: $\boxed{6}$ $\boxed{\text{MATH}}$ <PRB> <4:!> $\boxed{\text{ENTER}}$

## Histograms

In order to obtain the graph of a histogram, follow these steps:

1. Enter data into appropriate lists.
2. Clear or turn off all functions in the <y=> screen so that they will not appear on the graph of the histogram.
3. Turn off other plots in the PLOT screen so that they will not appear.
4. Clear any previous drawings that remain.
5. Set the viewing window so that the appropriate range of the variable and the frequencies will appear on the screen.
6. Define the histogram.
7. Display the graph.

Now for more detail of the steps:

1. **Enter data.** See earlier section.

    In the examples, the list name $L_1$ is used for the variable list and $L_2$ is used for the frequency list. For a single list of data, use 1 for each frequency.

2. **Clear <y=> screen.**

    Select the <y=> screen and clear or turn off the functions.

3. **Turn off other plots,** say PLOT 3 with:

    $\boxed{\text{STAT PLOT}}$ <PLOT 3> <OFF> $\boxed{\text{ENTER}}$

4. **Clear drawings**

    Turn off Plot and <y=> functions.

    $\boxed{\text{DRAW}}$ <1:CLRDRAW> $\boxed{\text{ENTER}}$ $\boxed{\text{ENTER}}$

5. **Set the viewing window.**

    Set XMIN, XMAX, XSCL, YMIN, YMAX, YSCL using the window screen used to set the screen when graphing functions. Set XMIN and XMAX so all scores are in the interval (XMIN, XMAX). This interval will be the $x$-axis on the graph. XSCL is the width of a bar on the histogram and determines the interval length of each category. YMIN and YMAX determine the range of frequencies (the $y$-scale).

6. **Define the histogram.**

    Three plots are allowed. This example will use PLOT 1.

    Press $\boxed{\text{STAT PLOT}}$ <1:PLOT1> $\boxed{\text{ENTER}}$

You will see a screen similar to the following.

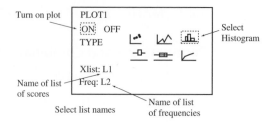

On that screen select as shown in the figure: <ON>, histogram for type, $L_1$ for Xlist, and $L_2$ for Freq. Press [ENTER] after each selection.

7. **Display the Histogram.**

Go to home screen and press [GRAPH].

**Box Plot.**   You can graph the box plot of data. Follow the steps for a histogram except for step 6. Substitute the following in step 6:

6. Select the box plot type. The symbol is ⊢═══⊣ instead of histogram.

## Graphing Calculator Programs

Some problems, such as finding the reduced echelon form of a matrix or the simplex solution of a linear programming problem, can be lengthy and tedious. Programs are available in which the calculator performs the calculations. The programs are listed in the Student Solutions Manual and on the text companion website. Some of the programs are included at the end of sections in the text. In the list below, the section where the program appears in the text is given after the name of the program. The programs available include the following:

**PIVOT (Section 2.3).**   The program pivots on the matrix entry specified by the user. This program is useful in reducing a matrix or performing the pivots in a simplex solution.

**SMPLX (Section 4.2).**   The user specifies the pivot row and column in a simplex tableau, and the program finds the next tableau.

**BINOM.**   For a given $N$, $p$, and $x$, this calculates the binomial probability

$$P(X = x) = C(N, x)p^x(1 - p)^{N-x}$$

**ANNA.**   This calculates the amount of an annuity when the periodic payment, periodic interest rate, and number of periods are given.

**PAYANN.**   This computes the periodic payment of an annuity that will yield a specified amount at a future date.

**PVAL.**   This computes the present value of an annuity with specified periodic payments.

**LNPAY.**   This finds the monthly payments required to amortize a loan.

**ANNGRO (Section 5.3).** This shows the growth of an annuity by computing the amount of annuity period by period.

**AMLN (Section 5.4).** This computes the amortization schedule of a loan, that is, the interest paid, principal paid, and balance of the loan for each month.

**BIOD (Section 8.6).** This computes the binomial probability distribution:

$$P(X = x) = C(N, x)p^x(1 - p)^{N-x} \quad (\text{for } x = 0, 1, 2, \ldots, N)$$

**BIOH (Section 8.6).** This computes the binomial probability distribution and displays it as a histogram.

**NORML (Section 8.7).** This finds the area under a normal curve between specified limits.

# USING EXCEL

Instructions for using Excel can be found at the end of some sections to show how Excel may be applied. Here is a list of the topics and the sections in which they occur.

*Continued*

# ANSWERS TO SELECTED ODD-NUMBERED EXERCISES

# CHAPTER 1

## Using Your TI Graphing Calculator, Section 1.1

**1.** $y = 3, 27, 51$     **3.** $y = 9, 9, 14, 30$     **5.** $y = 0, -0.25, -1.6$, and $8.9231$

## Using Excel, Section 1.1

**1.** =A4+B4 in C4     **3.** =C4+C5 in C6     **5.** =B2*B3 in B4     **7.** =(B1+B2)/2 in B3     **9.** =2.1*A5-1.8 in B5
**11.** =1.5*A1+3.25 in B1, drag through B6

## Using Your TI Graphing Calculator, Section 1.2

**1.** $y = 5x + 4$          **3.** $y = -1.4x + 8.2$     **5.** $5x - 2y = 12$     **7.** $2.4x + 5.3y = 15.6$

  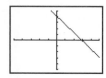  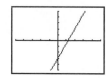  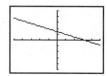

**9.** $18.59$     **11.** $35.54, -7.52, 26.06$

## Using Excel, Section 1.2

**1.** $0.25$     **3.** $2.73$     **5.** $y = -0.67x + 4.33$     **7.** $y = -0.56x + 4.24$
**1.** 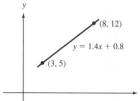     **3.**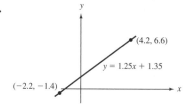

## Using Your TI Graphing Calculator, Section 1.3

**1.** $(4, 7)$     **3.** $(4.27, 0.91)$     **5.** $(2, 6)$

## Using Excel, Section 1.3

**1.** $-35, 215, 590, 1015$, and $1490$     **3.** $-1396; 16,709; 52,919; 77,059; 127,753$     **5.** $(44.13, 2144.66)$

## Review Exercises, Chapter 1

**1.** **(a)** 16   **(b)** 2   **(c)** 12.5   **(d)** $\dfrac{7b - 3}{2}$     **3.** 22     **5.** **(a)** \$4.20   **(b)** 2.75 pounds

**(a)** $f(x) = 29.95x$   **(b)** $f(x) = 2.25x + 40$

**9. (a)**

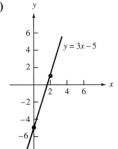

**(b)**

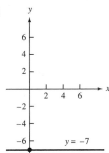

**(c)**

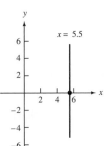

**(d)**

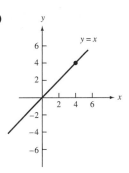

**11.** **(a)** Slope is $-2$, $y$-intercept is 3.   **(b)** Slope is $\frac{2}{3}$, $y$-intercept is $-4$.   **(c)** Slope is $\frac{5}{4}$, $y$-intercept is $\frac{3}{2}$.   **(d)** Slope is $-\frac{6}{7}$, $y$-intercept is $-\frac{5}{7}$.   **13.** **(a)** $-\frac{6}{5}$ **(b)** 3 **(c)** $\frac{5}{2}$   **15.** **(a)** $y = -\frac{3}{4}x + 5$ **(b)** $y = 8x - 3$ **(c)** $y = -2x + 9$ **(d)** Horizontal line, $y = 6$ **(e)** $y = -\frac{1}{6}x + \frac{23}{6}$ **(f)** Vertical line, $x = -2$ **(g)** $y = \frac{4}{3}x + \frac{13}{3}$ or $4x - 3y = -13$
**17.** **(a)** $y = 2$ **(b)** $x = -4$ **(c)** $x = 5$ **(d)** $y = 6$   **19.** The lines are not parallel.   **21.** The lines are not parallel.
**23.** The lines are not parallel.   **25.** $C(x) = 36x + 12{,}800$   **27.** **(a)** \$4938 **(b)** 655 bags   **29.** **(a)** $R(x) = 11x$
**(b)** $C(x) = 6.5x + 675$ **(c)** $x = 150$   **31.** 365 watches   **33.** **(a)** $BV = -2075x + 17{,}500$ **(b)** \$2075 **(c)** \$7125
**35.** $BV = -296x + 1540$   **37.** $k = 5$   **39.** $BV = -3800x + 22{,}000$   **41.** $C(x) = 0.67x + 480$   **43.** The second
plan is better when sales exceed 20,000 items.   **45.** $k = 26$   **47.** \$1015   **49.** Plan 1 is better when sales are less than
\$2667. Plan 2 is never better. Plan 3 is better when sales are larger than \$2667.   **51.** **(a)** $y = 120x + 8400$ **(b)** During the
14th month   **53.** $y = 500x - 1700$   **55.** **(a)** $y = 0.196x + 11.43$ **(b)** 13.19 **(c)** During 2008–2009
**57.** **(a)** $y = 70.83x + 564$ **(b)** \$1697.28 **(c)** About 2022

# CHAPTER 2

## Using Your TI Graphing Calculator, Section 2.2

**1.** $(1, -2, 3)$    **3.** $(6, -1, 5)$

## Using Excel, Section 2.2

**1.** $(1, -2, 3)$   **3.** $(6, -1, 5)$   **5.** $(2, -3, 5, 7)$

## Using Your TI Graphing Calculator, Section 2.3

**1.** $(2, -3, 4)$   **3.** $(-34.74, 20.55, -9.98)$   **5.** $x_1 = 12.3 - 0.31x_4$   **1.** $\begin{bmatrix} 1 & 0 & 0 & 1.3 & 2.5 \\ 0 & 1 & 0 & 0.7 & 0.5 \\ 0 & 0 & 1 & -1.2 & 0.5 \end{bmatrix}$
**3.** $(-2, 5, -3)$   **5.** $(6.1, 11.3, 2.7)$   $\quad x_2 = 7.2 + 0.40x_4$
$\quad x_3 = 9.5 - 0.68x_4$

## Using Excel, Section 2.3

**1.** $(-5, 2, 6)$   **3.** $(-34.74, 20.55, -9.98)$

## Using Your TI Graphing Calculator, Section 2.4

**1.** **(a)** $\begin{bmatrix} 4 & 1 & 3 \\ 10 & 13 & 2 \end{bmatrix}$ **(b)** $\begin{bmatrix} 2 & 6 & 4 \\ 8 & 10 & 14 \end{bmatrix}$ **(c)** $\begin{bmatrix} -2 & 5 & 1 \\ -2 & -3 & 12 \end{bmatrix}$ **(d)** $\begin{bmatrix} 9 & 5 & 8 \\ 24 & 31 & 11 \end{bmatrix}$

## Using Excel, Section 2.4

**1.** $\begin{bmatrix} 9 & 5 & -2 \\ 2 & 14 & 11 \\ -2 & 9 & 7 \end{bmatrix}$    **3.** $\begin{bmatrix} 4 & 12 & -8 \\ 20 & 36 & 28 \\ -16 & 0 & 24 \end{bmatrix}$    **5.** $\begin{bmatrix} 52 & 24 & -8 \\ 2 & 66 & 52 \\ -4 & 54 & 30 \end{bmatrix}$

## Using Your TI Graphing Calculator, Section 2.5

**1.** $\begin{bmatrix} 2 & 10 \\ 29 & -11 \end{bmatrix}$    **3.** $\begin{bmatrix} 13 & 7 \\ 12 & -10 \end{bmatrix}$    **5.** $A^2 = \begin{bmatrix} 2 & 3 & 2 \\ 6 & 7 & 6 \\ 8 & 6 & 8 \end{bmatrix}$    $A^3 = \begin{bmatrix} 10 & 9 & 10 \\ 26 & 25 & 26 \\ 28 & 30 & 28 \end{bmatrix}$    $A^4 = \begin{bmatrix} 38 & 39 & 38 \\ 102 & 103 & 102 \\ 116 & 114 & 116 \end{bmatrix}$

## Using Excel, Section 2.5

**1.** $\begin{bmatrix} 12 & 15 \\ 22 & 29 \end{bmatrix}$    **3.** $\begin{bmatrix} -3 & 7 & 17 \\ 8 & 7 & 13 \end{bmatrix}$

## Using Your TI Graphing Calculator, Section 2.6

**1.** $\begin{bmatrix} 0.4 & 1.4 & -1 \\ -0.8 & -0.8 & 1 \\ -0.2 & -1.2 & 1 \end{bmatrix}$    **3.** No inverse

## Using Excel, Section 2.6

**1.** $\begin{bmatrix} -17 & -24 & 15 \\ -26 & -37 & 23 \\ 41 & 58 & -36 \end{bmatrix}$    **3.** $\begin{bmatrix} -85 & -120 & 75 \\ -26 & -37 & 23 \\ 32.8 & 46.4 & -28.8 \end{bmatrix}$    **5.** $A$ has no inverse.

## Using Your TI Graphing Calculator, Section 2.8

**1.** $y = 1.64x - 2.71$    **3.** $y = 0.50x + 4.13$

## Using Excel, Section 2.8

**1.** $y = 1.2x + 0.8$    **3.** $y = 0.99x - 24.14$

## Review Exercises, Chapter 2

**1.** $\left(\frac{1}{4}, \frac{17}{8}\right)$    **3.** $(6, -4)$    **5.** $\left(2, 0, \frac{1}{3}\right)$    **7.** $(-6, 2, -5)$    **9.** No solution    **11.** $x = -z, y = 1 - z$

**13.** $x_1 = -56 + 29x_4, x_2 = 23 - 12x_4, x_3 = -13 + 8x_4$    **15.** $x_1 = 3 + \left(\frac{1}{2}\right)x_3, x_2 = \frac{2}{3} + \left(\frac{4}{3}\right)x_3$    **17.** $\frac{3}{4}$

**19.** $\begin{bmatrix} -3 & -2 \\ 6 & 7 \end{bmatrix}$    **21.** $\begin{bmatrix} 11 & -3 \\ 7 & -1 \\ 3 & 0 \end{bmatrix}$    **23.** $[3]$    **25.** Cannot multiply them    **27.** $\begin{bmatrix} -\frac{5}{2} & 3 \\ \frac{7}{2} & -4 \end{bmatrix}$

**29.** $\begin{bmatrix} -2 & -6 & 15 \\ 0 & -1 & 2 \\ 1 & 2 & -5 \end{bmatrix}$    **31.** $\begin{bmatrix} 6 & 4 & -5 & | & 10 \\ 3 & -2 & 0 & | & 12 \\ 1 & 1 & -4 & | & -2 \end{bmatrix}$    **33.** $\begin{bmatrix} 1 & 0 & -\frac{3}{5} & 0 \\ 0 & 1 & \frac{9}{5} & 0 \\ 0 & 0 & 0 & 1 \end{bmatrix}$

**35.** 5 free throws, 11 two-pointers, and 3 three-pointers   **37.** $20,000 in bonds, $30,000 in stocks   **39.** 120 of High-Tech, 68 of Big Burger   **41.** 630 at plant A and 270 at plant B   **43.** 7400   **45.** $y = 2.3x + 4.1$   **47.** 224 of 1.5-pound and 168 of 2.5-pound fruitcakes

---

# CHAPTER 3

## Using Your TI Graphing Caculator, Section 3.2

**1.** Corners: $(0, 0), (8, 2), (3, 6), (0, 7), (9.33, 0)$   **3.** Corners: $(0, 0), (0, 11), (10, 9), (21, 4), (27, 0)$

## Using Excel, Section 3.2

**1.** Intersection at $(7, 5)$   **3.** The first two intersect at $(10, 12)$.   The first and third intersect at $(14, 4)$.   The second and third intersect at $(5, 10)$.   **5.** Intersection at $(2.3, 6.7)$

## Using Your TI Graphing Calculator, Section 3.3

**1.** $\{0, 704, 1251, 1035\}$   **3.** $\{0, 77, 114, 91, 20\}$   **5.** $\{117, 91, 147, 201\}$

## Using Excel, Section 3.3

**1.** $\{0, 704, 1251, 1035\}$   **3.** $\{0, 77, 114, 91, 20\}$   **5.** $\{117, 91, 147, 201\}$

## Review Exercises, Chapter 3

**1.**

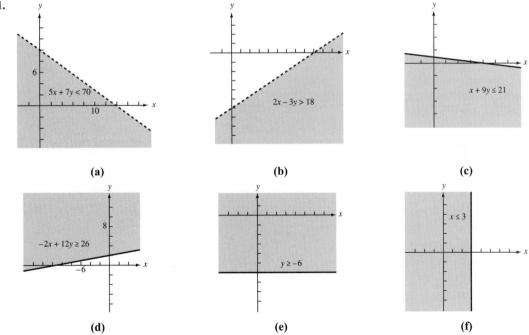

(a)   (b)   (c)

(d)   (e)   (f)

**3.** Corners are $(-9, -5)$, $(-1, -5)$, and $(3, -1)$.

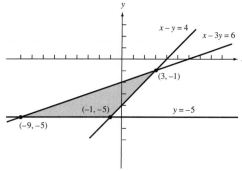

**5.** Corners are $(-4, 2)$, $(4, 8)$, $(5, 0)$, and $(-2, 0)$.

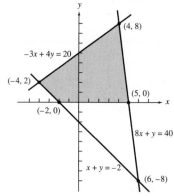

**7.** Corners are $(-\frac{4}{3}, -\frac{2}{3})$, $(0, 2)$, $(2, 2)$, and $(2, 1)$.

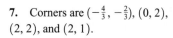

**9.** Maximum $z$ is 22 at $(2, 3)$.    **11.** **(a)** Minimum $z$ is 34 at $(2, 6)$.    **(b)** Minimum $z$ is 92 at $(2, 6)$, $(8, 1)$ and points in between.    **13.** Maximum $z = 290$ at $(20, 30)$.

**15.** **(a)** $65x + 105y \leq 700$    **(b)**

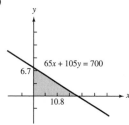

**17.** Let $x$ = number of adult tickets
$y$ = number of children's tickets
$$x + y \leq 275$$
$$7.50x + 4.00y \geq 1100$$
$$x \geq 0, y \geq 0$$

**19.** 640 bars of standard and 280 bars of premium yield a revenue of \$85,600.

**21.** Let $x_1$ = number of A-teams
$x_2$ = number of B-teams
$x_3$ = number of C-teams

Maximize number of inoculations $z = 175x_1 + 110x_2 + 85x_3$, subject to:
$$x_1 + x_2 + x_3 \leq 75 \text{ (number of doctors)}$$
$$3x_1 + 2x_2 + x_3 \leq 200 \text{ (number of nurses)}$$
$$x_1 \geq 0, x_2 \geq 0, x_3 \geq 0$$

**23.** Let $x_1$ = number of Type I pattern
$x_2$ = number of Type II pattern
$x_3$ = number of Type III pattern
$x_4$ = number of Type IV pattern

Maximize profit $z = 48x_1 + 45x_2 + 55x_3 + 65x_4$, subject to:
$$40x_1 + 25x_2 + 30x_3 + 45x_4 \leq 1250 \text{ (number tulips)}$$
$$25x_1 + 50x_2 + 40x_3 + 45x_4 \leq 1600 \text{ (number daffodils)}$$
$$6x_1 + 4x_2 + 8x_3 + 2x_4 \leq 195 \text{ (number boxwood)}$$
$$x_1 \geq 0, x_2 \geq 0, x_3 \geq 0, x_4 \geq 0$$

# CHAPTER 4

## Using Excel, Section 4.2

**1.** Maximum $z = 117$ at $(6, 5)$     **3.** Maximum $z = 696$ at $(12, 12, 24)$

## Using Excel, Section 4.4

**1.** Maximum $z = 308$ at $(14, 30, 0)$.     **3.** Maximum $z = 13$ at $(1, 2.5)$.

## Review Exercises, Chapter 4

**1.** $6x_1 + 4x_2 + 3x_3 + s_1 = 220$
$x_1 + 5x_2 + x_3 + s_2 = 162$
$7x_1 + 2x_2 + 5x_3 + s_3 = 139$

**3.** $6x_1 + 5x_2 + 3x_3 + 3x_4 + s_1 = 89$
$7x_1 + 4x_2 + 6x_3 + 2x_4 + s_2 = 72$

**5.** $10x_1 + 12x_2 + 8x_3 + s_1 = 24$
$7x_1 + 13x_2 + 5x_3 + s_2 = 35$
$-20x_1 - 36x_2 - 19x_3 + z = 0$

**7.** $3x_1 + 7x_2 + s_1 = 14$
$9x_1 + 5x_2 + s_2 = 18$
$x_1 - x_2 + s_3 = 21$
$-9x_1 - 2x_2 + z = 0$

**9.** $x_1 + x_2 + x_3 + s_1 = 15$
$2x_1 + 4x_2 + x_3 + s_2 = 44$
$-6x_1 - 8x_2 - 4x_3 + z = 0$

**11.** **(a)** 6 in row 2, column 2   **(b)** 5 in row 2, column 4

**13.** **(a)** $x_1 = 0, x_2 = 80, s_1 = 0, s_2 = 42, z = 98$   **(b)** $x_1 = 73, x_2 = 42, x_3 = 15, s_1 = 0, s_2 = 0, s_3 = 0, z = 138$

**15.** **(a)**
$$\begin{bmatrix} 11 & 5 & 3 & 1 & 0 & 0 & 0 & 142 \\ -3 & -4 & -7 & 0 & 1 & 0 & 0 & -95 \\ 2 & 15 & 1 & 0 & 0 & 1 & 0 & 124 \\ \hline -3 & -5 & -4 & 0 & 0 & 0 & 1 & 0 \end{bmatrix}$$
**(b)**
$$\begin{bmatrix} 7 & 4 & 1 & 0 & 0 & 28 \\ -1 & -3 & 0 & 1 & 0 & -6 \\ \hline 14 & 22 & 0 & 0 & 1 & 0 \end{bmatrix}$$

**17.** **(a)**
$$\begin{bmatrix} -15 & -8 & 1 & 0 & 0 & 0 & -120 \\ 10 & 12 & 0 & 1 & 0 & 0 & 120 \\ -15 & -5 & 0 & 0 & 1 & 0 & -75 \\ \hline -5 & -12 & 0 & 0 & 0 & 1 & 0 \end{bmatrix}$$
**(b)**
$$\begin{bmatrix} 14 & 9 & 1 & 0 & 0 & 0 & 126 \\ -10 & -11 & 0 & 1 & 0 & 0 & -110 \\ -5 & 1 & 0 & 0 & 1 & 0 & 9 \\ \hline 3 & 2 & 0 & 0 & 0 & 1 & 0 \end{bmatrix}$$

**19.** Maximum $z = 47$ at $(13, 0, 4)$.     **21.** Unbounded feasible region, no maximum

**23.** Maximum $z = 456$ at $(60, 132, 0)$.     **25.** **(a)** $x_3$   **(b)** $x_2$     **27.**
$$\begin{bmatrix} 3 & 4 & 5 \\ 1 & 0 & 7 \\ -2 & 6 & 8 \end{bmatrix} \begin{bmatrix} 4 & -5 \\ 3 & 0 \\ 2 & 12 \\ 1 & 9 \end{bmatrix}$$

**29.** Multiple solutions, minimum $z = 64$ at $(0, 6, 1)$ and $(0, 8, 0)$     **31.** Maximum $z = 92$ at $(4, 4, 3)$.

**33.** Maximum $z = 22$ at $(2, 3, 1)$.     **35.** Unbounded feasible region, no maximum     **37.** Minimum $z = 252$ at $(6, 4)$.     **39.** Maximum $z = 2700$ at $(0, 180)$.     **41.** Maximum $z = \frac{17}{4}$ at $(\frac{3}{4}, \frac{11}{4})$

**43.** Let $x_1$ = number of hunting jackets
        $x_2$ = number of all-weather jackets
        $x_3$ = number of ski jackets

**45.** $A$ can be any value from 6 through 20.

Maximize $z = 7.5x_1 + 9x_2 + 11x_3$, subject to
$3x_1 + 2.5x_2 + 3.5x_3 \le 3200$
$26x_1 + 20x_2 + 22x_3 \le 18{,}000$
$x_1 \ge 0, x_2 \ge 0, x_3 \ge 0$

$$\begin{bmatrix} 3 & 2.5 & 3.5 & 1 & 0 & 0 & 3{,}200 \\ 26 & 20 & 22 & 0 & 1 & 0 & 18{,}000 \\ \hline -7.5 & -9 & -11 & 0 & 0 & 1 & 0 \end{bmatrix}$$

# CHAPTER 5

## Using Your TI Graphing Calculator, Section 5.2

**1.**

| Year | Amount |
|------|--------|
| 0 | 200.00 |
| 1 | 210.20 |
| 2 | 220.92 |
| 3 | 232.19 |
| 4 | 244.03 |
| 5 | 256.47 |
| 6 | 269.55 |
| 7 | 283.30 |
| 8 | 297.75 |
| 9 | 312.94 |
| 10 | 328.89 |

**3.**   About 16 years

## Using Excel, Section 5.2

**1.**

| Year | Amount |
|------|--------|
| 0 | 500.00 |
| 1 | 530.00 |
| 2 | 561.80 |
| 3 | 595.51 |
| 4 | 631.24 |
| 5 | 669.11 |
| 6 | 709.26 |

**3.**

| Year | 6% | 5.50% |
|------|------|-------|
| 0 | 2000.00 | 2000.00 |
| 1 | 2120.00 | 2110.00 |
| 2 | 2247.20 | 2226.05 |
| 3 | 2382.03 | 2348.48 |
| 4 | 2524.95 | 2477.65 |
| 5 | 2676.45 | 2613.92 |

**5.**   Just over 9 years

## Using Your TI Graphing Calculator, Section 5.3

**1.**

| Month | Amount |
|-------|--------|
| 1 | 200.0 |
| 2 | 401.50 |
| 3 | 604.51 |
| 4 | 809.05 |
| etc., for 15 months | |

**3.**

| Year | Amount |
|------|--------|
| 1 | 1000.00 |
| 2 | 2072.00 |
| 3 | 3221.18 |
| 4 | 4453.11 |
| etc., for 10 years | |

## Using Excel, Section 5.3

**1.**

| Month | Amount |
|-------|--------|
| 1 | 1,500.00 |
| 2 | 3,085.50 |
| 3 | 4,761.37 |
| 4 | 6,532.77 |
| 5 | 8,405.14 |
| 6 | 10,384.23 |
| 7 | 12,476.13 |
| 8 | 14,687.27 |
| 9 | 17,024.45 |
| 10 | 19,494.84 |

**3.**

| Month | Amount |
|-------|--------|
| 1 | 75.00 |
| 2 | 150.34 |
| 3 | 226.01 |
| 4 | 302.03 |
| 5 | 378.39 |
| 6 | 455.09 |
| 7 | 532.14 |
| 8 | 609.54 |
| 9 | 687.28 |
| 10 | 765.37 |
| 11 | 843.82 |
| 12 | 922.61 |

## Using Your TI Graphing Calculator, Section 5.4

**1.**

| Month | Interest | Principal repaid | Balance |
|-------|----------|------------------|---------|
| 1 | 15.00 | 159.90 | 1840.10 |
| 2 | 13.80 | 161.10 | 1679.00 |
| 3 | 12.59 | 162.31 | 1516.69 |
| 4 | 11.38 | 163.52 | 1353.17 |
| etc., for 12 months | | | |

**3.**

| Month | Interest | Principal repaid | Balance |
|-------|----------|------------------|---------|
| 1 | 552.50 | 250.02 | 84,749.98 |
| 2 | 550.07 | 251.65 | 84,498.33 |
| 3 | 549.24 | 253.28 | 84,245.05 |
| 4 | 547.59 | 254.93 | 83,990.12 |
| etc., for 15 years | | | |

## Review Exercises, Chapter 5

**1.** $90   **3.** $960   **5.** $1350   **7.** $D = \$1530, PR = \$6970$   **9.** $1125.22   **11.** After 10 years, the value is $A = 1000(1.017)^{40} = 1962.63$, so it will not double in value.   **13.** 4.65%   **15.** 7.442% **17.** Effective rate of 5.6% = 5.72%. So, 5.6% compounded quarterly is better.   **19.** **(a)** $10,099.82   **(b)** $2099.82 **21.** **(a)** $7063.49   **(b)** $2063.49   **23.** $16,695.38   **25.** $33,648.57   **27.** The investment will not reach $3500 in 6 years.   **29.** $5637.09   **31.** $6942.02   **33.** $7245.15   **35.** $962,997   **37.** $3736.29 **39.** $4411.16   **41.** $33,648.57   **43.** After 5 years, the total payment is $297,189.48, so the term should be less than 5 years.   **45.** $276.76   **47.** $1183.24   **49.** $573,937.65   **51.** $17,706.09   **53.** Amount after 5 years, $1360.19; amount after 10 years, $1868.42   **55.** $3023.65   **57.** $39,443.62   **59.** $1157.29 **61.** The average annual rate of increase was 7.8%.

---

# CHAPTER 6

## Using Your TI Graphing Calculator, Section 6.4

**1.** 95,040   **3.** 201,600

## Using Excel, Section 6.4

**1.** 120   **3.** 1.15585 E+29

## Using Your TI Graphing Calculator, Section 6.5

**1.** 3003   **3.** 2,882,880   **5.** 72.833

## Using Excel, Section 6.5

**1.** 20   **3.** 1.80535 E+13

## Review Exercises, Chapter 6

**1.** **(a)** True   **(b)** False   **(c)** False   **(d)** False   **(e)** True   **(f)** False   **(g)** True   **(h)** False   **(i)** True   **(j)** False **(k)** True   **(l)** False   **(m)** True   **(n)** False   **(o)** False   **3.** **(a)** Equal   **(b)** Not equal   **(c)** Equal, both are empty **5.** 14 **7.**

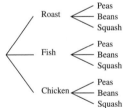

**9.** **(a)** 6,760,000   **(b)** 3,276,000   **11.** 3003   **13.** 3,121,200   **15.** 1440   **17.** **(a)** 990   **(b)** 1331   **19.** 6300 **21.** This totals 58, not 60.   **23.** **(a)** 20   **(b)** 6   **(c)** 23   **25.** 280   **27.** 10   **29.** **(a)** 8000   **(b)** 6840   **31.** C(15, 4) C(20, 4) C(25, 3) C(11, 1)   **33.** 15,120   **35.** 26,334   **37.** **(a)** 1680   **(b)** 126   **(c)** 5040   **(d)** 1   **(e)** 24   **(f)** 35 **(g)** 1680   **(h)** 630,630   **(i)** 280   **39.** 142,500   **41.** 14   **43.** 210   **45.** ∅, {red}, {white}, {blue}, {red, white}, {red, blue}, {white, blue}, {red, white, blue}   **47.** 5040   **49.** 85   **51.** **(a)** 120   **(b)** 20   **c)** 10,080

# CHAPTER 7

## Review Exercises, Chapter 7

**1.** No   **3.** $\frac{146}{360} = 0.406$   **5.** $\frac{17}{30} = 0.567$   **7.** $\frac{2}{15}$   **9.** $\frac{4}{13}$   **11.** $\frac{1}{32}$   **13.** **(a)** $\frac{6}{11} = 0.545$   **(b)** $\frac{133}{528} = 0.252$
**15.** **(a)** $\frac{3}{13} = 0.231$   **(b)** 0.769   **17.** Because $P(\text{even})P(10) = \frac{5}{169} \neq P(\text{even} \cap 10) = \frac{4}{52}$, even and 10 are not
independent events.   **19.** **(a)** $\frac{1}{1296}$   **(b)** $\frac{24}{1296} = \frac{1}{54}$   **(c)** $\frac{1}{72}$   **21.** **(a)** 0.036   **(b)** 0.196   **(c)** 0.084   **23.** **(a)** 0.290
**(b)** 0.234   **25.** **(a)** 0   **(b)** $\frac{1}{30}$   **(c)** $\frac{2}{30}$   **(d)** $\frac{15}{30}$   **(e)** $\frac{7}{30}$   **27.** 0.154   **29.** **(a)** 12   **(b)** 12   **31.** **(a)** 0.06
**(b)** 0.09   **33.** **(a)** $\frac{225}{420} = 0.536$   **(b)** $\frac{68}{420} = 0.162$   **(c)** $\frac{118}{242} = 0.488$   **(d)** $\frac{98}{225} = 0.436$
**35.**   $P(\text{seat belt not used and injuries}) = 0.344$   **37.** **(a)** 0.92   **(b)** 0.08
$P(\text{seat belt not used})P(\text{injuries}) = 0.204$
Dependent

# CHAPTER 8

## Using Excel, Section 8.1

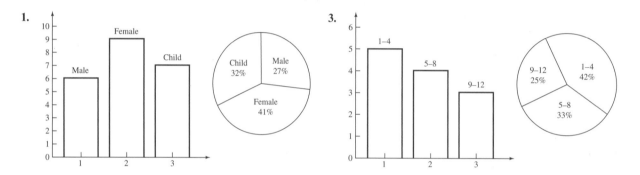

## Using Your TI Graphing Calculator, Section 8.2

**1.** Mean = 6.43, median = 7   **3.** Mean = 8.675, median = 7.9   **5.** 2.25

## Using Excel, Section 8.2

**1.** Mean = 6.43, median = 7   **3.** Mean = 8.675, median = 7.9

## Using Your TI Graphing Calculator, Section 8.3

**1.** Mean = 4.17, population standard deviation = 2.41, sample standard deviation = 2.64.
**3.** Mean = 8.18, population standard deviation = 2.66, sample standard deviation = 2.79.
**1.** Five-point summary = $\{2, 3.5, 6, 11.5, 14\}$

## Using Excel, Section 8.3

**1.** Range = 16, sample standard deviation = 5.4556, population standard deviation = 5.2017.
**3.** Range = 31, $s = 9.18$, $\sigma = 8.71$

## Using Your TI Graphing Calculator, Section 8.6

**1.** $p = 0.35, n = 4$

| X | P(X) |
|---|------|
| 0 | 0.1785 |
| 1 | 0.3845 |
| 2 | 0.3105 |
| 3 | 0.1115 |
| 4 | 0.0150 |

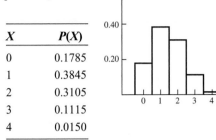

**3.** $p = 0.70, n = 5$

| X | P(X) |
|---|------|
| 0 | 0.0024 |
| 1 | 0.0284 |
| 2 | 0.1323 |
| 3 | 0.3087 |
| 4 | 0.3602 |
| 5 | 0.1681 |

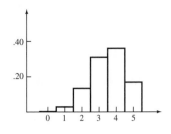

## Using Your TI Graphing Calculator, Section 8.7

**1.** 0.7011     **3.** 0.7612

## Using Excel, Section 8.7

**1.** 0.9088     **3.** 0.5074     **5.** 0.1499     **7.** 0.8783

## Review Exercises, Chapter 8

**1.**

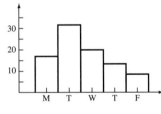

New accounts

**3. (a)** 6  **(b)** 10.5  **(c)** 2     **5.** 2.64     **7.** 26

**9.** 0, 1, 2, 3     **11.** 2.85     **13.** 2.67     **15.** 0.0227     **17.**

| X | P(X) |
|---|------|
| 0 | 0.3164 |
| 1 | 0.4219 |
| 2 | 0.2109 |
| 3 | 0.0469 |
| 4 | 0.0039 |

**19. (a)** 0.6368  **(b)** 0.0009  **(c)** 0.1669     **21.** 81st percentile

**23.**

| X | P(X) |
|---|-------|
| 1 | 0.025 |
| 2 | 0.200 |
| 3 | 0.4625 |
| 4 | 0.2688 |
| 5 | 0.0438 |

**25.**

| X | Number of outcomes |
|---|--------------------|
| 0 | 10 |
| 1 | 30 |
| 2 | 15 |
| 3 | 1 |

**27.**

| X | Number of ways x can occur |
|---|----------------------------|
| 1 | 1 |
| 2 | 1 |
| 3 | 4 |

**29.** 11.238 **31.** **(a)** 0.0616 **(b)** 0.0181 **(c)** 0.0159 **33.** $0.575 < x < 0.891$ **35.** 0.292 **37.** 0.65
**39.** **(a)** 0.89 **(b)** 0.25 **41.** 0.245

---

# CHAPTER 9

## Review Exercises, Chapter 9

**1.** **(a)** This game is not strictly determined. **(b)** This game is strictly determined. The $(2, 2)$ location is the saddle point. The value of the game is 4. **(c)** This game is strictly determined. The $(2, 2)$ location is the saddle point. The value of the game is 275. **(d)** This game is strictly determined. The $(3, 2)$ location is the saddle point. The value of the game is 2.
**3.** **(a)** This game is not strictly determined, so there is no solution. **(b)** Offense adopts strategy 2, and the defense adopts strategy 1. **5.** **(a)** 7.16 **(b)** 6.8 **(c)** 2.4 **7.** The farmer should plant 24% of the crop in the field and 76% in the greenhouse.

---

# CHAPTER 10

## Review Exercises, Chapter 10

**1.** **(a)** Statement **(b)** Not a statement **(c)** Not a statement **(d)** Statement **3.** **(a)** Rhonda is not sick today.
**(b)** Rhonda is sick today, and she has a temperature. **(c)** Rhonda is not sick today, and Rhonda does not have a temperature. **(d)** Rhonda is sick today, or she has a temperature **5.** **(a)** True **(b)** True **(c)** False **(d)** True
**7.** **(a)** $T$ **(b)** $T$ **(c)** $T$ **(d)** $F$ **9.** **(a)** True **(b)** False **11.** *Inverse*: "If I do not turn in my paper late, then I will not be penalized." *Converse*: "If I will be penalized, then I will turn in my paper late." *Contrapositive*: "If I am not penalized, then I did not turn in my paper late."

**13.**

| $p$ | $q$ | $\sim p$ | $p \wedge q$ | $\sim p \to (p \wedge q)$ |
|---|---|---|---|---|
| $T$ | $T$ | $F$ | $T$ | $T$ |
| $T$ | $F$ | $F$ | $F$ | $T$ |
| $F$ | $T$ | $T$ | $F$ | $F$ |
| $F$ | $F$ | $T$ | $F$ | $F$ |

**15.**

| $p$ | $q$ | $p \vee q$ | $p \wedge (p \vee q)$ |
|---|---|---|---|
| $T$ | $T$ | $T$ | $T$ |
| $T$ | $F$ | $T$ | $T$ |
| $F$ | $T$ | $T$ | $F$ |
| $F$ | $F$ | $F$ | $F$ |

As $p$ and $p \wedge (p \vee q)$ have the same truth values, they are logically equivalent.

**17.** Valid by indirect reasoning **19.** Valid by syllogism **21.** Valid by disjunctive syllogism

---

# APPENDIX A

## Section A.1

**1.** $-13$ **3.** 23 **5.** $-30$ **7.** $-35$ **9.** $-5$ **11.** $-7$ **13.** $-10$ **15.** $-8$ **17.** 3 **19.** $\frac{9}{5}$
**21.** $\frac{17}{12}$ **23.** $-\frac{11}{12}$ **25.** $\frac{13}{20}$ **27.** $-\frac{1}{35}$ **29.** $\frac{2}{3}$ **31.** $\frac{5}{14}$ **33.** $\frac{1}{15}$ **35.** $\frac{8}{15}$ **37.** $\frac{12}{35}$ **39.** $\frac{20}{33}$
**41.** $\frac{4}{3}$ **43.** $\frac{15}{8}$ **45.** $\frac{171}{40}$ **47.** $-6a - 22b$ **49.** $-10a - 50b$

## Section A.2

**1.** $-3$ **3.** 4 **5.** 2 **7.** $\frac{1}{2}$ **9.** $-\frac{7}{4}$ **11.** $\frac{20}{11}$ **13.** $\frac{15}{23}$ **15.** 6 **17.** **(a)** \$242 **(b)** \$412
**(c)** 950 miles **19.** **(a)** \$14.28 **(b)** \$44.10 **(c)** 61 pounds

## Section A.3

**1.**

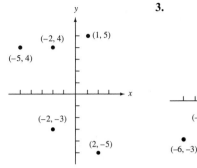

**3.**

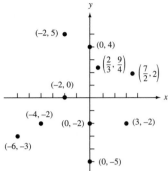

**5.** **(a)** $x$ is negative and $y$ is positive. **(b)** $x$ and $y$ are both negative. **(c)** $x$ is positive and $y$ is negative.

**7.**

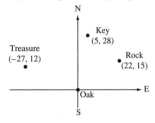

## Section A.4

**1.** **(a)** True because $9 - 3$ is positive **(b)** True because $4 - 0$ is positive **(c)** False because $-5 - 0$ is not positive
**(d)** True because $-3 - (-15)$ is positive. **(e)** True because $\frac{5}{6} - \frac{2}{3} = \frac{1}{6}$ is positive **3.** $x < \frac{9}{2}$ **5.** $x \geq -16$

**7.** $x > -3$ **9.** ⟶ 5 **11.** ⟵ $-9$ **13.** ⟵ $-4$

**15.** ⟵ $7/9$ **17.** $-7$ ⟶ $17/3$ **19.** $2$ ⟶ $8$ **21.** $(-\infty, -1]$

**23.** $\left(-\infty, \frac{1}{9}\right]$ **25.** $\left(-13, -\frac{17}{4}\right]$ **27.** $x \leq -\frac{1}{2}$ **29.** $-\frac{23}{3} > x \geq -11$ **31.** 8, 9, or 10 correct answers
**33.** 20, 21, 22, 23, 24, or 25

# INDEX